Electronic and Thermal Properties of Graphene

Electronic and Thermal Properties of Graphene

Special Issue Editor

Kyong Yop Rhee

MDPI • Basel • Beijing • Wuhan • Barcelona • Belgrade

Special Issue Editor
Kyong Yop Rhee
Department of
Mechanical Engineering,
Kyung Hee University
Korea

Editorial Office
MDPI
St. Alban-Anlage 66
4052 Basel, Switzerland

This is a reprint of articles from the Special Issue published online in the open access journal *Nanomaterials* (ISSN 2079-4991) from 2019 to 2020 (available at: https://www.mdpi.com/journal/nanomaterials/special_issues/electronic_graphene_nano).

For citation purposes, cite each article independently as indicated on the article page online and as indicated below:

LastName, A.A.; LastName, B.B.; LastName, C.C. Article Title. *Journal Name* **Year**, *Article Number*, Page Range.

ISBN 978-3-03936-400-8 (Hbk)
ISBN 978-3-03936-401-5 (PDF)

Contents

About the Special Issue Editor

Kyong Yop Rhee has been a professor of mechanical engineering at Kyung Hee University (South Korea) since 1999. His main research interests are nanocomposites, surface treatment, fracture, fatigue and composite materials. He has published 361 scientific papers in leading international journals, and he has led 67 R&D projects. His current h-index is 46. He earned his BS and MS degrees in Mechanical Engineering at Seoul National University (South Korea). He earned his PhD in Mechanical Engineering at Georgia Institute of Technology, after which he completed a two-year postdoctoral fellowship at Rutgers University.

 *nanomaterials*

Editorial

Electronic and Thermal Properties of Graphene

Kyong Yop Rhee

Department of Mechanical Engineering, College of Engineering, Kyung Hee University, Yongin 446-701, Korea; rheeky@khu.ac.kr

Received: 30 April 2020; Accepted: 5 May 2020; Published: 11 May 2020

Recently, the development of nanotechnology has bloomed in numerous industries. In this regard, graphene, a two-dimensional carbon nanomaterial, has been extensively researched, due to its high academic interests and commercial potential. Due to its unique structure and outstanding material properties, the research on graphene has enabled substantial progress in the development of electronics devices, thermoelectric appliances, optical devices, sensors, and energy harvesting applications. In addition, the mechanical properties, such as young modulus (1TPa), intrinsic strength (130 GPa), high stiffness, and fracture strain, make graphene a promising candidate for industries like nanocomposite coatings, paints, and bioapplications [1]. The development of graphene and graphene-based nanocomposites is always under intensive research, to provide the foundations for next generation devices. In this framework, the present special issue explored recent advances in the graphene based devices and their applications in next generation electronics. Here, we have discussed remarkable electronic and thermal properties of graphene such as Dirac fermions, high electrical conductivity, high seebeck coefficient, the quantum hall effect and thermoelectric effects [2]. This special issue compiles 19 articles dedicated to thermal, electrical and thermoelectric properties of graphene: 14 research articles and five review articles.

Graphene shows numerous novel and inimitable physical properties, such as the anomalous quantum hall effect, the ambipolar electric field effect, Klein tunneling, and ballistic transport, due to its exclusive crystal and electronic structure. However, the band gap characteristics of graphene limits its applications and commercial interests. The properties of graphene can be tuned by altering its nanostructure in terms of boundary configuration, structural defects, chemical doping, and the formation of heterogeneous structures to meet the demands of the electronic, photo thermal, thermoelectric and photoelectric fields [3]. High energy density and low cost are the prerequisite conditions for next generation transportation and grid storage. Lithium sulfur batteries are widely popular in this respect, however, due to their low active material utilization and poor life cycle, Hearin Jo et al. [4] use carbon coated separators with different ratios of carbon black and vapor grown carbon fibers to inhibit the migration of the soluble polysulfide, preventing it from reaching to the Li metal surface. Hong et al. [5] prepared carbon foam from carboxymethyl cellulose and compared enhancement in the thermal conductivity of neat carbon foam (CF), and carbon foam with Ag (CF-Ag), Al (CF-Al), CNT (CF-CNT), and graphene (CF-G). Their studies showed higher thermal conductivities of CF-Ag, CF-Al, CF-CNT, and CF-G, compared to CF. Among all the CF, the phase change temperature is the fastest for CF-G. In addition, the SEM analysis showed effective impregnation of erythritol into the pores of carbon foams, which minimized the latent heat loss during thermal cycling. Therefore, these CF impregnated with erythritol represents promising material for thermal energy storage applications. Jhao Yi Wu et al. [6] reported a facile process to synthesize few layer graphene (FLG) from graphite by a liquid exfoliation process. The similar D/G ratio of FLG and graphite demonstrates generation of very few structural defects. The FLG/polyvinylidene difluoride thin films showed superior electrical conductivity with great flexibility and mechanical strength under bending conditions, and have immense potential in conductive adhesive applications. The manuscript by Maoyuan Li [7] and Huaipeng Wang [1] discussed molecular dynamics simulations (MDS) to study the mechanical properties of graphene and carbon honeycomb (CHC). Maoyuan Li et al. investigated the effects of temperature, strain rate, and defects in the mechanical properties of graphene. Their results indicate a reduction in the Young modulus, fracture strength, and fracture strain with increases in the temperature. The existence of defects in graphene hampers the mechanical properties significantly, due to local stress concentration points, whereas the thermal conductivity showed a low temperature dependent behavior. In a similar study, Huaipeng Wang et al. presents an interesting insight between the covalent bonds of hinge atoms of CHC and their plastic behavior.

The special issue is completed with review papers [2,3,8,9], which compile recent research findings on the thermal, electrical, and thermoelectric properties of graphene and its composites. In addition, the application of graphene in the development of electrochemical biosensors was also discussed. Chao Lv et al. [10] discussed recent developments and the potential of graphene-based humidity sensors in their review paper.

In summary, this special issue of nanomaterials entitled *"Electronic and Thermal Properties of Graphene"* compiles a series of original research articles and recent review papers to provide depth information on the development of graphene based next generation devices. We hope that this special issue will provide an interesting insight into the latest perspectives in this rapidly evolving industry.

Funding: This research received no external funding.

Conflicts of Interest: The author declare no conflicts of interest.

References

1. Wang, H.; Cao, Q.; Peng, Q.; Liu, S. Atomistic study of mechanical behaviors of carbon honeycombs. *Nanomaterials* **2019**, *9*, 109. [CrossRef] [PubMed]
2. Sang, M.; Shin, J.; Kim, K.; Yu, K.J. Electronic and thermal properties of graphene and recent advances in graphene based electronics applications. *Nanomaterials* **2019**, *9*, 374. [CrossRef] [PubMed]
3. Wang, J.; Mu, X.; Sun, M. The thermal, electrical and thermoelectric properties of graphene nanomaterials. *Nanomaterials* **2019**, *9*, 218. [CrossRef] [PubMed]
4. Jo, H.; Oh, J.; Lee, Y.M.; Ryou, M.H. Effect of varying the ratio of carbon black to vapor-grown carbon fibers in the separator on the performance of Li–S batteries. *Nanomaterials* **2019**, *9*, 436. [CrossRef] [PubMed]
5. Kim, H.G.; Kim, Y.S.; Kwac, L.K.; Shin, H.J.; Lee, S.O.; Lee, U.S.; Shin, H.K. Latent heat storage and thermal efficacy of carboxymethyl cellulose carbon foams containing Ag, Al, carbon nanotubes, and graphene in a phase change material. *Nanomaterials* **2019**, *9*, 158. [CrossRef] [PubMed]
6. Wu, J.Y.; Lai, Y.C.; Chang, C.L.; Hung, W.C.; Wu, H.M.; Liao, Y.C.; Huang, C.H.; Liu, W.R. Facile and green synthesis of graphene-based conductive adhesives via liquid exfoliation process. *Nanomaterials* **2019**, *9*, 38. [CrossRef] [PubMed]
7. Li, M.; Deng, T.; Zheng, B.; Zhang, Y.; Liao, Y.; Zhou, H. Effect of defects on the mechanical and thermal properties of graphene. *Nanomaterials* **2019**, *9*, 347. [CrossRef] [PubMed]
8. Lee, J.H.; Park, S.J.; Choi, J.W. Electrical property of graphene and its application to electrochemical biosensing. *Nanomaterials* **2019**, *9*, 297. [CrossRef] [PubMed]
9. Cheng, J.; Fan, F.; Chang, S. Recent progress on graphene-functionalized metasurfaces for tunable phase and polarization control. *Nanomaterials* **2019**, *9*, 398. [CrossRef] [PubMed]
10. Lv, C.; Hu, C.; Luo, J.; Liu, S.; Qiao, Y.; Zhang, Z.; Song, J.; Shi, Y.; Cai, J.; Watanabe, A. Recent advances in graphene-based humidity sensors. *Nanomaterials* **2019**, *9*, 422. [CrossRef] [PubMed]

 *nanomaterials*

Article

Effect of Varying the Ratio of Carbon Black to Vapor-Grown Carbon Fibers in the Separator on the Performance of Li–S Batteries

Hearin Jo [1], Jeonghun Oh [1], Yong Min Lee [2] and Myung-Hyun Ryou [1,*

[1] Department of Chemical and Biological Engineering, Hanbat National University, 125 Dongseo-daero, Yuseong-gu, Daejeon 34158, Korea; hearin.jo124@gmail.com (H.J.); jeonghun.ohh@gmail.com (J.O.)

[2] Department of Energy Systems Engineering, Daegu Gyeongbuk Institute of Science and Technology (DGIST), 333 Techno Jungang-Daero, Daegu 42988, Korea; yongmin.lee@dgist.ac.kr

* Correspondence: mhryou@hanbat.ac.kr; Tel.: +82-42-821-1534

Received: 29 January 2019; Accepted: 12 March 2019; Published: 15 March 2019

Abstract: Lithium–sulfur (Li–S) batteries are expected to be very useful for next-generation transportation and grid storage because of their high energy density and low cost. However, their low active material utilization and poor cycle life limit their practical application. The use of a carbon-coated separator in these batteries serves to inhibit the migration of the lithium polysulfide intermediate and increases the recyclability. We report the extent to which the electrochemical performance of Li–S battery systems depends on the characteristics of the carbon coating of the separator. Carbon-coated separators containing different ratios of carbon black (Super-P) and vapor-grown carbon fibers (VGCFs) were prepared and evaluated in Li–S batteries. The results showed that larger amounts of Super-P on the carbon-coated separator enhanced the electrochemical performance of Li–S batteries; for instance, the pure Super-P coating exhibited the highest discharge capacity (602.1 mAh g^{-1} at 150 cycles) with a Coulombic efficiency exceeding 95%. Furthermore, the separators with the pure Super-P coating had a smaller pore structure, and hence, limited polysulfide migration, compared to separators containing Super-P/VGCF mixtures. These results indicate that it is necessary to control the porosity of the porous membrane to control the movement of the lithium polysulfide.

Keywords: carbon-coated separator; polysulfide; shuttle effect; lithium–sulfur batteries

---

1. Introduction

The continuously increasing worldwide demand for energy has resulted in energy storage systems becoming essential for the successful implementation of various electric devices such as electric vehicles, portable electronic devices, and energy storage systems [1–6]. Lithium–sulfur batteries have been regarded as promising candidates due to their high theoretical capacity (1675 mA·h·g^{-1}), low cost, and the environmentally friendly characteristics of sulfur. Despite the many advantages of sulfur cathodes, Li–S batteries have poor cycle performance due to following chronic drawbacks, which lead failure at successful commercialization [7–11]: (1) Sulfur has insulating properties that interfere with uniform electrons throughout the active materials during operation. As a result, poor utilization of the active sulfur material occurs during electrochemical reactions. (2) Lithium polysulfides (Li$_2$S$_x$, $4 \leq x \leq 8$), the intermediates of sulfur during the charging/discharging processes, are a fatal component impeding the cycle performance of Li–S batteries. Lithium polysulfide dissolves easily in the electrolyte from the sulfur cathode and consumes electrons directly from the both electrodes, cathodes, and anodes inside the battery system, unlike conventional battery systems, where electrons are consumed along the conductors. This series of internal cyclic electron consumption is called the "shutting effect" [12–14].

Various approaches have been proposed to overcome the drawbacks of the sulfur cathodes described above to improve Li–S batteries through the modification of the sulfur cathodes and electrolyte systems, and in this regard, advanced sulfur/carbon composites [3,15,16], functional polymeric binders [17,18], sulfur/metal–organic frameworks [19,20], solid electrolytes [21–23], and functional additives [24] have been studied.

Although the separators are one of the main components of the battery system (composed of anodes, cathodes, electrolytes, and separators), the significance of the separators for cycle performance of Li–S batteries was underestimated until a functional separator containing a carbon coating layer was proposed [25–28]. A carbon coating layer not only provides an electron path to the sulfur cathodes but also effectively reduces the migration of lithium polysulfides through the separators and leaves them on the sulfur cathode surface. As a result, the carbon-coated separators improve the cycle performance of the Li–S batteries by helping the sulfur cathodes to reuse lithium polysulfides during the repeated charging/discharging processes. Furthermore, from a practical point of view, the use of functional separators is not only economical but is also advantageous over other existing approaches associated with sulfur cathode and electrolyte modification, since it can be applied to a variety of existing technologies.

We have focused on the selection guidelines for carbon materials for Li–S batteries, and to do so we fabricated carbon-coated separators using two representative commercial carbon materials, carbon black (Super-P) and vapor-grown carbon fibers (VGCFs). We see that when Super-P and VGCFs are used as conductive additives for lithium–cobalt–oxide (LiCoO$_2$) cathodes, there is a synergy that leads to improved LiCoO$_2$ performance in unexpected combinations. Taking this into account, it is thought that the combination of Super-P and VGCFs for carbon-coated separator needs to be investigated for Li–S batteries, because changing the ratio of Super-P and VGCFs significantly changes the electrical conductivity and porosity of the coating separators.

The design of carbon-coated separators for Li–S batteries has to carefully consider the porosity of the coating layer. For instance, the use of carbon-coated separators which are not porous with a dense surface structure would prevent the liquid electrolytes containing the polysulfide from flowing through the layer fluently, thereby making it difficult for the polysulfide to migrate to the Li metal surface. From a kinetic point of view, however, limited movement of the liquid electrolyte inside the batteries may result in electrochemical performance degradation due to low reaction rates. On the other hand, highly porous, carbon-coated separators would be beneficial to the kinetic behavior of the Li–S battery system but would be vulnerable to polysulfide migration. In this study, the porosity of the carbon-coated separator was adjusted by using mixtures of different types of carbon materials, VGCFs and Super-P. Simply changing the VGCF to Super-P ratio in the carbon-coated separator enabled us to optimize the porosity of the separator as an effective approach to obtain a stable high-performance Li–S battery with exceptional rate capability.

2. Materials and Methods

2.1. Materials

Sulfur (100 mesh, Sigma–Aldrich, St. Louis, MO, USA), Ketjenblack (Ketjenblack®EC-600JD, AkzoNobel, Amsterdam, Netherlands), poly(vinylidene fluoride) (PVdF, KF-1300, Kureha, Iwaki, Japan, M_w = 350000), vapor-grown carbon fibers (VGCFs, Showa Denko K.K, Tokyo, Japan), Super-P (Li-conductive) (IMERYS, Paris, France), poly(vinylidene fluoride-co-hexafluoropropylene) (PVdF-HFP, Kynar Flex® 2801, Arkema Inc., Colombes, France), N-methyl-2-pyrrolidone (NMP, Sigma–Aldrich), 1,3-dioxolane (DOL, Sigma–Aldrich, St. Louis, MO, USA), 1,2-dimethoxyethane (DME, Sigma–Aldrich, St. Louis, MO, USA), LiTFSI (Enchem, Jecheon, Korea), Li metal foil (thickness = 200 μm, Honjo Metal Co., Osaka, Japan), and polypropylene (PP) separators (thickness = 25 μm, Celgard 2400, Celgard®, Charlotte, NC, USA) were used as separators.

2.2. Preparation of the VGCF and Super-P Carbon Composite

The VGCF and Super-P powder, both of which were purchased from commercial corporations, were first mixed together in a specific weight ratio and then dispersed in a poly(vinylidene fluoride-co-hexafluoropropylene) (PVdF-HFP, Kynar Flex® 2801, Arkema Inc., Colombes, France) solution (2 wt.% of PVdF-HFP dissolved in *N*-methyl-2-pyrrolidone (NMP, Sigma–Aldrich, St. Louis, MO, USA)) to form a uniform slurry.

2.3. Preparation of the Modified Separators

In this work, the separators were modified by fabricating them with three different coatings by using slurries with different compositions. The first (VGCF-coated separator) consisted of 70 wt.% VGCF and 30 wt.% PVdF-HFP; The second (VGCF and Super-P composite-coated separator) consisted of 35 wt.% VGCF, 35 wt.% Super-P, and 30 wt.% PVdF-HFP; the third (Super-P-coated separator) consisted of 70 wt.% Super-P and 30 wt.% PVdF-HFP. These three kinds of separators were prepared by modifying the conventional separator (PP separator, Celgard 2400) by directly applying a coating of the three slurries mentioned above on the PP separator using a gap-controlled doctor blade. After they were dried at 50 °C for 12 h in the oven, the three kinds of modified separators were punched into circular disks with a diameter of 18 mm. The fabricated layers of carbon coating had an average thickness of ~10 μm, which is the minimum value to ensure reasonable polysulfide inhibition behavior. The coating thickness issue will be further discussed in polysulfide diffusion experiments corresponding to Figure 2. The carbon loadings of the VGCF-coated separator, VGCF and Super-P composite-coated separator, and Super-P-coated separator were 0.18, 0.20, and 0.33 mg cm^{-2}, respectively.

2.4. Preparation of the Sulfur Cathode and Cell Assembly

A sulfur/carbon (S/C) (Ketjen black) composite (S/C = 80/20 in weight) was prepared with the melt diffusion method. A slurry consisting of the S/C composite (70 wt.%), vapor grown carbon fiber (20 wt.%), and PVdF (10 wt.%) as a binder, was poured onto aluminum foil. Then the coated foil was dried at 50 °C for 12 h. Finally, the sulfur cathode was roll-pressed and punched into circular disks with a diameter of 12 mm. The areal loading of sulfur for the as-prepared electrodes ranged from 1.3 mg·cm^{-2} to 1.5 mg·cm^{-2}. The electrochemical properties were tested by assembling 2032 coin-type half-cells using the sulfur electrodes, VGCF/Super-P coated separators, and Li metal as the counter electrode. The electrolyte was 1 M LiTFSI (lithium bis (trifluoromethanesulfonyl) imide) in DOL and DME (1:1 by volume) with 0.2 M LiNO$_3$ as an additive. To standardize the measurement protocol, the amount of electrolyte added to each cell was controlled to 200 μL. Cell assembly was carried out in an argon-filled glove box, and all capacity values were calculated based on the sulfur mass.

2.5. Electrochemical Testing

After assembly, the coin cells (sulfur/Li metal) were stored for 12 h before the electrochemical measurements. Cycle performance was evaluated by cycling the unit cells over different potential ranges (1.9–2.8 V versus Li/Li$^+$) in a constant current (CC) mode during both the charging and discharging processes at a constant current density C/2 (resp. 0.76 mA·cm^{-2} for sulfur) using a charge/discharge cycler (PNE Solution, Suwon, Korea) at 25 °C. The cycle performance was evaluated at 1 C (CC during the charge and discharge processes within the same potential ranges.) The rate capability was evaluated by increasing the discharge current densities from C/5 to 3 C (C/5, C/2, 1 C, 2 C, and 3 C). The cells were discharged in CC mode while maintaining a charging current density of C/2 in CC mode.

2.6. Characterization and Electrochemical Measurements

After the electrochemical investigations were performed, the fully charged cells up to 2.8 V versus Li/Li$^+$ were carefully disassembled in a dry Ar-filled glove box. The samples were washed several times with dimethyl carbonate (DMC, anhydrous, >99%, S Sigma–Aldrich, St. Louis, MO, USA), and then dried overnight under vacuum before observation. The samples were analyzed by performing field emission scanning electron microscopy with energy-dispersive X-ray analysis (FE-SEM/EDX, S-4800, Hitachi, Tokyo, Japan).

The Gurley number was evaluated using a densometer (4110N, Thwing-Albert, West Berlin, NJ USA) by measuring the time required for passing 100 mL of air through separators under 6.52 psi pressure [29].

3. Results and Discussion

3.1. Morphology and Physical Characterization of the Carbon-Coated Separator

For simplicity, the carbon-coated separator containing pure Super-P, the Super-P/VGCF combination, and pure VGCF are denoted as the Super-P, Super-P/VGCF, and VGCF separators, respectively.

Figure 1 shows the surface morphologies of each of these carbon-coated separators. The Super-P separator had a dense surface structure in powder form with an average particle size of ~40 nm. The larger size columnar VGCF particles (average particle diameter = ~150 nm, average length = 15 μm) resulted in the VGCF separator showing the most porous surface structure. Thus, the different dimensions and shapes of the Super-P and VGCF particles strongly influenced the surface morphology of the Super-P/VGCF separator, which was strongly dependent on the Super-P/VGCF ratio. The physical properties of each separator such as the Gurley number and surface resistance are listed in Table 1. All of the carbon-coated separators (the Super-P, Super-P/VGCF, and VGCF separators) had higher Gurley numbers than the bare uncoated separator (Celgard 2400). The additional carbon coating layer of ~9 μm with high tortuosity played the role of a gas barrier.

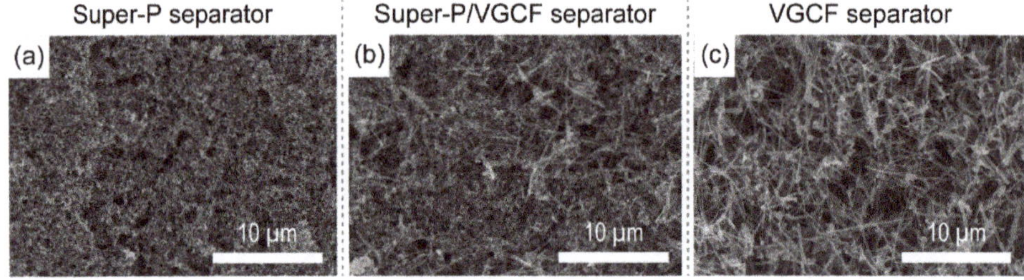

Figure 1. Scanning electron microscopy (SEM) images of the separator surface containing (**a**) Super-P; (**b**) Super-P/vapor-grown carbon fiber (VGCF) (5:5 by wt.%); and (**c**) VGCF.

Table 1. Physical properties of bare polypropylene (PP) and different conductive additives (Super-P, Super-P/VGCF, VGCF).

	Celgard 2400	Super-P	Super-P/VGCF	VGCF
Thickness (μm)	25	34	35	34
Gurley number (s·100 mL^{-1})	546.4	633.2	583.2	570.6
Surface resistance (mΩ·cm)	N.A.	286.1	165.2	84.3

On the other hand, VGCF exhibited the lowest surface resistance, which is in good agreement with our previous study [30]. Similar to the present study, the previous study investigated the effect of various types of conductive additives (pure Super-P, pure VGCF, and a mixture of Super-P and VGCF) on lithium–cobalt–oxide (LiCoO$_2$) cathodes. The LiCoO$_2$ cathodes containing pure VGCF revealed the lowest surface resistance because VGCF builds an "expressway" for electron transfer, which facilitates electron transfer across the cathode.

3.2. Polysulfide Suppression Behaviors of Carbon-Coated Separators

In general, separators for batteries are placed between the cathode and anode and are designed to have a highly porous structure to allow ion migration through the pores. Separators with high porosity containing more massive amounts of liquid electrolyte result in improved ion mobility [31,32]. However, a Li–S battery with highly porous separators can increase the mobility of polysulfides, resulting in more severe polysulfide shuttle behavior which can critically affect the cycle performance [33]. Taking this into account, the polysulfide inhibition behavior of carbon-coated separators prepared with various Super-P and VGCF ratios was investigated.

As shown in Figure 2, the polysulfide permeability of the separator was examined. The inner glass tube was filled with a mixture consisting of 15 mL of a solution of 0.43 M Li_2S_6 and 15 mL of DME/DOL (1:1, by vol.). The outer glass tube was filled with 30 mL of DME/DOL (1:1, by vol.). Because of the difference in polysulfide concentration in the two tubes, the polysulfide spread to the outer glass tube over time. The bare separator (Celgard 2400) showed the fastest polysulfide diffusion, whereas the Super-P separator exhibited the best protection against polysulfide diffusion. After 6 h, the solution in the outer glass tube wrapped with the bare separator turned brown, but the Super-P separator ensured that the solution in the outer glass tube remained transparent even after 12 h. On the other hand, when the carbon coating layer was less than 10 µm, the diffusion of the polysulfide could not be adequately suppressed even by the Super-P separator. Thus, an increase in the Super-P ratio more effectively inhibited the diffusion of polysulfide. These results indicate that Super-P effectively immobilizes polysulfide inside nano-sized porous structures.

Figure 2. Digital camera images of polysulfide diffusion experiments after (**a**) 2 h and (**b**) 6 h. The inner glass tube filled with a mixture of polysulfide (Li_2S_6) and control electrolyte (DME/DOL, 1:1 by vol.) was wrapped with various types of separators, while the outer glass tube was filled with control electrolyte. S:V indicates the weight ratio of Super-P to VGCF.

3.3. Electrochemical Performance

Figure 3a shows the galvanostatic discharge/charge potential profiles during pre-cycling for the Li–S battery containing the three types of carbon-coated separators (Super-P, Super-P/VGCF (5:5 by wt.%), and VGCF) measured at C/5. During the discharging process (lithiation), the upper discharge plateau near 2.4 V represents the conversion of sulfur (S_8) to soluble polysulfide (Li_2S_x, $4 \leq x \leq 8$), and the lower discharge plateau near 2.1 V represents the conversion of soluble polysulfide (Li_2S_x, $4 \leq x \leq 8$) to solid polysulfide (Li_2S_2/Li_2S) [34]. During the charging process, the first long and flat plateau near 2.2 V corresponds to the conversion of solid polysulfide (Li_2S_2/Li_2S) to soluble polysulfide (Li_2S_x, $4 \leq x \leq 8$), and the plateau near 2.35 V corresponds to the conversion of soluble polysulfide (Li_2S_x, $4 \leq x \leq 8$) to sulfur (S_8) [34]. Although the theoretical potentials of each plateau are 2.18 and 2.33 V versus Li/Li^+, the plateaus generally differed during charging and discharging because of the IR drop ascribed to the high internal resistance of Li–S batteries [35]. The Li–S batteries containing Super-P exhibited the highest initial discharge capacity (1219.5 $mA \cdot h \cdot g^{-1}$) with the highest Coulombic efficiency (100%). The first discharge capacity of each cell containing the carbon-coated separator exceeded that of the bare separator (Super-P = 1213.0 $mA \cdot h \cdot g^{-1}$, Super-P/VGCF = 1158.4 $mA \cdot h \cdot g^{-1}$, VGCF = 1120.7 $mA \cdot h \cdot g^{-1}$, bare PP = 1017.4 $mA \cdot h \cdot g^{-1}$). The hysteresis shown in Figure 3a usually can be observed from the initial discharge of other conversion electrode materials because this is attributed to the poor electrical contact of the initial grain boundaries between active materials and conducting carbon materials [36]. As can be seen in Figure 3b, the hysteresis observed under 2.0 V versus Li/Li^+ during pre-cycling (associated with Figure 3a) disappeared. Considering this, we can infer that the Li−S batteries were stabilized during pre-cycling.

As shown in Figure 3c, the cycle performance of the Li–S batteries containing the carbon-coated separators was evaluated at a discharging rate of 1 C. The unit cells containing larger amounts of Super-P showed a higher initial discharge capacity (after the first cycle, Super-P = 984.3 $mA \cdot h \cdot g^{-1}$, Super-P/VGCF = 903.0 $mA \cdot h \cdot g^{-1}$, VGCF = 795.9 $mA \cdot h \cdot g^{-1}$, bare PP = 690.4 $mA \cdot h \cdot g^{-1}$). The cycle performance of Li–S batteries was greatly improved when larger amounts of Super-P were used (after 150 cycles, Super-P = 602.1 $mA \cdot h \cdot g^{-1}$, Super-P/VGCF = 501.5 $mA \cdot h \cdot g^{-1}$, VGCF = 301.0 $mA \cdot h \cdot g^{-1}$, bare PP = 0 $mA \cdot h \cdot g^{-1}$). Because the Gurley number is defined by passing a specific amount of air through the medium, the exact correlation between the polysulfide and the separators cannot be clearly explained. Nonetheless, it is plausible that separators with a high Gurley number are beneficial in inhibiting polysulfide migration since the Gurley number reflects the tortuosity of the separators [32,37]. With this in mind, the improved cycle performance of the Li–S unit cells containing Super-P separators is reasonable.

The rate capabilities of the Li–S batteries containing the various carbon-coated separators were also evaluated by increasing the discharging current density step-wise from C/5 (0.27 $mA \cdot cm^{-2}$) to 3 C (4.02 $mA \cdot cm^{-2}$) every seven cycles. As shown in Figure 3d, the rate capabilities of the Li–S unit cells were significantly improved when more substantial amounts of Super-P were used (at the 35th cycle for the 3 C rate: Super-P = 659.8 $mA \cdot h \cdot g^{-1}$, Super-P/VGCF = 582.1 $mA \cdot h \cdot g^{-1}$, VGCF = 509.5 $mA \cdot h \cdot g^{-1}$, bare PP = 3.4 $mA \cdot h \cdot g^{-1}$). These results were unusual because the Super-P separators with the highest Gurley number, and with the highest tortuosity, exhibited the highest rate capabilities. Given the results, we can infer that, of the two main factors, the tortuosity of the separators and migration of polysulfide, the latter is more decisive in determining the cycle performance as well as the rate capabilities of Li–S batteries.

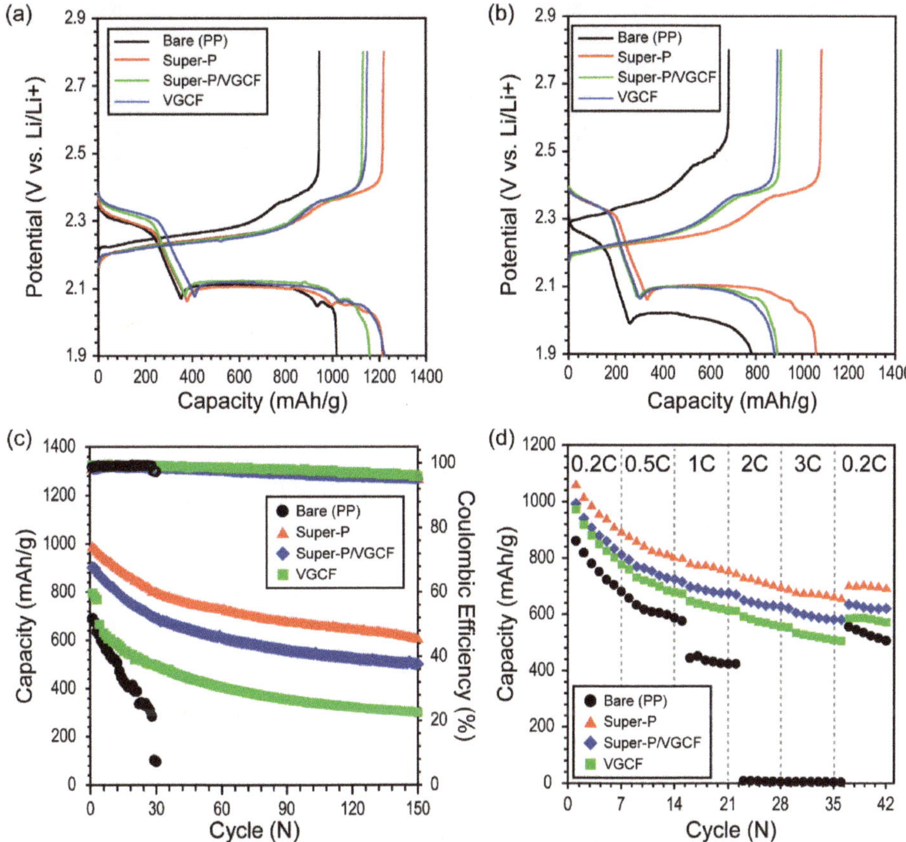

Figure 3. Charge/discharge potential profiles during (**a**) pre-cycling (current density = C/5) and (**b**) the first cycling for rate capability test (associated with Figure 3d); (**c**) cycle performance for the Li–S batteries containing various carbon-coated separators, respectively (current density = 1 C); and (**d**) rate cyclability (charging current density was varied from C/5 to 3 C, while the discharging current density was fixed at C/5). Super-P/VGCF consisted of Super-P:VGCF = 5:5 by wt.%.

3.4. Post-Mortem Analysis of Li–S Batteries after Cycling

After 20 cycles (corresponding to Figure 3c), the sulfur cathodes were retrieved from fully charged Li–S unit cells and the surface structures of the sulfur cathodes were observed using SEM. As shown in Figure 4, the morphological structure of the sulfur cathodes varied depending on the type of coating that was used on the separator. Deep holes were observed across the entire surface of the sulfur cathode in the Li–S unit cells containing bare separators (Figure 4a) and VGCF (Figure 4d). On the other hand, the sulfur cathodes of the Li–S unit cells containing larger amounts of Super-P showed a dense structure with fewer pores. Because polysulfide intermediates are highly soluble in electrolytes, these results are plausible because Super-P-rich separators more efficiently hinder the movement of polysulfides [28,38].

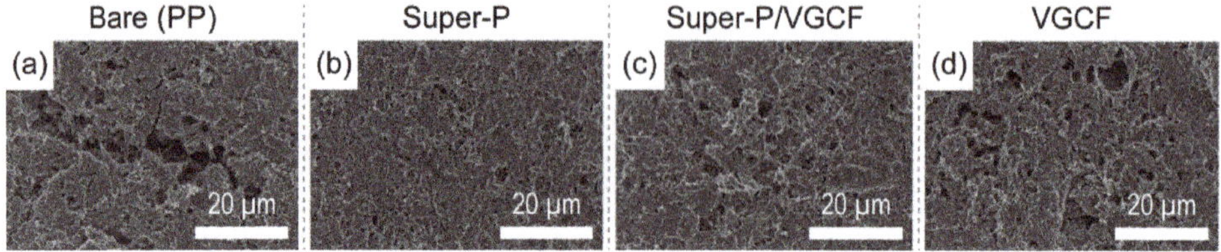

Figure 4. SEM images of the surfaces of (**a**)–(**d**) sulfur cathodes after 20 cycles (corresponding to Figure 3c).

The surface structure of the cathode side of each separator was also observed using SEM and EDX. As shown in Figure 5, the morphological structures of the separators are almost similar to those shown in Figure 1. In contrast, the elemental composition determined by EDX showed that the Super-P-rich separator contained larger amounts of the element sulfur (Super-P = 6.54 wt.%, Super-P/VGCF = 2.00 wt.%, and VGCF = 1.08 wt.%). On the other hand, as shown in Figure 6, after exposure to the same experimental conditions, the Li metal surface was observed using SEM and EDX. Again, although the morphological structure was almost the same, the Li metal recovered from the disassembled Li–S unit cells containing Super-P-rich separators contained a smaller amount of the element sulfur on the surface (Super-P = 6.36 wt.%, Super-P/VGCF = 13.28 wt.% and VGCF = 16.14 wt.%). The sulfur element was originated from the sulfur-containing species, namely polysulfide and LiTFSI. The relationship between polysulfide and LiTFSI for electrochemical decomposition during discharge has not yet been clearly understood. Nonetheless, it can be easily deduced that the elemental change of the Li metal surface depends mainly on the amount of polysulfide. This is because the amount of LiTFSI is the same in all cases because the same amount of liquid electrolyte is used, but the amount of polysulfide changes during operation.

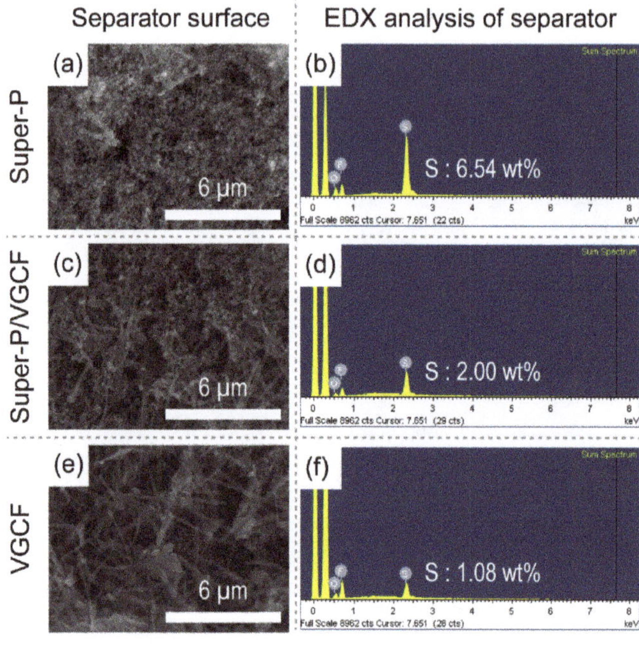

Figure 5. (**a,c,e**) SEM images and (**b,d,f**) EDX elemental analysis of the surfaces of the separators on the cathode side after 20 cycles (corresponding to Figure 3c).

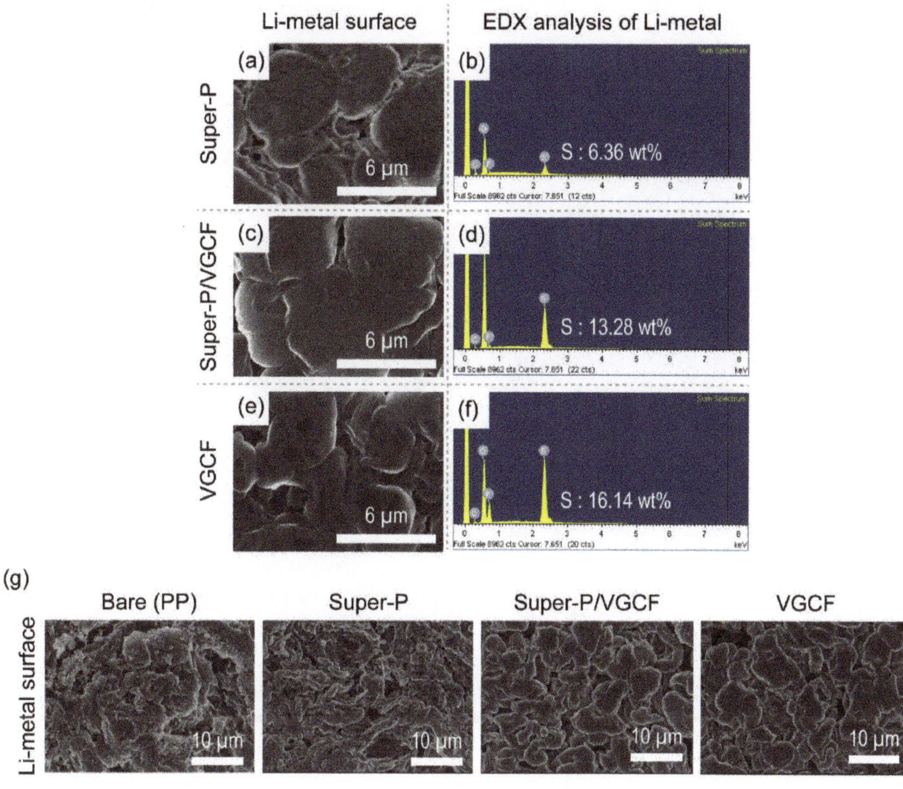

Figure 6. (**a,c,e**) SEM images and (**b,d,f**) EDX elemental analysis of the Li metal after 20 cycles (corresponding to Figure 3c); (**g**) SEM images of the Li metal after 50 cycles (corresponding to Figure 3c).

4. Conclusions

In this study, the effect of the Super-P/VGCF ratio of the carbon-coated separators on the electrochemical performance of Li–S batteries was investigated. Although the Super-P-rich separator exhibited the highest tortuosity with the highest Gurley number, Li–S unit cells containing the Super-P-rich separator showed superior cycle performance and rate capabilities compared to the other types of separators. This implies that polysulfide shuttling is the main factor determining the performance of Li–S batteries rather than the dynamic behavior of separators. Furthermore, we demonstrated that the migration of the soluble polysulfide was efficiently inhibited by the Super-P-rich separators, which prevented the polysulfide from reaching the Li metal surface (Figure 7). Consequently, manipulating the porosity of the porous membrane to control the migration of soluble lithium polysulfide is of key importance for the development of Li–S batteries.

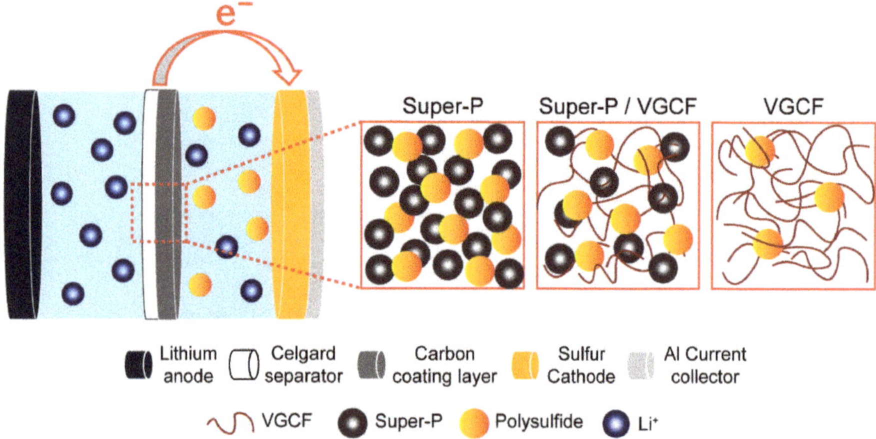

Figure 7. Schematic representation of polysulfide immobilization by Super-P-rich separators in Li–S cells. The black and yellow spheres represent lithium and sulfur particles, respectively.

Author Contributions: H.J. and J.O. both conducted and equally contributed to the experiments and writing of the manuscript. M.-H.R. and Y.M.L. both equally supervised and guided the experiments and refined the manuscript.

Funding: This work was supported by the Basic Science Research Program through the National Research Foundation of Korea (NRF) funded by the Ministry of Education (NRF-2016R1D1A3B03933293). This research was supported by the Basic Science Research Program through the National Research Foundation of Korea (NRF) funded by the Ministry of Education (No.2018M3A7B4071066). This research was supported by the Industrial Technology Innovation Project of Korea Industrial Technology Evaluation and Management Institute (No. 20001370).

Conflicts of Interest: The authors declare no conflicts of interest.

References

1. Elazari, R.; Salitra, G.; Garsuch, A.; Panchenko, A.; Aurbach, D. Sulfur-impregnated activated carbon fiber cloth as a binder-free cathode for rechargeable Li-S batteries. *Adv. Mater.* **2011**, *23*, 5641–5644. [CrossRef] [PubMed]

2. Guo, B.; Li, Y.; Yao, Y.; Lin, Z.; Ji, L.; Xu, G.; Liang, Y.; Shi, Q.; Zhang, X. Electrospun $Li_4Ti_5O_{12}$/C composites for lithium-ion batteries with high rate performance. *Solid State Ionics* **2011**, *204*, 61–65. [CrossRef]

3. Ji, X.; Lee, K.T.; Nazar, L.F. A highly ordered nanostructured carbon–sulphur cathode for lithium–sulphur batteries. *Nat. Mater.* **2009**, *8*, 500. [CrossRef] [PubMed]

4. Ji, X.; Nazar, L.F. Advances in Li–S batteries. *J. Mater. Chem.* **2010**, *20*, 9821–9826. [CrossRef]

5. Lu, S.; Cheng, Y.; Wu, X.; Liu, J. Significantly improved long-cycle stability in high-rate Li–S batteries enabled by coaxial graphene wrapping over sulfur-coated carbon nanofibers. *Nano Lett.* **2013**, *13*, 2485–2489. [CrossRef] [PubMed]

6. Yang, Z.; Zhang, J.; Kintner-Meyer, M.C.; Lu, X.; Choi, D.; Lemmon, J.P.; Liu, J. Electrochemical energy storage for green grid. *Chem. Rev.* **2011**, *111*, 3577–3613. [CrossRef] [PubMed]

7. Chen, C.; Fu, K.; Lu, Y.; Zhu, J.; Xue, L.; Hu, Y.; Zhang, X. Use of a tin antimony alloy-filled porous carbon nanofiber composite as an anode in sodium-ion batteries. *RSC Adv.* **2015**, *5*, 30793–30800. [CrossRef]

8. Ge, Y.; Jiang, H.; Zhu, J.; Lu, Y.; Chen, C.; Hu, Y.; Qiu, Y.; Zhang, X. High cyclability of carbon-coated TiO_2 nanoparticles as anode for sodium-ion batteries. *Electrochim. Acta* **2015**, *157*, 142–148. [CrossRef]

9. Peng, H.-J.; Liang, J.; Zhu, L.; Huang, J.-Q.; Cheng, X.-B.; Guo, X.; Ding, W.; Zhu, W.; Zhang, Q. Catalytic self-limited assembly at hard templates: A mesoscale approach to graphene nanoshells for lithium–sulfur batteries. *ACS Nano* **2014**, *8*, 11280–11289. [CrossRef] [PubMed]

10. Qiu, Y.; Li, W.; Zhao, W.; Li, G.; Hou, Y.; Liu, M.; Zhou, L.; Ye, F.; Li, H.; Wei, Z. High-rate, ultralong cycle-life lithium/sulfur batteries enabled by nitrogen-doped graphene. *Nano Lett.* **2014**, *14*, 4821–4827. [CrossRef] [PubMed]

11. Xu, G.-L.; Xu, Y.-F.; Fang, J.-C.; Peng, X.-X.; Fu, F.; Huang, L.; Li, J.-T.; Sun, S.-G. Porous graphitic carbon loading ultra high sulfur as high-performance cathode of rechargeable lithium–sulfur batteries. *ACS Appl. Mater. Interface* **2013**, *5*, 10782–10793. [CrossRef] [PubMed]

12. Lim, J.; Pyun, J.; Char, K. Recent approaches for the direct use of elemental sulfur in the synthesis and processing of advanced materials. *Angew. Chem. Int. Ed.* **2015**, *54*, 3249–3258. [CrossRef] [PubMed]

13. Qie, L.; Manthiram, A. A facile layer-by-layer approach for high-areal-capacity sulfur cathodes. *Adv. Mater.* **2015**, *27*, 1694–1700. [CrossRef] [PubMed]

14. Wu, F.; Shi, L.; Mu, D.; Xu, H.; Wu, B. A hierarchical carbon fiber/sulfur composite as cathode material for Li–S batteries. *Carbon* **2015**, *86*, 146–155. [CrossRef]

15. Liang, C.; Dudney, N.J.; Howe, J.Y. Hierarchically structured sulfur/carbon nanocomposite material for high-energy lithium battery. *Chem. Mater.* **2009**, *21*, 4724–4730. [CrossRef]

16. Ma, G.; Wen, Z.; Jin, J.; Lu, Y.; Rui, K.; Wu, X.; Wu, M.; Zhang, J. Enhanced performance of lithium sulfur battery with polypyrrole warped mesoporous carbon/sulfur composite. *J. Power Sources* **2014**, *254*, 353–359. [CrossRef]

17. Li, G.; Cai, W.; Liu, B.; Li, Z. A multi functional binder with lithium ion conductive polymer and polysulfide absorbents to improve cycleability of lithium–sulfur batteries. *J. Power Sources* **2015**, *294*, 187–192. [CrossRef]

18. Wang, J.; Yao, Z.; Monroe, C.W.; Yang, J.; Nuli, Y. Carbonyl-β-cyclodextrin as a novel binder for sulfur composite cathodes in rechargeable lithium batteries. *Adv. Funct. Mater.* **2013**, *23*, 1194–1201. [CrossRef]

19. Li, M.-T.; Sun, Y.; Zhao, K.-S.; Wang, Z.; Wang, X.-L.; Su, Z.-M.; Xie, H.-M. Metal–organic framework with aromatic rings tentacles: High sulfur storage in Li–S batteries and efficient benzene homologues distinction. *ACS Appl. Mater. Interface* **2016**, *8*, 33183–33188. [CrossRef] [PubMed]

20. Zheng, J.; Tian, J.; Wu, D.; Gu, M.; Xu, W.; Wang, C.; Gao, F.; Engelhard, M.H.; Zhang, J.-G.; Liu, J. Lewis acid–base interactions between polysulfides and metal organic framework in lithium sulfur batteries. *Nano Lett.* **2014**, *14*, 2345–2352. [CrossRef] [PubMed]

21. Hayashi, A.; Ohtomo, T.; Mizuno, F.; Tadanaga, K.; Tatsumisago, M. All-solid-state Li/S batteries with highly conductive glass–ceramic electrolytes. *Electrochem. Commun.* **2003**, *5*, 701–705. [CrossRef]

22. Tao, X.; Liu, Y.; Liu, W.; Zhou, G.; Zhao, J.; Lin, D.; Zu, C.; Sheng, O.; Zhang, W.; Lee, H.-W. Solid-state lithium–sulfur batteries operated at 37C with composites of nanostructured $Li_7La_3Zr_2O_{12}$/carbon foam and polymer. *Nano Lett.* **2017**, *17*, 2967–2972. [CrossRef] [PubMed]

23. Wu, X.-L.; Zong, J.; Xu, H.; Wang, W.; Liu, X.-J. Effects of LAGP electrolyte on suppressing polysulfide shuttling in Li–S cells. *RSC Adv.* **2016**, *6*, 57346–57356. [CrossRef]

24. Lin, Z.; Liu, Z.; Fu, W.; Dudney, N.J.; Liang, C. Phosphorous pentasulfide as a novel additive for high-performance lithium-sulfur batteries. *Adv. Funct. Mater.* **2013**, *23*, 1064–1069. [CrossRef]

25. Balach, J.; Jaumann, T.; Klose, M.; Oswald, S.; Eckert, J.R.; Giebeler, L. Mesoporous carbon interlayers with tailored pore volume as polysulfide reservoir for high-energy lithium–sulfur batteries. *J. Phys. Chem. C* **2015**, *119*, 4580–4587. [CrossRef]

26. Chung, S.-H.; Manthiram, A. A hierarchical carbonized paper with controllable thickness as a modulable interlayer system for high performance Li–S batteries. *Chem. Commun.* **2014**, *50*, 4184–4187. [CrossRef] [PubMed]

27. Chung, S.H.; Manthiram, A. Bifunctional separator with a light-weight carbon-coating for dynamically and statically stable lithium-sulfur batteries. *Adv. Funct. Mater.* **2014**, *24*, 5299–5306. [CrossRef]

28. Yao, H.; Yan, K.; Li, W.; Zheng, G.; Kong, D.; Seh, Z.W.; Narasimhan, V.K.; Liang, Z.; Cui, Y. Improved lithium–sulfur batteries with a conductive coating on the separator to prevent the accumulation of inactive S-related species at the cathode–separator interface. *Energy Environ. Sci.* **2014**, *7*, 3381–3390. [CrossRef]

29. Yeon, D.; Lee, Y.; Ryou, M.-H.; Lee, Y.M. New flame-retardant composite separators based on metal hydroxides for lithium-ion batteries. *Electrochim. Acta* **2015**, *157*, 282–289. [CrossRef]

30. Cho, I.; Choi, J.; Kim, K.; Ryou, M.-H.; Lee, Y.M. A comparative investigation of carbon black (Super-P) and vapor-grown carbon fibers (VGCFs) as conductive additives for lithium-ion battery cathodes. *RSC Adv.* **2015**, *5*, 95073–95078. [CrossRef]

31. Yang, M.; Hou, J. Membranes in lithium ion batteries. *Membranes* **2012**, *2*, 367–383. [CrossRef] [PubMed]

32. Zhang, S.S. A review on the separators of liquid electrolyte Li-ion batteries. *J. Power Sources* **2007**, *164*, 351–364. [CrossRef]

33. Freitag, A.; Stamm, M.; Ionov, L. Separator for lithium-sulfur battery based on polymer blend membrane. *J. Power Sources* **2017**, *363*, 384–391. [CrossRef]

34. Jayaprakash, N.; Shen, J.; Moganty, S.S.; Corona, A.; Archer, L.A. Porous hollow carbon@sulfur composites for high-power lithium–sulfur batteries. *Angew. Chem. Int. Ed.* **2011**, *50*, 5904–5908. [CrossRef] [PubMed]

35. He, M.; Yuan, L.-X.; Zhang, W.-X.; Hu, X.-L.; Huang, Y.-H. Enhanced cyclability for sulfur cathode achieved by a water-soluble binder. *J. Phys. Chem. C* **2011**, *115*, 15703–15709. [CrossRef]

36. Zhang, S. Understanding of sulfurized polyacrylonitrile for superior performance lithium/sulfur battery. *Energies* **2014**, *7*, 4588–4600. [CrossRef]

37. Arora, P.; Zhang, Z. Battery separators. *Chem. Rev.* **2004**, *104*, 4419–4462. [CrossRef] [PubMed]

38. Seh, Z.W.; Li, W.; Cha, J.J.; Zheng, G.; Yang, Y.; McDowell, M.T.; Hsu, P.-C.; Cui, Y. Sulphur–TiO$_2$ yolk–shell nanoarchitecture with internal void space for long-cycle lithium–sulphur batteries. *Nat. Commun.* **2013**, *4*, 1331. [CrossRef] [PubMed]

Article

Facile Preparation and Characterization of Carbon Fibers with Core-Shell Structure from Graphene-Dispersed Isotropic Pitch Compounds

Dong Hun Lee [1], Yong-Hwan Choi [1,2], Kyong Yop Rhee [3], Kap Seung Yang [4,5,* and Byung-Joo Kim [1,*]

[1] R&D division, Korea Institute of Carbon Convergence Technology, Jeonju-si, Jeollabuk-do 54853, Korea; leedonghun002@gmail.com (D.H.L.); ebaek241@naver.com (Y.-H.C.)
[2] Department of Organic Materials & Fiber Engineering, Chonbuk National University, Jeonju-si, Jeollabuk-do 54896, Korea
[3] Department of Mechanical Engineering, College of Engineering, Kyung Hee University, Yongin-si, Gyeonggi-do 17104, Korea; rheeky@khu.ac.kr
[4] Carbon Composite Laboratory, R&D Center, HPK Inc., Hwaseong-si, Gyeonggi-do 18487, Korea
[5] Department of Polymer & Fiber System Engineering, Chonnam National University, Gwangju 61186, Korea
* Correspondence: ksyang@chonnam.ac.kr (K.S.Y.); kimbj2015@gmail.com (B.-J.K.); Tel.: +82-62-5301774 (K.S.Y.); +82-63-2193710 (B.-J.K.)

Received: 14 February 2019; Accepted: 21 March 2019; Published: 3 April 2019

Abstract: In this study, isotropic pitch-based carbon fibers were prepared from a mixture of petroleum residue and graphene nanoplatelets with different contents. The softening point and synthetic yield of synthesized isotropic pitches were analyzed and compared to characterize the nature of the pitches. The surface and thermal characteristics of the fibers were observed using scanning electron microscopy and thermogravimetric analysis (TGA), respectively. From the results, it was observed that the prepared carbon fibers had an interesting core-shell structure. In the TGA analysis with air, the carbon fiber having 0.1 wt.% of graphene showed a higher residue yield than that of the sample having 1.0 wt.% of graphene. This result can be explained due to the graphene being placed on the surface region of the carbon fibers and directly helping to increase the surface area of the carbon fibers, resulting in rapid oxidation due to the enhanced contact area with oxygen.

Keywords: pyrolysis fuel oil (PFO); isotropic pitch; graphene; carbon fiber

---

1. Introduction

Carbon fiber (CF) can be prepared from various precursors, such as gases (benzene [1], ethane [2], and methane [3]), polymers (cellulose (rayon), polyacrylonitrile (PAN), polyvinylchloride, and phenol resin [4]), and pitches (isotropic and mesophase pitch). Carbon fiber is considered a useful reinforcement for composites due to its excellent properties, such as its high modulus, dimensional stability, and excellent thermal and electrical conductivities [5–9]. Commercial production has been achieved from only three kinds of precursors: PAN, rayon, and pitches [10]. The pitch-based CFs can be further classified into two types: isotropic pitch and anisotropic (or mesophase) pitch [11,12]. Isotropic pitch-based CFs are widely used for general performance applications because isotropic pitch is easy to spin, and its physicochemical properties can be easily controlled during the carbonization process [13]. Graphene has become one of the most important nanomaterials, and many researchers have been continuing various studies using it in various fields [14–16]. Recently, a variety of studies to combine graphene and other materials are underway [17–20] in attempts to unlock synergetic effects of two materials. From the point of view of mechanical applications, Ji et al. [17] have reviewed graphene/polymer composite fibers. These composite fibers usually exhibit enhanced mechanical, thermal, conductive, and antibacterial properties. In the aspect of energy storage applications, He et al. [18] have made microporous carbon/graphene composites using coal tar pitch and graphene oxide via KOH activation. These composites exhibit a high specific capacitance, good rate performance, and excellent cycle stability due to the good electrical properties that result from the addition of the proper content of graphene. Cheng et al. and Ma et al. [19,20] have studied the co-carbonization behavior of petroleum pitch/graphene oxides. Such composites form different anisotropic structures depending on the content of graphene. However, the present works focus on the properties of graphene/polymer composite fibers or electrical properties of graphene/pitch based porous materials; there has been no report on surface morphology change of graphene-dispersed isotropic pitch-based fiber. In this study, graphene-dispersed isotropic pitch was synthesized, and CF was prepared using it. Various contents of graphene were added to the pitch during the synthesis step to observe the effects of graphene content on the morphology of the CFs. The morphology change and surface properties of the CFs with addition of graphene were observed using scanning electron microscopy (SEM), transmission electron microscopy (TEM), atomic force microscopy (AFM), X-ray photoelectron spectroscopy (XPS), Raman spectroscopy, and X-ray diffraction analysis (XRD); the thermal and electrical properties were also characterized.

2. Materials and Methods

2.1. Materials

The isotropic pitch was synthesized from pyrolysis fuel oil (PFO)-based materials (PFO$^{\#}$), which were thermally fractionated at 360 °C. The phase of PFO$^{\#}$ is solid and has heavy molecular weight fractions of PFO via thermal separation. XGnP®Graphene Nanoplatelets Grade C650 (XG Sciences Inc., Lansing, MI, USA) was selected for the graphene source. N-methyl-2-pyrrolidone anhydrous grade (NMP, Sigma-Aldrich Co., St. Louis, USA) was used as a dispersion agent for the graphene.

2.2. Synthesis of Isotropic Pitch Using PFO$^{\#}$

Heat treatment was applied for the synthesis of isotropic pitches without any catalyst. A 500 mL Pyrex flask containing 30 g of PFO$^{\#}$ was placed in the reactor; it was heated to 280 or 300 °C at a 5 °C/min heating rate, held at that temperature for a certain time, and finally cooled to room temperature. While the heat treatment was underway, the reactant was bubbled with nitrogen at 800 mL/min and stirred at 400 rpm at the same time.

2.3. Synthesis of Isotropic Pitches Using PFO$^{\#}$ and Graphene

Graphene was put into beakers with various PFO$^{\#}$ weight%'s with 30 g of NMP and sonicated for 120 min to disperse the material. The 500 mL flask containing 30 g of PFO$^{\#}$ was placed in the reactor and the dispersed graphene solution was poured into the flask. A two-step heat treatment was used for the synthesis. First, the mixture was heated to 300 °C at 5 °C/min, held for 360 min, and cooled down. Then, it was heated again to 350 °C at 5 °C/min and held at that temperature for 240 min and before cooling to room temperature.

2.4. Carbon Fiber Preparation

Synthesized pitches (with or without graphene) were deposited in a stainless-steel chamber equipped with a spinneret (0.5×0.5 mm, L/D = 1) and spun into fiber at 300 m/min speed while applying pressurized nitrogen gas (Figure 1). The spinning temperature was controlled in a range 40 to 50 °C higher than the temperature of the softening point of the pitches [21]. The spun fibers were stabilized in an oven under air atmosphere, followed with a two-step program of heating at 1 °C/min from room temperature to 300 °C and holding for 60 min at the same temperature. The stabilized fibers were carbonized in a horizontal furnace heated under a nitrogen atmosphere at 5 °C/min up to 1000 °C, at which temperature, they were held for 60 min.

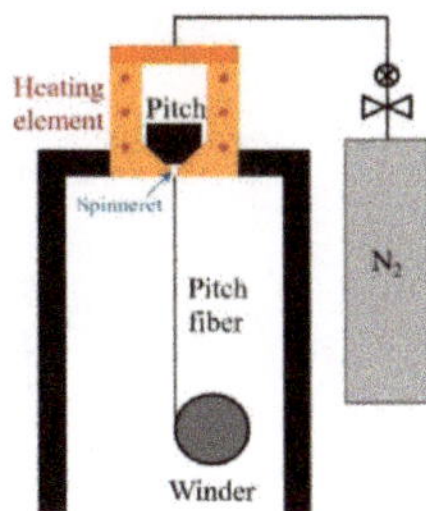

Figure 1. Melt-spinning instrument scheme.

2.5. Characterization of Materials and Products

The yield of synthesized pitches was calculated using Equation (1), and the softening points of synthesized pitches were measured using a Mettler FP 90(Mettler Toredo, Columbus, Ohio, USA) by following the directions of the ASTM (American Society for Testing and Materials) D3104 standard test method.

$$\text{Yield}(\%) = \frac{Weight\ of\ the\ product}{Weight\ of\ the\ reactant} \times 100 \tag{1}$$

The thermal properties of the samples were analyzed using an SDTA 841e – TGA analyzer (Mettler Co.) in the temperature range of 30 °C to 800 °C at a heating rate of 10 °C/min in an air atmosphere.

The surface structure and functional groups of the fibers were analyzed using SEM (S-4700, Hitachi, Tokyo, Japan), AFM (XE-70, Park systems. Co., Suwon, Korea), XPS (PHI 5000 Versa Probes II, ULVAC-PHI, Inc., Kanagawa, Japan), and TEM (JEM-2010, JEOL, Co., Tokyo, Japan). In order to prepare the TEM sample, the carbon fibers were molded into an epoxy resin. The samples were cut using a focused ion beam (FIB, JIB-4601F, JEOL, Co., Tokyo, Japan). The microstructural property was analyzed using Raman spectroscopy (LabRAM HR800, Horiba, Co., Kyoto, Japan) and XRD (X′pert pro Powder, Malvern PANalytical., Eindhoven, Netherlands). The electrical resistivity of the carbon fibers was measured using a Loresta GP resistivity meter (MCP-T610, Mitsubishi Chemical Co., Kanagawa, Japan) connected with a four-point-probe (MCP-TP03P, Mitsubishi Chemical Co., Kanagawa, Japan). A tubular furnace with an alumina tube was used for the oxidation of the carbon fibers to observe different weight loss behaviors via surface oxidation due to the different morphologies of the fibers. Material was heated to 500 °C at 20 °C/min and held at the same temperature for 40 min with 100 mL/min of nitrogen flow. Then, the atmospheric gas was shifted from nitrogen to air. Finally, system was held for 20 min in the same conditions for both samples.

3. Results and Discussion

The sample preparation ID of PFO#280(240)-800 indicates PFO# was manufactured using the following heat treatment conditions: '280' and '(240)' are the holding temperature and time; '800' is the flow rate of nitrogen atmosphere. The data about synthesis yield and softening points are listed in Table 1. The PFO# was observed as a solid petroleum residue with a softening point in the range of 162.0 to 172.7 °C. As the reaction temperature and time increased, the yield and the softening point changed. When the reaction temperature increased, the yield showed a decrease, and the softening point exhibited an increase. In the case of the increase in holding time, two factors (softening point and synthesis yield) showed the same trend when the reaction temperature increased. This is probably due to the decrease in the low molecular weight fractions using a heat treatment and the increase in average molecular weight. When graphene was added, it was possible through the heat treatment to produce a pitch having a softening point similar to that in the case of the material not containing graphene.

Table 1. Yields and Softening Points of Pitches Synthesized as Functions of Thermal Treatment Conditions and Graphene Addition.

Sample	Yield (%)	Softening Point (°C)
PFO[#]	-	167.4 ± 5.4
PFO[#] 280(240)-800	83.3	190.3 ± 0.2
PFO[#] 300(120)-800	48.3	274.4 ± 6.0
PFO[#] 300(240)-800	45.0	319.5 ± 8.1
PFO[#]/0.1 graphene 300(360)350(240)-800	50.6	285.6 ± 4.9
PFO[#]/1.0 graphene 300(360)350(240)-800	54.5	277.6 ± 4.8

Note: Each sample was measured two times.

Figure 2 shows the TGA and derivative of the TGA (DTG) data of PFO[#]-based fibers (nitrogen atmosphere, 50 mL/min of feeding rate). PFO[#]/graphene-based fibers exhibited a lower weight loss than that of the PFO[#]-based fibers. The weight of the PFO[#]-based fibers decreased rapidly around 500 °C. However, the PFO[#]/graphene-based fibers showed better thermal resistance (they showed a similar drop point around 550–600 °C). It seems that the addition of graphene significantly increased the thermal stability of the pitch fiber.

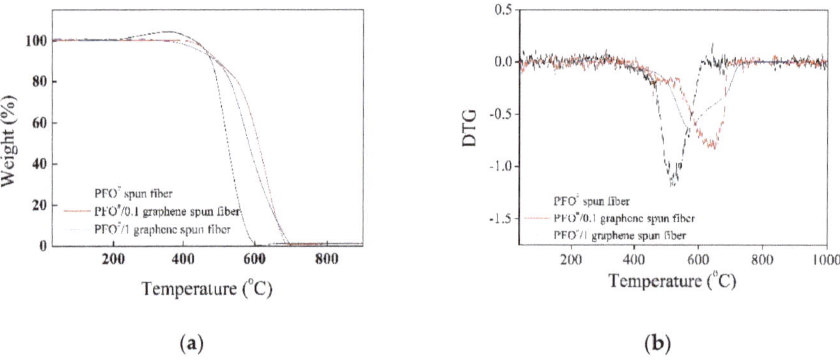

(a)   (b)

Figure 2. Thermal analysis of PFO[#]-based fibers before and after graphene addition; (a) TGA, (b) DTG.

Generally, pitch-based spun fibers are highly brittle and difficult to handle. Thus, in this work, they were stabilized in an oven under an ambient condition by heating them from room temperature to 300°C at a heating rate of 1 °C/min and then maintaining this temperature for 60 min [22]. Finally, the stabilized fibers were carbonized by heating up to 1000 °C at a heating rate of 5 °C/min in an inert atmosphere. The mass change of the pitch-based fibers is listed in Table 2. It is found that a significant weight was gained in the stabilized fibers (≈10.5%) because oxygen atoms diffuse into the fibers to induce thermosetting [23]. This crosslinking reaction helps the fibers maintain their shape and allows them to be carbonized in the following process without inter-filament fusing [24]. It is interesting to note that the fibers having graphene exhibit lower stabilization yield than that of the neat pitch-based fiber. This indicates that graphene is not oxidized or is less oxidized at the stabilization condition. Moreover, it can be recognized that the graphene played the role of a gas barrier, preventing oxygen diffusion in the fiber due to the severe difference of the stabilization yield.

Table 2. Process Yields of PFO#-based Fibers as Functions of Thermal Treatment Conditions and Graphene addition.

Sample	Stabilization Yield (%)
PFO# 300(240)-800	110.5
PFO#/0.1 graphene 300(360)350(240)-800	109.6
PFO#/1.0 graphene 300(360)350(240)-800	107.7

Figure 3 presents SEM images of the PFO#-based carbon fibers. The neat pitch-based carbon fiber (Figure 3a) had smooth morphology and exhibits a range of diameters of 20.7–23.7 μm. There were no distinctive defects on the surfaces of the fibers. However, some voids were observed in the SEM micrographs of the surface and cross-sectional area of the PFO#/graphene-based CFs, and the skin area of the fibers showed a unique core-shell structure. It is considered that the graphene added during the synthesis of pitch did not directly participate in the reaction, but was placed at the outside of the fiber during the spinning process as a result of the different rheological properties of the pure pitch and the pitch/graphene-mixed domain (during spinning, the core areas of the fibers were exposed to high flow compared to the outer side; this can cause extraction of graphene added to the outer side, resulting in the formation of a core-shell structure).

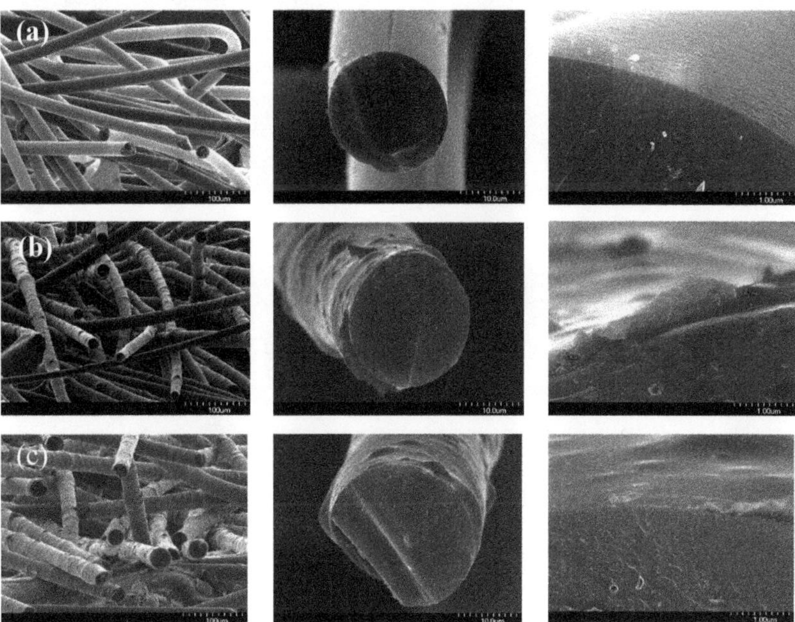

Figure 3. SEM images of PFO#-based carbon fibers; (a) PFO#, (b) PFO#/0.1 graphene, and (c) PFO#/1.0 graphene.

Longitudinal TEM images of PFO#-based carbon fibers are shown in Figure 4. Concerning the longitudinal TEM images, the structure of the PFO#-based carbon fiber displayed a random microtexture of carbon microcrystallite. The structure of the PFO#/0.1graphene-based carbon fiber displayed a random microtexture of carbon microcrystallite with crumpled graphene sheets. In the high magnification images, the graphene sheet displayed a well-aligned carbon crystallite microtexture.

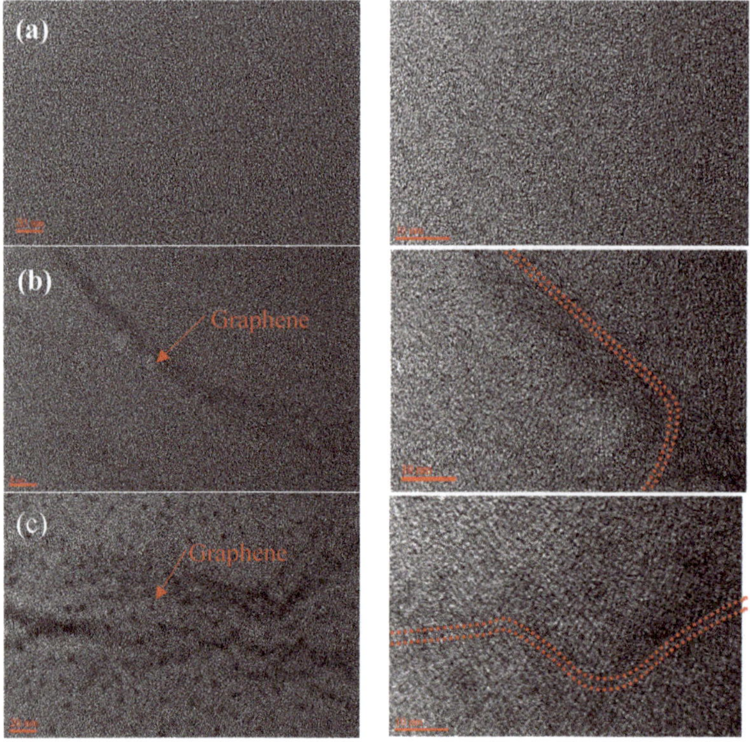

Figure 4. TEM images of PFO#-based carbon fibers: (**a**) PFO# and (**b,c**) PFO#/0.1 graphene.

Figure 5 confirms that the electrical conductivity of PFO#-based carbon fibers showed a slight increase with the addition of the graphene. The electrical conductivity of the carbon fiber was considered to be affected by the graphene, which oriented in the longitudinal direction as shown in Figure 4.

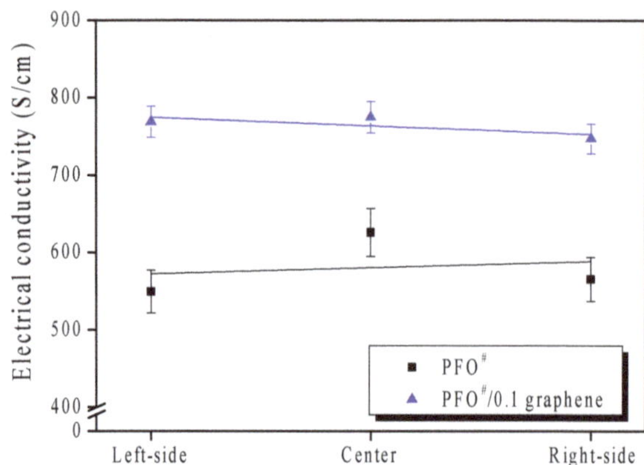

Figure 5. Electrical conductivity of PFO#-based carbon fibers.

AFM topographic images of PFO#-based carbon fibers and height profiles derived from AFM are presented in Figure 6. An absolute comparison using AFM was impossible due to the difference in diameter of the fibers. However, the AFM images show particles on the surface after addition of graphene, and the height profiles were also significantly changed.

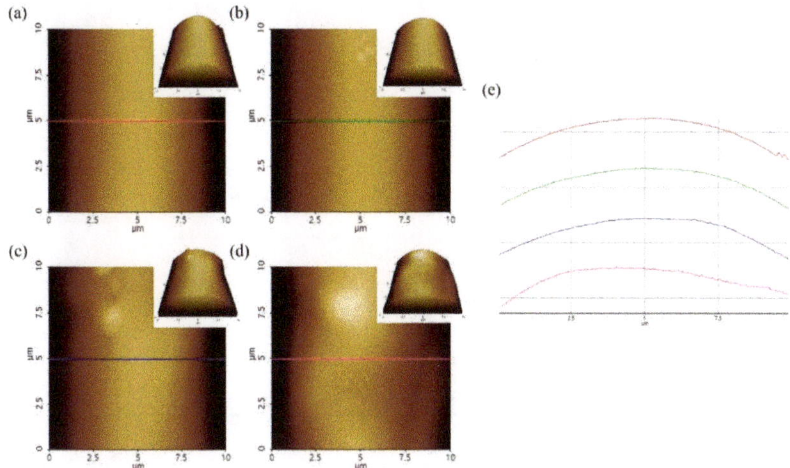

Figure 6. AFM surface morphology of PFO#-based carbon fibers; (**a,b**) PFO#, (**c,d**) PFO#/0.1graphene, and (**e**) height pattern along the color lines.

The diffraction patterns of the PFO#-based carbon fibers show two broad (002) diffraction peaks at around 2θ = 23°, with another (10*l*) diffraction peak at around 2θ = 44°, indicating a graphitic structure [25,26]. The inter-planar spacing d_{002} was calculated using the Bragg equation from the intensity of the diffraction peak. Lateral size (L_a) and the stacking height (L_c) of the crystallite were determined using Equation (2,3) [25,26]:

$$L_a = 1.84\lambda / A_a \cos \beta_a \tag{2}$$

$$L_c = 0.89\lambda / A_c \cos \beta_c \tag{3}$$

where λ is the wavelength of X-rays used, A_α and A_c are the full half width of the (10*l*) and (002) peaks, and β_α and β_c are the corresponding scattering angles.

Figure 7 and Table 3 present the XRD results of the PFO#-based carbon fibers. It can be seen that the XRD peak intensities of carbon fibers with 0.1 wt.% of graphene were higher than those without graphene, and the addition of graphene reduces the lateral size (L_a) but increased the stacking height (L_c). It was also interesting to note that the centers of the (002) and (101) peaks were shifted to the right side, indicating that graphitization was accelerated by the addition of graphene to the pitch.

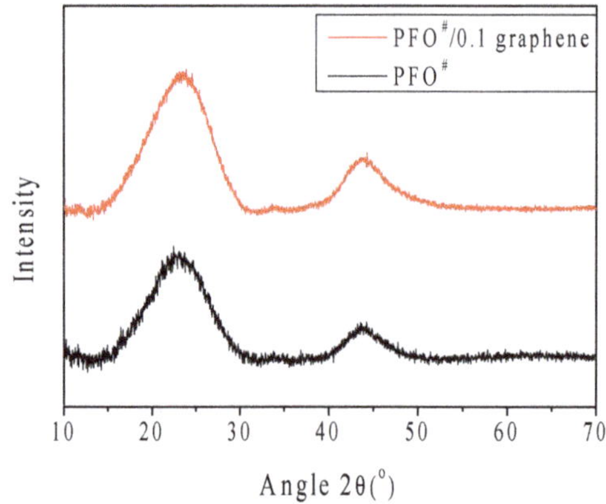

Figure 7. XRD patterns of PFO[#]-based carbon fibers; PFO[#], PFO[#]/0.1 graphene.

Table 3. Crystallite size results of PFO[#]-based carbon fibers: PFO[#], PFO[#]/0.1 graphene.

Sample	002 Peak				10*l* Peak		
	2θ	FWHM	d_{002} (Å)	L_c (Å)	2θ	FWHM	L_a (Å)
PFO[#]	22.95 ± 0.02	7.54	3.73	13.25	44.05 ± 0.06	4.97	35.25
PFO[#]/0.1 Graphene	23.08 ± 0.01	7.73	3.83	17.95	44.24 ± 0.03	6.32	27.74

Figure 8 and Table 4 present the Raman spectroscopy results of the PFO[#]-based carbon fibers. Strong peaks can be seen in the range of 1300–1600 cm^{-1}. These Raman spectra exhibited two large peaks: one near 1330 cm^{-1}, which was a D peak from amorphous structures of carbon, and another near 1580 cm^{-1}, which was a G peak from the graphitic structures of carbon [27,28]. The D-band peak position of carbon fibers shifted from 1346 to 1359 cm^{-1}, and the R-value (defined as I_D/I_G) decreased from 2.39 to 2.28. These results can be explained as showing that there was a shift toward a higher wavelength number, which meant greater ordering of the PFO[#]/0.1 graphene-based carbon fiber [29].

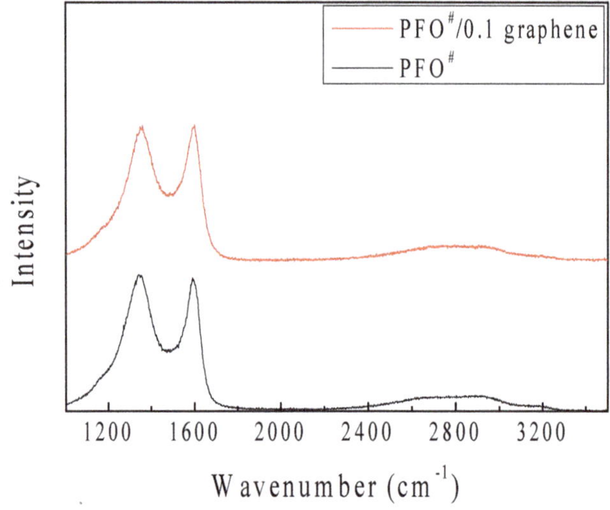

Figure 8. Raman spectra of PFO[#]-based carbon fibers; PFO[#], PFO[#]/0.1 graphene.

Table 4. Raman Spectra Results of PFO# and PFO#/0.1 Graphene-Based Carbon Fiber.

Sample	Peak Position		Peak Intensity		R-Value (I_D/I_G)
	D Band	G Band	D Band	G Band	
PFO#	1346.28 ± 0.76	1588.45 ± 0.50	574.28	570.73	2.39
PFO#/0.1 Graphene	1359.43 ± 0.84	1591.63 ± 0.52	384.16	396.21	2.28

In the wide-scan XPS spectra (Figure 9a), the surface of PFO#-based carbon fibers can be seen to contain the expected two elements (C and a small amount of O), because carbon fibers were formed via carbonization at 800 °C in the nitrogen condition. The C_{1s} spectra (Figure 9b) were used to analyze the types and amounts of functional groups of carbon. These results were fitted into five peaks at 284.3, 285.2, 286.7, 288.6, and 290.5 eV, which correspond to C–C (aromatic), C–C or C–H (aliphatic), C–O, –C=O, and –COOH [30–33], respectively. The area ratios of aromatic carbon bonds/other carbon bonds, which depends on the degree of disorder of the carbon materials [31], were calculated from Figure 9b and are shown in Table 5. It is clear that the ratio for PFO#/0.1 graphene-based carbon fibers was larger than that of PFO#-based carbon fibers. This result seems to derive from the more ordered structure shown in the Raman analysis.

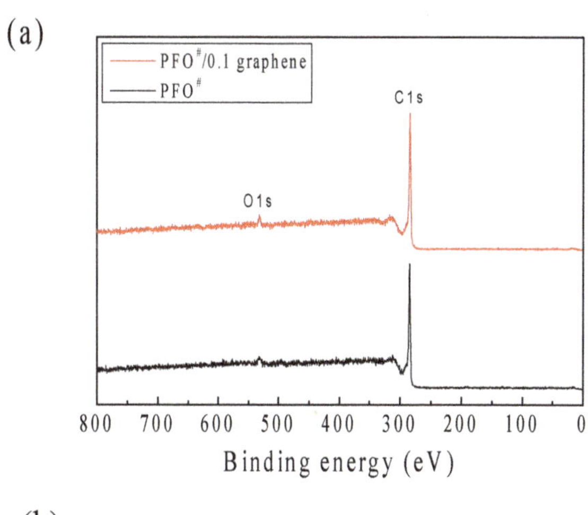

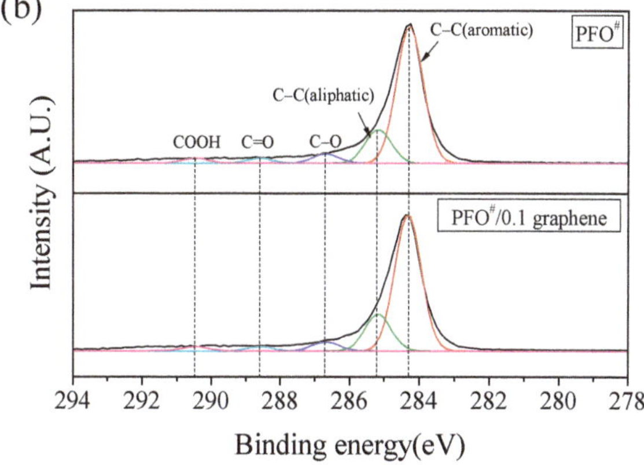

Figure 9. (**a**) Wide-scan XPS spectra and (**b**) C_{1s} XPS spectra of PFO#-based carbon fibers.

Table 5. Area Ratios of Chemical Bonding Peaks from C_{1s} XPS Spectra.

	Peak Area		Area Ratio
	Aromatic Carbon	**Other Carbon**	
PFO$^{\#}$	5135.04	3225.53	1.59
PFO$^{\#}$/0.1 graphene	4596.05	2515.76	1.83

Table 6 and Figure 10 present the oxidation results of each carbon fiber. After oxidation, the surface of the carbon fibers was distinctively changed. The surface of PFO$^{\#}$-based carbon fibers (Figure 10a) was smoother than that shown in Figure 3a, but in the PFO$^{\#}$/graphene-based carbon fibers (Figure 10b,c) the destruction of the core-shell structure during the oxidation process can be observed. It appears that graphene, which was a shell, competitively reacted with oxygen atoms, and this was related to a decrease in oxidation yield when the graphene content increased (Table 6).

Table 6. Yields of Carbon Fibers after Oxidation at 500 °C in Air.

Sample	Yield (%)
PFO$^{\#}$ 300(240)-800	89.1
PFO$^{\#}$/0.1 graphene 300(360)350(240)-800	86.6
PFO$^{\#}$/1.0 graphene 300(360)350(240)-800	52.1

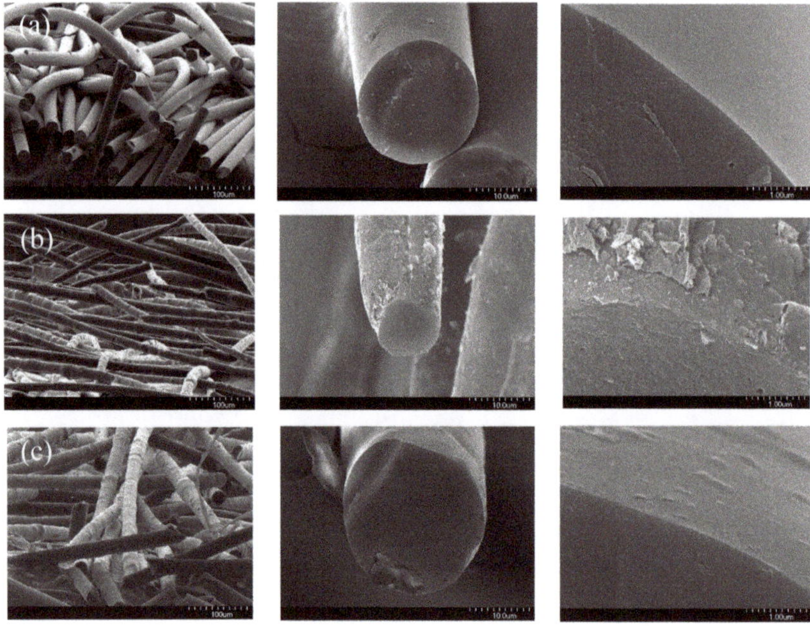

Figure 10. Oxidized PFO$^{\#}$-based carbon fibers; (**a**) PFO$^{\#}$, (**b**) PFO$^{\#}$/0.1 graphene, and (**c**) PFO$^{\#}$/1.0 graphene.

These oxidation results are direct proof that the core-shells were formed by graphene addition to the pitch precursors. Moreover, it was possible to obtain functional fibers with high specific surface area using this core-shell structure.

The carbon fibers fabricated in this work were identified as having a higher surface area and better electric conductivity than virgin carbon fibers. Usually, a high surface area of fillers can cause good mechanical adhesion with a polymer matrix in a composite system. Recently, research to apply carbon fibers has been used to fabricate polymer-matrix 3D printing wires [34,35]. However, low adhesion strength between carbon fibers (or other carbonaceous materials) and thermoplastic matrix has been a critical obstacle to commercialization. The high surface area of carbon fibers can be an excellent solution to enhance the mechanical interfacial strength between fibers and thermoplastic matrices.

Moreover, with the development of the shale gas industry, the importance of proton-exchange-membrane-fuel-cells (PEMFC) has been steadily increasing. The gas-dispersion-layer (GDL) in a PEMFC is normally made of carbon fibers due to their excellent mechanical properties, chemical resistance, and electric conductivity. Carbon fibers with a core-shell structure, like those prepared in this work, can be candidate materials for GDL [36–38]. Additionally, this kind of carbon fiber can be easily activated by further oxidation in steam and to have porous structures that will maintain higher electric conductivity compared to that of conventionally activated carbon fibers. This means that carbon fibers with core-shell structures can be used as electrode materials for energy storage devices such as fuel cells and supercapacitors.

4. Conclusions

In this study, using various thermal treatments, isotropic pitches/graphene precursors were prepared from compounds of petroleum residue and graphene-dispersed solution. As the reaction temperature and time increased, the yield decreased, and the softening point increased. The addition of graphene increased the thermal stability, surface roughness, and ordered structures of the pitch fibers. Isotropic pitch-based carbon fibers, which were prepared from petroleum residue and graphene-dispersed solution, showed core-shell structures. When 0.1 wt.% and 1.0 wt.% of graphene were added, the oxidation yields were 86.6% and 52.1%, respectively, which seem to be due to the acceleration of the oxidation reaction due to the presence of graphene at high temperature. Carbon fibers with a core-shell structure, which have a high surface area, could be used in C/C composite materials, and would result in good adhesion with the polymer matrix. Carbon fibers with more ordering of structures via the addition of graphene lead to improved electrical conductivity, such that they could be used in electrode materials to obtain enhanced electrical conductivity.

Author Contributions: D.H.L. and Y.-H.C. analyzed the data for the work, acquired the data for the work, drafted the work, gave final approval of the version to be published, and agreed to be accountable for all aspects of the work in questions related to its accuracy. K.Y.R. and K.S.Y. designed the concept of the work, and agreed to be accountable for all aspects of the work in questions related to its accuracy. B.-J.K. analyzed the data for the work, revised it critically for important intellectual content, gave final approval of the version to be published, and agreed to be accountable for all aspects of the integrity, investigation and resolution of the work.

Funding: This work was supported by the Technology Innovation Program (10083586, Development of Petroleum-based Graphite Fibers with Ultra-High Thermal Conductivity), funded by the Ministry of Trade, Industry & Energy (MOTIE, Korea).

Conflicts of Interest: The authors declare no conflict of interest.

References

1. Jayasankar, M.; Agarwal, N.; Chand, R.; Gupta, S.K.; Kunzru, D. Vapor grown carbon fibers from benzene pyrolysis: Filament length distributions. *Carbon* **1996**, *34*, 127–134. [CrossRef]
2. Thornton, J.M.; Walker, S.G. Catalytic carbon deposition on three-dimensional carbon fiber preforms using alkane gas feedstocks. *New Carbon Mater.* **2009**, *24*, 251–259. [CrossRef]
3. Sacco, A.; Thacker, P.; Chang, T.N. The initiation and growth of filamentous carbon from alpha-iron in H_2, CH_4, H_2O, CO_2, and Co Gas-Mixtures. *J. Catal.* **1984**, *85*, 224–236. [CrossRef]
4. Chen, Y.J. *Activated Carbon Fiber and Textiles*; Woodhead Publishing: Sawston, UK, 2017; p. 189.
5. Northolt, M.G.; Veldhuizen, L.H.; Jansen, H. Tensile deformation of carbon fibers and the relationship with the modulus for shear between the basal planes. *Carbon* **1991**, *29*, 1267–1279. [CrossRef]
6. Kumar, S.; Anderson, D.P.; Crasto, A.S. Carbon fibre compressive strength and its dependence on structure and morphology. *J. Mater. Sci.* **1993**, *28*, 423–439. [CrossRef]

7. Edie, D.D.; Fain, C.C.; Robinson, K.E.; Harper, A.M.; Rogers, D.K. Ribbon-shape carbon fibers for thermal management. *Carbon* **1993**, *31*, 941–949. [CrossRef]

8. Hong, S.H.; Korai, Y.; Mochida, I. Development of mesoscopic textures in transverse cross-section of mesophase pitch-based carbon fibers. *Carbon* **1999**, *37*, 917–930. [CrossRef]

9. Hong, S.H.; Korai, Y.; Mochida, I. Mesoscopic texture at the skin area of mesophase pitch-based carbon fiber. *Carbon* **2000**, *38*, 805–815. [CrossRef]

10. Peebles, L.H. *Carbon Fibers: Formation, Structure, and Properties*; CRC Press: Boca Raton, FL, USA, 1995; pp. 3–42.

11. Wazir, A.H.; Kakakhel, L. Preparation and characterization of pitch-based carbon fibers. *New Carbon Mater.* **2009**, *24*, 83–88. [CrossRef]

12. Mora, E.; Blanco, C.; Prada, V.; Santamaria, R.; Granda, M.; Menendez, R. A study of pitch-based precursors for general purpose carbon fibres. *Carbon* **2002**, *40*, 2719–2725. [CrossRef]

13. Lee, H.M.; Kwac, L.K.; An, K.H.; Park, S.J.; Kim, B.J. Electrochemical behavior of pitch-based activated carbon fibers for electrochemical capacitors. *Energy Convers. Manag.* **2016**, *125*, 347–352. [CrossRef]

14. Geim, A.K. Graphene: Status and prospects (Review). *Science* **2009**, *324*, 1530–1534. [CrossRef] [PubMed]

15. Rao, C.N.R.; Sood, A.K.; Subrahmanyam, K.S. Govindaraj, Graphene: The New Two-Dimensional Nanomaterial. *Angew. Chem. Int. Ed.* **2009**, *48*, 7752–7777. [CrossRef]

16. Xu, Z.; Gao, C. Graphene fiber: A new trend in carbon fibers. *Mater. Today* **2015**, *18*, 480–492. [CrossRef]

17. Ji, X.; Xu, Y.; Zhang, W.; Cui, L.; Liu, J. Review of functionalization, structure, and properties of graphene/polymer composite fibers. *Compos. Part A* **2016**, *87*, 29–45. [CrossRef]

18. He, X.; Wang, J.; Xu, G.; Yu, M.; Wu, M. Synthesis of microporous carbon/graphene composites for high-performance supercapacitors. *Diam. Relat. Mater.* **2016**, *66*, 119–125. [CrossRef]

19. Cheng, Y.; Yang, L.; Fang, C.; Guo, X. Co-carbonization behavior of petroleum pitch/graphene oxide: Influence on structure and mechanical property of resultant cokes. *J. Anal. Appl. Pyrolysis* **2016**, *122*, 387–394. [CrossRef]

20. Ma, Y.; Ma, C.; Sheng, J.; Zhang, H.; Wang, R.; Xie, Z.; Shi, J. Nitrogen-doped hierarchical porous carbon with high surface area derived from graphene oxide/pitch oxide composite for supercapacitors. *J. Colloid Interface Sci.* **2016**, *461*, 96–103. [CrossRef] [PubMed]

21. Yang, J.; Shi, K.; Li, X.; Yoon, S.H. Preparation of isotropic spinnable pitch and carbon fiber from biomass tar through the co-carbonization with ethylene bottom oil. *Carbon Lett.* **2018**, *25*, 89–94.

22. Lee, D.H.; Choi, J.S.; Oh, y. S.; Kim, Y.A.; Yang, K.S.; Ryu, H.J.; Kim, Y.J. Catalytic hydrogenation-assisted preparation of melt spinnable pitches from petroleum residue for making mesophase pitch based carbon fibers. *Carbon Lett.* **2017**, *24*, 28–35.

23. Donnet, J.B.; Bansal, R.C. *Carbon Fibers*; Marcel Dekker: New York, NY, USA, 1990; p. 55.

24. Zhu, J.; Park, S.W.; Joh, H.I.; Kim, H.C.; Lee, S.H. Preparation and characterization of isotropic pitch-based carbon fiber. *Carbon Lett.* **2013**, *14*, 94–98. [CrossRef]

25. Takagi, H.; Maruyama, K.; Yoshizawa, N.; Yamada, Y.; Sato, Y. XRD analysis of carbon stacking structure in coal during heat treatment. *Fuel* **2004**, *83*, 2427–2433. [CrossRef]

26. Lu, L.; Sahajwalla, V.; Kong, C.; Harris, D. Quantitative X-ray diffraction analysis and its application to various coals. *Carbon* **2001**, *39*, 1821–1833. [CrossRef]

27. Endo, M.; Kim, C.; Karaki, T.; Kasai, T.; Matthews, M.J.; Brown, S.D.M.; Dresselhaus, M.S.; Tamaki, T.; Nishimura, Y. Structural characterization of milled mesophase pitch-based carbon fibers. *Carbon* **1998**, *36*, 1633–1641. [CrossRef]

28. Zhu, Y.; Zhao, X.; Gao, L.; Cheng, J. Pyrolysis kinetics and microstructure of thermal conversion products on toluene soluble component from two kinds of modified pitch. *Carbon Lett.* **2018**, *28*, 38–46.

29. Roh, J.S.; Kim, S.H. Structural Study of the Oxidized High Modulus Carbon Fiber using Laser Raman Spectroscopy. *Carbon Lett.* **2009**, *10*, 38–42. [CrossRef]

30. Kim, B.H.; Lee, D.H.; Yang, K.S.; Lee, B.C.; Kim, Y.A.; Endo, M. Electron Beam Irradiation-Enhanced Wettability of Carbon Fibers. *ACS Appl. Mater. Interfaces* **2011**, *3*, 119–123. [CrossRef] [PubMed]

31. Boudou, J.P.; Paredes, J.I.; Cuesta, A.; Martìnez-Alonso, A.; Tascón, J.M.D. Oxygen plasma modification of pitch-based isotropic carbon fibres. *Carbon* **2003**, *41*, 41–56. [CrossRef]

32. Moreno-Castilla, C.; López-Ramón, M.V.; Carrasco-Marín, F. Changes in surface chemistry of activated carbons by wet oxidation. *Carbon* **2000**, *38*, 1995–2001. [CrossRef]

Nanomaterials **2019**, *9*, 521

33. Estrade-Szwarckopf, H. XPS photoemission in carbonaceous materials: A "defect" peak beside the graphitic asymmetric peak. *Carbon* **2004**, *42*, 1713–1721. [CrossRef]
34. Zhong, J.; Zhou, G.-X.; He, P.-G.; Yang, Z.-H.; Ja, D.-C. 3D printing strong and conductive geo-polymer nanocomposite structures modified by graphene oxide. *Carbon* **2017**, *117*, 421–426. [CrossRef]
35. Zhang, H.; Yang, D.; Sheng, Y. Performance-driven 3D printing of continuous curved carbon fibre reinforced polymer composites: A preliminary numerical study. *Compos. Part B* **2018**, *151*, 256–264. [CrossRef]
36. Indayaningsih, N.; Priadi, D.; Zulfia, A. Suprapedi, Analysis of coconut carbon fibers for gas diffusion layer material. *Key Eng. Mater.* **2011**, *462–463*, 937–942. [CrossRef]
37. Kannan, M.A.; Munukutla, L. Carbon nano-chain and carbon nano-fibers based gas diffusion layers for proton exchange membrane fuel cells. *J. Power Sources* **2007**, *167*, 330–335. [CrossRef]
38. Tötzke, C.; Manke, I.; Hartnig, C.; Kuhn, R.; Riesemeier, H.; Banhart, J. Investigation of carbon fiber gas diffusion layers by means of synchrotron X-ray tomography. *ECS Trans.* **2011**, *41*, 379–386.

 *nanomaterials*

Article

Introducing Well-Defined Nanowrinkles in CVD Grown Graphene

Tim Verhagen [1,*, Barbara Pacakova [2,3], Martin Kalbac [2] and Jana Vejpravova [1,***

[1] Department of Condensed Matter Physics, Faculty of Mathematics and Physics, Charles University, Ke Karlovu 5, 121 16 Prague 2, Czech Republic

[2] J. Heyrovsky Institute of Physical Chemistry of the CAS, v.v.i., Dolejskova 2155/3, CZ-182 23 Prague 8, Czech Republic; barbara.pacakova@ntnu.no (B.P.); martin.kalbac@jh-inst.cas.cz (M.K.)

[3] Department of Physics, Norwegian University of Science and Technology (NTNU), Høgskoleringen 5, NO-7491 Trondheim, Norway

* Correspondence: verhagen@mag.mff.cuni.cz (T.V.); jana@mag.mff.cuni.cz (J.V.); Tel.: +420-9515-2735 (J.V.)

Received: 31 January 2019; Accepted: 23 February 2019; Published: 4 March 2019

Abstract: The control of graphene's topography at the nanoscale level opens up the possibility to greatly improve the surface functionalization, change the doping level or create nanoscale reservoirs. However, the ability to control the modification of the topography of graphene on a wafer scale is still rather challenging. Here we present an approach to create well-defined nanowrinkles on a wafer scale using nitrocellulose as the polymer to transfer chemical vapor deposition grown graphene from the copper foil to a substrate. During the transfer process, the complex tertiary nitrocellulose structure is imprinted into the graphene area layer. When the graphene layer is put onto a substrate this will result in a well-defined nanowrinkle pattern, which can be subsequently further processed. Using atomic force and Raman microscopy, we characterized the generated nanowrinkles in graphene.

Keywords: CVD graphene; transfer; ruga; wrinkle; ripple; Raman spectroscopy; AFM

---

1. Introduction

The transfer of either exfoliated or chemical vapor deposition (CVD) grown graphene to another substrate is one of the key steps during the fabrication of graphene-based devices. The creation of rugae [1], a single state of a corrugated material configuration in which graphene can be mostly observed as exhibiting either wrinkles, ripples, folds, wrinklons or crumples, during the transfer is generally seen as unwanted, as these topographic features deteriorate the properties of graphene [2,3]. Several techniques were developed to improve the transfer of graphene as perfect as possible. Especially, the introduction of an all-dry transfer (whether or not using h-BN and annealing) resulted in devices with a minimum of these unwanted features [4,5].

However, those "unwanted topographic features" can be very suitable to further explore graphene's potential. One of the most significant changes due to the presence of these rugae is a change in the pyramidalization angle, which describes the angle between the σ- and π-orbitals of graphene, and which is a measure for the chemical reactivity of carbon allotropes [6–8]. As the pyramidalization angle is highly dependent on the local curvature, a larger reactivity can be expected around the rugae. The chemical reactivity of the rugae can be further controlled via the doping level [9]. As they are locally delaminated from the substrate, the influence of the substrate doping is reduced. Furthermore, rugae can act as a generator for pseudo magnetic fields [10] or for anisotropic transport properties [11].

Besides changing these fundamental properties of graphene, a homogeneous, local deformation of graphene can act as a trap for liquids and gases, as graphene is an impenetrable membrane for most liquids and gases [12]. Besides acting as a storage space, these nanoreservoirs could also be an ideal chemical reactor for two dimensional (2D) chemistry [13] or used as a model system to study 2D confined liquids or the effect of a 2D liquid–membrane interface [14].

As shown recently by Hallam et al. [15], the strain in the graphene layer which is induced during the after-growth cooling is released during the removal of the Cu foil during the Cu etching step. Depending on the stiffness of the polymer, the strain within the graphene layer can be released via rugae. For CVD transferred graphene, these rugae most often consist of networks of wrinkles which have a width ranging between one and hundreds of nm, a height below 15 nm, a length (much) larger than 100 nm and an aspect ratio larger than 10. Although it is straightforward to induce some of these topographic corrugations during the transfer of graphene, it is a challenge to create these topographic features homogeneously in the whole transferred graphene layer.

The actual type of corrugation that is created during the transfer can be predicted if the properties of the used polymer for transfer are known. Using rugae mechanics, it can be shown that the shear modulus of both graphene and the polymer, the adhesion energy, the mismatch strain between the polymer and graphene, the orientation of graphene's grain and defects edges and the structure of the used polymer are the dominant factors that determine the types of rugae that can be observed when the graphene is transferred to its final substrate [16,17]. For many polymers these parameters are known, and this information can be used to fine-tune the final rugae landscape.

In this article we show that by transferring graphene using nitrocellulose as a transfer polymer, we can induce a homogeneous landscape of nanoscale wrinkles in the whole graphene sheet, in which topography is induced from the structure of the used nitrocellulose polymer. This easy and straightforward method dramatically improves the production of large homogeneous areas of graphene with nanoscale wrinkles, which are created in a controllable and reproducible way. The ability to create large areas of graphene with the desired topographical properties improves the study of the influence of rugae on the electrical, optical and chemical properties of graphene, as there is no need to extensively search for large, homogeneous areas with the desired type of rugae.

2. Materials and Methods

Graphene samples were grown by the CVD method as reported previously [18]. In brief, the polycrystalline copper foil was heated to 1000 °C and annealed for 20 min under a flow of 50 standard cubic centimeters per minute (sccm) H_2. The copper foil was exposed to 30 sccm CH_4 and 50 sccm H_2 for 20 min, after which the copper foil was cooled to room temperature.

On top of the foil, a nitrocellulose layer (NC) (Collodion solution for microscopy, 2% in amyl acetate, Sigma Aldrich 09817, St. Louis, Missouri, MO, USA), was spincoated with 2700 revolutions per minute (rpm) for 30 s and air dried, which results in a thickness of the NC of approximately 50 nm. The transfer of the 1×1 cm^2 graphene/NC layer is the same as is used for the transfer with PMMA. In short; the Cu foil was etched away using $FeCl_3$ (copper etchant type CE-100, Transene Company Inc., Danvers, MA, USA), and the graphene/NC sample was fished using a Si/SiO$_2$ wafer from the solution and subsequently washed several times using highly purified water. Finally, the graphene/NC layer was fished using the final substrate, which was carefully cleaned using acetone and isopropanol, and dried with nitrogen gas. Residual NC can be easily removed from the graphene by washing the sample several times with methanol. The samples were dried with nitrogen gas and no further annealing steps were performed.

Prepared samples were characterized by atomic force microscopy (AFM) and Raman spectroscopy. The Raman spectra were obtained at a WITec alpha300R spectrometer (WITec Wissenschaftliche Instrumente und Technologie GmbH, Ulm, Germany) equipped with a piezo stage. Raman spectral maps were measured with 2.33-eV (532-nm) laser excitation, a laser power of approximately 1 mW, a grating of 600 lines/mm and lateral steps of 1 μm in both directions. The laser was focused on the sample with a 100× objective. The D, G and 2D modes of the graphene monolayer were analyzed using pseudo-Voigt peak profiles, where one, three (G_1, G_2 and D') and two ($2D_1$ and $2D_2$) peak profiles were used to fit the D, G and 2D modes respectively. The G_1, G_2, $2D_1$ and $2D_2$ modes were used to account for the delaminated graphene, which was created by the NC transfer [19].

AFM was measured using PeakForce quantitative nano-mechanical mapping (PeakForceQNM) and imaging in tapping mode, using a Dimension Icon microscope (Bruker Inc., Billerica, MA, USA). Images were captured using Bruker Scanasyst-Air probes (k = 0.4 N/m, f_0 = 70 kHz, nominal tip radius = 2 nm) at room temperature in air. High quality images were processed in the standard way using Gwyddion software [20], applying line-by-line first order leveling and scar removal.

To determine the overall orientation of wrinkles, Gwyddion's 2D FFT analysis (Gwyddion, Czech Metrological Institute, Brno, Czech Republic) is used. Grain analysis was used to evaluate the surface coverage by the winkles (A_{wr}), and to determine their geometry and width (w_{wr}). Surface characterization was used for the estimation of the roughness (R_a), root-mean-square roughness (R_q), and surface area (SA). The profile height distribution was analyzed to determine the median peak-to-valley distance (R_{tm}), of the wrinkle graphene layer topography, which was used as the wrinkle height. The topography AFM image was cut into individual lines using home-built procedures, in order to identify all the minima and maxima attributed to the highest parts of the wrinkles and the lowest parts in the inter-wrinkle valleys. Then, the median difference of all maxima and minima was calculated as $R_{tm} = p_\mu - v_\mu$, where p_μ is the median wrinkle height and v_μ is the median valley depth, calculated over all the lines in the image.

3. Results

Figure 1a shows an AFM image of the CVD grown graphene transferred with NC on Si/SiO_2 substrate. It can be clearly seen that the graphene is not flat, but a clear, fine landscape of graphene wrinkles is present. Graphene is wrinkled very homogeneously, as shown from the 2D FFT in Figure 1c, within the whole graphene layer and consisted everywhere of a fine structure composed of connected wrinkles.

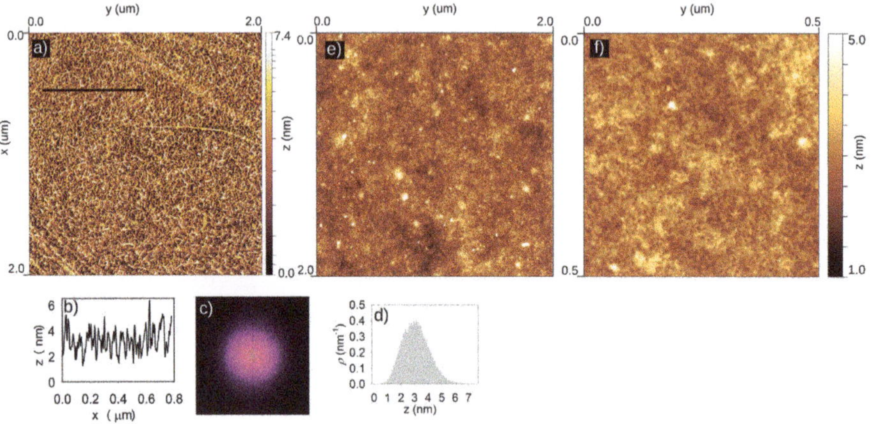

Figure 1. The atomic force microscopy (AFM) topography image of the wrinkled NC transferred graphene (**a**) with the cross-section (black line) in panel (**b**). The 2D FFT image of the AFM topography image (**c**), showing no orientation preference of wrinkles, and the height histogram of the topography image (**d**). (**e**) Large-scale AFM image of NC on a Si/SiO_2 substrate. (**f**) A close up AFM image of the NC on a Si/SiO_2 substrate as shown in (**e**).

Figure 1e shows an AFM image of a NC layer directly spincoated on a Si/SiO_2 substrate without graphene. On the 2×2 um^2 scale, the NC layer was relatively smooth. In Figure 1f a close up of this layer is shown where the very fine, complex tertiary structure of the NC is visible with a height of less than 1 nm, which resembles the landscape of graphene wrinkles presented in Figure 1a.

Homogeneously distributed wrinkles were found to be uniformly covering almost 65% of graphene sheet area (see Table 1). The large-scale wrinkles and folds, which are often present when graphene is transferred with PMMA [11], were rarely observed. If there were large folds, they were wrinkled in the same way as the whole graphene layer. On the large scale, there was no orientation preference, indicating an isotropic compression of the graphene layer, and wrinkles were oriented equally in all directions, as can be seen from the 2D FFT image (Figure 1c). If there was some preferential orientation direction of the wrinkles, the 2D FFT image would not be spherical, but asymmetric in the direction of preferential orientation. Furthermore, the intensity of the 2D FFT sphere was homogeneous and no bright rings in the sphere are visible. This indicates that the distance between two wrinkles varied from wrinkle to wrinkle and no repeating wrinkle pattern was present. The wrinkle height, which is represented by the R_{tm} value, reached 2.5 nm and wrinkle width varies from 6 to 15 nm.

Table 1. The basic characteristics of the NC transferred graphene, determined from the AFM images. Roughness (R_a), root-mean-square roughness, (R_q), wrinkle height (R_{tm}), surface area (SA), area covered with the wrinkles (A_{wr}) and width, (w_{wr}).

R_a (nm)	R_q (nm)	R_{tm} (nm)	SA (%)	A_{wr} (%)	w_{wr} (nm)
0.83	1.03	2.5	105	65	6–15

Raman Spectroscopy

Figure 2 shows a typical Raman spectrum of a CVD graphene monolayer transferred with NC. While the G and 2D modes of graphene are clearly visible, almost no D mode is visible indicating that no significant amount of defects were created in graphene during the NC transfer. Furthermore, the G and 2D modes are rather asymmetric, and therefore fitted with a G_2 and $2D_2$ sub-band. The presence of these sub-bands confirms the results of the AFM analysis, indicating that a significant part of the graphene was delaminated [19]. Finally, the intensity of the G and 2D modes is approximately the same, which was caused by the relatively high doping level of the graphene, as can be seen in Figure 3e [21].

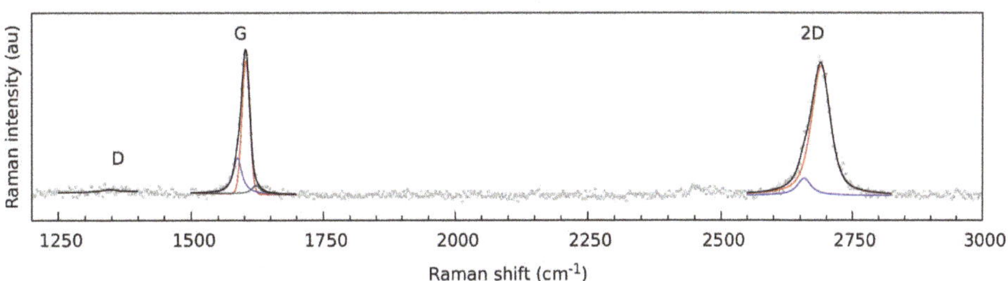

Figure 2. Typical Raman spectrum of graphene, which is transferred using NC. The experimental data are marked by gray dots and fits of the individual bands are shown in red for the G_1 and $2D_1$ components, blue for the G_2 and $2D_2$ components and grey for the D′ component, respectively. The resulting curve (sum of the individual components) is represented by a solid black line.

The Raman data were further analyzed in order to get deeper insight in the charge and strain management in the graphene monolayer with regularly distributed nanowrinkles. Raman spectral maps of the Raman shift of the G_1 and $2D_1$ modes, and the full width at half maximum (FWHM) of the $2D_1$ mode are shown in Figure 3a–c. The Raman shift of the G_1 and $2D_1$ modes was not homogeneous, and well-defined regions with a smaller or larger Raman shift are visible. These regions were caused by the different Cu grains of the Cu foil, onto which the graphene was grown. As shown before [21], the actual strain and doping varies depending on the crystal orientation of the Cu grain in the Cu foil. This variation in strain and doping was visible as a variation in the actual Raman shift and FWHM of the G and 2D modes. Using well known relations describing the relation between strain, doping and the measured Raman shifts of the G and 2D modes, an actual estimation of the strain and doping can be made [22,23]. As a first estimate, a biaxial strain distribution was assumed. As shown in Figure 3d, also in the calculated strain map, clear regions with a different strain were be observed, whereas in the doping map shown in Figure 3e these regions showed a much more homogeneous doping.

The FWHM of the 2D mode, which acts as a probe of the nanometer-scale flatness [24] is shown in Figure 3c. The FWHM of the $2D_1$ mode of the NC transferred graphene was rather homogeneous and there was no obvious dependence visible, as could be seen for the Raman shift of the G_1 and $2D_1$ modes. This is an indication that during the NC transfer, the wrinkles as shown in the AFM images in Figure 1 were homogeneously induced in the graphene layer. Figure 3f shows the distribution of the FWHM of the $2D_1$ mode, which shows a narrow distribution of FWHM values around 48 cm^{-1}, where extremely flat graphene can have a FWHM of the 2D mode well below 20 cm^{-1} [5].

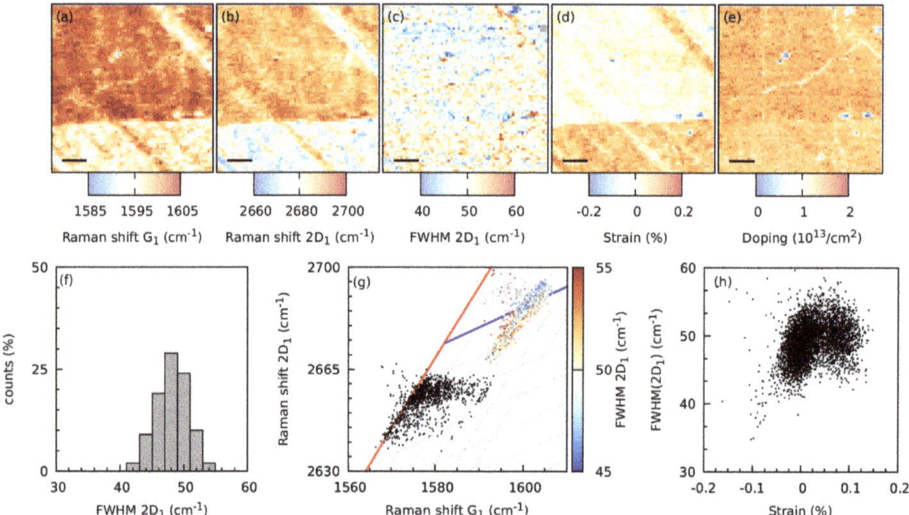

Figure 3. Raman spectral maps of the fitted Raman shift of the G_1 (**a**) and $2D_1$ (**b**) modes, the full width at half maximum (FWHM) of the $2D_1$ mode (**c**) and the calculated strain (**d**) and doping (**e**) for graphene transferred with NC. (**f**) A histogram of the fitted FWHM of the $2D_1$ mode. (**g**) Correlation plot of the Raman shift of the $2D_1$ and $2D_2$ modes as a function of the Raman shifts of the G_1 and G_2 modes, respectively. The ($G_2 - 2D_2$) data points are shown in black, whereas the ($G_1 - 2D_1$) data points are colored, which represents the FWHM of the $2D_1$ mode. (**h**) The correlation between the FWHM of the $2D_1$ mode versus the calculated, biaxial strain.

However, the correlation between the calculated strain and FWHM of the $2D_1$, as shown in Figure 3h, shows that there might be a correlation between the strain and the FWHM of the $2D_1$ mode. For increasing compressive strain, the FWHM of the 2D mode slightly decreased, whereas for increasing tensile strain, the FWHM of the 2D mode has a larger scatter and was not increasing.

This trend is also visible in the correlation plot between the Raman shift of the $G_1 - 2D_1$ (colored points) and $G_2 - 2D_2$ (black points) modes, as shown in Figure 3g, where the FWHM of the $2D_1$ mode is indicated via a color code. The attached graphene to the substrate (colored points) was relatively doped and a large strain variation was present. Interestingly, these points were clustered close to the blue line indicating graphene with a strain value close to zero, but with an increasing doping level, had a lower FWHM of the $2D_1$ mode than points that had some tensile strain.

As the FWHM of the $2D_1$ mode of the NC transferred graphene seemed to be strain dependent, this suggests that there is also the uniaxial strain component present within the graphene layer [25]. Although all the wrinkles are oriented randomly with respect to each other, it is not surprising that graphene feels locally around a single wrinkle this uniaxial strain component due to the large amount of one dimensional wrinkles. Subsequently, Raman spectroscopy will probe the contribution of all the different local strain components [24]. No significant influence of the doping level on the FWHM of the $2D_1$ mode is expected, as the FWHM did not change significantly when the doping level was changed.

The delaminated graphene (black points) had a very low doping level and showed the same large strain variation as observed for the attached graphene.

4. Discussion

The rugae landscape in CVD grown graphene that is transferred to another substrate can be controlled by the used polymer. Up to now, the choice of polymer was often influenced by the possibility to completely remove the polymer after the transfer, so that as clean as possible graphene could be obtained. Here, we show that NC transferred graphene, besides utilizing a very clean transfer method [15], also results in the creation of a very fine pattern of wrinkles in the whole transferred graphene sample, which is evidenced by AFM and an extreme broadening of the 2D mode in Raman spectra.

The relatively rigid NC preserves the strain in graphene that is built up during the CVD growth as can be seen in the maps of the Raman shift of the G and 2D modes (Figure 3). Areas with a well-defined Raman shift are clearly recognizable. The variation of the Raman shift in the different areas is mostly caused by the variation in strain, as can be seen in the Figure 3, which strongly depends on the Cu grain orientations in the Cu foil where graphene was grown on.

The structure of NC shows at the nanoscale a complex network of nanoscale wrinkles, as is visible in Figure 1f. When the Cu foil is etched away below the graphene/NC stack, the rigid NC polymer layer is able to preserve the compressive, growth-induced strain as was already observed by Hallam et al. [15]. However on the nanoscale, some strain is released locally in the graphene layer and thereby graphene locally copies the complex tertiary NC structure and forms a complex nanowrinkle network, as shown in Figure 1a.

The observed wrinkles in the AFM images are not caused by polymer residuals. The strong asymmetric G and 2D modes indicates that a significant part of the graphene layer is delaminated [15]. Furthermore, the FWHM of the 2D mode acts as a probe of the nanometer scale flatness [24]. The extremely broad FWHM of the 2D mode of the NC transferred graphene films indicates that graphene is not flat on the nanometer scale, and the wrinkles observed with AFM are thus caused solely by the graphene layer.

5. Conclusions

We have shown that when NC is used as the polymer for the transfer of CVD grown graphene from the Cu foil to another substrate, randomly distributed regular nanowrinkles are imprinted in the graphene layer. Our approach opens up many new possibilities to control and exploit the properties of graphene with nanometer spatial resolution, like the doping level or chemical functionalization. Moreover, it allows to discover the potential of nanoscale reservoirs on wafer scale.

Author Contributions: All authors contributed to conceptualization, methodology; all authors commented on the manuscript; other contributions: investigation, T.V. and B.P.; formal analysis, T.V. and B.P.; data curation, T.V. and B.P.; writing—original draft preparation, T.V.; writing—review and editing, T.V. and J.V.; visualization, T.V. and B.P.; supervision, J.V. and M.K.; project administration, J.V. and M.K.; funding acquisition, J.V. and M.K.

Funding: This research was funded by the European Research Council, ERC Starting Grant TSuNAMI, number 716265 and Ministry of Education, Youth and Sports of the Czech Republic under Operational Programme Research, Development and Education, project Carbon allotropes with rationalized nanointerfaces and nanolinks for environmental and biomedical applications (CARAT), number CZ.02.1.01/0.0/0.0/16_026/0008382.

Conflicts of Interest: The authors declare no conflict of interest.

References

1. Deng, S.; Berry, V. Wrinkled, rippled and crumpled graphene: An overview of formation mechanism, electronic properties, and applications. *Mater. Today* **2016**, *19*, 197. [CrossRef]
2. Zhao, G.; Li, X.; Huang, M.; Zhen, Z.; Zhong, Y.; Chen, Q.; Zhao, X.; He, Y.; Hu, R.; Yang, T.; et al. The physics and chemistry of graphene-on-surfaces. *Chem. Soc. Rev.* **2017**, *46*, 4417. [CrossRef] [PubMed]
3. Chen, M.; Haddon, R.C.; Yana, R.; Bekyarova, E. Advances in transferring chemical vapour deposition graphene: A review. *Mater. Horiz.* **2017**, *4*, 1054. [CrossRef]
4. Wang, L.; Meric, I.; Huang, P.Y.; Gao, Q.; Gao, Y.; Tran, H.; Taniguchi, T.; Watanabe, K.; Campos, L.M.; Muller, D.A.; et al. One-dimensional electrical contact to a two-dimensional material. *Science* **2013**, *342*, 614. [CrossRef] [PubMed]
5. Banszerus, L.; Schmitz, M.; Engels, S.; Dauber, J.; Oellers, M.; Haupt, F.; Watanabe, K.; Taniguchi, T.; Beschoten, B.; Stampfer, C. Ultrahigh-mobility graphene devices from chemical vapor deposition on reusable copper. *Sci. Adv.* **2015**, *1*, e1500222. [CrossRef] [PubMed]
6. Park, S.; Srivastava, D.; Cho, K. Generalized, Chemical reactivity of curved Surfaces: Carbon nanotubes. *Nano. Lett.* **2003**, *3*, 1273. [CrossRef]
7. Haddon, R.C. π-Electrons in three dimensions. *Acc. Chem. Res.* **1988**, *21*, 243. [CrossRef]
8. Niyogi, S.; Hamon, M.A.; Hu, H.; Zhao, B.; Bhowmik, P.; Sen, R.; Itkis, M.E.; Haddon, R.C. Chemistry of Single-Walled Carbon Nanotubes. *Acc. Chem. Res.* **2002**, *35*, 1105. [CrossRef] [PubMed]
9. Wang, Q.H.; Jin, Z.; Kim, K.K.; Hilmer, A.J.; Paulus, G.L.; Shih, C.C.-J.; Ham, M.-H.; Sanchez-Yamagishi, J.D.; Watanabe, K.; Taniguchi, T.; et al. Understanding and controlling the substrate effect on graphene electron-transfer chemistry via reactivity imprint lithography. *Nat. Chem.* **2012**, *7*, 724. [CrossRef] [PubMed]
10. Levy, N.; Burke, S.A.; Meaker, K.L.; Panlasigui, M.; Zettl, A.; Guinea, F.; Castro Neto, A.H.; Crommie, M.F. Strain-induced pseudo magnetic fields greater than 300 tesla in graphene nanobubbles. *Science* **2010**, *329*, 544. [CrossRef] [PubMed]
11. Zhu, W.; Low, T.; Perebeinos, V.; Bol, A.A.; Zhu, Y.; Yan, H.; Tersoff, J.; Avouris, P. Structure and electronic transport in graphene wrinkles. *Nano Lett.* **2012**, *12*, 3431. [CrossRef] [PubMed]
12. Khestanova, E.; Guinea, F.; Fumagalli, L.; Geim, A.K.; Grigorieva, I.V. Universal shape and pressure inside bubbles appearing in van der Waals heterostructures. *Nat. Commun.* **2016**, *7*, 2587. [CrossRef] [PubMed]
13. Vasu, K.S.; Prestat, E.; Abraham, J.; Dix, J.; Kashtiban, R.J.; Beheshtian, J.; Sloan, J.; Carbone, P.; Neek-Amal, M.; Haigh, S.J.; et al. Van der Waals pressure and its effect on trapped interlayer molecules. *Nat. Commun.* **2016**, *7*, 12168. [CrossRef] [PubMed]
14. Algara-Siller, G.; Lehtinen, O.; Wang, F.C.; Nair, R.R.; Kaiser, U.; Wu, H.A.; Geim, A.K.; Grigorieva, I.V. Square ice in graphene nanocapillaries. *Nature* **2015**, *519*, 443. [CrossRef] [PubMed]
15. Hallam, T.; Berner, N.C.; Yim, C.; Duesberg, G.S. Strain, bubbles, dirt, and folds: A study of graphene polymer-assisted transfer. *Adv. Mater. Interfaces* **2014**, *1*, 1400115. [CrossRef]
16. Diab, M.; Kim, K.-S. Ruga-formation instabilities of a graded stiffness boundary layer in a neo-Hookean solid. *Proc. R. Soc. A* **2014**, *470*, 20140218. [CrossRef]
17. Wang, Q.; Zhao, X. A three-dimensional phase diagram of growth-induced surface instabilities. *Sci. Rep.* **2015**, *5*, 8887. [CrossRef] [PubMed]
18. Kalbac, M.; Frank, O.; Kavan, L. Effects of heat treatment on raman spectra of two-layer 12C/13C grapheme. *Chem. Eur. J.* **2012**, *18*, 13877. [CrossRef] [PubMed]
19. Vejpravova, J.; Pacakova, B.; Endres, J.; Mantlikova, A.; Verhagen, T.; Vales, V.; Frank, O.; Kalbac, M. Graphene wrinkling induced by monodisperse nanoparticles: Facile control and quantification. *Sci. Rep.* **2015**, *5*, 15061. [CrossRef] [PubMed]
20. Necas, D.; Klapetek, P. Gwyddion: An open-source software for SPM data analysis. *Cent. Eur. J. Phys.* **2012**, *10*, 181. [CrossRef]
21. Frank, O.; Vejpravova, J.; Holy, V.; Kavan, L.; Kalbac, M. Interaction between graphene and copper substrate: The role of lattice orientation. *Carbon* **2014**, *68*, 440. [CrossRef]
22. Verhagen, T.G.A.; Drogowska, K.; Kalbac, M.; Vejpravova, J. Temperature-induced strain and doping in monolayer and bilayer isotopically labeled graphene. *Phys. Rev. B* **2015**, *92*, 125437. [CrossRef]
23. Lee, J.E.; Ahn, G.; Shim, J.; Lee, Y.S.; Ryu, S. Optical separation of mechanical strain from charge doping in grapheme. *Nat. Commun.* **2012**, *3*, 1024. [CrossRef] [PubMed]

24. Neumann, C.; Reichardt, S.; Venezuela, P.; Drogeler, M.; Banszerus, L.; Schmitz, M.; Watanabe, K.; Taniguchi, T.; Mauri, F.; Beschoten, B.; et al. Raman spectroscopy as probe of nanometre-scale strain variations in grapheme. *Nat. Commun.* **2015**, *6*, 8429. [CrossRef] [PubMed]
25. Frank, O.; Mohr, M.; Maultzsch, J.; Thomsen, C.; Riaz, I.; Jalil, R.; Novoselov, K.S.; Tsoukleri, G.; Parthenios, J.; Papagelis, K.; et al. Raman 2D-Band Splitting in Graphene: Theory and Experiment. *ACS Nano* **2011**, *5*, 2231. [CrossRef] [PubMed]

Article

Transparent Electrothermal Heaters Based on Vertically-Oriented Graphene Glass Hybrid Materials

Lingzhi Cui [1,2], Kejian Cui [2], Haina Ci [1,2,3], Kaiqiang Zheng [2], Huanhuan Xie [1,2], Xuan Gao [2], Yanfeng Zhang [1,2,4,*] and Zhongfan Liu [1,2,*]

1   Center for Nanochemistry (CNC), Beijing National Laboratory for Molecular Sciences, College of Chemistry and Molecular Engineering, Peking University, Beijing 100871, China; cuilz-cnc@pku.edu.cn (L.C.); cihn-cnc@pku.edu.cn (H.C.); xiehh-cnc@pku.edu.cn (H.X.)
2   Beijing Graphene Institute, Beijing 100091, China; cuikj@bgi-graphene.com (K.C.); zhengkq@bgi-graphene.com (K.Z.); gaoxuan@bgi-graphene.com (X.G.)
3   Academy for Advanced Interdisciplinary Studies, Peking University, Beijing 100871, China
4   Department of Materials Science and Engineering, College of Engineering, Peking University, Beijing 100871, China
*   Correspondence: yanfengzhang@pku.edu.cn (Y.Z.); zfliu@pku.edu.cn (Z.L.); Tel.: +86-010-62767065 (Y.Z.); +86-010-62757157 (Z.L.)

Received: 24 February 2019; Accepted: 29 March 2019; Published: 6 April 2019

Abstract: Transparent heating devices are widely used in daily life-related applications that can be achieved by various heating materials with suitable resistances. Herein, high-performance vertically-oriented graphene (VG) films are directly grown on soda-lime glass by a radio-frequency (rf) plasma-enhanced chemical vapor deposition (PECVD) method, giving reasonable resistances for electrothermal heating. The optical and electrical properties of VG films are found to be tunable by optimizing the growth parameters such as growth time, carrier gas flow, etc. The electrothermal performances of the derived materials with different resistances are thus studied systematically. Specifically, the VG film on glass with a transmittance of ~73% at 550 nm and a sheet resistance of ~3.9 KΩ/□ is fabricated into a heating device, presenting a saturated temperature up to 55 °C by applying 80 V for 3 min. The VG film on the glass at a transmittance of ~43% and a sheet resistance of 0.76 KΩ/□ exhibits a highly steady temperature increase up to ~108 °C with a maximum heating rate of ~2.6 °C/s under a voltage of 60 V. Briefly, the tunable sheet resistance, good adhesion of VG to the growth substrate, relative high heating efficiency, and large heating temperature range make VG films on glass decent candidates for electrothermal related applications in defrosting and defogging devices.

Keywords: graphene; transparent heater; PECVD

---

1. Introduction

Electrically driven transparent heaters have attracted a lot of attention in broad areas containing defogging in automobiles, outdoor displays, heating retaining windows, and other heating systems [1]. To date, the customarily employed materials like strips of metal alloys (such as Fe-Cr-Al or Ni-Cr) are often used in commercial heaters. However, they present several drawbacks like a heavy weight, opacity, and low heating efficiency [2,3]. Therefore, the conductive metal-oxide represented by indium tin oxide (ITO) has been commercially used for fabricating conventional transparent heaters for its extra-high transparency and conductivity [4]. However, disadvantages lie in its intolerance to acids or bases, fragile under bending deformation, as well as the limited reserves of indium, etc. Consequently, the use of many alternative materials have been attempted to serve as heating media. Metal nanowires such as silver nanowires have been developed to be used in the stretchable, wearable transparent heaters [5], while facing the challenge of easy oxidation in the atmosphere. Carbon nanotubes (CNTs) also possess impressive electrical, optical, and thermal properties, and the related heating device was first fabricated by Han and co-workers with the achievement of a remarkable heating performance [6]. However, drawbacks lie in the multi-step process and the necessity to use harsh chemicals, since CNTs need to be purified, dispersed, sonicated and transferred onto targeted substrates.

Graphene, a two-dimensional layered material which possesses sp^2-hybridized carbons, has attracted a lot of attention since it was first obtained in 2004 [7]. The excellent electrical [8], optical [9], mechanical [10] and thermal conductive [11] properties endow graphene with versatile applications in varied fields [12], including sensors [13], transparent electrodes [14,15], field-effect transistors [16] and so on. Up to now, various methods have been developed to achieve graphene, such as mechanical exfoliation [7], reduction of graphene oxide [15], anodic bonding [17], epitaxial growth [18] and chemical vapor deposition [19]. Graphene films assembled by chemically reduced graphene oxides (RGO) were firstly used to fabricate transparent heating devices and displayed decent performances [20]. However, the treatment of reduced graphene (RGO) needs high-temperature annealing over 800 °C. Transfer of graphene on metals onto targeting substrates is another promising method to fabricate transparent heating devices [21–23]. But it always requires a complicated transfer process, which customarily introduces breakage, wrinkle, and contaminations in the derived graphene. Alternatively, vertically-oriented graphene (VG) has many unique properties, containing free-standing, non-agglomerated morphology, a large amount of exposed edges, etc. [24–26]. More importantly, the extremely high in-plane conductivity of graphene can be effectively used, avoiding the sheet-to-sheet resistance. The VG films are thus proposed to serve as perfect heating media. In particular, the combination of VG films and traditional functional substrates like glass should be more promising for related applications.

Herein, VG films were directly grown on soda-lime glass by using a low temperature (~600 °C) radio frequency (rf) plasma-enhanced chemical vapor deposition (PECVD) route using CH_4 as the precursor. The optical and electrical properties of VG films were precisely controlled by alternating the growth parameters, especially the growth time. Accordingly, the resulted VG films are featured with a broad range of tunable transparency levels at 20–80% and sheet resistance levels of 250–5600 $\Omega \cdot sq^{-1}$. The as-grown VG films on glass hybrids were then manufactured to transparent heating templates for the functions of defogging and defrosting, showing relatively high heating efficiency, uniform temperature distribution, low electrical consumption and a wide range of heating temperature. The highly tunable range of sheet resistance, high chemical stability and good adhesion of graphene to the growth substrates were considered to be the unique properties of the current system, compared with other materials such as metal alloys related or RGO-related heating media in the electrothermal devices.

2. Materials and Methods

2.1. Synthesis of Vertically-Oriented Graphene

Uniform VG films were directly grown on soda-lime glass by an rf-PECVD method. In a typical PECVD process, the used soda-lime glass substrate was cleaned with deionized water, ethanol and acetone before loading into the quartz tube placed inside a three-zone furnace. CH_4 was selected as the carbon source. A copper coil was used to assist the generation of plasma, which was placed ~40 cm away from the glass substrate. A vacuum pump was used to keep the system at low pressure (~15 Pa). The typical growth temperature was set at ~600 °C. The thicknesses of the VG samples were controlled by varying the growth time, plasma generating power, flow rate of the precursor gas, and growth temperature. After the optimization of the other parameters at a steady condition, the growth time is selected as the fundamental parameter.

2.2. Characterizations

The samples were systematically characterized using Raman spectroscopy (LabRAM HR Evolution, Horiba, Paris, France; 532 nm laser excitation, 100× objective lens), scanning electron microscopy (SEM, FEI Quattro S, IBM, Waltham, MA, USA, operating at 1 kV), four-probe resistance measurement meter (RTS-4, Guangzhou 4-probe Tech Co. Ltd., Guangzhou, China), atomic force microscopy (AFM, BRUKER Dimension Icon, Karlsruhe, Germany), infrared imaging system (FLIR A615, Täby, Sweden) and UV-Vis spectroscopy (Perkin-Elmer Lambda 950 spectrophotometer, Waltham, MA, USA).

3. Results

3.1. Structural Characterizations of Vertically-Oriented Graphene Films

The growth mechanism of VG on the soda-lime glass by the rf-PECVD method is summarized in Figure 1a. In the beginning, the carbon source CH_4 is disassociated into highly reactive carbon species (CH_x, C_2H_y, and atomic C and H) by the radio-frequency discharge. These active species are first adsorbed onto the glass surface at the temperature of ~600 °C (right below the softening point of soda-lime glass). The irregular cracks, impurities and dangling bonds on the substrate surface serve as nucleation sites, inducing the formation of a horizontal carbon buffer layer. The carbon species then migrate on the buffer layer and form carbon nanosheets or islands with open edges with the increasing growth time, under the effects of localized electric field and the internal stress. After that, carbon nanosheets grow vertically, and the growth is terminated until the closure of open edges, which is determined by the competition between the etching effect in the plasma and the materials deposition [27].

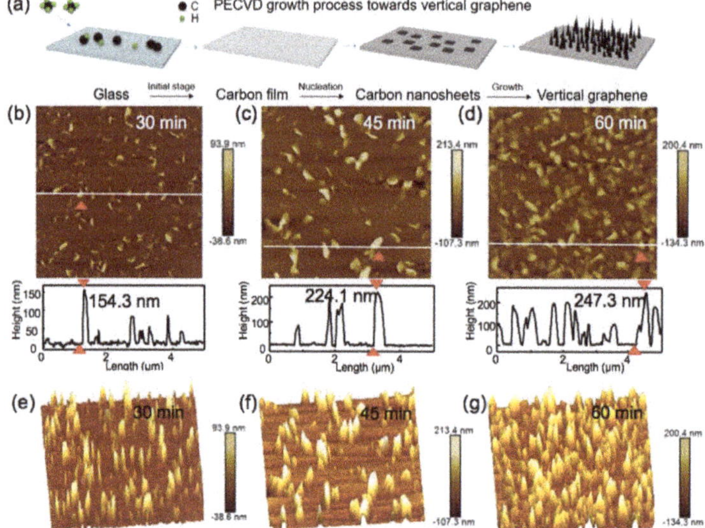

Figure 1. Plasma-enhanced chemical vapor deposition (PECVD) growth towards the formation of vertically-oriented graphene (VG). (**a**) Schematic illustration of VG growth by the radio-frequency PECVD (rf-PECVD) method. (**b–d**) Typical 2D atomic force microscopy (AFM) height images (5 μm × 5 μm) of directly-grown VG films on soda-lime glass at various growth time: (**b**) 30 min, (**c**) 45 min, (**d**) 60 min, respectively, with the identical growth conditions (10 standard cubic centimeter per minute (sccm) CH_4, 600 °C, 300 W). Insets show the section view of the images along the white-dashed line direction. The corresponding 3D AFM height images are shown in (**e–g**).

To verify the generation of VG, various comparative experiments with varying growth time were carried out under the identical PECVD growth condition (10 standard cubic centimeter per minute (sccm) CH_4, 300 W, 600 °C, 15 Pa). The surface morphologies of the VG films deposited on soda-lime glass were first characterized by atomic force microscopy (AFM). Typical 2D and 3D AFM images of VG films on soda-lime glass with different growth times of 30, 45 and 60 min are shown in Figure 1b–g. The height profiles of these samples are plotted along the lines in Figure 1b–d. The red arrows in each image mark the ridge differences. At the initial growth stage (within the early 30 min), a polycrystalline carbon layer was formed on the glass surface Figure 1b,e, and more detailed images are shown in Figure S1. Individual graphene nanosheets of ~154.3 nm high also appeared on the carbon layer. With increasing growth time, this vertical growth was then facilitated by the strain energy at the curved edges, and the nearby VG domains merged with each other and reached ~224 nm at ~45 min (Figure 1c,f). After 60 mins' growth, the corrugation of the VG nanosheets reached ~247 nm and the density of nanosheets increased gradually, as shown in Figure 1d,g.

The corresponding growth-time-dependent SEM images are also presented in Figure S2. In the early stage of growth (20 min), a polycrystalline carbon buffer layer was formed on the glass surface, accompanied by the formation of a few nanosheets oriented perpendicularly to the glass surface (Figure S2a). As the growth progressed, nucleation sites initialize on the boundaries and dangling bonds induced by ion bombardment, leading to the growth of vertically-oriented graphene (Figure S2b,c). With the increase of growth time to 60 min, the randomly distributed vertical graphene nanosheets cover the whole surface and keep growing (Figure S2d). When the growth time reached 90 min, much larger graphene nanosheets formed with the evolution of network structures (Figure S2e). The cross-section image suggests that these vertically-oriented graphene sheets are around 1 μm high (Figure S2f).

The preference of graphene to grow vertically rather than parallel to the substrate surface can be explained as follows: (1) the irregular cracks and dangling bonds on the substrate surface serve as nucleation sites, and a buffer layer is formed [28]; (2) the edges of buffer layer or defects curve upward, and vertical growth of graphene happens in these curved edges [27]; (3) the diffusion of carbon cations along the vertical nanowalls is enhanced by the electric field in the sheath layer in plasma, as well as by the stronger localized electric field due to the sharp features of VG edges [29–31].

Figure 2a shows the photographs of the VG films on soda-lime glass samples for 20, 30, 45, 60 and 90 min at 600 °C under the system pressure of 15 Pa. Copper tapes were attached to both sides of the glass for fabricating transparent heating devices, as displayed in the latter part of the work (Figure S3). These samples are marked as 20#, 30#, 45#, 60#, 90#, according to their different growth time. Evidently, the grey contrast of the VG films on glass samples (superimposed on a scenic photo) from left to right increases, which is caused by the increase of the density of VG nanosheets and the lateral expansion of nanosheets. This observation is also supported by the corresponding Raman and transmittance measurements (Figure 2b,c). Raman spectra were also performed on the above mentioned VG nanosheets directly grown on glass. All the spectra are standardized by G peak intensity for comparison. The D, G, and 2D peaks are fitted with the Lorentzian function and the D′ peak fitted by a Fano line shape. Here, the heights of specific peaks are used to indicate the peaks intensities, which are denoted as I_D, I_G, $I_{D'}$, I_{2D} for the D, G, D′, and 2D peaks, respectively.

The I_{2D}/I_G intensity ratio increases from ~0.47 to 0.94 with increasing growth time from 20 to 90 min, as clearly shown in Figure S4a. In addition, the I_D/I_G intensity ratio decreases from 2.1 to 1.2 (Figure S4b), demonstrating the improvement of crystal quality and the expansion of lateral length of the VG nanosheets with an increasing growth time from 20 to 90 min. This can be explained by the empirical equation proposed by Cancado et al. [32], which describes the relation among the in-plane sp^2 crystallite size L_a, the excitation energy of laser source E_L, and I_D/I_G:

$$L_a(nm) = 560/E_L^4 (I_D/I_G)^{-1}$$

Moreover, D′ band appears around 1610 cm^{-1}, corresponding to the highest frequency feature in the density of states [33]. The intensity ratios of $I_D/I_{D'}$ were ~5.49, 4.33 and 6.58 at the first growth stage for 20, 30, 45 min, respectively, which indicates the occurrence of obvious vacancy defects [34]. After 45 min of growth, the intensity ratios of $I_D/I_{D'}$ were estimated to be ~2.89 and 2.73, suggesting the evolution of abundant grain boundary defects [34,35]. The details are shown in Table S1. Specifically, the transparencies of the VG films on glass (at 550 nm) are tunable from 80%, 73%, 63%, and 43% to 20%, by varying the growth time of VG from 20, 30, 45, and 60 to 90 min, respectively (Figure 2c). The sheet resistance also varies significantly with the values of ~5.60, 3.90, 0.90, 0.76 and 0.26 kΩ/□, accordingly, as shown in Figure 2d. Briefly, this series of results show the wide tunability of the optical and electrical properties of the synthesized VG films on soda-lime glass.

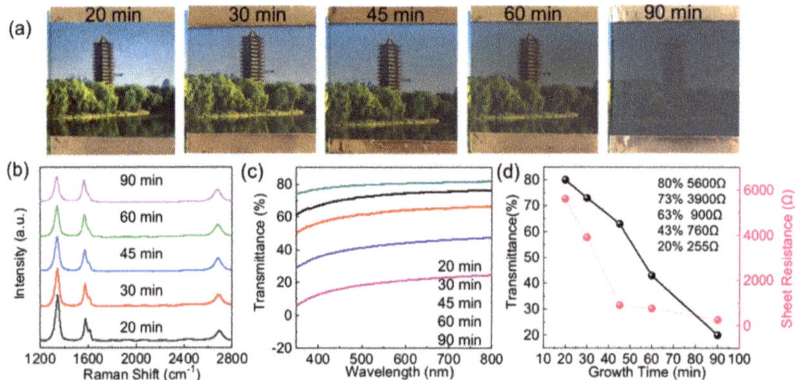

Figure 2. (**a**) Photographs of 5 cm × 5 cm PECVD-derived VG films on the soda-lime glass at 600 °C with varying the growth time from 20, 30, 45, 60 to 90 min, respectively. (**b**) Corresponding Raman spectra of the VG nanosheets films, respectively. (**c**) Corresponding transmittance spectra in the wavelength range of 300–800 nm for VG films on glass, respectively. (**d**) Corresponding sheet resistances and UV-vis transmittance spectra at 550 nm for the VG nanosheets on the glass, respectively.

3.2. Heating Performances of VG-Based Transparent Heater Devices

The unique advantages of the VG nanosheets on soda-lime glass, including the one-step growth strategy, comparatively low growth temperature, reduced sheet resistance, excellent uniformity of transmittance and sheet resistance, and more importantly, its combination with soda-lime glass, make it readily practical for the fabrication of transparent heating devices. Herein, sample 45# (5 cm × 5 cm, 63%, 900 Ω/□) was fabricated into a heater, as it has relatively high transparency and low sheet resistance. Sheet resistance mapping of sample 45# was first performed to address the excellent uniformity (Figure S5). Figure 3 indicates the infrared images of the VG film of 45# by applying different voltages (20, 40, 60, 80 V), respectively. As expected, the final equilibrium temperature increases obviously as the input voltage goes up. And the time consumed to reach the equilibrium temperature is shorter for the sample at higher input voltage. Although there is no inter-digital electrode loaded on the graphene film, obvious heat radiation can be observed at a low input voltage of 20 V. Overall, the temperature distribution is rather homogeneous according to the infrared thermal images showing a uniform color contrast. This is attributed to the uniform sheet resistance/electrical conductivity, as well as the excellent thermal conductivity of the as-grown VG films on glass. It should also be noticed that the size of VG on the glass sample is only limited by the size of the furnace, and it can be enlarged to several inch scales by a large furnace or a different growth route [25]. Therefore, the VG films on glass hybrids are desirable for a wide range of applications in electrothermal heating related aspects.

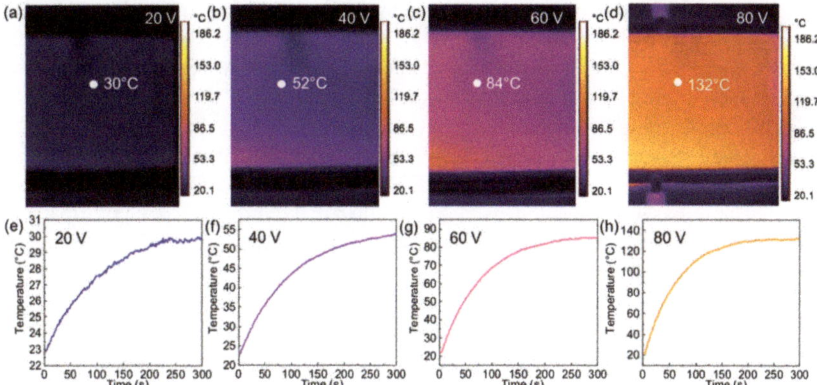

Figure 3. The infrared thermal images of 45# VG films on glass (5 cm × 5 cm, 63%, 900 Ω/□) under different applied voltages, under (**a**) 20 V, (**b**) 40 V, (**c**) 60 V and (**d**) 80 V, respectively. Time versus temperature profiles of the samples with respect to different applied voltages are listed in (**e**) 20 V, (**f**) 40 V, (**g**) 60 V and (**h**) 80 V, respectively.

The as-obtained samples were then fabricated into heaters for electrothermal performance tests. Copper tapes were attached to both sides of the glass as electrodes. An alternating current (ac) was applied to the VG films on soda-lime glass platforms (valid heating area: 5×5 cm^2). Their electrothermal performances were studied under atmospheric conditions, as demonstrated in Figure 4. Upon connecting electrical power, the surface temperature monotonically rises over time until a stable temperature is obtained. With increasing the applied voltage, the steady temperature of a given film surges. Moreover, the longer the growth time (with respect to lowered resistance), the higher the steady temperature was attained at a given voltage. Among these, the VG film obtained by 20 min (20#, 80%, 5600 Ω/□) has a relatively higher sheet resistance, which shows the lowest steady temperature and the lowest heating rate given the same applied voltage.

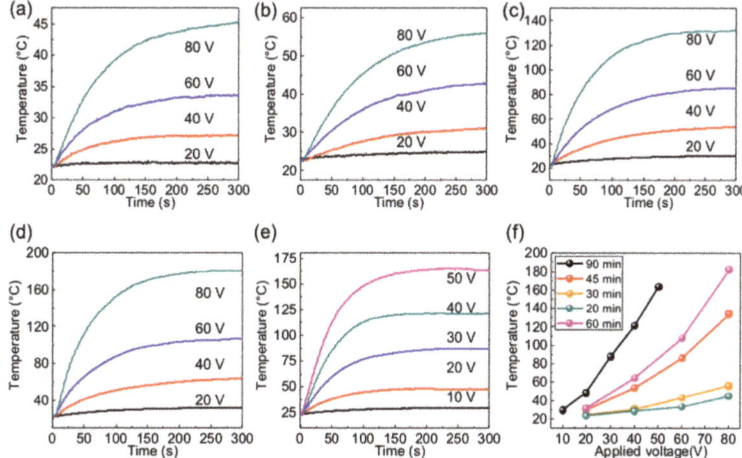

Figure 4. Electrothermal performances of VG films on glass (5 cm × 5 cm) under varied growth times from 20, 30, 45, and 60 min to 90 min, respectively. (**a**–**e**) Time versus temperature profiles under different applied voltages. (**f**) Steady-state temperature versus the applied voltage plots for the different samples.

This can be explained by the relative low transition efficiency from electrical energy to Joule-heating for the current sample. Since $P = U^2/R$, where P is power, U is the applied voltage, R is the resistance. Sample 20# (80%, 5600 $\Omega/\square$) has a relatively higher sheet resistance, which shows the lowest steady temperature (~32 °C) and the lowest heating rate (0.23 °C/s) given the same applied voltage (60 V) (Figure 3a and the Supporting Information (SI), Figure S6). Therefore, the sample marked with 20# can be used as transparent heating device considering its low electricity consumption (e.g., maximum input power density less than 0.045 W cm^{-2}) and good transparency ~80% over the whole vision region. The input power density is lower than that of the RGO-based transparent heater at the same sheet resistance, but with approximately equal transparency under the same applied voltage [15]. In comparison, as shown in Figure 4c–e, the 60# and 90# samples (43%, 760 $\Omega/\square$; 20%, 255 $\Omega/\square$) show higher steady temperature and heating rates, due to their lower sheet resistances with regard to that of the 45# sample (63%, 900 $\Omega/\square$).

Notably, all of the samples can reach the maximum temperature in less than 3 min, highly indicating their rather high electrothermal heating performances. Specifically, the 45# sample (43%, 760 $\Omega/\square$) can reach a maximum temperature of ~45 °C under 40 V, and ~130 °C under 80 V. With regard to the 90# sample, the steady temperature can reach ~160 °C under 50 V, exhibiting the maximum heating rates of ~2.6 °C/s at an applied voltage of 40 V (Figure S7). 50 V is the cut off voltage for 90# sample as the temperature can reach over 200 °C when 80 V is applied, which causes copper tape oxidization. Compared to 20# (80%, 5600 $\Omega/\square$), the 30# (73%, 3900 $\Omega/\square$), 45# (63%, 900 $\Omega/\square$), 60# (43%, 760 $\Omega/\square$), 90# (20%, 255 $\Omega/\square$) samples can achieve higher input power at the same applied voltage due to their improved electrical conductivities. But the maximum input power density is still less than 0.4 W cm^{-2} (see Figure S8). Under the same input power density ~0.4 W cm^{-2}, the VG-based heating device can obtain a relatively high heating temperature (~160 °C), twice as high as that of the RGO-based heating devices (~80 °C) [15]. For 45#, 60#, 90# samples, improved electrothermal performances can be realized, while their transparencies were reduced dramatically. Due to their different electrothermal properties, the applications of the hybrid materials can be very different. The performances of the reported electrothermal heaters are summarized in Table S2. Notably, under the same power density, the max heating rate achieved in this work is comparable, or even larger than that reported previously. As for the low transparency films, they can be used as heating devices which do not require high transparency, such as antifogging mirrors in bathrooms and vehicles.

There are also some applications requiring high transparency and excellent heating performance, such as transparent heaters on building windows or cars. The 20# (80%, 5900 $\Omega/\square$) and 30# (73%, 3900 $\Omega/\square$) samples with high transparency can be candidates for the above application scenarios. VG has good uniformity at both high and low sheet resistances (30#), as demonstrated in Figure 5a (size: 5 cm × 5 cm, transmittance: ~73%, sheet resistance: ~3900 $\Omega/\square$). Detailed statistical distributions of the sample are shown in Figure 5b. With 40 V input voltage, the VG glass shows a uniform temperature distribution of 39.0 ± 1.0 °C, as shown in the infrared thermal image in Figure 5c. It is known that one of the probable applications of the heating performance for VG glass is the defroster device. Herein, the sample 30# (VG films on glass: 73%, 3900 $\Omega/\square$) was fabricated into a defroster device in Figure 5d,e, which possessed a good defrosting performance with a completion time of ~120 s under 40 V. In general, the VG directly grown on soda-lime glass by the rf-PECVD method possesses excellent graphene thickness uniformity, high transparency, homogeneous sheet resistance, thus affording advanced defrosting potential in outdoor displays or other equipment requiring transparent, conductive heating films for defogging or temperature maintenance.

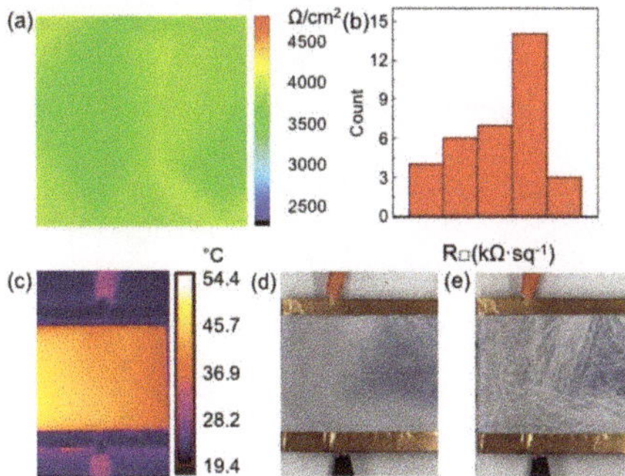

Figure 5. (**a**) Spatial distribution of the sheet resistance of the 30# sample (transmittance ~73%). The map is composed of 36 points, collected from 5 cm × 5 cm graphene glass. (**b**) The statistical distribution of the sheet resistance. (**c**) Infrared temperature image on the VG glass (5 cm × 5 cm) fabricated by the rf-PECVD route. Defrosting results (**d**) before, and (**e**) after heating the 30# VG film by an applied voltage of 40 V.

4. Conclusions

We report the direct synthesis of high-performance VG films on soda-lime glass by a rf-PECVD method. Considering the tunable transparency and resistance, such directly grown VG films on soda-lime glass are exploited to fabricate electrical thermal heating devices. The electrothermal performance was studied in terms of applied voltage and heating rate. Moreover, prototype devices for defogging or defrosting were also fabricated by these VG films on glass hybrids, which possess homogeneous sheet resistance, low electrical consumption, and excellent heating efficiency. In essence, this work provides a reliable method for the production of VG on the glass, as well as presenting a visual demonstration of its applications in electrothermal related fields, especially in traffic transportation, home appliances, or any other industrial fields.

Supplementary Materials: The following are available online at http://www.mdpi.com/2079-4991/9/4/558/s1, Figure S1: Detailed AFM image of VG film obtained in 30 mins (1.6 μm × 1.6 μm), Figure S2: (a–e) SEM images of graphene nanowalls directly grown on the soda-lime glass for growth periods of 20, 30, 45, 60 and 90 min, respectively, with plasma power 300 W at 600 °C. (f) Cross section SEM image of graphene nanowalls through 90 min growth, Figure S3: Schematically illustration of the structure of the heating devices, Figure S4: (a,b) I_{2D}/I_G and I_D/I_G ratio of different growth time, Figure S5: Sheet resistance mapping of sample 45# indicates the excellent uniformity, Figure S6: Heating rates of (a) 20#, (b) 30#, (c) 45#, (d) 60# at 60 V, Figure S7: Heating rates of (a) 20#, (b) 30#, (c) 45#, (d) 60#, (e) 90# at 40 V, Figure S8: (a) steady-state temperatures vs input power; (b) steady-state temperature vs power density. Table S1: The intensity ratios of I_D/I_G, $I_{D'}/I_G$, $I_D/I_{D'}$ for growth time of 20, 30, 45, 60 and 90 min, Table S2: Heaters fabricated by carbon-based films and their performances.

Author Contributions: L.C. and H.C. synthesized the samples, L.C. characterized the samples and analyzed the data. K.C. and K.Z. contributed to analysis tools and software. L.C. designed the experiments. L.C. wrote the paper. X.G. and H.X. help review the article. Y.Z. and Z.L. guide and help review the article.

Funding: This work was supported by the National Key Research and Development Program of China (No. 2016YFA0200103), the National Natural Science Foundation of China (NSFC) (Nos. 51432002, 51520105003), Beijing Municipal Science & Technology Commission (No. Z181100004818002).

Conflicts of Interest: The authors declare no conflict of interest.

References

1. Kim, J.H.; Ahn, B.D.; Kim, C.H.; Jeon, K.A.; Kang, H.S.; Lee, S.Y. Heat generation properties of Ga doped ZnO thin films prepared by rf-magnetron sputtering for transparent heaters. *Thin Solid Films* **2008**, *516*, 1330–1333. [CrossRef]
2. Wu, Z.P.; Wang, J.N. Preparation of large-area double-walled carbon nanotube films and application as film heater. *Phys. E Low-dimens. Syst. Nanostr.* **2009**, *42*, 77–81. [CrossRef]

3. Janas, D.; Koziol, K.K. A review of production methods of carbon nanotube and graphene thin films for electrothermal applications. *Nanoscale* **2014**, *6*, 3037–3045. [CrossRef] [PubMed]

4. Ederth, J.; Johnsson, P.; Niklasson, G.A.; Hoel, A.; Hultåker, A.; Heszler, P.; Granqvist, C.G.; van Doorn, A.R.; Jongerius, M.J.; Burgard, D. Electrical and optical properties of thin films consisting of tin-doped indium oxide nanoparticles. *Phys. Rev. B* **2003**, *68*, 155410. [CrossRef]

5. Hong, S.; Lee, H.; Lee, J.; Kwon, J.; Han, S.; Suh, Y.D.; Cho, H.; Shin, J.; Yeo, J.; Ko, S.H. Highly Stretchable and Transparent Metal Nanowire Heater for Wearable Electronics Applications. *Adv. Mater.* **2015**, *27*, 4744–4751. [CrossRef]

6. Yoon, Y.-H.; Song, J.-W.; Kim, D.; Kim, J.; Park, J.-K.; Oh, S.-K.; Han, C.-S. Transparent Film Heater Using Single-Walled Carbon Nanotubes. *Adv. Mater.* **2007**, *19*, 4284–4287. [CrossRef]

7. Novoselov, K.S.; Geim, A.K.; Morozov, S.V.; Jiang, D.; Zhang, Y.; Dubonos, S.V.; Grigorieva, I.V.; Firsov, A.A. Electric Field Effect in Atomically Thin Carbon Films. *Science* **2004**, *306*, 666–669. [CrossRef]

8. Novoselov, K.S.; Geim, A.K.; Morozov, S.V.; Jiang, D.; Katsnelson, M.I.; Grigorieva, I.V.; Dubonos, S.V.; Firsov, A.A. Two-dimensional gas of massless Dirac fermions in graphene. *Nature* **2005**, *438*, 197–200. [CrossRef]

9. Nair, R.R.; Blake, P.; Grigorenko, A.N.; Novoselov, K.S.; Booth, T.J.; Stauber, T.; Peres, N.M.R.; Geim, A.K. Fine Structure Constant Defines Visual Transparency of Graphene. *Science* **2008**, *320*, 1308. [CrossRef] [PubMed]

10. Lee, C.; Wei, X.; Kysar, J.W.; Hone, J. Measurement of the Elastic Properties and Intrinsic Strength of Monolayer Graphene. *Science* **2008**, *321*, 385–388. [CrossRef]

11. Nika, D.L.; Pokatilov, E.P.; Askerov, A.S.; Balandin, A.A. Phonon thermal conduction in graphene: Role of Umklapp and edge roughness scattering. *Phys. Rev. B* **2009**, *79*, 155413. [CrossRef]

12. Huang, X.; Yin, Z.; Wu, S.; Qi, X.; He, Q.; Zhang, Q.; Yan, Q.; Boey, F.; Zhang, H. Graphene-Based Materials: Synthesis, Characterization, Properties, and Applications. *Small* **2011**, *7*, 1876–1902. [CrossRef] [PubMed]

13. Shao, Y.; Wang, J.; Wu, H.; Liu, J.; Aksay, I.A.; Lin, Y. Graphene Based Electrochemical Sensors and Biosensors: A Review. *Electroanalysis* **2010**, *22*, 1027–1036. [CrossRef]

14. Wang, X.; Zhi, L.; Müllen, K. Transparent, Conductive Graphene Electrodes for Dye-Sensitized Solar Cells. *Nano Lett.* **2008**, *8*, 323–327. [CrossRef] [PubMed]

15. Becerril, H.A.; Mao, J.; Liu, Z.; Stoltenberg, R.M.; Bao, Z.; Chen, Y. Evaluation of Solution-Processed Reduced Graphene Oxide Films as Transparent Conductors. *ACS Nano* **2008**, *2*, 463–470. [CrossRef] [PubMed]

16. Das, A.; Pisana, S.; Chakraborty, B.; Piscanec, S.; Saha, S.K.; Waghmare, U.V.; Novoselov, K.S.; Krishnamurthy, H.R.; Geim, A.K.; Ferrari, A.C.; et al. Monitoring dopants by Raman scattering in an electrochemically top-gated graphene transistor. *Nat. Nanotechnol.* **2008**, *3*, 210. [CrossRef]

17. Moldt, T.; Eckmann, A.; Klar, P.; Morozov, S.V.; Zhukov, A.A.; Novoselov, K.S.; Casiraghi, C. High-yield production and transfer of graphene flakes obtained by anodic bonding. *ACS Nano* **2011**, *5*, 7700–7706. [CrossRef]

18. Berger, C.; Song, Z.; Li, X.; Wu, X.; Brown, N.; Naud, C.; Mayou, D.; Li, T.; Hass, J.; Marchenkov, A.N.; et al. Electronic Confinement and Coherence in Patterned Epitaxial Graphene. *Science* **2006**, *312*, 1191–1196. [CrossRef]

19. Sun, Z.; Yan, Z.; Yao, J.; Beitler, E.; Zhu, Y.; Tour, J.M. Growth of graphene from solid carbon sources. *Nature* **2010**, *468*, 549. [CrossRef]

20. Sui, D.; Huang, Y.; Huang, L.; Liang, J.; Ma, Y.; Chen, Y. Flexible and Transparent Electrothermal Film Heaters Based on Graphene Materials. *Small* **2011**, *7*, 3186–3192. [CrossRef]

21. Lee, B.J.; Jeong, G.H. Fabrication of defrost films using graphenes grown by chemical vapor deposition. *Curr. Appl. Phys.* **2012**, *12*, S113–S117. [CrossRef]

22. Bae, J.J.; Lim, S.C.; Han, G.H.; Jo, Y.W.; Doung, D.L.; Kim, E.S.; Chae, S.J.; Huy, T.Q.; Luan, N.V.; Lee, Y.H. Heat dissipation of transparent graphene defoggers. *Adv. Funct. Mater.* **2012**, *22*, 4819–4826. [CrossRef]

23. Kang, J.; Kim, H.; Kim, K.S.; Lee, S.K.; Bae, S.; Ahn, J.H.; Kim, Y.J.; Choi, J.B.; Hong, B.H. High-performance graphene-based transparent flexible heaters. *Nano Lett.* **2011**, *11*, 5154–5158. [CrossRef]

24. Qi, Y.; Deng, B.; Guo, X.; Chen, S.; Gao, J.; Li, T.; Dou, Z.; Ci, H.; Sun, J.; Chen, Z.; et al. Switching Vertical to Horizontal Graphene Growth Using Faraday Cage-Assisted PECVD Approach for High-Performance Transparent Heating Device. *Adv. Mater.* **2018**, *30*, 1704839. [CrossRef]

25. Ci, H.; Ren, H.; Qi, Y.; Chen, X.; Chen, Z.; Zhang, J.; Zhang, Y.; Liu, Z. 6-inch uniform vertically-oriented graphene on soda-lime glass for photothermal applications. *Nano Res.* **2018**, *11*, 3106–3115. [CrossRef]

26. Sun, J.; Chen, Y.; Cai, X.; Ma, B.; Chen, Z.; Priydarshi, M.K.; Chen, K.; Gao, T.; Song, X.; Ji, Q.; et al. Direct low-temperature synthesis of graphene on various glasses by plasma-enhanced chemical vapor deposition for versatile, cost-effective electrodes. *Nano Res.* **2015**, *8*, 3496–3504. [CrossRef]

27. Malesevic, A.; Vitchev, R.; Schouteden, K.; Volodin, A.; Zhang, L.; von Tendeloo, G.; Vanhulsel, A.; Haesendonck, C.V. Synthesis of few-layer graphene via microwave plasma-enhanced chemical vapour deposition. *Nanotechnology* **2008**, *19*, 305604. [CrossRef] [PubMed]

28. Yang, C.Y.; Bi, H.; Wan, D.Y.; Huang, F.Q.; Xie, X.M.; Jiang, M.H. Direct PECVD growth of vertically erected graphene walls on dielectric substrates as excellent multi-functional electrodes. *J. Mater. Chem. A* **2013**, *1*, 770–775. [CrossRef]

29. Zhao, J.; Shaygan, M.; Eckert, J.; Meyyappan, M.; Rümmeli, M.H. A growth mechanism for free-standing vertical graphene. *Nano Lett.* **2014**, *14*, 3064–3071. [CrossRef]

30. Ostrikov, K.K.; Levchenko, I.; Xu, S. Self-organized nanoarrays: Plasma-related controls. *Pure Appl. Chem.* **2008**, *80*, 1909–1918. [CrossRef]

31. Neyts, E.C.; van Duin, A.C.; Bogaerts, A. Insights in the plasma-assisted growth of carbon nanotubes through atomic scale simulations: Effect of electric field. *J. Am. Chem. Soc.* **2011**, *134*, 1256–1260. [CrossRef] [PubMed]

32. Cancado, L.G.; Takai, K.; Enoki, T.; Endo, M.; Kim, Y.A.; Mizusaki, H. General equation for the determination of the crystallite size L-a of nanographite by Raman spectroscopy. *Appl. Phys. Lett.* **2006**, *88*, 163106. [CrossRef]

33. Ni, Z.H.; Fan, H.M.; Feng, Y.P.; Shen, Z.X.; Yang, B.J.; Wu, Y.H. Raman spectroscopic investigation of carbon nanowalls. *J. Chem. Phys.* **2006**, *124*, 204703. [CrossRef]

34. Eckmann, A.; Felten, A.; Mishchenko, A.; Britnell, L.; Krupke, R.; Novoselov, K.S.; Casiraghi, C. Probing the nature of defects in graphene by Raman spectroscopy. *Nano Lett.* **2012**, *12*, 3925–3930. [CrossRef]

35. Wu, J.B.; Lin, M.L.; Cong, X.; Liu, H.N.; Tan, P.H. Raman spectroscopy of graphene-based materials and its applications in related devices. *Chem. Soc. Rev.* **2018**, *47*, 1822–1873. [CrossRef] [PubMed]

 *nanomaterials*

Review

Recent Advances in Graphene-Based Humidity Sensors

Chao Lv [1,2], Cun Hu [1,3], Junhong Luo [1], Shuai Liu [3], Yan Qiao [4], Zhi Zhang [1], Jiangfeng Song [1], Yan Shi [1], Jinguang Cai [1,*] and Akira Watanabe [2,*]

[1] Institute of Materials, China Academy of Engineering Physics, Jiangyou 621908, China; lvchao219@foxmail.com (C.L.); hucun402@163.com (C.H.); luojunhong@caep.cn (J.L.); zhangzhi@caep.cn (Z.Z.); songjiangfeng@caep.cn (J.S.); shiyan@caep.cn (Y.S.)

[2] Institute of Multidisciplinary Research for Advanced Materials, Tohoku University, 2-1-1 Katahira, Aoba-ku, Sendai 980-8577, Japan

[3] School of Chemistry and Chemical Engineering, Southwest Petroleum University, Chengdu 610500, China; shuailiu@swpu.edu.cn

[4] College of Chemistry and Molecular Engineering, Zhengzhou University, Zhengzhou 450001, China; yanqiao@zzu.edu.cn

[*] Correspondence: caijinguang@foxmail.com (J.C.); watanabe@tagen.tohoku.ac.jp (A.W.)

Received: 31 January 2019; Accepted: 3 March 2019; Published: 12 March 2019

Abstract: Humidity sensors are a common, but important type of sensors in our daily life and industrial processing. Graphene and graphene-based materials have shown great potential for detecting humidity due to their ultrahigh specific surface areas, extremely high electron mobility at room temperature, and low electrical noise due to the quality of its crystal lattice and its very high electrical conductivity. However, there are still no specific reviews on the progresses of graphene-based humidity sensors. This review focuses on the recent advances in graphene-based humidity sensors, starting from an introduction on the preparation and properties of graphene materials and the sensing mechanisms of seven types of commonly studied graphene-based humidity sensors, and mainly summarizes the recent advances in the preparation and performance of humidity sensors based on pristine graphene, graphene oxide, reduced graphene oxide, graphene quantum dots, and a wide variety of graphene based composite materials, including chemical modification, polymer, metal, metal oxide, and other 2D materials. The remaining challenges along with future trends in high-performance graphene-based humidity sensors are also discussed.

Keywords: humidity sensors; graphene; graphene oxide; reduced graphene oxide; chemical modified graphene; graphene/polymer; graphene quantum dots; graphene/metal oxide; graphene/2D materials

---

1. Introduction

Humidity sensors are a common type of sensors in our daily life, and play a significant role in numerous application fields ranging from humidity control for various kinds of industrial processing, agricultural moisture monitoring, and medical fields to weather forecasting, indoor humidity sensing, and domestic machine controlling [1–4], and the corresponding research has continued for more than 100 years since the 18th century. Simply, humidity sensors show the humidity by converting the amount of water molecules in the environment into a measurable signal. According to the change of the physical parameters after interacting with water molecules, humidity sensors can be categorized into many types, such as the capacitive type, resistive type, impedance type, optic-fiber type, quartz crystal microbalance (QCM) type, surface acoustic wave (SAW) type, resonance type, and so on [3]. Many materials sensitive to water molecules have been developed as sensing materials in humidity sensors, including ceramics, such as Al_2O_3, SiO_2, and spinel compounds [2]; semiconductors, such as TiO_2 [5,6], SnO_2 [7–10], ZnO [11–14], In_2O_3, Si [15], and perovskite compounds [16,17]; polymers, such as polyelectrolytes [18,19], conducting and semiconducting polymers [20], and hydrophilic polymers [21–24]; 2D materials, such as MoS_2 [25–27], WS_2 [28–30], and black phosphorus [31–34]; and carbon materials, such as porous carbon [35], carbon nanotubes [36,37], and graphene [38,39].

Among them, graphene is a unique material, which is an atomically thin, planar membrane of carbon atoms arranged in a honeycomb lattice. The unique atomic arrangement brings unique electronic structures, and extraordinary physical and chemical properties, which promote graphene materials' applications in many fields, including electronics, optoelectronics, spintronics, catalysts, energy generation and storage, molecular separation, and chemical sensors [38–47]. Graphene and graphene-based materials have shown great potential for the detection of various kinds of gases due to their ultrahigh specific surface areas, extremely high electron mobility at room temperature, and low electrical noise due to the quality of their crystal lattice and very high electrical conductivity [48]. Although many excellent review articles have summarized the progresses on graphene-based chemical, gas, and tactile sensors [38,39,45,46,48–54], and several review articles have addressed the mechanisms, materials, and development of humidity sensors [1–4,55,56], there are still no specific reviews on the progresses of graphene-based humidity sensors. This review focuses on the recent advances in graphene-based humidity sensors, which is divided into three main parts: Preparation and properties of graphene, sensing mechanisms of graphene-based humidity sensors, and advances in the humidity sensors based on graphene and graphene composite materials. The first part will briefly introduce three commonly used methods to prepare pristine graphene and graphene oxide, as well as the properties of different kinds of graphene materials. The second part will explain the sensing mechanisms of seven types of commonly studied humidity sensors, including field-effect transistor (FET)-type, resistive, impedance, capacitive, surface acoustic wave (SAW), quartz crystal microbalance (QCM), and optical fiber humidity sensors, and the progresses of SAW, QCM, and optical fiber humidity sensors based on graphene materials will be comprehensively reviewed in this part. The third part mainly focuses on the progresses of graphene-based humidity sensors operating in electronic modes, such as field-effect transistor (FET)-type, resistive, impedance, and capacitive humidity sensors, which are classified into eight sub-parts according to the properties and composites of graphene materials. Finally, the remaining challenges along with future trends in high-performance and practical graphene-based humidity sensors are also discussed.

2. Preparation and Properties of Graphene

2.1. Preparation of Graphene

The preparation of graphene materials is critical for practical applications. Various methods have been developed to synthesize graphene materials with different structures since the first preparation was accomplished by mechanical exfoliation of graphite in 2004. In this method, Scotch tapes were generally employed to stick the graphite crystals or graphite flakes, and repeatedly peeled over a long period to obtain the graphene layers. This process can also be conducted with a silicon wafer with an oxidized layer. The products are monolayer or few-layer graphene with perfect atomic arrangement, which can be used to fabricate principle electronic devices, such as gas sensors or electronic biosensors [48,57]. However, this method has an extremely low productivity, thus it is difficult to be applied widely [58].

Fortunately, the chemical vapor deposition (CVD) technique, which was widely used in the preparation of one-dimensional (1D) structures, has been developed and extensively explored to grow high-quality graphene with superior properties [59]. Like the growth process of 1D nanostructures, the growth of graphene includes three main steps, the dissolution of carbon atoms into the metal substrate at higher temperature, the precipitation of carbon from the substrate in the cooling stage caused by the reduced solubility, and the formation of graphene layers on the metal surface [60]. Some metals, such as Ni, Cu, Fe, Co, Ir, Pt, and Ni-Mo alloy, have been developed as a catalyst and substrate to grow graphene layers. The growth of single-layer or multilayer graphene films can be controlled by using metals with different solubility of carbon atoms. For example, multilayer graphene films are usually prepared on Fe, Ni, and Co substrates with relatively high solubility [61], while single-layer graphene films can grow on Pt and Cu substrates due to the relatively low solubility (Figure 1a). To use the graphene layers in electronic devices, the effective transfer technology of the chemical wet-etching method was developed to dissolve the metal substrate and transfer the graphene films, and this process does not produce defects or degrade graphene properties. This process was further developed in a roll-to-roll way for continuous and large-area production (Figure 1b) [62]. Nevertheless, more facile, effective, and low-cost technologies are urgently required for large-scale commercial applications.

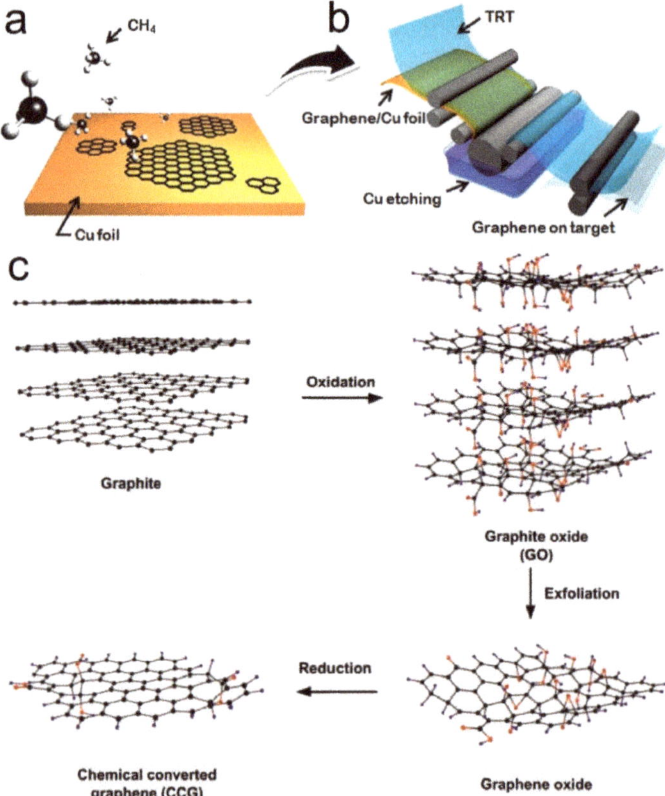

Figure 1. Schematics of the synthesis mechanism of chemical vapor deposition (CVD) graphene on Cu foil (**a**) and the roll-to-roll transfer process of graphene (**b**). Reproduced with permission from [42]. Copyright Wiley-VCH, 2016. Schematic of the preparation of graphene oxide and reduced graphene oxide by the reduction of graphene oxide (**c**). Reproduced with permission from [63]. Copyright Wiley-VCH, 2011.

Another common strategy of the solution-processed approach, which is also called the Hummers method, has been widely used in the massive and reproducible preparation of graphene materials [40]. This approach involves two main steps, the oxidation and exfoliation of graphite in liquid solution and the formation of graphene by reduction. In the first step, the graphite powder is oxidized by strong acids and oxidants, followed by exfoliation under sonication, producing hydrosoluble graphene oxide (GO) (Figure 1c) [63]. GO with different layers can be obtained by purification and separation by centrifugation. Then, reduced graphene oxide (rGO) can be obtained through chemical reduction, thermal reduction, microwave irradiation, electrochemical reduction, photo reduction, or laser-induced reduction. Compared to the CVD method, which mainly grows single-layer or multilayer graphene films, this method can produce GO or rGO with abundant functional groups in a powder form at a very large scale and with relatively low costs [52,64]. However, this method has some shortages, such as environmental pollution, explosion hazards, and a long period during the oxidation process, and the resultant graphene materials have lots of defects.

2.2. Properties of Graphene

The unique atomic thickness and honeycomb lattice structure of sp^2-bonded carbon atoms, endow graphene with excellent physicochemical, electronic, optical, thermal, and mechanical properties [65]. Graphene can be regarded as a single layer of graphite, but it exhibits a semi-metal feature with a unique zero band gap, different from the metallic behavior of graphite. The charge carriers of graphene have zero rest mass near their Dirac point with a high carrier density of up to 10^{13} cm^{-2}, thus graphene exhibits a significant high room-temperature carrier mobility of ~10,000 cm^2 V^{-1} s^{-1} and a low temperature carrier mobility of 200,000 cm^2 V^{-1} s^{-1}, corresponding to a resistivity of 10^{-6} Ω [65,66]. Due to the perfect atomic arrangement, graphene possesses excellent mechanical properties with fracture strains of up to 25% and a Young's modulus of ~1.1 TPa [67], even higher than steel's tensile strength, showing large potential for applications in flexible, stretchable, and wearable electronics. Single-layer graphene has a very high transparency of up to 97.7% with an extremely low theoretical sheet resistance of 30 Ω sq^{-1}, showing significant promise in high-performance optic and electronic devices. Additionally, graphene has a very high specific surface area of up to ~2600 m^2/g [38]. The extremely high electron mobility, the ultrahigh specific surface area to adsorb molecules, and inherently low electronical noise due to the high-quality crystal lattice and very high electrical conductivity make graphene highly sensitive to changes in its chemical environment, thus graphene is recognized as an ideal candidate for the ultrahigh sensitivity detection of gases in various environments [48].

Some of the above properties are only shown in pristine perfect single-layer graphene, which is quite different from graphene materials, such as GO or rGO, prepared by the solution-phase method. In contrast to CVD growth, the solution-phase approach mostly introduces plenty of oxygenic functional groups, such as carboxyl, epoxides, and hydroxyl groups, and defects exist on the graphene basal plane and edges [45]. Thus, as-prepared GO is an electrical insulator with no conductivity, and reduction is necessary to make it conductive in most practical applications. The functional groups, defects, and other unavoidable contamination may degenerate the physical properties. Generally, the conductivity, transparency, and mechanical properties of rGO are several orders lower than those of pristine graphene [40]. Nevertheless, GO and rGO show other advantages. For example, the abundant oxygen containing groups endow GO or rGO with various stimuli-responsive behaviors in aqueous solution. GO and rGO can be further chemically modified to improve their properties for additional functions [45]. The electronic properties of rGO can be modulated by heteroatom doping, and other properties can also be realized by covalently grafting functional groups. In addition, non-covalent interactions between GO or rGO and other components can be employed to prepare graphene-based composite structures for more extensive applications [40]. Furthermore, GO or rGO can be dispersed in solutions and assembled into 1D fibers, 2D films, and 3D porous structures, which diversifies their application areas. Therefore, there is plenty of room to develop and improve the performance of gas sensors, including humidity sensors, based on GO and rGO materials.

3. Mechanisms of Graphene-Based Humidity Sensors

Principally, water molecules in the gas environment will adsorb onto the graphene surfaces in a graphene-based humidity sensor, which causes changes of some properties of the graphene materials, corresponding to the humidity change. Various kinds of graphene-based humidity sensors have been developed according to different sensing mechanisms or sensor configurations. This section briefly introduces seven types of sensing mechanisms commonly applied in graphene-based humidity sensors, and the progress of graphene-based humidity sensors working in the last three mechanisms, i.e., SAW, QCM, and optical fiber, is also included.

3.1. Field-Effect Transistor (FET) Humidity Sensors

FET configuration has been widely applied in many gas sensing devices due to the simple preparation process, high sensitivity, and portability, as well as the easy miniaturization to the nanoscale [57]. A typical graphene-based FET humidity sensor is comprised of three main parts, graphene as the channel material, two conductive electrodes as the source and drain electrodes, and a gate electrode with a thin dielectric layer (Figure 2). A bias voltage is applied on the gate electrode through the dielectric layer, which can modulate the conductivity of the graphene channel. Humidity sensing is conducted by measuring the current change of the graphene channel before and after exposing it to a humidity environment under a constant applied gate voltage. When adsorbing the water molecules, the electronic structure of the sensing graphene will be changed, leading to a change of the conductivity. In this type of humidity sensor, the sensing material, graphene, can be easily functionalized with different groups or complexed with other components to improve the sensitivity of the humidity sensors. Notably, many gases, including CO, NO, NH_3, NO_2, H_2, SO_2, H_2S, and ethanol, can be detected by such a type of gas sensor due to the different interactions between the gas molecules and sensing materials [48,68].

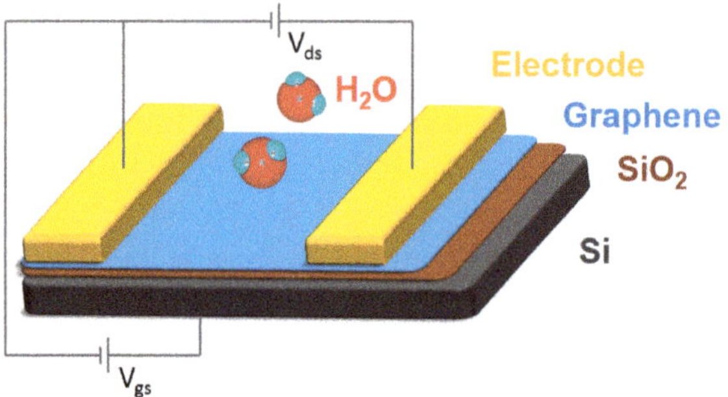

Figure 2. Schematic of a graphene-based field-effect transistor (FET) humidity sensor with source-drain voltage, V_{ds}, and gate voltage, V_{gs}, control.

3.2. Resistive Humidity Sensors

Resistive humidity sensors are one of the most widely investigated types of humidity sensing devices because of their many intrinsic advantages, such as the facile fabrication, simplicity of operation, cost effectiveness, reusability, low driven power, and easy miniaturization. The sensing mechanism is based on the change of the electrical resistance of graphene materials caused by the water molecules adsorbed on the surfaces, and the humidity can be effectively determined by measuring the resistance change (Figure 3) [46,69]. Equivalently, the humidity sensor can also detect the current change. The configuration of resistive humidity sensors is relatively simple, which consists of sensing materials located between two conductive electrodes on an inert substrate. The electrodes always employ an interdigitated structure to improve the sensing area and thus the sensitivity. The sensing graphene materials can easily be coated onto the substrate with patterned electrodes through spin-coating, drop-casting, spray-coating, dip-coating, printing, or laser transfer deposition. Generally, the sensitivity of such types of humidity sensors can be improved by increasing the materials, surface areas, as well as the functional groups sensitive to water molecules through constructing porous structures, surface modification, or compositing with other sensitive components. Sensors in this type are also widely employed to detect different kinds of gases, such as CO, H_2, NO_x, SO_x, O_2, Cl_2, and organic vapors, by changing the sensing materials [38,46,48,52,70–72].

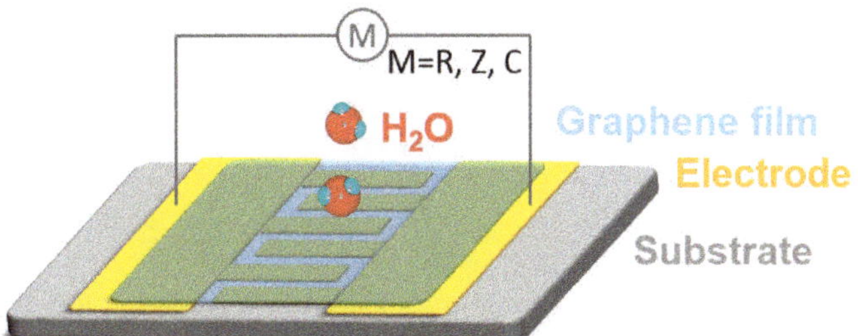

Figure 3. Schematic of a graphene-based humidity sensor in a resistive, impedance, or capacitive working mode depending on the measuring parameter.

3.3. Impedance Humidity Sensors

Impedance humidity sensors are a type of mixed potential sensors, which work based on impedance change. The humidity sensor is applied with a sinusoidal voltage in the frequency domain, and the impedance is calculated by measuring the current. In such a type of humidity sensor, an impedance spectroscopy is utilized to measure the response of the humidity sensor over frequencies ranging from subhertz to megahertz [73]. There is an advantage in such-type humidity sensors that it has the potential to accurately detect a relatively low humidity, but generally, it requires professional high-performance impedance spectroscopy. The structure of the impedance humidity sensor is similar to the resistive one, in which sensing materials are deposited on two conductive electrodes with gaps on an inert substrate (Figure 3). In impedance humidity sensors, GO and rGO based materials rather than pristine graphene are utilized as sensing materials, because the abundant functional groups on the surfaces as well as porous structures can provide more active sites for water molecules, which induces a simultaneous resistance and capacitance change of the graphene materials, i.e., comprehensive impedance change [74]. Similarly, GO or rGO can be composited with other active components to further improve the sensitivity. Impedance sensors have also been developed for the efficient detection of various gases, including NO_x, CO, and hydrocarbons, with low concentrations [69].

3.4. Capacitive Humidity Sensors

Capacitive humidity sensors utilize the special properties of GO materials, which are insulating, but show proton conductivity when the surfaces adsorb water molecules, with the proton conductivity being related to the water concentration in the environment. This induces the change of the capacitance [75]. The sensor detects the humidity by measuring the capacitance of the device. Therefore, GO can work as a water molecule-sensitive dielectric in a double-layer electrochemical capacitor to detect humidity in the atmosphere. A typical capacitive humidity sensor is generally the in-plane type to expose the maximum active surface area, which consists of two parts, interdigitated conductive electrodes and GO-based sensing materials as the dielectric (Figure 3) [76]. In capacitive humidity sensors, interdigitated conductive electrodes can be metal materials, as well as rGO, which can form all-graphene based humidity sensors. For the dielectric, GO can be composited with other proton conductive components to improve the sensitivity and response [77].

3.5. Surface Acoustic Wave (SAW) Humidity Sensors

Principally, SAW humidity sensors are a type of microelectromechanical system, which detect the humidity change depending on the modulation of surface acoustic waves. An SAW sensor basically comprises a piezoelectric substrate, an input interdigitated transducer (IDT) on one side, a second output IDT on the other side of the substrate, and a delay line in the space between the two IDTs

(Figure 4) [78,79]. The input IDT produces a mechanical SAW under the sinusoidal electrical signal using the piezoelectric effect of the piezoelectric substrate, and the SAW propagates across the decay line, where the SAW is influenced by the environment. The SAW is converted into an electric signal by the piezoelectric effect in the output IDT, and any changes made to the mechanical wave are reflected in the output electric signal. The sensing can be accomplished by measuring changes in the amplitude, phase, frequency, or time-delay between the input and output electrical signals. Any changes of the physical or chemical properties of the sensor surface, such as the mass, conductance, or viscoelasticity, affect the acoustic wave velocity or attenuation [80]. Graphene-based SAW humidity sensors generally employ graphene materials as the acoustic layer and utilize the mass change induced SAW change to detect the humidity because hydrophilic GO or GO-based composites enhance the water adsorption capability through the oxygen-rich groups, strengthening the mass loading effect. Based on the above mechanism, SAW sensors towards other gases or targets can also be designed by modulating the properties of the sensing materials and their interaction with sensing targets.

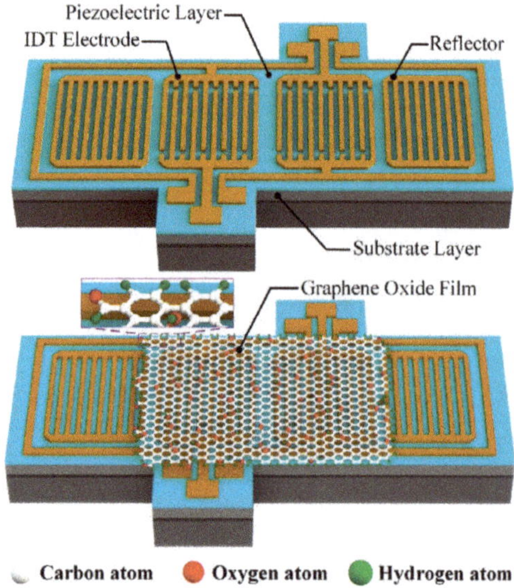

Figure 4. Schematic view of a surface acoustic wave sensor with a clean surface and covered with a graphene oxide (GO) film. Reproduced with permission from [79], Copyright Elsevier B.V., 2017.

Balashov et al. investigated the sensing performance of SAW humidity sensors by coating a submicro-thick film of GO or a polyvinyl alcohol (PVA) thin film on the decay line, and the GO-based humidity sensor showed a sensitivity of 1.54 kHz/% RH (relative humidity), much higher than the 0.47 kHz/% RH and 0.13 kHz/% RH for the PVA coated and uncoated one, respectively [78]. The SAW humidity sensor prepared on the $LiNbO_3$ surface coated with a CVD-growth graphene layer showed a frequency downshift of 1.38 kHz/% RH in the RH range lower than 50% due to the mass loading and a frequency downshift of 2.6 kHz/% RH in the RH range higher than 50% due to the change in the elastic properties of graphene film [81]. The SAW humidity sensor based on GO coated $LiNbO_3$ plate showed a super high sensitivity by employing a high-order lamb wave with a large coupling constant, standard $LiNbO_3$ plate, and graphene oxide sorbent film, showing a minimal detectable level as low 0.03% RH, and reproducibility of $\pm$2.5% [82]. Xuan et al. demonstrated SAW humidity sensors based on a ZnO piezoelectric thin film on a glass substrate with GO as the sensing layer, which exhibited a high sensitivity from 0.5% RH to the 85% RH range with very fast responses (rise time of 1 s and fall time of 19 s) [83]. Then, they realized a high-performance flexible SAW humidity sensor with the same GO sensing layer on a piezoelectric ZnO thin film deposited on a flexible polyimide substrate [84]. Similarly, Kuznetsova et al. reported an SAW humidity sensor based on a GO film/ZnO film/Si substrate with an enhanced sensitivity of about 91 kHz/% RH and a linear response towards relative humidity in the range of 20–98% [85]. Recently, Xie's group proposed an SAW humidity sensor based on an AlN/Si (doped) layered structure with a GO sensing layer for a high sensitivity and low temperature coefficient of frequency [79]. The humidity sensor coated with the GO layer exhibited an improved sensitivity of up to 42.08 kHz/% RH in the RH higher than 80% and worked well at both low (<10% RH) and high (>90% RH) humidity. Moreover, it showed low hysteresis, outstanding short-term repeatability, long-term stability, and improved thermal stability.

3.6. Quartz Crystal Microbalance (QCM) Humidity Sensors

The QCM technique has been widely used to analyze the mass, membrane structure, molecular interaction, and viscoelasticity changes on the surface of electrodes. QCMs coated with water-sensitive materials can act as sensors for humidity detection, which have the advantages of being ultra-sensitive towards water molecules adsorbed on the crystal surface and a digital frequency output. A QCM humidity sensor can be directly connected to a digital control system to simplify the signal processing circuit. A typical QCM humidity sensor consists of a bare quartz crystal resonator deposited with a humidity sensing material on its electrode, forming a composite resonator (Figure 5) [86]. The composite resonator shows an oscillating frequency shift when the water molecules adsorb or desorb on the sensing material, reflecting the ambient humidity change. There are two types of interaction between water molecules and sensing materials in a QCM humidity sensor, the mass change due to the adsorption and desorption on the surface (mass effect) and the viscoelastic change of the adsorbed film caused by the penetration of water molecules (viscosity effect) [87].

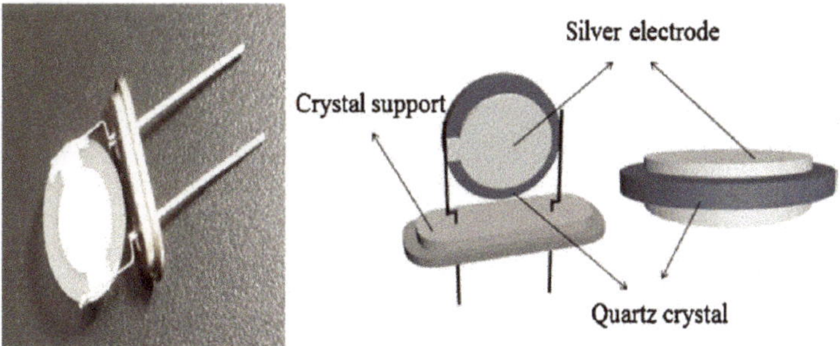

Figure 5. Photograph and structure of the quartz crystal microbalance (QCM) sensor. Reproduced with permission from [86], Copyright Elsevier B.V., 2017.

GO is recognized as a promise humidity sensing material because of its abundant hydrophilic oxygen-containing functional groups, such as hydroxyl, carboxyl, and epoxy groups. Yao et al. demonstrated the first GO thin film-coated QCM humidity sensors, which exhibited excellent humidity sensitivity properties with a sensitivity of up to 22.1 Hz/% RH and a linear frequency response in the wide detection range of 6.4–93.5% RH. They found that, at the low RH range (<54.3%), mass changes caused by water molecules' adsorption/desorption accounted for the frequency response of the sensor, while in the high RH range (>54.3%), both the water adsorption/desorption mass change and expansion stress of the GO thin film induced by the swelling effect caused the frequency change [88]. They also demonstrated that the GO coated QCM humidity sensor showed better stability than the PEG-coated one [89]. The incorporation of other components into the GO film has also been widely investigated to improve the humidity sensing performance by addressing the mass effect or viscosity effect. Other carbon materials, such as multi-walled carbon nanotubes (MWCNTs), nanodiamond, or fullerene, were introduced into the GO film to prevent the viscosity effect, because the intercalation may expand the GO film and improve water molecule diffusion [87,90,91]. Many hygroscopic or hydrophilic components, such as poly (diallyl dimethyl ammonium chloride) [18], poly ethyleneimine and protonated poly ethylenimine [92,93], polyethylene oxide [94], polyaniline [86], SnO_2 [95], and ZnO [96], have been introduced into GO or rGO films by the direct mixing method or layer-by-layer assembly, which can provide more active sites for water adsorption to improve the humidity sensing performance, including the sensitivity, response/recovery time, hysteresis, repeatability, selectivity, and long-term stability. Additionally, diamine- and β-cyclodextrin-functionalized graphene oxide films exhibited a good response to the low humidity region due to the strong sensitivity of the $-CONHC_2H_4NH_2$ groups to water molecules [97].

3.7. Optical Fiber Humidity Sensors

Optical fiber humidity sensors detect changes of the optical properties inside the fiber caused by water molecules, such as changes of the transmitted optical power, dielectric properties, or the refractive index. Compared to electronic humidity sensors, optical fiber humidity sensors show several advantages, such as the possibility of working in harsh conditions, including flammable environments; higher temperature and pressure ranges; and electromagnetic immunity. However, some facts, including the fabrication repeatability and the high cost of the optical equipment, have prevented the commercial applications of optical fiber humidity sensors. According to the working principle, optical fiber humidity sensors can be classified into several groups, such as humidity sensors based on the optical absorption of materials; humidity sensors based on fiber Bragg gratings and long-period fiber gratings; humidity sensors based on interference (Fabry-Pérot, Sagnac, Mach-Zehnder, Michelson, and modal interferometers); humidity sensors based on micro-tapers, micro-ring, micro-knot resonators, and whispering galleries modes; and humidity sensors based on electromagnetic resonances, specifically lossy mode resonances [98]. The classification is shown in Figure 6. More details can be found in a comprehensive and specialized review on optical fiber humidity sensors summarized by Ascorbe et al. [98], thus this review will not describe the detailed mechanisms.

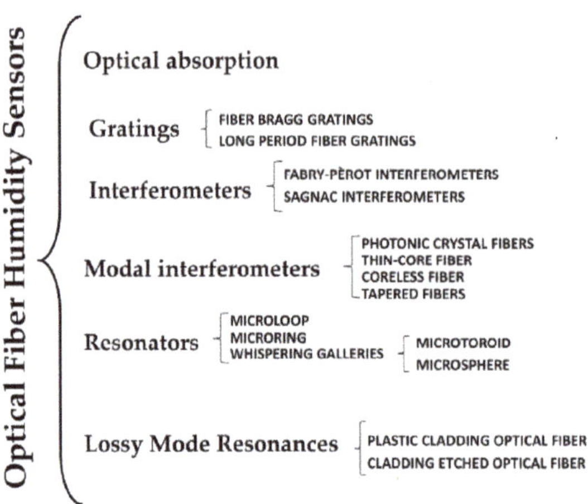

Figure 6. Classification of optical fiber humidity sensors [98].

Graphene materials have the potential to be applied in optical fiber humidity sensors due to the good adsorption towards water molecules, which will induce changes of the optical properties in the optical fiber. For example, Xiao et al. reported an rGO based optical fiber humidity sensor by coating an rGO film on the polished surface of a side-polished fiber, which achieved a power variation of up to 6.9 dB in the high relative humidity range (70–95%) and displayed a linear response with a correlation coefficient of 98.2%, sensitivity of 0.31 dB/% RH, response speed of faster than 0.13% RH/s, and good repeatability in the 75–95% RH [99]. Recently, they demonstrated an improved optical fiber humidity sensor based side-polished single-mode fiber coated with GO film, which exhibited a better performance [100]. The water adsorbed on the GO film caused the dielectric property change, which could influence the TE-mode absorption at 1550 nm of the polymer channel waveguide, constructing the relation between optical absorption and relative humidity [101]. Wang et al. developed humidity sensors based on tilted fiber Bragg grating coated with GO film, which showed a maximum sensitivity of 0.129 dB/% RH with a linear correlation coefficient of 99% under the RH range of 10–80% due to the dependence of the cladding mode resonances on the water absorption and desorption on the GO film [102]. They also investigated optical fiber humidity sensors based on an in-fiber Mach-Zehnder interferometer coated with GO or GO/PVA composite, showing high sensitivity, good stability, and linearity [103,104]. An optical fiber humidity sensor based on a Fabry–Perot resonator was also demonstrated by depositing rGO on the other surface of a hollow core fiber, and the light leakage at the resonant wavelength, which depended on the refractive index of rGO, could be measured to reflect the humidity [105]. Optical fibers coated with graphene materials can be applied not only in humidity sensors, but also in chemical and biological sensors [106].

4. Humidity Sensors Based on Graphene Materials

As seen in the previous sections, water molecules in the environment can induce physical and electronic changes in graphene-based sensing materials. The detection of humidity is thus possible by monitoring a variety of properties, such as the resistance, impedance, capacitance, mass, and surface acoustic wave. Previous studies reporting SAW, QCM, and optical fiber sensors were covered in Sections 3.5–3.7. This section describes recent advances in graphene-based humidity sensors based on electronic properties using measuring techniques, such as FET, resistance, impedance, and capacitance, and it is organized to illustrate the utilization of the different sensor types for the eight sub-classes of graphene materials.

4.1. Humidity Sensors Based on Pristine Graphene

Novoselov et al. demonstrated the first gas sensors of the FET type based on pristine graphene prepared by micromechanical cleavage of graphite at the surface of oxidized Si wafers, which exhibited a significant response to not only water molecules, but also NH_3, NO_2, and CO gases (Figure 7a) [57]. The gas sensor showed different responses to the gases due to the electronic properties, where NO_2 and H_2O act as acceptors, and NH_3 and CO are donors. The exceptionally low-noise electronic property endows the graphene-based sensors with a high sensitivity. Then, the bandgap of graphene was tuned by exposing it to the humidity environment, which was increased with the amount of water molecules adsorbed on the graphene surface, thus the resistivity of the graphene increased with the humidity increasing (Figure 7b) [107]. Smith demonstrated a high-performance humidity sensor based on the electrical resistance change of CVD-grown single-layer graphene placed on the top SiO_2 layer of an Si wafer (Figure 7c), which showed response and recovery times of less than 1 s due to the fast adsorption and desorption of the water molecules from the graphene surface (Figure 7d) [108]. They also revealed that the sensitivity of the resistance of a graphene patch to water vapor resulted from the interaction between the water electrostatic dipole moment and the impurity bands in the substrate according to the simulations. Popov's study revealed that water molecules adsorbed at different defected places showed different effects on the resistance change of the graphene due to the different interaction mechanisms [109]. Adsorption at grain boundary defects is assumed to lead to an increase in film resistivity due to the donor property of water and the p-type conductivity of graphene, while adsorption at edge defects in multilayer graphene films leads to the formation of conductive chains with ionic conductivity. Son et al. found that physical defects in the graphene could hardly increase the humidity sensing performance, while the distinct changes were observed with chemical

Nanomaterials **2019**, *9*, 422

defects by controlling the thickness and the coverage area of the poly(methyl methacrylate) (PMMA) on the graphene surface [110].

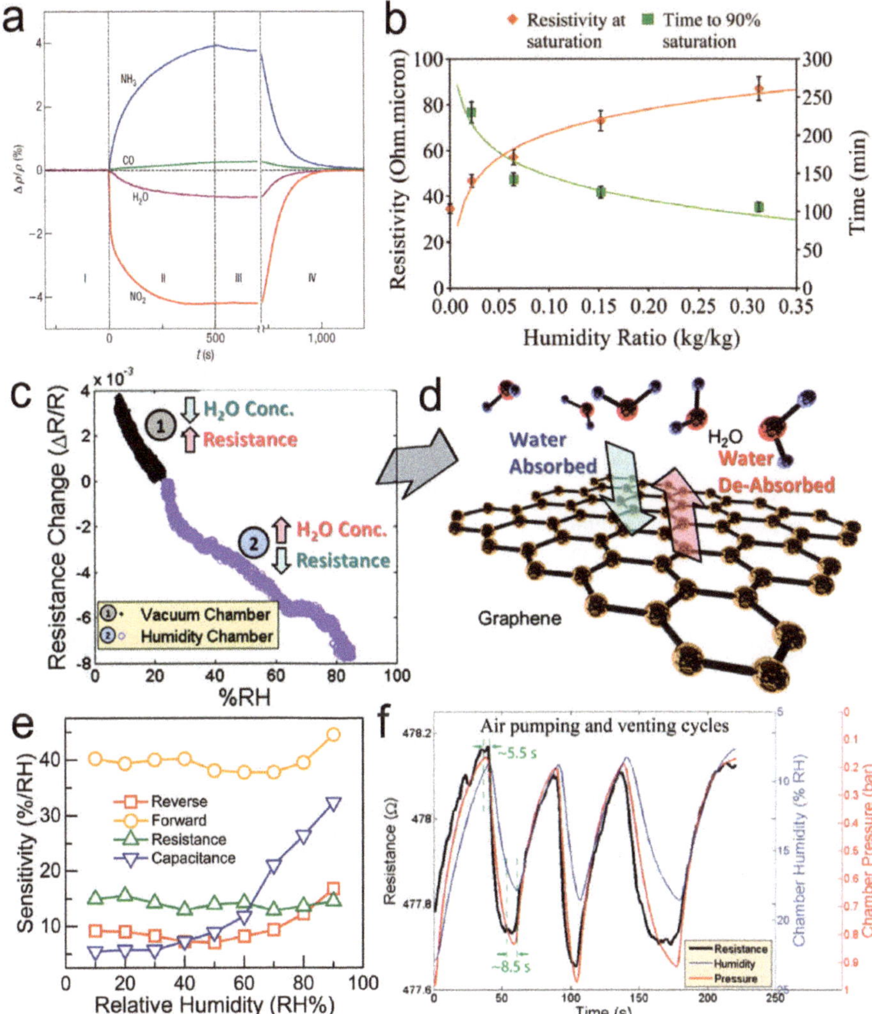

Figure 7. Changes in resistivity caused by graphene's exposure to various gases diluted in concentration to 1 ppm (**a**). Reproduced with permission from [57], Copyright Nature Publishing Group, 2007. The values of maximum resistivity observed, and the time taken to achieve 90% of the saturation value for different values of absolute humidity (**b**). Reproduced with permission from [107], Copyright Wiley-VCH, 2010. Resistance change in the graphene device versus the relative humidity (%RH) for a device placed in a vacuum chamber (1) and the same device placed in a humidity chamber (2) (**c**) and the interaction of water molecules with the graphene surface (**d**). Reproduced with permission from [108], Copyright Royal Society of Chemistry, 2015. Sensitivity of the humidity sensors under current and capacitance modes (**e**). Reproduced with permission from [111], Copyright Wiley-VCH, 2017. Resistance response of the double-layer graphene device in comparison with the %RH response from a commercial humidity sensor as well as the response from a commercial pressure sensor during three consecutive cycles of pumping air from the environment into and out of the vacuum chamber (**f**). Reproduced with permission from [112], Copyright Elsevier B.V., 2017.

Shehzad et al. demonstrated a high-performance multimode humidity sensor by constructing a graphene/Si Schottky junction. The intrinsic properties were influenced by the adsorbent water molecules. The device could detect humidity when it was both forward and reverse biased, as well as in the resistive and capacitive mode, the corresponding sensitivity of which reached 17%, 45%, 26%, and 32% per relative humidity (% RH), respectively (Figure 7e) [111]. Fan et al. investigated the gas sensing behaviors of a double-layer graphene gas sensor, the resistance of which showed a fast response and recovery towards humidity (Figure 7f), but their experiments and theoretical calculations indicated that the resistance response to the humidity of double-layer graphene was lower than that of single-layer graphene [112]. Recently, Zhu's group reported a high-performance humidity sensor based on wrinkled graphene [113], the wrinkled morphology of which could effectively prevent the aggregation of water microdroplets and thus improve the evaporation compared to flat pristine graphene (Figure 8a,b). The device exhibited an ultrafast response to the humidity with a short response time of down to 12.5 ms, which can be used to monitor sudden changes in respiratory rate and depth (Figure 8c). Additionally, the application of graphene woven fabrics prepared with CVD growth have also been demonstrated in simultaneous sensing of humidity and temperature with high sensitivity [114]. Notably, a recent study indicated that a relative humidity of over 50% may modify the interlayer interaction, thus affecting the properties of bilayer graphene [115], therefore, the repeatability and stability of the humidity sensors based on multilayer graphene materials should be paid attention to. The sensing performance of humidity sensors based on pristine graphene materials is summarized in Table 1, which indicates that, although several humidity sensors showed a fast response and recovery, the sensitivity was relatively low and the long-term stability was not investigated.

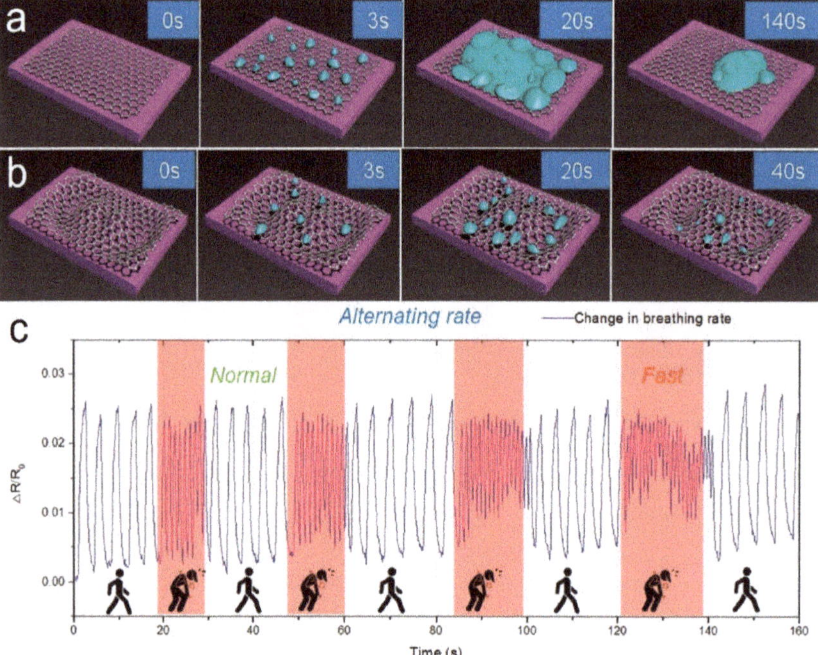

Figure 8. Schematic illustration of the water adsorption and desorption on surfaces of CVD-growth flat graphene (**a**) and wrinkled graphene (**b**), and breathing signals recorded by wrinkled graphene sensors with alternating breathing speed and depth during physical activity (**c**). Reproduced with permission from [113], Copyright Wiley-VCH, 2018.

Table 1. Summary of the sensing performances of humidity sensors based on pristine graphene.

Material	Preparation Method	Type	Operating Temperature	Sensitivity	Selectivity	Humidity Range	Response/Recovery Time	Stability	Ref.
Graphene	micromechanical cleavage of graphite	FET	R.T.	—	Low (NH_3, CO, NO_2)	1 ppm	Hundreds of seconds	—	[57]
Graphene	CVD	Resistive	R.T.	0.31%	High (Ar, N_2, O_2)	1–96%	0.6 s/0.4 s	—	[108]
Multilayer graphene	CVD	Resistive	25 °C	10–17%	—	15–80%	<1 s	—	[109]
Graphene/Si Schottky junction	CVD	Resistive Capacitive	10–90 °C	45% 32%	High (N_2, O_2, Ar, CO_2)	10–90%	8 s/19 s 4 s/10 s	—	[111]
Double-layer graphene	CVD	Resistive	25 °C	~0.784–0.933%	Low (CO_2)	20–100%	Hundreds of milliseconds	—	[112]
Wrinkled graphene	CVD	Resistive	20–80 °C	—	—	11–95%	12.5 ms	—	[113]

4.2. Humidity Sensors Based on Graphene Oxide

Humidity sensors based on GO materials have been widely studied due to the simple, low-cost, and large-scale preparation of GO, and high proton-conductive sensitivity to water molecules. Based on the change of the proton conductivity, GO humidity sensors generally operate by detecting the capacitance or impedance signals [116,117]. Bi et al. constructed a GO humidity sensor by depositing a GO film on microscale interdigitated electrodes (Figure 9a), the capacitance change of which towards humidity was measured with an LCR meter (Figure 9b) [76]. The capacitance of the humidity sensor was related with the relative humidity in the gas environment, as well as the frequency (Figure 9c). This device exhibited a highly improved sensitivity at 15–95% compared to conventional capacitive humidity sensors, and a fast response time of 10.5 s and recovery time of 41 s. Subsequently, Borini et al. investigated the influence of the GO layer thickness on the response of the humidity sensor and demonstrated that a thickness of 15 nm of the GO film resulted in an ultrafast response to a modulated humid flow at the tens-of-microseconds scale while maintaining a full scale output of over an order of magnitude (Figure 9d,e) [76]. Ho et al. reported a stretchable and multimodal all graphene electronic skin by spray coating rGO or GO on a patterned CVD graphene on poly(dimethylsiloxane) (PDMS) to realize different functions, where the device made of GO on the graphene structure worked in a capacitive mode and showed a response to the humidity change (Figure 9f,g) [118]. Recently, Park et al. investigated the correlation between the sensitivity and the sorption/desorption hysteresis of thin film GO humidity sensors working in the conductance mode (Figure 9h), which indicated that the sensors made at pH 3.3 showed a lower sensitivity and hysteresis-induced error while those made at pH 9.5 showed both increased sensitivity and hysteresis-induced error (Figure 9i) [119]. They proposed that the enhanced sensitivity and the hysteresis of these sensors were based on the molecular interactions between the increased water and charged groups in GO at high pH, suggesting a trade-off relationship between sensitivity and hysteresis. In addition, various approaches have been employed to improve the sensing performance or application area of GO based humidity sensors, such as free-standing GO foam to increase the active site [120], ultralarge GO to improve the overall proton conductivity [121], silk fiber coated with GO to take advantage of silk fiber's flexibility [122], computer-aided design [123], investigation of the influence of the structure and coating methods [124], heteroatom-doping on GO to improve the response [125], as well as microstructure related synergic sensing for high-performance humidity sensors [126].

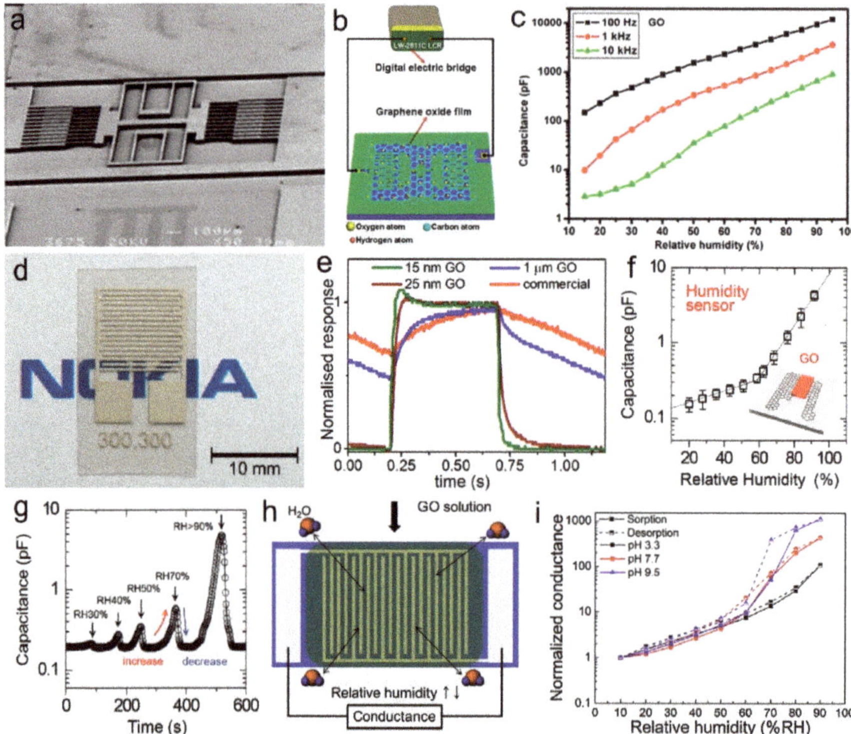

Figure 9. SEM image of the device (**a**), schematic diagram of the humidity testing system graphene oxide film as a humidity sensing material was placed on the two sets of interdigitated electrodes (**b**), and the output capacitances of sensors as a function of RH (**c**). Adapted from [76]. Photograph of a sprayed GO sensing element (**d**), and normalized response of the different sensors to a modulated humid air flow at 1 Hz (**e**). Reproduced with permission from [74], Copyright American Chemical Society, 2013. GO-based humidity sensor performance (capacitance vs RH) (**f**) and real-time RH sensing of the GO-based humidity sensor at specific RH (**g**). Reproduced with permission from [118], Copyright Wiley-VCH, 2016. The humidity sensing layer through drop-casting of GO followed by the measurement of relative humidity via conductance (**h**) and normalized conductance of three different humidity sensors as a function of RH (**i**). Reproduced with permission from [119], Copyright Elsevier B.V., 2017.

Generally, noble metal (Au, Ag, etc.) interdigitated patterns prepared by the lithographical technique are used as electrodes to form GO based humidity sensors. However, Au and Ag materials are relatively expensive, and the lithographical preparation needs professional instruments and a complex process. The laser direct writing technique, which is a noncontact, fast, single-step fabrication technique with no need for masks, postprocessing, or a complex clean room, and is compatible with current electronic product lines for commercial use, has been employed to fabricate energy storage devices, electrically conductive circuits, sensors, as well as self-powered integrated devices [127–133]. Graphene oxide can easily be reduced by laser irradiation, producing rGO with improved conductivity, which acts as conductive electrodes, while the GO can work as a water-sensitive solid electrolyte [134]. Ajayan's group prepared an interdigitated micro- supercapacitor (MSC) on a hydrated GO film (Figure 10a), and demonstrated that the proton conductivity of GO was related to the water concentration in the environment (Figure 10b). An et al. prepared a highly flexible humidity sensor based on an rGO/GO/rGO structure patterned by a fiber femtosecond laser (Figure 10c), which could reduce the GO to rGO with program-controlled patterns (Figure 10d) [135]. The impedance of the humidity sensor showed changes towards the relative humidity at different frequencies in a large range, indicating a high sensitivity (Figure 10e). Moreover, this device also exhibited a fast response time of 1.8 s and recovery time of 11.5 s (Figure 10f). Interestingly, this humidity sensor could be arranged in the substrate forming pixels, which could work as a noncontact e-skin with a high-spatial-resolution sensing capability.

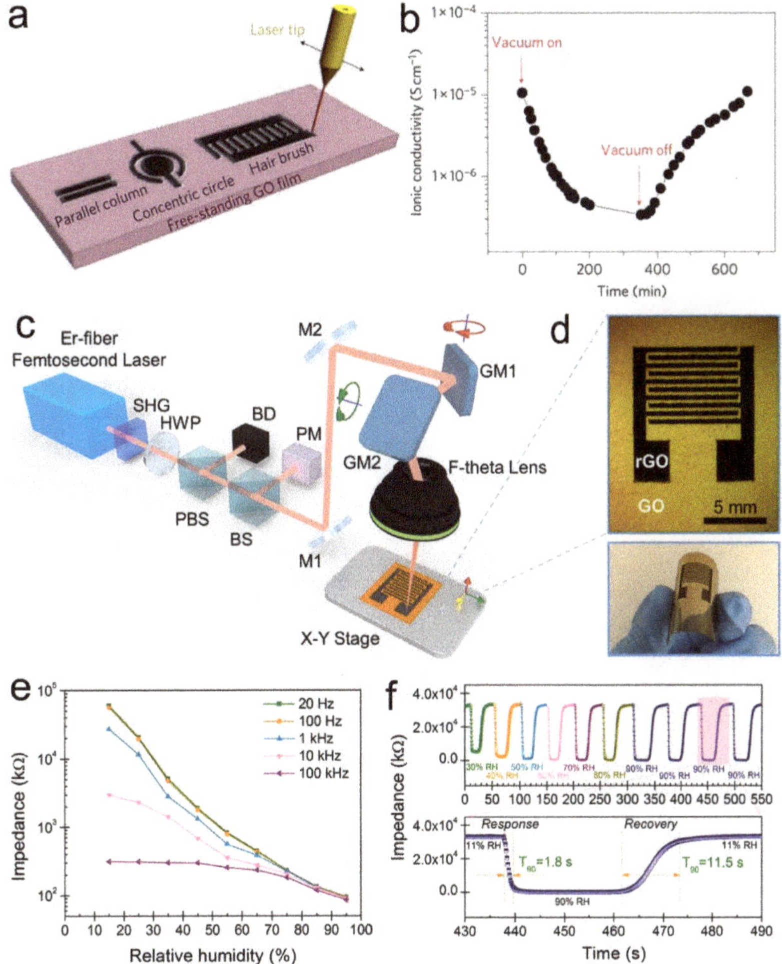

Figure 10. Schematic of CO_2 laser-patterning of free-standing hydrated GO films to fabricate rGO–GO–rGO devices with in-plane geometry (**a**) and the dependence of ionic conductivity on the exposure time to vacuum and air (**b**). Reproduced with permission from [75], Copyright Macmillan Publishers Limited, 2011. Schematic image of the laser direct writing system for the single-step fabrication of all-graphene noncontact sensors (**c**), optical images of the sensor where the rGO electrodes appear in black and the brown thin film corresponds to the GO sensing material (**d**), plots of impedance as a function of RH at different operation frequencies (**e**), and (**f**) upper panel: real-time moisture sensing with different RH ranges (all starting from 11% RH) and repeated RH detection between 11% RH and 90% RH for four cycles, and lower panel: response−recovery curve of the sensor with RH switching between 11% and 90%. Reproduced with permission from [135], Copyright American Chemical Society, 2017.

Recently, we reported the facile preparation of humidity sensors based on an rGO/GO/rGO structure on a flexible PET sheet through laser direct writing with a semiconductor diode laser (Figure 11a) [131]. After laser irradiation, the color of GO was turned to grey from black (Figure 11b), and the laser irradiated part expanded due to the evolved gases induced by laser heating (Figure 11c). Raman spectra, X-ray diffraction (XRD), and X-ray photoelectron spectroscopy (XPS) characterizations clearly showed that the GO was reduced to rGO after laser irradiation and became electrically conductive (Figure 11d). Instead of measuring the impedance or resistance of the rGO/GO/rGO structure, we proposed a novel alternating current (ac) detection mode by connecting the interdigitated structure to a simple circuit (Figure 11e), which showed an obvious response to the relative humidity (RH), ranging from 6.3% RH to 100% RH (Figure 11f). Compared to the low and unstable response in the direct current (dc) mode, the sensor working in this ac detection mode exhibited a dramatically enhanced sensitivity by about 45 times (Figure 11h). Moreover, the device showed a fast response time (1.9 s) and recovery time (3.9 s) (Figure 11g). The sensor also exhibited outstanding cycling stability, flexibility, and long-term stability (>1 year), as well as good reproducibility of the device preparation. The sensor could work on an iPhone to conduct the humidity sensing with an oscilloscope application on iPhone OS (iOS). Furthermore, an electronic circuit simulation method was used to fit the output waves, showing potential promise in real-time monitoring on a smartphone based on the Internet of things and big data technologies. The sensing performance of GO-based humidity sensors is summarized in Table 2, which suggests that such types of humidity sensors generally show a high sensitivity and fast response, as well as potential long-term stability. However, such types of humidity sensors are generally based on changes of the proton conductivity when adsorbing or desorbing water molecules, thus the selectivity may be disturbed by the gases of proton donors or acceptors.

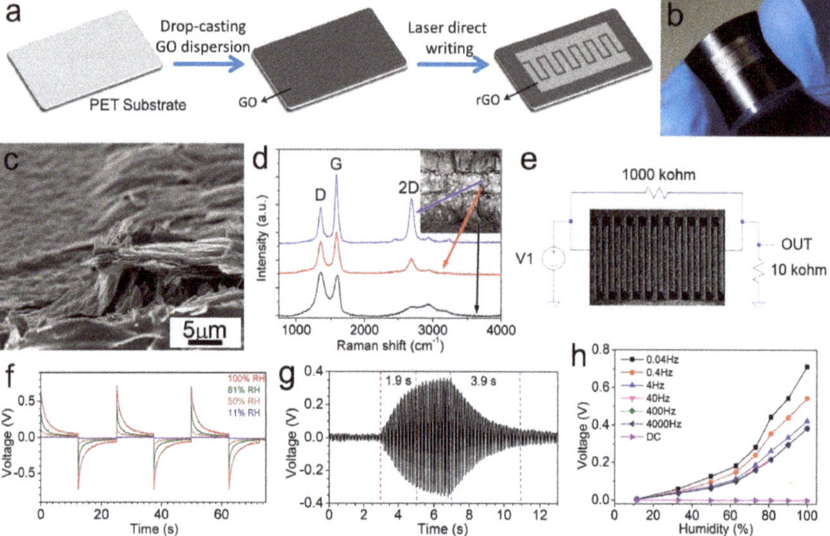

Figure 11. Schematic illustration for the preparation of rGO patterns by laser direct writing (**a**), interdigitated pattern of rGO/GO/rGO prepared on a flexible poly(ethylene terephthalate) (PET) film (**b**), SEM image of the structures obtained by laser direct writing (**c**), Raman spectra at different positions of a laser-irradiated line (**d**), electronic circuit of the rGO/GO/rGO humidity sensor (**e**), output waves of the humidity sensor responding to a rectangular alternating current (ac) wave with a (peak) pk−pk voltage of 1 V at 0.04 Hz (**f**), response and recovery of the humidity sensor at 40 Hz (**g**), and the change of the sensing peak voltages toward RH at different frequencies (**h**). Reproduced with permission from [131], Copyright American Chemical Society, 2018.

4.3. Humidity Sensors Based on Reduced Graphene Oxide

After reduction, the insulating graphene oxide is converted into conductive rGO, the conductivity of which is sensitive to water molecules due to the defects and remnant oxygen-containing groups, and rGO also shows some capacitive behavior because of the uncomplete reduction. Till now, rGO materials prepared with various reduction methods have been employed for humidity sensing. For example, Guo et al. realized the simultaneous reduction, patterning, and nanostructuring of graphene oxide on flexible PET substrates with a two-beam-laser interference method [136]. The as-prepared humidity sensor exhibited a high sensitivity and fast response and recovery performance due to the improved water molecules' adsorption provided by laser induced hierarchical graphene nanostructures. Humidity sensors prepared by layer-by-layer covalent anchoring of a GO film followed by a partially reduced process exhibited good sensitivity in the range of 30% to 90% RH with negligible hysteresis (<2.5% RH), short response time of 28 s and recovery time of 48 s, and good long-term stability [137]. Phan et al. investigated the influence of the reduction degree or the quantity of oxygen functional groups of the GO on humidity sensing using rapid thermal annealing [138]. They indicated that, as the annealing temperature increased, the resistivity decreased, and the GO film lost its capability to adsorb water molecules, thus the response of the humidity sensor decreased. However, a trade-off existed between the response and the long-term stability, which was quite poor in the as-deposited GO film. Recently, Shojaee et al. reported their study on the influence of the reduction degree of GO nanosheets on the humidity sensing performance by a hydrothermal reduction method, the reaction time of which accounted for the reduction degree [139]. The found that humidity sensors based on rGO with a moderate reduction exhibited an optimized sensitivity and response, because the sensitivity was attributed to the oxygen function groups while the response was attributed to the restoration of the sp^2 carbon network. A transparent humidity sensor made of rGO stripes driven by convective self-assembly exhibited a reversible response to humidity in the range of 10–70% RH [140]. In addition, rGO materials obtained by photo reduction, such as sunlight or flash, have also been applied in high-performance rGO-based humidity sensors [141,142]. Papazoglou et al. demonstrated an in-situ sequential laser transfer and laser reduction method to fabricate rGO-based humidity sensors using picosecond laser pulses, and this laser printed rGO humidity sensor showed a fast response time of less than 1 min in the water concentration of 1700–20,000 ppm with a limit of detection of 1700 ppm [143].

Table 2. Summary of the sensing performance of GO-based humidity sensors.

Material	Preparation Method	Type	Operating Temperature	Sensitivity/Response	Selectivity	Humidity Range	Response/Recovery Time	Long-Term Stability	Ref.
GO	Drop casting	Capacitive	25 °C	37,800%	—	15–95%	10.5 s/41 s	30 days	[76]
GO	Drop casting	Impedance	10–40 °C	—	—	10–90%	~30 ms	72 h	[74]
GO	Spary coating	Capacitive	22–90 °C	—	—	20–90%	—	500 cycles	[118]
GO	Drop casting	Conductance	25 °C	12.3 ± 2.2 µS/%RH (pH 3.3) 12.3 ± 2.2 µS/%RH (pH 9.5)	—	10–90%	2.2 s/1.6 s (pH 2.8) 91.8 s/11.3 s (pH 9.3)	—	[119]
GO foam	Dry in a frame	Impedance	25 °C	33,254%	—	11–95%	50 s/79 s	—	[120]
Ultralarge GO	Drop casting	Conductance	20 °C	4339 ± 433	—	7–100%	0.2s /0.7 s	5 days	[121]
GO	Simulation	Capacitive	25 °C	7680 pF/%RH	—	0–100%	<0.5 s	—	[123]
Li-doped GO	Drop casting	Resistive	25 °C	3038%	—	11–97%	4 s/25 s	—	[125]
GO	Spary coating	Capacitive	R.T.	—	—	12–97%	<0.1 s	—	[126]
rGO/GO/rGO	Laser direct writing	Impedance	23 °C	—	—	11–95%	1.8 s/11.5 s	30 days	[135]
rGO/GO/rGO	Laser direct writing	Voltage	R.T.	142.5	High (H$_2$, hexane, ethanol)	6.3–100%	1.9 s/3.9 s	> 1 year	[131]

Because of the importance of humidity sensing in daily life, wearable or portable humidity sensors based on rGO materials have attracted much research attention. To make the device wearable, one of the strategies is to prepare the humidity sensor in a fiber form. For example, Qu's group demonstrated multi-stimuli sensitive sensor based on double-helix core-sheath rGO-based microfibers, which showed a high current response to small perturbations induced by temperature variations, mechanical interactions, and relative humidity changes [144]. Recently, Choi et al. developed a unique humidity sensing layer with nitrogen-doped rGO fibers on colorless polyimide film, and tiny Pt nanoparticles were deposited on the surface of rGO, which acted as dissociation catalysts for humidity sensing [145]. The rGO fiber could detect a wide humidity range from 6.1% RH to 66.4% RH, with a 1.36-fold sensitivity at 66.4% RH of pure rGO fiber. Natural fibers, such as silk fibers or spider silk fibers, have been employed as supports to construct wearable devices due to their excellent mechanical properties, superior skin affinity, and biodegradable properties. Recently, Li et al. developed a flexible humidity sensor based on silk fabrics coated with Ni and GO nanosheets, which showed a fast response to the humidity change and could be used for human respiration monitoring [146]. Li et al. demonstrated a biomimetic electric multi stimuli sensing device by combining layer-by-layer deposition of graphene and supercontraction of spidroin fibers, which exhibited a rapid response and repeatability towards RH > 20% [147]. Ma et al. proposed a strong, tough, lightweight printable biopapers by introducing silk interlayers into graphene oxide films combining a seriography-guided reduction technique, and the resultant rGO/silk patterns could work as resistive moisture sensors to detect humidity within 3.0 s, showing potential for applications in wearable electronics [148].

The controllable preparation of homogeneous rGO thin film at a high efficiency is the basis for practical applications. Zhang's group proposed an effective and reproducible way to assemble rGO ultrathin films with a controllable thickness [149]. The preparation involved the vacuum filtration of rGO suspension and exfoliation at the liquid/air interface (Figure 12a). The ultrathin film on a PET substrate showed high transparency and flexibility with layer structured rGO (Figure 12b,c). The preparation showed a good reproducibility according to the transmittance of all 13 samples (Figure 12d). The device fabricated based on rGO ultrathin films and the Au electrode showed a fast response to RH from 4.3% to 75.7% with repeated response (Figure 12e,f). Furthermore, a flexible matrix panel was prepared based on the rGO ultrathin film and Au electrode arrays, which exhibited excellent noncontact humidity sensing performance with high sensitivity, high spatial resolution, and fast response (Figure 12g). Additionally, wearable humidity sensors can also be prepared based on a porous graphene network with a high response for detecting finger humidity, speaking, whistle rhythm, and respiration monitoring [150]. The sensing performance of rGO-based humidity sensors is summarized in Table 3, which indicates that rGO-based humidity sensors generally work based on impedance or resistance change, and generally the sensitivity is not so high because of the reduced functional oxygen-containing groups compared to GO. Therefore, combining rGO and other sensitive materials may be an effective way to improve the sensitivity.

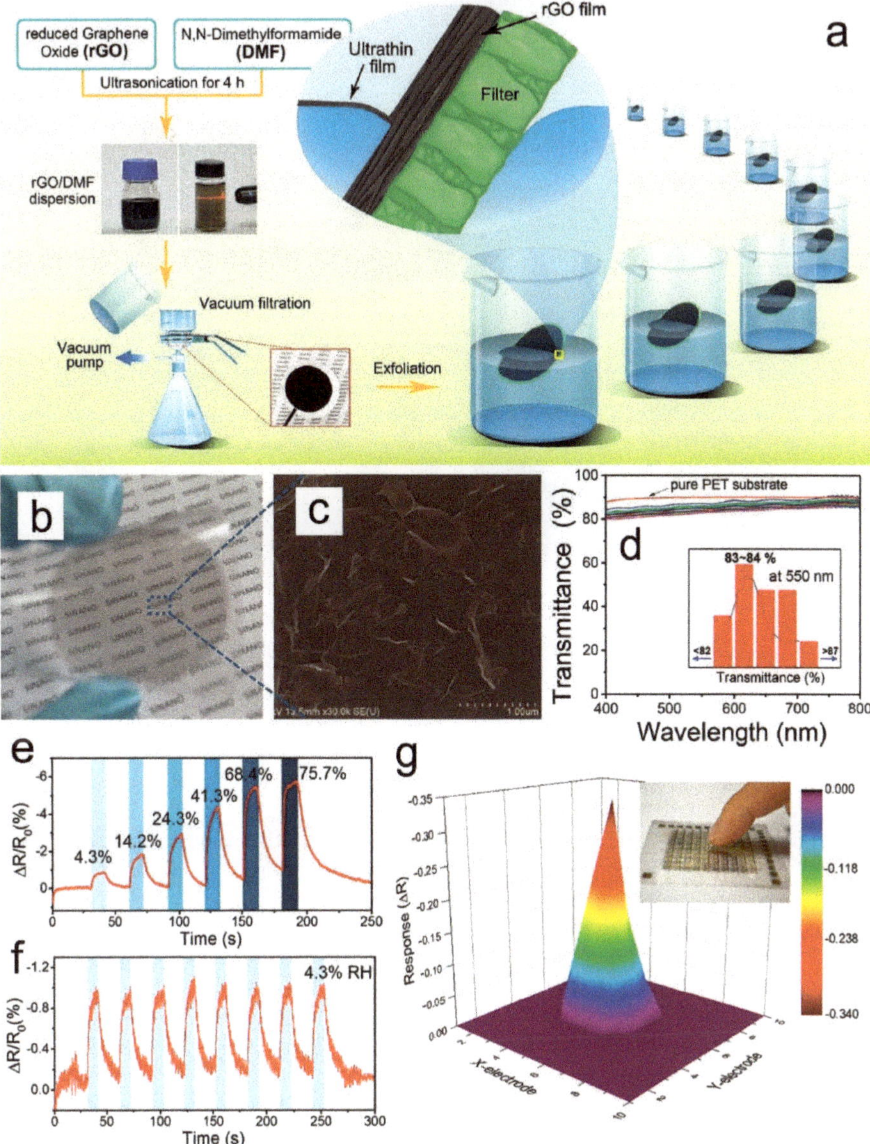

Figure 12. Schematic of the reproducible exfoliation process for the fabrication of r-GO ultrathin films (**a**), a typical photograph (**b**) and SEM image of an r-GO ultrathin film on a PET substrate (**c**), transmittance spectra of 13 r-GO ultrathin films on PET substrates (**d**), real-time response of one sensor in the flexible matrix device to RH from 4.3% to 75.7% (**e**), real-time-repeated response of the sensor at 4.3% RH for eight cycles (**f**), and the 3D mapping of matrix device when the fingertip approaches the relative center area of the device (**g**). Reproduced with permission from [149], Copyright Wiley-VCH, 2014.

Table 3. Summary of the sensing performance of humidity sensors based on reduced graphene oxide or graphene quantum dots.

Material	Preparation Method	Type	Operating Temperature	Sensitivity/Response	Selectivity	Humidity Range	Response/Recovery Time	Long-Term Stability	Ref.
rGO	Two-beam-laser interference reduction	Capacitive	25 °C	—	—	11–95%	3 s/10 s	—	[136]
rGO	LBL-anchored and chemical reduction	Impedance	15–35 °C	0.0423 log Z/%RH	—	30–90%	28 s/48 s	42 days	[137]
rGO	Rapid thermal annealing	Resistive	25 °C	35.3–0.075%	High (Acetone, ethanol, toluene, NH_3, H_2, CO, NO_2, C_2H_2)	20–95%	—	5 weeks	[138]
partially reduced GO	Hydrothermal reduction	Resistive	R.T.	3.3–105%	—	20–85%	4.2 s/3.6 s	14 days	[139]
rGO	Chemical reduction	Resistive	25 °C	17.6	—	10–70%	—	—	[140]
rGO	Sunlight reduction	Impedance	25 °C	three orders of magnitude	—	11–95%	16 s/47 s	30 days	[141]
rGO	Flash reduction	Impedance	20 °C	—	—	11–95%	1 s/24 s	30 days	[142]
N-doped rGO fiber	Thermal annealing	Resistive	R.T.	0.32–4.51%	—	6.1–99.9%	—	30 cycles	[145]
rGO-silk	Seriography-guided reduction	Resistive	25 °C	—	—	~20–97%	3 s/~1 min	—	[148]
rGO	Exfoliation at liquid/air interface	Resistive	R.T.	~6%	—	4.3–75.7%	4 s/10 s	—	[149]
GQDs	Hydrothermal	Resistive	24 °C	~390	High (CO, H_2, CH_4, CO_2)	1–100%	12 s/43 s	—	[151]
GQDs	Pyrolysis of citric acid	Resistive	R.T.	—	—	15–80%	~5 s	—	[152]

4.4. Humidity Sensors Based on Graphene Quantum Dots

Graphene quantum dots (GQDs) have similar structures and properties as layered graphene in the 2D layer, but their electronic structure is governed by the edge electronic states and size (quantum confinement), which can be manipulated to control their properties. Sreeprasad et al. demonstrated the electron-tunneling modulation in a percolating network of graphene quantum dots selectively assembled on a polyelectrolyte microfiber, which was employed to construct a humidity sensor, where the water-mass transferred from the polymer and electrons transported through the GQDs [153]. The reduction of 0.36 nm in the tunneling barrier width between GQDs increased the conductivity of the device by 43-fold. Ruiz et al. synthesized GQDs through pyrolysis of citric acid drop-casted on a metallic interdigitated microelectrode, forming a resistive humidity sensor [152]. The sensor showed an exponential dependence of sensitivity with the RH in the range of 15–80%, and a fast response time estimated at around 5 s. The capillary condensation of water molecules on the GQD surfaces accounted for the sensing performance. Then, Alizadeh et al. prepared a GQD humidity sensor by using a similar approach [151], which exhibited outstanding sensitivity to the variation of environment humidity with a response time of about 10 s. However, they found there were two different sensing mechanisms existing between 0–52% and 52–97%. In the low RH range, the adsorption of water molecules increased the hole-type carriers, thus decreasing the electrical resistance, while in the high RH range, the improved ionic proton transportation devoted by the adsorbed water molecules induced the resistance decrease of the sensor. Hosseini et al. demonstrated the first flexible humidity sensor based on GQDs, which were prepared using a facile hydrothermal method. The as-prepared GQDs were drop-casted on an interdigitated microelectrode on a flexible polyimide substrate (Figure 13a), and the GQD film showed a porous structure (Figure 13b). The sensor showed an exponential behavior response towards humidity in the RH range of 12–100%, with a response and recovery time of 12 and 43 s, respectively (Figure 13c). The device exhibited a fast response when exposed to a flow of exhaled breath with about 90% RH (Figure 13d) [151]. From the performance summary in Table 3, it can be found that GQDs-based humidity sensors have some advantages, such as high response, high selectivity, and relatively fast response and recovery. More studies should be done to develop high-performance humidity sensors based on GQDs.

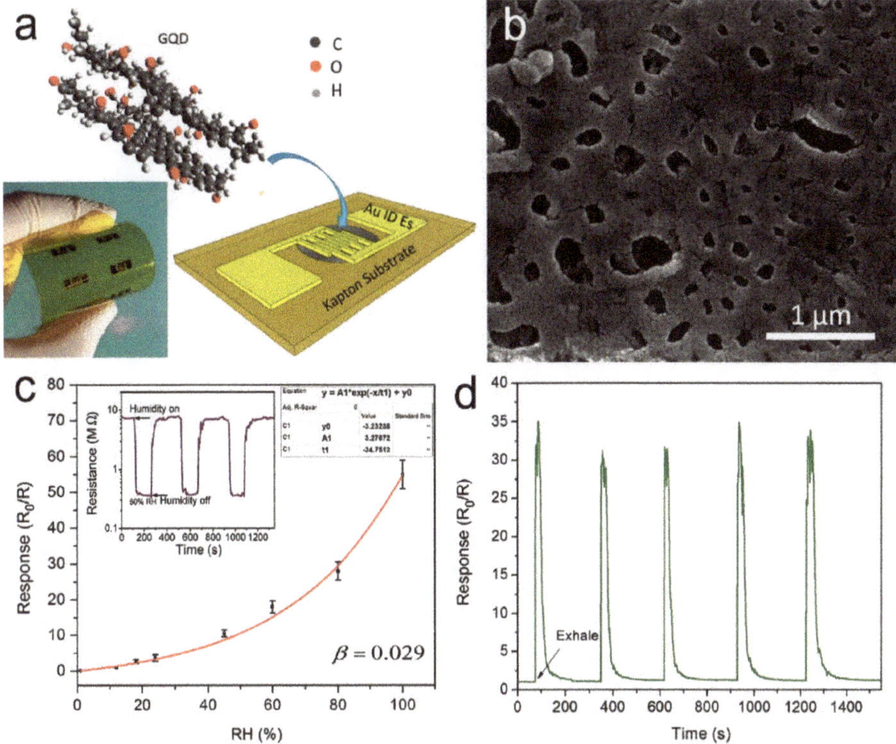

Figure 13. Schematic illustration of the fabricated flexible sensor (inset is an optical image of a flexible array of sensors) (**a**), SEM image of the drop-casted sensing layer (**b**), the sensor response to different levels of humidity (inset shows sensor resistance as a function of time upon exposure to 60% RH during subsequent cycles) (**c**), and response of the fabricated sensor to human breath (**d**). Reproduced with permission from [151], Copyright The Royal Society of Chemistry, 2017.

4.5. Humidity Sensors Based on Chemical Modified Graphene

Generally, pristine graphene or reduced graphene oxide without proper modification exhibits a relatively slow response and low sensitivity towards humidity changes, probably due to the small change in resistance, the decrease of oxygen-containing functional groups, or the aggregation during the chemical reduction process. Hence, chemical modification on graphene surfaces provides a pathway to improve the humidity sensing performance. Earlier, Huang et al. demonstrated the adding sugar through the solvothermal method could effectively tune the mount of the oxygenated group on graphene surfaces, which contributed to the improvement of the humidity sensing performance [154]. A humidity sensor based on amine rich polyethyleneimine (PEI) functionalized CVD-growth graphene showed an improved sensitivity, fast recovery, and good repeatability due to the electron transfer from amine groups in the polymer to graphene [155]. Chen et al. reported an ultra-strong PEI-GO nanocomposite by using PEI modified GO and glycerol diglycidyl ether, forming cross-linking networks [156]. Synergistic reinforcement of mechanical interlocking and hydrogen bonding led to a dramatic increase in the tensile strength and Young's modulus by 98.3% and 87%, respectively, at 7.5 wt% GO loading of PEI and the composite film showed robust humidity sensing performance over the RH range of 40–90%. Both Su's group and Lee's group studied amine-modified graphene oxide as sensing materials in humidity sensors, and demonstrated that the amine modification could improve the sensitivity towards humidity change, but may bring hysteresis-induced errors due to the interaction between water molecules and amine groups [157,158]. Wang et al. proposed supramolecularly modified graphene naphthalene-1-sulfonic acid sodium salt and silver nanoparticles (Ag-NA-rGO), and the resulting supramolecular composite-based humidity sensor exhibited an excellent sensing performance between 11% and 95%, including an ultrafast response and recovery time of <1 s, and a high sensitivity and stability, which was attributed to the large surface area and wide interlayer spacing in the supramolecular composite [159]. Some chemical modification of porous graphene oxide (pGO), such as phenyl, dodecyl, or ethanol, can decrease the humidity sensing sensitivity, but improve the sensitivity to other molecules [160].

In order to address the low resistance change of graphene towards humidity, Ali et al. introduced methyl red molecules to modify graphene surfaces and the composite was deposited over the interdigitated electrodes with the electrohydrodynamic technique (Figure 14a). The electrical resistance of the humidity sensor varied inversely over a wide range from 11 to 0.4 MΩ towards the RH content from 5% to 95%, and the humidity sensor showed 96.36% resistive and 2,869,500% capacitive sensitivity (Figure 14b), with a fast response and recovery times of 0.251 and 0.35 s, respectively [161]. The highly-improved humidity sensing performance was attributed to the water adsorption on methyl red, and induced a decrease of the overall film resistance and the paths between graphene flakes built by water adsorbed methyl red. Tao et al. proposed the preparation of hydrophobin (HFBI) protein wrapped rGO by a one-step exfoliation and functionalization, which was casted onto the interdigitated electrode (Figure 14c) [162]. Humidity sensors based on HFBI modified rGO showed a highly improved sensitivity compared to that based on pristine rGO (Figure 14d,e). Chen et al. employed a similar supramolecular assembly method to modify rGO with functional organic molecule pyranine, which has a pyrene ring decorated with hydrophilic sulfonic groups and can assemble with rGO through π-π interactions (Figure 14f) [163]. A humidity sensor based on this composite exhibited an excellent sensing performance, including a high sensitivity between 11% and 95% RH, fast response time of <2 s, small hysteresis of 8% RH, and good repeatability and stability (Figure 14g,h). Additionally, tannic acid modified rGO can be incorporated in PVA to improve its humidity sensing properties [164]. Recently, a typical metal organic framework (MOF), copper benzene-1,3,5-tricarboxylate (Cu-BTC), was incorporated and deposited on GO film to form a resistive humidity sensor, which showed an improved humidity sensing performance due to the facilitated water molecules captured by Cu-BTC from the environment [165]. The sensing performances of humidity sensors based on chemical modified graphene are summarized in Table 4, wherein humidity sensors based on graphene/methyl-red or pyranine-rGO showed excellent performance, but further studies on the stability and selectivity are required.

4.6. Humidity Sensors Based on Graphene/Polymer Composites

It is well-known that electric humidity sensors output electrical signals by sensing the variation of the moisture content by adsorbing/desorbing water molecules in the environment on the sensitive layer. To construct humidity sensors, polymers are one of the chosen sensitive materials, because they have some advantages, such as low cost, good sensitivity, fast response, flexibility, easy processability, etc. However, generally, they lack conductivity and show large hysteresis due to the cluster of water adsorbed inside bulk polymers, which may cause deformation and instability of the sensing polymer layer, eventually reducing the lifetime of the sensor. Therefore, the combination of polymers and graphene can harness the advantages of both to improve the humidity sensing performance and this has been widely investigated.

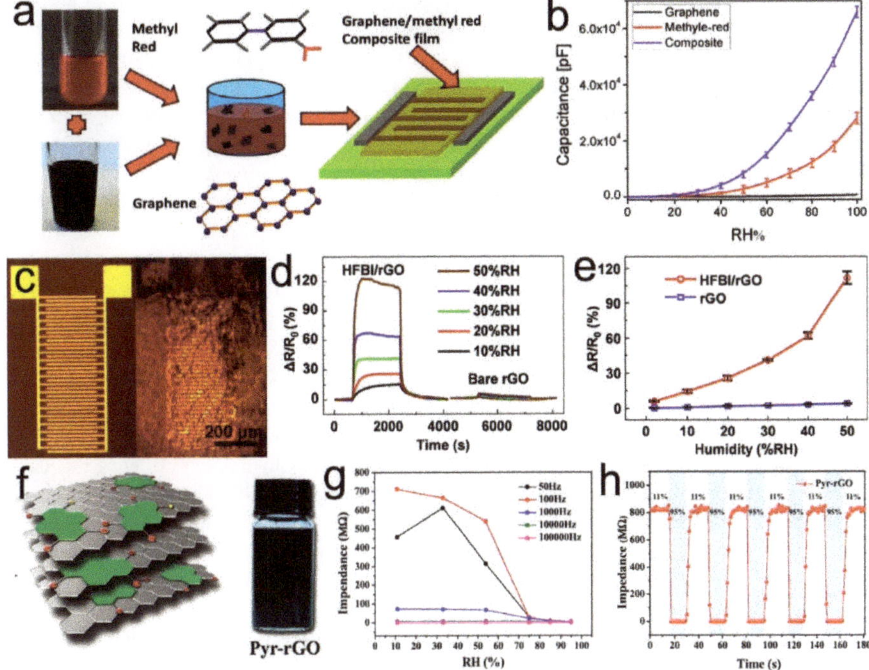

Figure 14. Schematic representation of the graphene/methyl-red composite based humidity sensor (**a**), and capacitance versus relative humidity (% RH) characteristics curves of the graphene/methyl-red composite, methyl-red only, and graphene only based humidity sensors measured in the humidity chamber at a 1 kHz frequency (**b**). Reproduced with permission from [161], Copyright, 2016 Elsevier B.V. Optical images of the comb electrode before and after coating a layer of hydrophobin (HFBI) wrapped rGO flakes (**c**), real-time responses of HFBI wrapped rGO sensor (left) and bare rGO sensor (right) to RH ranging from 2% to 50%, respectively (**d**), and relative resistance change upon the exposure to water molecules at the maximum equilibrium vs. RH values (**e**). Reproduced with permission from [162], Copyright Elsevier B.V., 2017. Schematic of the supramolecular assembly of Pyr-rGO sheets with the corresponding physical image in dispersion (**f**), the impedance curves of Pyr-rGO based humidity sensors measured at different frequencies under different RH levels (**g**), and the five-cycle response-recovery curve of Pyr-rGO (**h**). Adapted from [163].

Table 4. Summary of the sensing performance of humidity sensors based on chemical modified graphene.

Material	Preparation Method	Type	Operating Temperature	Sensitivity/Response	Selectivity	Humidity Range	Response/Recovery Time	Stability	Ref.
Graphene/methyl red	Electro-hydrodynamic	Resistive Capacitive	R.T.	96.36% (R) 2,869,500% (C)	—	5–95%	0.251 s/0.35 s	—	[161]
HFBI-rGO	Drop casting	Resistive	R.T.	2.24/RH	Medium (Ethanol, acetone, hexane)	2–50%	130 s/200 s	—	[162]
Pyranine-rGO	Supramolecular assembly	Impedance	25 °C	6000	—	11–95%	<2 s/<6 s	100 cycles	[163]
PEI-GO	Solution casting	Impedance	30 °C	—	—	40–90%	—	—	[156]
Diamine-GO	Brush coating	Impedance	15–35 °C	0.0545 log Z/% RH	—	20–90%	52 s/72 s	54 days	[157]
GO-NH$_2$	Drop casting	Conductance	25 °C	870 ± 90	—	5–95%	—	—	[158]
Ag-NA-rGO	Drop casting	Impedance	25 °C	600	—	11–95%	1 s/1 s	110 days	[159]

Cellulose is an abundant, renewable, and natural material on earth, which is a colorless, odorless, and nontoxic solid polymer, and possesses the advantages of a high mechanical strength, hydrophilicity, relative thermo-stability, biocompatibility, light weight, low price, and is eco-friendly. Moreover, cellulose is a high dielectric material and insulator, therefore, the incorporation of cellulose into graphene may be of benefit to the humidity sensing performance. For example, Kim's group first employed cellulose nanocrystals to modify rGO, and the resultant composite, m-r(CNC/GO), exhibited an improved sensitivity by 5 times compared to that of pristine rGO, which was attributed to its high surface-to-volume ratio and charge-storage capacity at junctions related to hydrophilic functional groups, such as carboxyl groups [166]. Then, they demonstrated that the incorporation of GO can highly improve the humidity sensor based on cellulose nanocrystal (CNC) due to the increased water molecule adsorption; the water uptake of the cellulose/GO composite in 90% RH was almost two times higher than the pristine cellulose film [167]. Nanocellulose was also used to assist graphene dispersion in the PVA nanocomposite (PVA/NFC/rGO) for humidity sensing, and the abundant hydroxyl groups on the surface of nanocellulose formed hydrogen bonds with water molecules for an improved sensitivity and decreased hysteresis [168]. Recently, Chen et al. prepared a cellulose/graphene composite film using an eco-friendly process by dispersing GO and cellulose homogeneously followed by in situ chemical reduction of GO to rGO, endowing the film with high conductivity and good mechanical properties [169]. The composite film was used as a sensing material to different external stimuli, such as temperature, stress/strain, liquids, as well as humidity in a resistance change. The swelling of the cellulose matrix by adsorbing water molecules increased the distance between the rGO sheets, thus increasing the resistivity of the material.

Lignin is also an abundant polymer in nature and the largest biomass source with an aromatic skeleton. Technical lignin from biomass can be sulfonated into water-soluble sodium lignosulfonate (LS) with abundant oxygen containing groups, which can provide a large number of moisture acceptors for humidity sensing. The hydrophobic phenylpropane units of LS can be spontaneously cross-linked into 3D networks, which is beneficial for the fast swelling and shrinking during adsorption and desorption of water molecules. Nevertheless, LS is electrically insulating, preventing it from being applied in resistive humidity sensors. Recently, Chen et al. designed an LS/rGO composite-based resistive humidity sensor using LS as the moisture sensing layers and rGO as the resistance transduction layers [170]. The LS/rGO composite film showed good flexibility (Figure 15a) and exhibited a more compact interlayered structure compared to pure rGO film (Figure 15b,c), and this alternative multilayered structure could provide water molecules as accepted sites as well as resistant transduction. The LS/rGO humidity sensor showed an apparent response to the RH from 22% to 97% with a dependent resistance increase with the increasing RH (Figure 15d). It also showed a relatively fast response to the RH change (Figure 15e) and a long-term stability lasting for 30 days (Figure 15f). The sensing mechanism may be attributed to the reduction of the hole density in the p-type rGO layers caused by physisorbed water molecules in low humidity and the interlayer swelling effect of LS networks in high humidity (Figure 15g).

Polyelectrolytes are always employed in humidity sensors due to their function groups being sensitive to water molecules. However, their impedance at low relative humidity levels is too high to be measured exactly due to their conductivity, while graphene materials have the advantage of a high surface area and relatively high conductivity. Therefore, it is an effective strategy to improve the sensing performance by combining polyelectrolytes and graphene materials. Li et al. first prepared two composites of polyelectrolytes, i.e., cationic poly(diallydimethylammonium chloride) (PDDA) and anionic sodium poly(4-styrenesulfonate) (PSSNa), and rGO as sensing materials [171]. Both the polyelectrolytes and polyelectrolyte/rGO showed a high response towards RH in the range of 10–90%. However, at lower RH, the polyelectrolytes showed an impedance that was too high to detect. With an increase of the rGO concentration in the polyelectrolytes, the impedance at low humidity was largely decreased by ~45 times. Both composites exhibited high sensitivity and good sensing linearity in their impedance response to humidity in the RH range of 0.2–30%. Typically, PDDA/rGO showed a higher sensitivity of 1000% with respect to the 300% for PSSNa/rGO. Subsequently, they prepared another impedance-type polyelectrolyte/graphene humidity sensor by sequentially depositing the thin films of cross linked and quaternized poly(4-vinylpyridine) (QC-P4VP) and rGO onto interdigitated gold electrodes [172]. The QC-P4VP/rGO humidity sensor showed much lower impedance than the QC-P4VP humidity sensor in the low RH levels and could detect ultra-low RH of 0.18% with a high response with an increase of 500% between 1.1% and 0.18% RH. Moreover, it showed a small hysteresis of ~4.5% RH and a fast response time of 21 s and recovery time of 78 s. Zhang et al. reported rGO/PDDA and GO/PDDA composite humidity sensors by using a layer-by-layer self-assembly approach. The rGO/PDDA exhibited a resistive type humidity sensor, which showed a stable and fast response to the RH in the RH range of 11–97% with a fair sensitivity, and the sensing mechanism was attributed to the p-type semiconducting properties of rGO at low RH, and interlayer swelling of rGO/PDDA film at high RH rather than the ionic conductivity [173]. Then, they demonstrated an ultrahigh performance humidity sensor based on GO/PDDA working in a capacitive type. The humidity sensor showed an unprecedented response of up to 265,640% in the RH range of 11–97% with a short response and recovery time of within 1 s. Thus, it can be used to sense human breath. The excellent sensing performance was attributed to enhanced proton transportation and water molecule permeation in the mesoporous film with adsorbed water molecules [174].

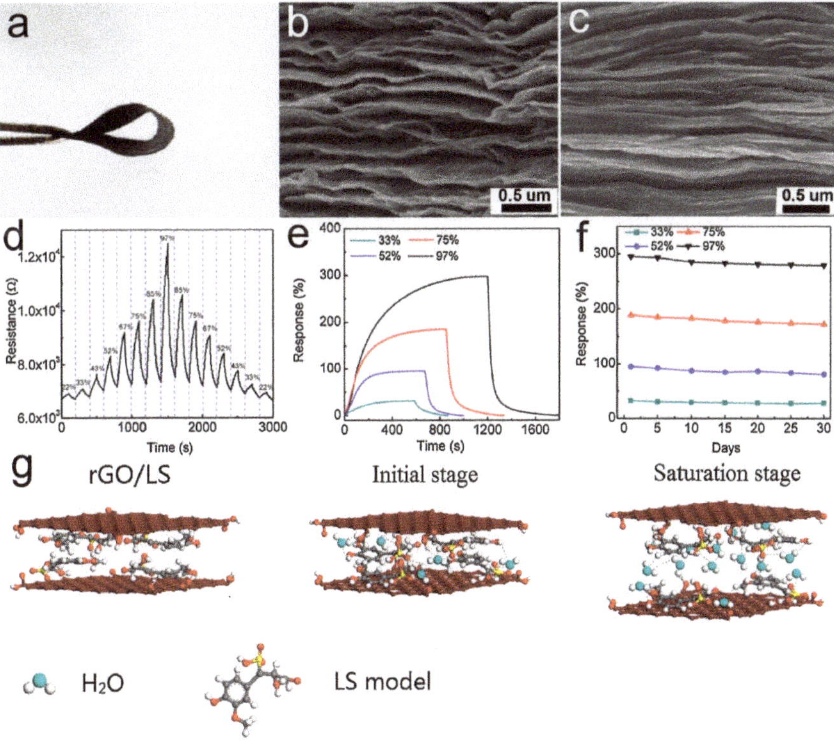

Figure 15. Photographs of a flexible rGO/LS-1 thin film (**a**), cross-sectional SEM images of rGO (**b**) and rGO/LS-1 (**c**) thin-film, real-time resistance measurement of the rGO/LS thin-films under switching RH (**d**), time-dependent humidification-dehumidification curves to a relative humidity pulse between 0% and 33%, 52%, 75%, and 97%, respectively (**e**), long-term stability of rGO/LS thin-film at 33%, 52%, 75%, and 97% RH (**f**), and schematic image of the humidity sensing mechanism of an rGO/LS thin-film at the initial stage and the saturation stage under a humidity environment, respectively (**g**). Reproduced with permission from [170], Copyright Elsevier B.V., 2017.

Dopamine is a small molecule containing catechol and amine groups, and its polymerized form, known as poly(dopamine) (PDA), is similar to the adhesive proteins of mussels, which can be used to modify the materials to bind strongly to most organic and inorganic surfaces. Hwang et al. demonstrated the incorporation of PDA-GO into PVA to synthesize graphene-reinforced PVA composite films, during which PDA reduced GO to rGO [175]. This film showed reinforcement and toughening mechanical properties, as well as an apparent response to RH in the range of 40–100% due to the proton generation on the surface of rGO and PVA when adsorbing water molecules. Recently, He et al. reported a high performance humidity sensor for wearable devices through a bioinspired atomic-precise tunable graphene/PDA heterogeneous sensing junction [176]. The strategy for preparing the bioinspired graphene nanochannels confined poly(dopamine) (GNCP) sensor is schematically shown in Figure 16a, which indicates the dispersion of PDA modified rGO was drop-cased on the flexible polyimide substrate with an interdigitated Au electrode. After slowly evaporating the solvent, the rGO/PDA nanosheets spontaneously self-assembled into a 40 nm thick rGO/PDA film where monolayer graphene alternated with PDA molecular layers with abundant graphene-polymer heterogeneous sensing junctions (Figure 16b–d). The humidity sensing range of the rGO/PDA depended on the PDA content (Figure 16e). The typical composite of GNCP-4 was highly sensitive to variation of RH and improved by almost 4 orders of magnitude with little hysteresis when the RH changed from 0% to 97% (Figure 16f,g). This device also showed a very fast response and recovery, with a response time of about 20 ms and recovery time of 17 ms (Figure 16h,i). A large number of hydroxyl and amino groups in the PDA polymer acting as proton donors and accepters could provide a possible proton transport mechanism by proton-hopping among the carboxyl and amino. Furthermore, this humidity sensor could detect humidity fluctuation information for voiceprint recognition anticounterfeiting (Figure 17), and monitor human respiratory, as well as be applied in noncontact human skin activity real-time monitoring devices.

In addition, some other polymers composited with graphene materials have also been reported, such as poly(N-vinyl pyrrolidone) (PVP), polypyrrole (PPy), poly(vinyl alcohol) (PVA), polyvinylidene fluoride (PVDF), polyurethane (PU), and Nafion. For example, an rGO/PVP composite based resistive humidity sensor was highly sensitive to RH in the range of 30–90% with a response time of ~3 s. The water molecules adsorbed by PVP assisted the charge transfer on the rGO/PVP composite sheets, thus increasing the sensitivity and response [177]. The PVP could also assist the dispersion of graphene to form graphene/PVP ink, which was inkjet-printed on the interdigitated electrode to form a humidity sensor, and further integrated into a complementary metal oxide semiconductor (CMOS) to fulfill the humidity sensing [178]. Su et al. prepared an rGO/PVP resistive humidity sensor, which showed high sensitivity towards RH in the range of 7–97.3% with a response and recovery time of 2.8 and 3.5 s, respectively. Interestingly, such a humidity sensor was integrated with a triboelectric nanogenerator to form a self-powered humidity sensor [179]. Hernández-Rivera et al. reported a capacitive humidity sensor based on an electrospun PVDF/graphene membrane, the sensing principle of which was based on the dielectric constant change of membranes due to the water vapor inside fibrous structure, and the enhanced response of the device may be caused by the improved hydrophobicity brought by the incorporation of graphene into the PVDF [180]. Lin et al. prepared a graphene/PPy composite by a chemical oxidative polymerization method with graphene, which acted as the sensing materials in the impedance humidity sensor. The humidity sensor based on 10% graphene/PPy material showed optimized sensing properties in the RH range of 12–90% with a higher sensitivity of 138 and small humidity hysteresis of <0.16%, as well as fast response and recovery times of 15 and 20 s, respectively [181]. Trung et al. reported a high-performance humidity sensor based on a rGO/PU composite sensing layer and an elastomeric conductive electrode, which exhibited fast response and recovery times of 3.5 and 7 s, respectively. Furthermore, this device remained almost unchanged under stretching up to a strain of 60% and after 10,000 stretching cycles at a 40% strain in the presence of humidity, meaning it can easily be attached to a finger to monitor humidity [182]. Recently, Leng demonstrated that the incorporation of Nafion into diamine modified GO (MGO) could improve the linearity of the sensor compared to MGO [183]. Additionally, fullerene and multiwalled carbon nanotubes have been introduced to composites with graphene to form humidity sensors with improved performances by increasing the accessible surface area of the composite material and accelerating the diffusion of the water molecules [184,185]. Plenty of humidity sensors based on graphene/polymer composites have been reported, the performances of which are summarized in Table 5. The combination with polymer can endow the composite with not only improved sensing performances, such as higher sensitivity and lower detection limit, but also additional functions, such as strength, flexibility, adhesion, and stability. Nevertheless, as we all know, the aging of polymers will be an issue for polymer-incorporated humidity sensors during their practical use. Therefore, accelerated aging tests on humidity sensors based on graphene/polymer composites are suggested.

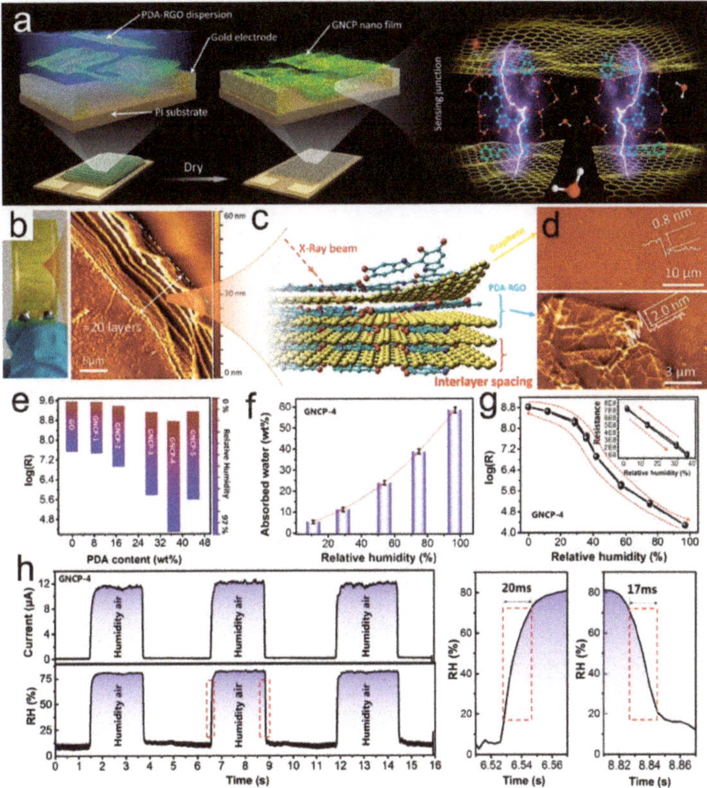

Figure 16. Schematic fabrication layer-by-layer stacking process of the graphene nanochannels confined poly(dopamine) (GNCP) high order superlattice sensing junction structure (**a**), photograph of a drop-casted GNCP sensing element and atomic force microscope (AFM) image of nano-size layered structure poly(dopamine) (PDA)/graphene film on it (**b**), illustration of superlattices composed of alternating atomic scale PDA/graphene layers (**c**), AFM image of graphene before (up) and after PDA modification (down) (**d**), the RH dependent resistance response range of PDA/graphene sensors as a function of PDA content (**e**), histogram plots of absorbed water of PDA/graphene at different humidity atmospheres (**f**), the derived RH dependent resistance changes of PDA/graphene and its magnified curve of the low RH region from 0% to 35% (**g**), and the changes in the measured current from the film at 1 V as RH was switched between dry air (RH ≈ 10%) and humidity air (RH ≈ 80%) (right) (**h**). Estimated results showed the ultrafast response (20 ms) and recovery (17 ms) times (left). Reproduced with permission from [176], Copyright American Chemical Society, 2018.

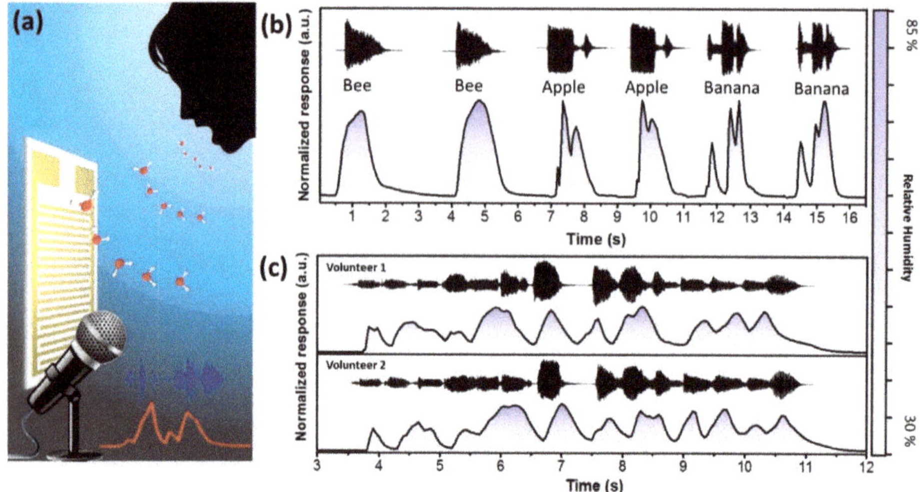

Figure 17. (**a**) Schematic illustration of a humidity sensor for human exhaled air detection during speaking. (**b**) Repeated responses of a PDA/graphene sensor to three different words. (**c**) Responses of a PDA/graphene sensor to the song "Twinkle Twinkle Little Star" sung by two different volunteers. Reproduced with permission from [176], Copyright American Chemical Society, 2018.

Table 5. Summary of the sensing performance of humidity sensors based on graphene/polymer composites.

Material	Preparation Method	Type	Operating Temperature	Sensitivity/Response	Selectivity	Humidity Range	Response/Recovery Time	Stability	Ref.
m-r(CNC/GO)	Layer-by-layer spraying	Resistive	25 °C	~55	—	20–100%	~10 s/~5 s	—	[166]
CNC/GO	Pour drying	Capacitive	25–45 °C	547	—	25–90%	—	—	[167]
PVA/NFC/rGO	Pour drying	Resistive	R.T.	0.347/%RH	—	30–98%	<2 min	—	[168]
rGO/cellulose	Pour casting	Resistive	23 °C	~40%	—	35–90%	—	—	[169]
rGO/LS	Vacuum filtration	Resistive	25 °C	298%	—	22–97%	100 s/100 s	30 days	[170]
PDA/rGO	Drop casting	Resistive	25 °C	20,000	—	0–97%	20 ms/17 ms	60 days	[176]
PDA/PVA/rGO	Pour drying	Resistive	30 °C	—	—	40–100%	—	—	[175]
PDDA/rGO PSSNa/rGO	Dip coating	Impedance	22–25 °C	1000% (0.2–30%) 300% (0.2–30%)	—	0.2–90%	16 s/24 s 38 s/70 s	—	[171]
QC-P4VP/rGO	Dip-coating	Impedance	20–40 °C	500% (0.18–2.1%)	—	0.18–98%	21 s/78 s	—	[172]
PDDA/rGO	Layer-by-layer self-assembly	Resistive	25 °C	8.69–37.43%	—	11–97%	108–147 s/ 94–133 s	60 days	[173]
PDDA/GO	Layer-by-layer self-assembly	Capacitive	25 °C	1552.3 pF/% RH	—	11–97%	1 s/1 s	60 days	[174]
rGO/PVP	Spin coating	Resistive	R.T.	—	—	30–90%	3 s/3 s	—	[177]
Graphene/PVP	Ink-jet printing	Resistive	R.T.	0.3–0.21%/%RH	—	10–80%	6–16 s/ 60–300 s	28 days	[178]
rGO/PVP	Spary coating	Resistive	20–60 °C	~7	High (HCHO, H_2S, NH_3, acetone, H_2)	7.0–97.3%	2.8s/3.5s	28 days	[179]
PVDF/graphene	Electro-spinning	Capacitive	25 °C	0.0463pF/% RH	—	40–90%	~1000 s/21.3 s	—	[180]
Graphene/PPy	Dip coating	Impedance	R.T.	138	—	12–90%	15 s/20 s	—	[181]
rGO/PU	Spin coating	Resistive	R.T.	—	—	10–70%	3.5 s/7 s	—	[182]
MGO/Nafion	Drop casting	Impedance	R.T.	—	—	11–97%	100–300 s/ 100–300 s	2 months	[183]
GO/C60	Drop casting	Capacitive	R.T.	2770%	—	11–97%	8 s/7 s	21 days	[184]
GO/MWCNT	Drop casting	Capacitive	25 °C	7980 pF/% RH	—	11–97%	5 s/2.5 s	—	[185]

4.7. Humidity Sensors Based on Graphene/Metal or Metal Oxide Composites

Metal oxide nanostructures, such as SnO_2, CuO, ZnO, and TiO_2, showed humidity sensing performance due to their high specific surface area, diverse morphology, vacancies, and defects, but a lack of conductivity and slow electron diffusion limited the response towards humidity while pristine GO or rGO exhibited relatively poor sensing properties toward humidity, including low sensitivity and irreversibility. Thus the composite of graphene and metal oxide structures may improve the performance of the humidity sensor. Graphene/SnO_2 composite was widely studied in humidity sensing, and the incorporation of graphene could highly improve the humidity sensing performance. For example, graphene coated SnO_x/carbon fiber (graphene/SnO_x/CF) showed a sensitivity of 6.22, more than 2 times more than the 2.71 for the uncoated one [186]. Xu et al. demonstrated a humidity sensor based on GO wrapped SnO_2@graphene, which showed a very high sensitivity of up to 32 MΩ/% RH, fast response and recovery time of <1 s, and good stability, which could be attributed to the improved conductivity by graphene and oxygen-rich groups (hydroxyl and epoxy groups) from GO [187]. Zhang et al. fabricated SnO_2 nanoparticles/rGO composites by using a facile one-step hydrothermal route, and deposited the composite on microelectrodes forming the humidity sensor (Figure 18a) [77]. The device showed highly improved sensitivity compared to the pristine rGO (Figure 18b), as well as fast response and recovery (Figure 18c), which may be attributed to the active sites, such as vacancies and defects, brought by SnO_2 nanoparticles and the heterojunction created at the interface of the two materials. They also investigated the humidity sensing performance of the SnO_2 nanoparticles/rGO based humidity sensor working in a resistive mode [188]. Reduced GO was also employed to improve the sensing performance of the humidity sensor based on Fe-doped SnO_2 [189]. For other metal oxide/graphene composites, Wang et al. demonstrated that a humidity sensor based on rGO/CuO composite exhibited a relatively good humidity sensing performance, including high sensitivity and fast response, which was attributed to the formation of an rGO−CuO Schottky junction [190]. It has also been demonstrated that the incorporation of graphene shows improvements on the sensing performance of humidity sensors based on ZnO or TiO_2 nanoparticles [191–194].

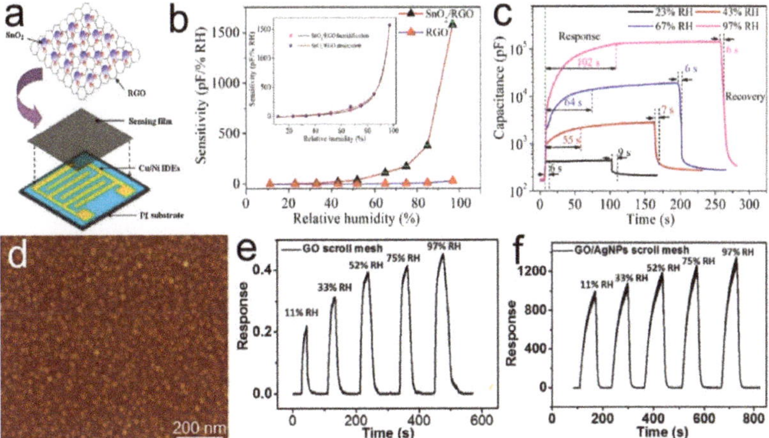

Figure 18. Schematic illustration of as-fabricated sensor prototype (**a**), sensitivity comparison between SnO_2/rGO composite and rGO towards humidity (**b**), and response and recovery curves of the SnO_2/rGO composite sensor towards an RH pulse from dry air to other RH levels (**c**). Reproduced with permission from [77], Copyright Elsevier B.V., 2015. AFM images of GO-Ag sheets (**d**), time-dependent response and recovery curve of an rGO scroll meshes-based device (**e**) and an rGO-Ag scroll meshes-based device (**f**) at different humidity. Reproduced with permission from [195], Copyright The Royal Society of Chemistry, 2017.

Several works have studied the combination of metal nanoparticles and graphene applied in humidity sensors. For example, Liu et al. prepared Ag nanoparticles encapsulated GO scrolls to form GO-Ag scrolls by a molecular combing method, where Ag nanoparticles uniformly deposited on the GO layer (Figure 18d) [195]. After reduction by hydrazine, the humidity sensor based on rGO-Ag scroll meshes exhibited a 3 orders of magnitude response towards humidity compared to that of rGO scroll meshes (Figure 18e,f). The excellent sensitivity was attributed to the enhanced conductivity of rGO-Ag scroll meshes induced by the encapsulation of Ag nanoparticles. Su et al. demonstrated the self-assembly of Au nanoparticles on the surface provided conduction pathways, thus improving the sensitivity and linearity of the sensing film [196]. Interestingly, Yeo et al. realized the suppression of humidity dependence of the rGO sensor by incorporating Cu nanoparticles to decrease the electrical resistances to detect other gases [197]. The sensing performance of humidity sensors based on graphene/metal oxide or metal nanoparticles is summarized in Table 6.

4.8. Humidity Sensors Based on Graphene/2D Materials

Recently, other 2D materials, such as transition metal dichalcogenides, like MoS_2 and WS_2 and black phosphorus, have attracted increasing attention for ultrasensitive sensor applications due to their unique structure and electronic properties. Several 2D materials, such as MoS_2, WS_2, and black phosphorus (BP), have been employed for composites with graphene as humidity sensing materials for humidity sensors with improved performance. Burman et al. developed the first MoS_2/GO nanocomposite-based humidity sensor, which exhibited a high sensing response lying between 55 times at 35% RH and 1600 times at 85% RH, and the high response was attributed to the high proton conductivity in the water layer for both MoS_2 and GO [198]. Recently, Park et al. reported chemoresistive humidity sensors based on rGO/MoS_2 composites, which were prepared by simple ultrasonication without additives and additional heating followed by drop-casting on the interdigitated electrodes (Figure 19a) [199]. The rGO/MoS_2 humidity sensor exhibited a 200 times higher response to humidity at room temperature, compared to the pristine rGO humidity sensor (Figure 19b), and showed a dependent response towards humidity change (Figure 19c). The electronic sensitization due to p–n heterojunction formation and porous structures between rGO and MoS_2 accounted for the remarkable improvement in the sensing performance of the rGO/MoS_2 humidity sensor (Figure 19d,e). They also demonstrated a high-performance humidity sensor based on rGO/MoS_2 hybrid composites synthesized by the hydrothermal method [200]. In addition, WS_2/GO nanohybrids humidity sensors have also been demonstrated with a high response up to 65.8 times at 40% RH and 590 times at 80% RH and a fast response time of 25 s and a recovery time of 29 s, which was attributed to the oxygen linking activities at the GO/WS_2 interface for better proton conductivity [201]. BP materials showed ultra-sensitive humidity sensing performance due to the natural adsorption of water molecules induced by the specific 2D layer-crystalline structure, but lack repeatability due to the instability of BP with water molecules. Phan et al. introduced graphene into BP forming BP/graphene composite to overcome this limitation, and the stability of the humidity sensor was improved by the BP/graphene interface. The humidity sensor exhibited a linear response in the range of 15–70% RH with a sensitivity of 43.4% at 70% RH, and a fast response time of 9 s and a recovery time of 30 s [202]. Humidity sensors based on graphene/2D materials are listed in Table 6, which indicates that graphene/MoS_2 composites have large potential in high-performance humidity sensors, but the response and recovery times still need to be reduced.

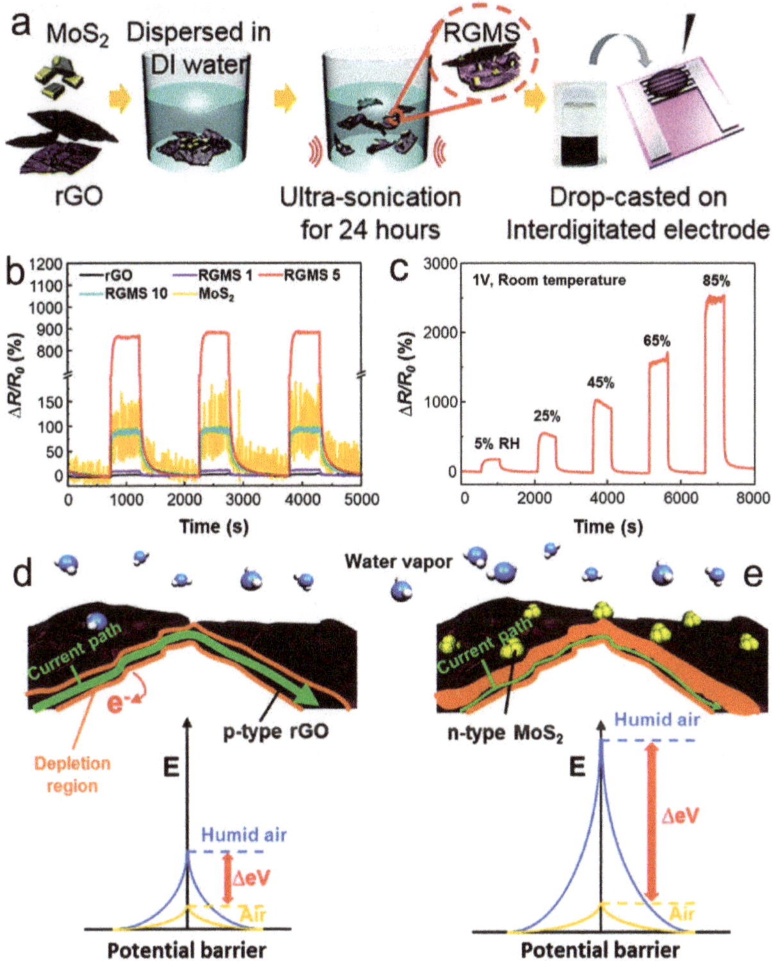

Figure 19. Fabrication procedure of the rGO@MoS₂ humidity sensor (**a**), response curves of rGO, RGMS 1, RGMS 5, RGMS 10, and MoS₂ to 50% RH at 25 °C (**b**), response curves to different RHs of 5%, 25%, 45%, 65%, and 85% at 1 V (**c**), and Schematic of the mechanism with enhanced humidity sensing properties of the enhanced depletion region on (**d**) bare rGO and (**e**) RGMS. Reproduced from [199] with permission from The Royal Society of Chemistry.

Table 6. Summary of the sensing performance of humidity sensors based on graphene/metal, metal oxide, or 2D material composites.

Material	Preparation Method	Type	Operating Temperature	Sensitivity/Response	Selectivity	Humidity Range	Response/Recovery Time	Stability	Ref.
Graphene/SnOx/CF	Electro-spinning	Resistive	20.5 °C	6.22	—	30–80%	8 s/6 s	—	[186]
SnO_2@G-GO	Electro-spinning	Impedance	20 °C	32 MΩ/% RH	—	30–90%	1 s/1 s	3 cycles	[187]
SnO_2/rGO	Hydrothermal	Resistive	R.T.	15.19–45.02%	—	11–97%	<100 s/<100 s	60 days	[188]
SnO_2/rGO	Hydrothermal	Capacitive	25 °C	1604.89 pF/%RH	—	11–97%	6–102 s/6–9 s	—	[77]
rGO/Fe:SnO_2	Electrostatic interaction	Resistive	R.T.	3.23	—	0–100%	—	—	[189]
CuO/rGO	Hydrothermal	Impedance	25 °C	22,700	—	11–98%	2 s/17 s	—	[190]
Graphene/TiO_2	Sol–gel	Impedance	25 °C	151	High (NH_3, NO_2, NO, CO)	12–90%	128 s/68 s	—	[191]
ZnO/GO	Layer-by-layer self-assembly	Capacitive	R.T.	17785.6 pF/%RH	—	0–97%	~50 s/~20 s	30 days	[192]
Graphene/ZnO	Spin coating	Impedance	R.T.	—	—	0–85%	1 s/2 s	—	[193]
TiO_2/GO	Drop casting	Resistive	R.T.	> 10^6	High (H_2, CO, CH_4, NO_2)	9–90%	~70 s	2 months	[194]
rGO–Ag scroll	Molecular combing	Resistive	R.T.	908–1243	—	11–97%	50 s/13 s	30 days	[195]
Au/GO/silica	Sol-gel	Impedance	15–35 °C	−0.0281 log Z/%RH	—	20–90%	119 s/125 s	15 days	[196]
Cu-embeded rGO fiber	Wet spinning	Resistive	20 °C	0.94	—	34–75%	—	—	[197]
rGO/MoS_2	Mixing and drop casting	Resistive	27 °C	2494.25%	High (H_2, CH_3COCH_3, NO_2, NH_3)	5–85%	6.3 s/30.8 s	20 months	[199]
MoS_2/GO	Mixing and drop casting	Resistive	R.T.	1600	—	25–85%	43 s/37 s	90 days	[198]
rGO/MoS_2	Hydrothermal	Resistive	R.T.	~25%	High (NO_2, NH_3, H_2, C_2H_5OH)	10–90%	30 s/253 s	—	[200]
WS_2/GO	Mixing and drop casting	Resistive	25 °C	590 0.044/%RH	—	40–80%	25 s/29 s	28 days	[201]
Black P/ graphene	CVD-G electrospray BP	Resistive	R.T.	43.4%	—	15–70%	9 s/30 s	2 weeks	[202]

5. Summary and Perspective

Various effective approaches have been devoted to developing graphene-based humidity sensors with high sensing performance. First, humidity sensors based on pristine graphene show high sensitivity with a detecting limit as low as 1 ppm and fast response, however, it suffers from low selectivity and relatively slow recovery. Chemical modification on the graphene surface may be a possible way to improve the selectivity and recovery. Second, graphene oxide-based humidity sensors possess high sensitivity and fast response in the high humidity range, but it is difficult for them to detect low humidity conditions of less than 5% RH. The swelling effect and relatively low stability possibly exist at high RH ranges, and the capacitive or impedance working mode generally needs relatively complicated circuits. Third, reduced graphene oxide provides an effective strategy to construct resistive humidity sensors with simple and easy fabrication and measurements and low power consumption, but it also reduces the available functionional groups for adsorbing water molecules, thus decreasing the sensitivity and sensing humidity range. Fourth, various kinds of materials, such as chemicals, polymers, metal, metal oxide, and 2D materials, have been incorporated into graphene oxide or reduced graphene oxide film to improve the humidity sensing performance, and some of them show excellent performances, such as wide sensing RH range, high sensitivity, small hysteresis, and fast response, but they may suffer from reproducibility and long-term stability.

Table 7 provides the performance of several graphene-based humidity sensors with superior comprehensive characteristics. All of them have high sensitivity, fast response, and broad humidity range, but there are still some challenges in the preparation and practical applications of the graphene-based humidity sensors. First, the design and development of humidity sensors based on graphene-materials with a comprehensively high performance, including high sensitivity, high selectivity, wide humidity range, fast response and recovery, and small hysteresis, is still required, and the long-term stability must be especially addressed. Second, the reproducibility for graphene-based materials is a big challenge, because the preparation process, especially the oxidation degree and reduction degree of graphene, is quite difficult to control precisely, thus resulting in materials with different sensing performance. Third, although several studies involve detection in a low humidity range of less than 2% RH, it still requires studies addressing the sensing in the low humidity range. Fourth, besides sensing performance, it is time to consider issues involved in the commercialization of graphene-based humidity sensors, such as large-scale manufacturing, integration, encapsulation and packaging, repeatability, long-term stability in practical use, anti-scratch and anti-chemical characteristics, calibration-free characteristics, etc. Fortunately, Goldsmith et al. have demonstrated a process on industrially manufactured graphene-based digital biosensors [203], but the commercialization still has a long way to go. Furthermore, most current studies on graphene-based humidity sensors have not considered the power consumption issue, which is very important for the future of wearable electronics.

Table 7. Performance of the selected graphene-based humidity sensors with superior characteristics.

Material	Type	Response	Selectivity	Humidity Range	Limit of Detection	Response Time	Recovery Time	Stability	Ref.
GO	Impedance	$\sim10^2$ (1)	—	10–90%	10% (2)	~30 ms	~30 ms	72h	[74]
Ultralarge GO	Conductance	4339 ± 433	—	7–100%	7% (2)	0.2 s	0.7 s	5 days	[121]
rGO/GO/rGO	Voltage	142.5	High (H_2, hexane, ethanol)	6.3–100%	6.3% (2)	1.9 s	3.9 s	> 1 year	[131]
rGO	Impedance	$\sim10^3$ (1)	—	11–95%	11% (2)	1 s	24 s	30 days	[142]
Graphene/methyl red	Resistive Capacitive	96.36% (R) 2869500% (C)	—	5–95%	5% (2)	0.251 s	0.35 s	—	[161]
PDA/rGO	Resistive	20000	—	0–97%	10% (1)	20 ms	17 ms	60 days	[176]
rGO/MoS$_2$	Resistive	2494.25%	High (H_2, CH_3COCH_3, NO_2, NH_3)	5–85%	5% (2)	6.3 s	30.8 s	20 months	[199]
PDDA/rGO PSSNa/rGO	Impedance	1000% (0.2–30%) 300% (0.2–30%)	—	0.2–90%	0.2%	16 s 38 s	24 s 70 s	—	[171]
QC-P4VP/rGO	Impedance	500% (0.18–2.1%)	—	0.18–98%	0.18%	21 s	78 s	—	[172]

(1) Estimated from the response curve; (2) the lowest RH in the sensing measurements.

Author Contributions: Conceptualization, J.C. and A.W.; resources, C.L., C.H., J.L., and Y.Q.; writing—original draft preparation, C.L., J.C., C.H., J.L., S.L., and Y.Q.; writing—review and editing, Y.S., J.S., Z.Z., and A.W.; visualization, C.L., C.H., Y.Q., and J.C.; project administration, Y.S., J.S., and Z.Z.; funding acquisition, J.C. and A.W.

Funding: This work was funded by National Natural Science Foundation of China (No. 21603201), Institute of Materials, China Academy of Engineering Physics (item No. TP02201303), and a Grant-in-Aid for Scientific Research on Innovative Areas "New Polymeric Materials Based on Element Blocks (No. 2401)" (JSPS KAKENHI Grant Number JP24102004) and JSPS KAKENHI Grant Number JP15H04132.

Conflicts of Interest: The authors declare no conflict of interest.

References

1. Seiyama, T.; Yamazoe, N.; Arai, H. Ceramic humidity sensors. *Sens. Actuat.* **1983**, *4*, 85–96. [CrossRef]
2. Chen, Z.; Lu, C. Humidity Sensors: A Review of Materials and Mechanisms. *Sens. Lett.* **2005**, *3*, 274–295. [CrossRef]
3. Blank, T.A.; Eksperiandova, L.P.; Belikov, K.N. Recent Trends of Ceramic Humidity Sensors Development: A Review. *Sens. Actuat. B Chem.* **2016**, *228*, 416–442. [CrossRef]
4. Peng, Y.; Zhao, Y.; Chen, M.-Q.; Xia, F. Research Advances in Microfiber Humidity Sensors. *Small* **2018**, *14*, 1800524. [CrossRef] [PubMed]
5. Li, Z.; Zhang, H.; Zheng, W.; Wang, W.; Huang, H.; Wang, C.; MacDiarmid, A.G.; Wei, Y. Highly Sensitive and Stable Humidity Nanosensors Based on LiCl Doped TiO_2 Electrospun Nanofibers. *J. Am. Chem. Soc.* **2008**, *130*, 5036–5037. [CrossRef] [PubMed]
6. Sun, A.; Huang, L.; Li, Y. Study on humidity sensing property based on TiO_2 porous film and polystyrene sulfonic sodium. *Sens. Actuat. B Chem.* **2009**, *139*, 543–547. [CrossRef]
7. Kuang, Q.; Lao, C.; Wang, Z.L.; Xie, Z.; Zheng, L. High-Sensitivity Humidity Sensor Based on a Single SnO_2 Nanowire. *J. Am. Chem. Soc.* **2007**, *129*, 6070–6071. [CrossRef]

8. Feng, H.; Li, C.; Li, T.; Diao, F.; Xin, T.; Liu, B.; Wang, Y. Three-dimensional hierarchical SnO_2 dodecahedral nanocrystals with enhanced humidity sensing properties. *Sens. Actuat. B Chem.* **2017**, *243*, 704–714. [CrossRef]

9. Li, W.; Liu, J.; Ding, C.; Bai, G.; Xu, J.; Ren, Q.; Li, J. Fabrication of Ordered SnO_2 Nanostructures with Enhanced Humidity Sensing Performance. *Sensors* **2017**, *17*, 2392. [CrossRef]

10. Malik, R.; Tomer, V.K.; Chaudhary, V.; Dahiya, M.S.; Sharma, A.; Nehra, S.P.; Duhan, S.; Kailasam, K. An excellent humidity sensor based on In–SnO_2 loaded mesoporous graphitic carbon nitride. *J. Mater. Chem. A* **2017**, *5*, 14134–14143. [CrossRef]

11. Gu, L.; Zheng, K.; Zhou, Y.; Li, J.; Mo, X.; Patzke, G.R.; Chen, G. Humidity sensors based on ZnO/TiO_2 core/shell nanorod arrays with enhanced sensitivity. *Sens. Actuat. B Chem.* **2011**, *159*, 1–7. [CrossRef]

12. Ates, T.; Tatar, C.; Yakuphanoglu, F. Preparation of semiconductor ZnO powders by sol–gel method: Humidity sensors. *Sens. Actuat. A Phys* **2013**, *190*, 153–160. [CrossRef]

13. Jagtap, S.; Priolkar, K.R. Evaluation of ZnO nanoparticles and study of $ZnO–TiO_2$ composites for lead free humidity sensors. *Sens. Actuat. B Chem.* **2013**, *183*, 411–418. [CrossRef]

14. Narimani, K.; Nayeri, F.D.; Kolahdouz, M.; Ebrahimi, P. Fabrication, modeling and simulation of high sensitivity capacitive humidity sensors based on ZnO nanorods. *Sens. Actuat. B Chem.* **2016**, *224*, 338–343. [CrossRef]

15. Kano, S.; Kim, K.; Fujii, M. Fast-Response and Flexible Nanocrystal-Based Humidity Sensor for Monitoring Human Respiration and Water Evaporation on Skin. *ACS Sens.* **2017**, *2*, 828–833. [CrossRef] [PubMed]

16. Manikandan, V.; Sikarwar, S.; Yadav, B.C.; Mane, R.S. Fabrication of tin substituted nickel ferrite (Sn-$NiFe_2O_4$) thin film and its application as opto-electronic humidity sensor. *Sens. Actuat. A Phys.* **2018**, *272*, 267–273. [CrossRef]

17. Tripathy, A.; Pramanik, S.; Manna, A.; Bhuyan, S.; Azrin Shah, F.N.; Radzi, Z.; Abu Osman, A.N. Design and Development for Capacitive Humidity Sensor Applications of Lead-Free Ca,Mg,Fe,Ti-Oxides-Based Electro-Ceramics with Improved Sensing Properties via Physisorption. *Sensors* **2016**, *16*, 1135. [CrossRef]

18. Yao, Y.; Ma, W. Self-Assembly of Polyelectrolytic/Graphene Oxide Multilayer Thin Films on Quartz Crystal Microbalance for Humidity Detection. *IEEE Sens. J.* **2014**, *14*, 4078–4084. [CrossRef]

19. Su, P.-G.; Li, W.-C.; Tseng, J.-Y.; Ho, C.-J. Fully transparent and flexible humidity sensors fabricated by layer-by-layer self-assembly of thin film of poly(2-acrylamido-2-methylpropane sulfonate) and its salt complex. *Sens. Actuat. B Chem.* **2011**, *153*, 29–36. [CrossRef]

20. Morais, R.M.; dos Santos Klem, M.; Nogueira, G.L.; Gomes, T.C.; Alves, N. Low Cost Humidity Sensor Based on PANI/PEDOT:PSS Printed on Paper. *IEEE Sens. J.* **2018**, *18*, 2647–2651. [CrossRef]

21. Xiao, X.; Zhang, Q.-J.; He, J.-H.; Xu, Q.-F.; Li, H.; Li, N.-J.; Chen, D.-Y.; Lu, J.-M. Polysquaraines: Novel humidity sensor materials with ultra-high sensitivity and good reversibility. *Sens. Actuat. B Chem.* **2018**, *255*, 1147–1152. [CrossRef]

22. Park, H.; Lee, S.; Jeong, H.S.; Jung, H.U.; Park, K.; Lee, G.M.; Kim, S.; Lee, J. Enhanced Moisture-Reactive Hydrophilic-PTFE-Based Flexible Humidity Sensor for Real-Time Monitoring. *Sensors* **2018**, *18*, 921. [CrossRef] [PubMed]

23. Zhuang, Z.; Qi, D.; Ru, C.; Pan, J.; Zhao, C.; Na, H. Fast response and highly sensitive humidity sensors based on $CaCl_2$-doped sulfonated poly (ether ether ketone)s. *Sens. Actuat. B Chem.* **2017**, *253*, 666–676. [CrossRef]

24. Sajid, M.; Siddiqui, G.U.; Kim, S.W.; Na, K.H.; Choi, Y.S.; Choi, K.H. Thermally modified amorphous polyethylene oxide thin films as highly sensitive linear humidity sensors. *Sens. Actuat. A Phys.* **2017**, *265*, 102–110. [CrossRef]

25. Tan, Y.; Yu, K.; Yang, T.; Zhang, Q.; Cong, W.; Yin, H.; Zhang, Z.; Chen, Y.; Zhu, Z. The combinations of hollow MoS_2 micro@nano-spheres: One-step synthesis, excellent photocatalytic and humidity sensing properties. *J. Mater. Chem. C* **2014**, *2*, 5422–5430. [CrossRef]

26. Zhao, J.; Li, N.; Yu, H.; Wei, Z.; Liao, M.; Chen, P.; Wang, S.; Shi, D.; Sun, Q.; Zhang, G. Highly Sensitive MoS_2 Humidity Sensors Array for Noncontact Sensation. *Adv. Mater.* **2017**, *29*, 1702076. [CrossRef]

27. Debasree, B.; Sumita, S.; Panchanan, P.; Prasanta Kumar, G. Pt decorated MoS_2 nanoflakes for ultrasensitive resistive humidity sensor. *Nanotechnology* **2018**, *29*, 115504.

28. Ravindra Kumar, J.; Prasanta Kumar, G. Liquid exfoliated pristine WS_2 nanosheets for ultrasensitive and highly stable chemiresistive humidity sensors. *Nanotechnology* **2016**, *27*, 475503.

29. Guo, H.; Lan, C.; Zhou, Z.; Sun, P.; Wei, D.; Li, C. Transparent, flexible, and stretchable WS_2 based humidity sensors for electronic skin. *Nanoscale* **2017**, *9*, 6246–6253. [CrossRef] [PubMed]

30. Zhang, D.; Cao, Y.; Li, P.; Wu, J.; Zong, X. Humidity-sensing performance of layer-by-layer self-assembled tungsten disulfide/tin dioxide nanocomposite. *Sens. Actuat. B Chem.* **2018**, *265*, 529–538. [CrossRef]

31. Yasaei, P.; Behranginia, A.; Foroozan, T.; Asadi, M.; Kim, K.; Khalili-Araghi, F.; Salehi-Khojin, A. Stable and Selective Humidity Sensing Using Stacked Black Phosphorus Flakes. *ACS Nano* **2015**, *9*, 9898–9905. [CrossRef] [PubMed]

32. Cho, S.-Y.; Lee, Y.; Koh, H.-J.; Jung, H.; Kim, J.-S.; Yoo, H.-W.; Kim, J.; Jung, H.-T. Superior Chemical Sensing Performance of Black Phosphorus: Comparison with MoS_2 and Graphene. *Adv. Mater.* **2016**, *28*, 7020–7028. [CrossRef] [PubMed]

33. Zhu, C.; Xu, F.; Zhang, L.; Li, M.; Chen, J.; Xu, S.; Huang, G.; Chen, W.; Sun, L. Ultrafast Preparation of Black Phosphorus Quantum Dots for Efficient Humidity Sensing. *Chem.-Eur. J.* **2016**, *22*, 7357–7362. [CrossRef] [PubMed]

34. Miao, J.; Cai, L.; Zhang, S.; Nah, J.; Yeom, J.; Wang, C. Air-Stable Humidity Sensor Using Few-Layer Black Phosphorus. *ACS Appl. Mater. Interfaces* **2017**, *9*, 10019–10026. [CrossRef]

35. Afify, A.S.; Ahmad, S.; Khushnood, R.A.; Jagdale, P.; Tulliani, J.-M. Elaboration and characterization of novel humidity sensor based on micro-carbonized bamboo particles. *Sens. Actuat. B Chem.* **2017**, *239*, 1251–1256. [CrossRef]

36. Zhou, G.; Byun, J.-H.; Oh, Y.; Jung, B.-M.; Cha, H.-J.; Seong, D.-G.; Um, M.-K.; Hyun, S.; Chou, T.-W. Highly Sensitive Wearable Textile-Based Humidity Sensor Made of High-Strength, Single-Walled Carbon Nanotube/Poly(vinyl alcohol) Filaments. *ACS Appl. Mater. Interfaces* **2017**, *9*, 4788–4797. [CrossRef] [PubMed]

37. Yeow, J.T.W.; She, J.P.M. Carbon nanotube-enhanced capillary condensation for a capacitive humidity sensor. *Nanotechnology* **2006**, *17*, 5441. [CrossRef]

38. Varghese, S.S.; Lonkar, S.; Singh, K.K.; Swaminathan, S.; Abdala, A. Recent advances in graphene based gas sensors. *Sens. Actuat. B Chem.* **2015**, *218*, 160–183. [CrossRef]

39. Singh, E.; Meyyappan, M.; Nalwa, H.S. Flexible Graphene-Based Wearable Gas and Chemical Sensors. *ACS Appl. Mater. Interfaces* **2017**, *9*, 34544–34586. [CrossRef] [PubMed]

40. Chen, D.; Feng, H.; Li, J. Graphene Oxide: Preparation, Functionalization, and Electrochemical Applications. *Chem. Rev.* **2012**, *112*, 6027–6053. [CrossRef]

41. Wang, H.; Maiyalagan, T.; Wang, X. Review on Recent Progress in Nitrogen-Doped Graphene: Synthesis, Characterization, and Its Potential Applications. *ACS Catal.* **2012**, *2*, 781–794. [CrossRef]

42. Jang, H.; Park, Y.J.; Chen, X.; Das, T.; Kim, M.-S.; Ahn, J.-H. Graphene-Based Flexible and Stretchable Electronics. *Adv. Mater.* **2016**, *28*, 4184–4202. [CrossRef] [PubMed]

43. Ji, L.; Meduri, P.; Agubra, V.; Xiao, X.; Alcoutlabi, M. Graphene-Based Nanocomposites for Energy Storage. *Adv. Energy Mater.* **2016**, *6*, 1502159. [CrossRef]

44. Kim, H.; Ahn, J.-H. Graphene for flexible and wearable device applications. *Carbon* **2017**, *120*, 244–257. [CrossRef]

45. Yu, X.; Cheng, H.; Zhang, M.; Zhao, Y.; Qu, L.; Shi, G. Graphene-based smart materials. *Nat. Rev. Mater.* **2017**, *2*, 17046. [CrossRef]

46. Nag, A.; Mitra, A.; Mukhopadhyay, S.C. Graphene and its sensor-based applications: A review. *Sens. Actuat. A Phys* **2018**, *270*, 177–194. [CrossRef]

47. Yao, Y.; Ping, J. Recent advances in graphene-based freestanding paper-like materials for sensing applications. *TrAC Trend. Anal. Chem.* **2018**, *105*, 75–88. [CrossRef]

48. Yavari, F.; Koratkar, N. Graphene-Based Chemical Sensors. *J. Phys. Chem. Lett.* **2012**, *3*, 1746–1753. [CrossRef]

49. Park, S.; An, J.; Suk, J.W.; Ruoff, R.S. Graphene-Based Actuators. *Small* **2010**, *6*, 210–212. [CrossRef]

50. Shao, Y.; Wang, J.; Wu, H.; Liu, J.; Aksay, I.A.; Lin, Y. Graphene Based Electrochemical Sensors and Biosensors: A Review. *Electroanalysis* **2010**, *22*, 1027–1036. [CrossRef]

51. Liu, Y.; Dong, X.; Chen, P. Biological and chemical sensors based on graphene materials. *Chem. Soc. Rev.* **2012**, *41*, 2283–2307. [CrossRef] [PubMed]

52. Yuan, W.; Shi, G. Graphene-based gas sensors. *J. Mater. Chem. A* **2013**, *1*, 10078–10091. [CrossRef]

53. Chen, S.; Jiang, K.; Lou, Z.; Chen, D.; Shen, G. Recent Developments in Graphene-Based Tactile Sensors and E-Skins. *Adv. Mater. Technol.* **2017**, *3*, 1700248. [CrossRef]

54. Tung, T.T.; Nine, M.J.; Krebsz, M.; Pasinszki, T.; Coghlan, C.J.; Tran, D.N.H.; Losic, D. Recent Advances in Sensing Applications of Graphene Assemblies and Their Composites. *Adv. Funct. Mater.* **2017**, *27*, 1702891. [CrossRef]

55. Farahani, H.; Wagiran, R.; Hamidon, M. Humidity Sensors Principle, Mechanism, and Fabrication Technologies: A Comprehensive Review. *Sensors* **2014**, *14*, 7881. [CrossRef]

56. Sikarwar, S.; Yadav, B.C. Opto-electronic humidity sensor: A review. *Sens. Actuat A Phys* **2015**, *233*, 54–70. [CrossRef]

57. Schedin, F.; Geim, A.K.; Morozov, S.V.; Hill, E.W.; Blake, P.; Katsnelson, M.I.; Novoselov, K.S. Detection of individual gas molecules adsorbed on graphene. *Nat. Mater.* **2007**, *6*, 652–655. [CrossRef]

58. Yi, M.; Shen, Z. A review on mechanical exfoliation for the scalable production of graphene. *J. Mater. Chem. A* **2015**, *3*, 11700–11715. [CrossRef]

59. Li, X.; Cai, W.; An, J.; Kim, S.; Nah, J.; Yang, D.; Piner, R.; Velamakanni, A.; Jung, I.; Tutuc, E.; et al. Large-Area Synthesis of High-Quality and Uniform Graphene Films on Copper Foils. *Science* **2009**, *324*, 1312–1314. [CrossRef] [PubMed]

60. Zhang, Y.; Zhang, L.; Zhou, C. Review of Chemical Vapor Deposition of Graphene and Related Applications. *Acc. Chem. Res.* **2013**, *46*, 2329–2339. [CrossRef]

61. Kim, K.-S.; Lee, H.-J.; Lee, C.; Lee, S.-K.; Jang, H.; Ahn, J.-H.; Kim, J.-H.; Lee, H.-J. Chemical Vapor Deposition-Grown Graphene: The Thinnest Solid Lubricant. *ACS Nano* **2011**, *5*, 5107–5114. [CrossRef] [PubMed]

62. Bae, S.; Kim, H.; Lee, Y.; Xu, X.; Park, J.-S.; Zheng, Y.; Balakrishnan, J.; Lei, T.; Ri Kim, H.; Song, Y.I.; et al. Roll-to-roll production of 30-inch graphene films for transparent electrodes. *Nat. Nanotechnol.* **2010**, *5*, 574–578. [CrossRef] [PubMed]

63. Bai, H.; Li, C.; Shi, G. Functional Composite Materials Based on Chemically Converted Graphene. *Adv. Mater.* **2011**, *23*, 1089–1115. [CrossRef] [PubMed]

64. Zhu, Y.; Murali, S.; Cai, W.; Li, X.; Suk, J.W.; Potts, J.R.; Ruoff, R.S. Graphene and Graphene Oxide: Synthesis, Properties, and Applications. *Adv. Mater.* **2010**, *22*, 3906–3924. [CrossRef] [PubMed]

65. Soldano, C.; Mahmood, A.; Dujardin, E. Production, properties and potential of graphene. *Carbon* **2010**, *48*, 2127–2150. [CrossRef]

66. Abergel, D.S.L.; Apalkov, V.; Berashevich, J.; Ziegler, K.; Chakraborty, T. Properties of graphene: A theoretical perspective. *Adv. Phys.* **2010**, *59*, 261–482.

67. Lee, C.; Wei, X.; Kysar, J.W.; Hone, J. Measurement of the Elastic Properties and Intrinsic Strength of Monolayer Graphene. *Science* **2008**, *321*, 385–388. [CrossRef] [PubMed]

68. He, Q.; Wu, S.; Yin, Z.; Zhang, H. Graphene-based electronic sensors. *Chem. Sci.* **2012**, *3*, 1764–1772. [CrossRef]

69. Yang, S.; Jiang, C.; Wei, S.-H. Gas sensing in 2D materials. *Appl. Phys. Rev.* **2017**, *4*, 021304. [CrossRef]

70. Meng, F.-L.; Guo, Z.; Huang, X.-J. Graphene-based hybrids for chemiresistive gas sensors. *TrAC Trends Anal. Chem.* **2015**, *68*, 37–47. [CrossRef]

71. Shafiei, M.; Bradford, J.; Khan, H.; Piloto, C.; Wlodarski, W.; Li, Y.; Motta, N. Low-operating temperature NO_2 gas sensors based on hybrid two-dimensional SnS_2-reduced graphene oxide. *Appl. Surf. Sci.* **2018**, *462*, 330–336. [CrossRef]

72. Piloto, C.; Shafiei, M.; Khan, H.; Gupta, B.; Tesfamichael, T.; Motta, N. Sensing performance of reduced graphene oxide-Fe doped WO_3 hybrids to NO_2 and humidity at room temperature. *Appl. Surf. Sci.* **2018**, *434*, 126–133. [CrossRef]

73. Rheaume, J.M.; Pisano, A.P. A review of recent progress in sensing of gas concentration by impedance change. *Ionics* **2011**, *17*, 99–108. [CrossRef]

74. Borini, S.; White, R.; Wei, D.; Astley, M.; Haque, S.; Spigone, E.; Harris, N.; Kivioja, J.; Ryhänen, T. Ultrafast Graphene Oxide Humidity Sensors. *ACS Nano* **2013**, *7*, 11166–11173. [CrossRef] [PubMed]

75. Gao, W.; Singh, N.; Song, L.; Liu, Z.; Reddy, A.L.M.; Ci, L.; Vajtai, R.; Zhang, Q.; Wei, B.; Ajayan, P.M. Direct Laser Writing of Micro-Supercapacitors on Hydrated Graphite Oxide Films. *Nat. Nanotechnol.* **2011**, *6*, 496–500. [CrossRef] [PubMed]

76. Bi, H.; Yin, K.; Xie, X.; Ji, J.; Wan, S.; Sun, L.; Terrones, M.; Dresselhaus, M.S. Ultrahigh Humidity Sensitivity of Graphene Oxide. *Sci. Rep.* **2013**, *3*, 2714. [CrossRef]

77. Zhang, D.; Chang, H.; Li, P.; Liu, R.; Xue, Q. Fabrication and characterization of an ultrasensitive humidity sensor based on metal oxide/graphene hybrid nanocomposite. *Sens. Actuat. B Chem.* **2016**, *225*, 233–240. [CrossRef]

78. Balashov, S.M.; Balachova, O.V.; Filho, A.P.; Bazetto, M.C.Q.; de Almeida, M.G. Surface Acoustic Wave Humidity Sensors Based on Graphene Oxide Thin Films Deposited with the Surface Acoustic Wave Atomizer. *ECS Trans.* **2012**, *49*, 445–450. [CrossRef]

79. Le, X.; Wang, X.; Pang, J.; Liu, Y.; Fang, B.; Xu, Z.; Gao, C.; Xu, Y.; Xie, J. A high performance humidity sensor based on surface acoustic wave and graphene oxide on AlN/Si layered structure. *Sens. Actuat. B Chem.* **2018**, *255*, 2454–2461. [CrossRef]

80. Wohltjen, H. Mechanism of operation and design considerations for surface acoustic wave device vapour sensors. *Sens. Actuat.* **1984**, *5*, 307–325. [CrossRef]

81. Guo, Y.J.; Zhang, J.; Zhao, C.; Hu, P.A.; Zu, X.T.; Fu, Y.Q. Graphene/LiNbO$_3$ surface acoustic wave device based relative humidity sensor. *Optik* **2014**, *125*, 5800–5802. [CrossRef]

82. Kuznetsova, I.E.; Anisimkin, V.I.; Gubin, S.P.; Tkachev, S.V.; Kolesov, V.V.; Kashin, V.V.; Zaitsev, B.D.; Shikhabudinov, A.M.; Verona, E.; Sun, S. Super high sensitive plate acoustic wave humidity sensor based on graphene oxide film. *Ultrasonics* **2017**, *81*, 135–139. [CrossRef] [PubMed]

83. Xuan, W.; He, M.; Meng, N.; He, X.; Wang, W.; Chen, J.; Shi, T.; Hasan, T.; Xu, Z.; Xu, Y.; et al. Fast Response and High Sensitivity ZnO/glass Surface Acoustic Wave Humidity Sensors Using Graphene Oxide Sensing Layer. *Sci. Rep.* **2014**, *4*, 7206. [CrossRef] [PubMed]

84. Xuan, W.; He, X.; Chen, J.; Wang, W.; Wang, X.; Xu, Y.; Xu, Z.; Fu, Y.Q.; Luo, J.K. High Sensitivity Flexible Lamb-Wave Humidity Sensors with a Graphene Oxide Sensing Layer. *Nanoscale* **2015**, *7*, 7430–7436. [CrossRef] [PubMed]

85. Kuznetsova, I.E.; Anisimkin, V.I.; Kolesov, V.V.; Kashin, V.V.; Osipenko, V.A.; Gubin, S.P.; Tkachev, S.V.; Verona, E.; Sun, S.; Kuznetsova, A.S. Sezawa wave acoustic humidity sensor based on graphene oxide sensitive film with enhanced sensitivity. *Sens. Actuat. B Chem.* **2018**, *272*, 236–242. [CrossRef]

86. Zhang, D.; Wang, D.; Li, P.; Zhou, X.; Zong, X.; Dong, G. Facile fabrication of high-performance QCM humidity sensor based on layer-by-layer self-assembled polyaniline/graphene oxide nanocomposite film. *Sens. Actuat. B Chem.* **2018**, *255*, 1869–1877. [CrossRef]

87. Ding, X.; Chen, X.; Chen, X.; Zhao, X.; Li, N. A QCM humidity sensor based on fullerene/graphene oxide nanocomposites with high quality factor. *Sens. Actuat. B Chem.* **2018**, *266*, 534–542. [CrossRef]

88. Yao, Y.; Chen, X.; Guo, H.; Wu, Z. Graphene oxide thin film coated quartz crystal microbalance for humidity detection. *Appl. Surf. Sci.* **2011**, *257*, 7778–7782. [CrossRef]

89. Yao, Y.; Chen, X.; Li, X.; Chen, X.; Li, N. Investigation of the stability of QCM humidity sensor using graphene oxide as sensing films. *Sens. Actuat. B Chem.* **2014**, *191*, 779–783. [CrossRef]

90. Li, X.; Chen, X.; Yao, Y.; Li, N.; Chen, X.; Bi, X. Multi-Walled Carbon Nanotubes/Graphene Oxide Composites for Humidity Sensing. *IEEE Sens. J.* **2013**, *13*, 4749–4756. [CrossRef]

91. Yao, Y.; Xue, Y. Impedance analysis of quartz crystal microbalance humidity sensors based on nanodiamond/graphene oxide nanocomposite film. *Sens. Actuat. B Chem.* **2015**, *211*, 52–58. [CrossRef]

92. Tai, H.; Zhen, Y.; Liu, C.; Ye, Z.; Xie, G.; Du, X.; Jiang, Y. Facile development of high performance QCM humidity sensor based on protonated polyethylenimine-graphene oxide nanocomposite thin film. *Sens. Actuat. B Chem.* **2016**, *230*, 501–509. [CrossRef]

93. Yuan, Z.; Tai, H.; Ye, Z.; Liu, C.; Xie, G.; Du, X.; Jiang, Y. Novel highly sensitive QCM humidity sensor with low hysteresis based on graphene oxide (GO)/poly(ethyleneimine) layered film. *Sens. Actuat. B Chem.* **2016**, *234*, 145–154. [CrossRef]

94. Wang, S.; Xie, G.; Su, Y.; Su, L.; Zhang, Q.; Du, H.; Tai, H.; Jiang, Y. Reduced graphene oxide-polyethylene oxide composite films for humidity sensing via quartz crystal microbalance. *Sens. Actuat. B Chem.* **2018**, *255*, 2203–2210. [CrossRef]

95. Zhang, D.; Wang, D.; Zong, X.; Dong, G.; Zhang, Y. High-performance QCM humidity sensor based on graphene oxide/tin oxide/polyaniline ternary nanocomposite prepared by in-situ oxidative polymerization method. *Sens. Actuat. B Chem.* **2018**, *262*, 531–541. [CrossRef]

96. Yuan, Z.; Tai, H.; Bao, X.; Liu, C.; Ye, Z.; Jiang, Y. Enhanced humidity-sensing properties of novel graphene oxide/zinc oxide nanoparticles layered thin film QCM sensor. *Mater. Lett.* **2016**, *174*, 28–31. [CrossRef]

97. Su, P.-G.; Lin, Y.-T. Low-humidity sensing properties of diamine- and β-cyclodextrin-functionalized graphene oxide films measured using a quartz-crystal microbalance. *Sens. Actuat. A Phys.* **2016**, *238*, 344–350. [CrossRef]

98. Ascorbe, J.; Corres, M.J.; Arregui, J.F.; Matias, R.I. Recent Developments in Fiber Optics Humidity Sensors. *Sensors* **2017**, *17*, 893. [CrossRef]

99. Xiao, Y.; Zhang, J.; Cai, X.; Tan, S.; Yu, J.; Lu, H.; Luo, Y.; Liao, G.; Li, S.; Tang, J.; et al. Reduced graphene oxide for fiber-optic humidity sensing. *Opt. Express* **2014**, *22*, 31555–31567. [CrossRef]

100. Huang, Y.; Zhu, W.; Li, Z.; Chen, G.; Chen, L.; Zhou, J.; Lin, H.; Guan, J.; Fang, W.; Liu, X.; et al. High-performance fibre-optic humidity sensor based on a side-polished fibre wavelength selectively coupled with graphene oxide film. *Sens. Actuat. B Chem.* **2018**, *255*, 57–69. [CrossRef]

101. Lim, H.W.; Yap, K.Y.; Chong, Y.W.; Ahmad, H. All-Optical Graphene Oxide Humidity Sensors. *Sensors* **2014**, *14*, 24329–24337. [CrossRef]

102. Wang, Y.; Shen, C.; Lou, W.; Shentu, F.; Zhong, C.; Dong, X.; Tong, L. Fiber optic relative humidity sensor based on the tilted fiber Bragg grating coated with graphene oxide. *Appl. Phys. Lett.* **2016**, *109*, 031107. [CrossRef]

103. Wang, Y.; Shen, C.; Lou, W.; Shentu, F. Polarization-dependent humidity sensor based on an in-fiber Mach-Zehnder interferometer coated with graphene oxide. *Sens. Actuat. B Chem.* **2016**, *234*, 503–509. [CrossRef]

104. Wang, Y.; Shen, C.; Lou, W.; Shentu, F. Fiber optic humidity sensor based on the graphene oxide/PVA composite film. *Opt. Commun.* **2016**, *372*, 229–234. [CrossRef]

105. Gao, R.; Lu, D.-F.; Cheng, J.; Jiang, Y.; Jiang, L.; Qi, Z.-M. Humidity sensor based on power leakage at resonance wavelengths of a hollow core fiber coated with reduced graphene oxide. *Sens. Actuat. B Chem.* **2016**, *222*, 618–624. [CrossRef]

106. Zhao, Y.; Li, X.-G.; Zhou, X.; Zhang, Y.-N. Review on the graphene based optical fiber chemical and biological sensors. *Sens. Actuat. B Chem.* **2016**, *231*, 324–340. [CrossRef]

107. Yavari, F.; Kritzinger, C.; Gaire, C.; Song, L.; Gulapalli, H.; Borca-Tasciuc, T.; Ajayan, P.M.; Koratkar, N. Tunable Bandgap in Graphene by the Controlled Adsorption of Water Molecules. *Small* **2010**, *6*, 2535–2538. [CrossRef]

108. Smith, A.D.; Elgammal, K.; Niklaus, F.; Delin, A.; Fischer, A.C.; Vaziri, S.; Forsberg, F.; Rasander, M.; Hugosson, H.; Bergqvist, L.; et al. Resistive graphene humidity sensors with rapid and direct electrical readout. *Nanoscale* **2015**, *7*, 19099–19109. [CrossRef]

109. Popov, V.I.; Nikolaev, D.V.; Timofeev, V.B.; Smagulova, S.A.; Antonova, I.V. Graphene-based humidity sensors: The origin of alternating resistance change. *Nanotechnology* **2017**, *28*, 355501. [CrossRef]

110. Son, Y.J.; Chun, K.-Y.; Kim, J.-S.; Lee, J.-H.; Han, C.-S. Effects of chemical and physical defects on the humidity sensitivity of graphene surface. *Chem. Phys. Lett.* **2017**, *689*, 206–211. [CrossRef]

111. Shehzad, K.; Shi, T.; Qadir, A.; Wan, X.; Guo, H.; Ali, A.; Xuan, W.; Xu, H.; Gu, Z.; Peng, X.; et al. Designing an Efficient Multimode Environmental Sensor Based on Graphene–Silicon Heterojunction. *Adv. Mater. Technol.* **2017**, *2*, 1600262. [CrossRef]

112. Fan, X.; Elgammal, K.; Smith, A.D.; Östling, M.; Delin, A.; Lemme, M.C.; Niklaus, F. Humidity and CO_2 gas sensing properties of double-layer graphene. *Carbon* **2018**, *127*, 576–587. [CrossRef]

113. Zhen, Z.; Li, Z.; Zhao, X.; Zhong, Y.; Zhang, L.; Chen, Q.; Yang, T.; Zhu, H. Formation of Uniform Water Microdroplets on Wrinkled Graphene for Ultrafast Humidity Sensing. *Small* **2018**, *14*, 1703848. [CrossRef]

114. Zhao, X.; Long, Y.; Yang, T.; Li, J.; Zhu, H. Simultaneous High Sensitivity Sensing of Temperature and Humidity with Graphene Woven Fabrics. *ACS Appl. Mater. Interfaces* **2017**, *9*, 30171–30176. [CrossRef]

115. Qadir, A.; Sun, Y.W.; Liu, W.; Oppenheimer, P.G.; Xu, Y.; Humphreys, C.J.; Dunstan, D.J. Effect of humidity on the interlayer interaction of bilayer graphene. *Phys. Rev. B* **2019**, *99*, 045402. [CrossRef]

116. Yao, Y.; Chen, X.; Zhu, J.; Zeng, B.; Wu, Z.; Li, X. The effect of ambient humidity on the electrical properties of graphene oxide films. *Nanoscale Res. Lett.* **2012**, *7*, 363. [CrossRef]

117. Jiawei, S.; Xiao, X.; Hengchang, B.; Haiyang, J.; Chongyang, Z.; Neng, W.; Jianqiu, H.; Meng, N.; Dan, L.; Litao, S. Solution-assisted ultrafast transfer of graphene-based thin films for solar cells and humidity sensors. *Nanotechnology* **2017**, *28*, 134004.

118. Ho, D.H.; Sun, Q.; Kim, S.Y.; Han, J.T.; Kim, D.H.; Cho, J.H. Stretchable and Multimodal All Graphene Electronic Skin. *Adv. Mater.* **2016**, *28*, 2601–2608. [CrossRef]

119. Park, E.U.; Choi, B.I.; Kim, J.C.; Woo, S.-B.; Kim, Y.-G.; Choi, Y.; Lee, S.-W. Correlation between the sensitivity and the hysteresis of humidity sensors based on graphene oxides. *Sens. Actuat. B Chem.* **2018**, *258*, 255–262. [CrossRef]

120. Feng, X.; Chen, W.; Yan, L. Free-standing dried foam films of graphene oxide for humidity sensing. *Sens. Actuat. B Chem.* **2015**, *215*, 316–322. [CrossRef]

121. Wee, B.-H.; Khoh, W.-H.; Sarker, A.K.; Lee, C.-H.; Hong, J.-D. A high-performance moisture sensor based on ultralarge graphene oxide. *Nanoscale* **2015**, *7*, 17805–17811. [CrossRef] [PubMed]

122. Han, I.K.; Kim, S.; Lee, G.I.; Kim, P.J.; Kim, J.-H.; Hong, W.S.; Cho, J.B.; Hwang, S.W. Compliment Graphene Oxide Coating on Silk Fiber Surface via Electrostatic Force for Capacitive Humidity Sensor Applications. *Sensors* **2017**, *17*, 407. [CrossRef]

123. Guo, R.; Tang, W.; Shen, C.; Wang, X. High sensitivity and fast response graphene oxide capacitive humidity sensor with computer-aided design. *Comput. Mater. Sci.* **2016**, *111*, 289–293. [CrossRef]

124. Leng, X.; Li, W.; Luo, D.; Wang, F. Differential Structure With Graphene Oxide for Both Humidity and Temperature Sensing. *IEEE Sens. J.* **2017**, *17*, 4357–4364. [CrossRef]

125. Rathi, K.; Pal, K. Impact of Doping on GO: Fast Response–Recovery Humidity Sensor. *ACS Omega* **2017**, *2*, 842–851. [CrossRef]

126. Wan, N.; Wang, T.; Tan, X.-Y.; Lu, S.; Zhou, L.-L.; Huang, J.-Q.; Pan, W.; Yang, Y.-M.; Shao, Z.-Y. Microstructure Related Synergic Sensoring Mechanism in Graphene Oxide Humidity Sensor. *J. Phys. Chem. C* **2018**, *122*, 830–838. [CrossRef]

127. Cai, J.; Lv, C.; Watanabe, A. Cost-Effective Fabrication of High-Performance Flexible All-Solid-State Carbon Micro-Supercapacitors by Blue-Violet Laser Direct Writing and Further Surface Treatment. *J. Mater. Chem. A* **2016**, *4*, 1671–1679. [CrossRef]

128. Cai, J.; Lv, C.; Watanabe, A. Laser Direct Writing of High-Performance Flexible All-Solid-State Carbon Micro-Supercapacitors for an On-Chip Self-Powered Photodetection System. *Nano Energy* **2016**, *30*, 790–800. [CrossRef]

129. Cai, J.; Lv, C.; Watanabe, A. High-Performance All-Solid-State Flexible Carbon/TiO$_2$ Micro-Supercapacitors with Photo-Rechargeable Capability. *RSC Adv.* **2017**, *7*, 415–422. [CrossRef]

130. Watanabe, A.; Cai, J. On Demand Process Based on Laser Direct Writing and the Sensor Application. *J. Photopolym. Sci. Tec.* **2017**, *30*, 341–343. [CrossRef]

131. Cai, J.; Lv, C.; Aoyagi, E.; Ogawa, S.; Watanabe, A. Laser Direct Writing of a High-Performance All-Graphene Humidity Sensor Working in a Novel Sensing Mode for Portable Electronics. *ACS Appl. Mater. Interfaces* **2018**, *10*, 23987–23996. [CrossRef]

132. Cai, J.; Lv, C.; Watanabe, A. Laser Direct Writing and Selective Metallization of Metallic Circuits for Integrated Wireless Devices. *ACS Appl. Mater. Interfaces* **2018**, *10*, 915–924. [CrossRef]

133. Cai, J.; Watanabe, A.; Lv, C. Laser direct writing of carbon-based micro-supercapacitors and electronic devices. *J. Laser Appl.* **2018**, *30*, 032603. [CrossRef]

134. Park, R.; Kim, H.; Lone, S.; Jeon, S.; Kwon, W.Y.; Shin, B.; Hong, W.S. One-Step Laser Patterned Highly Uniform Reduced Graphene Oxide Thin Films for Circuit-Enabled Tattoo and Flexible Humidity Sensor Application. *Sensors* **2018**, *18*, 1857. [CrossRef]

135. An, J.; Le, T.-S.D.; Huang, Y.; Zhan, Z.; Li, Y.; Zheng, L.; Huang, W.; Sun, G.; Kim, Y.-J. All-Graphene-Based Highly Flexible Noncontact Electronic Skin. *ACS Appl. Mater. Interfaces* **2017**, *9*, 44593–44601. [CrossRef]

136. Guo, L.; Jiang, H.-B.; Shao, R.-Q.; Zhang, Y.-L.; Xie, S.-Y.; Wang, J.-N.; Li, X.-B.; Jiang, F.; Chen, Q.-D.; Zhang, T.; et al. Two-beam-laser interference mediated reduction, patterning and nanostructuring of graphene oxide for the production of a flexible humidity sensing device. *Carbon* **2012**, *50*, 1667–1673. [CrossRef]

137. Su, P.-G.; Chiou, C.-F. Electrical and humidity-sensing properties of reduced graphene oxide thin film fabricated by layer-by-layer with covalent anchoring on flexible substrate. *Sens. Actuat. B Chem.* **2014**, *200*, 9–18. [CrossRef]

138. Phan, D.-T.; Chung, G.-S. Effects of rapid thermal annealing on humidity sensor based on graphene oxide thin films. *Sens. Actuat. B Chem.* **2015**, *220*, 1050–1055. [CrossRef]

139. Shojaee, M.; Nasresfahani, S.; Dordane, M.K.; Sheikhi, M.H. Fully integrated wearable humidity sensor based on hydrothermally synthesized partially reduced graphene oxide. *Sens. Actuat A Phys* **2018**, *279*, 448–456. [CrossRef]

140. Zaharie-Butucel, D.; Digianantonio, L.; Leordean, C.; Ressier, L.; Astilean, S.; Farcau, C. Flexible transparent sensors from reduced graphene oxide micro-stripes fabricated by convective self-assembly. *Carbon* **2017**, *113*, 361–370. [CrossRef]

141. Han, D.-D.; Zhang, Y.-L.; Ma, J.-N.; Liu, Y.; Mao, J.-W.; Han, C.-H.; Jiang, K.; Zhao, H.-R.; Zhang, T.; Xu, H.-L.; et al. Sunlight-Reduced Graphene Oxides as Sensitive Moisture Sensors for Smart Device Design. *Adv. Mater. Technol.* **2017**, *2*, 1700045. [CrossRef]

142. He, Y.; Liu, Y.; Ma, J.; Han, D.; Mao, J.; Han, C.; Zhang, Y. Facile Fabrication of High-Performance Humidity Sensors by Flash Reduction of GO. *IEEE Sens. J.* **2017**, *17*, 5285–5289. [CrossRef]

143. Papazoglou, S.; Petridis, C.; Kymakis, E.; Kennou, S.; Raptis, Y.S.; Chatzandroulis, S.; Zergioti, I. In-situ sequential laser transfer and laser reduction of graphene oxide films. *Appl. Phys. Lett.* **2018**, *112*, 183301. [CrossRef]

144. Zhao, F.; Zhao, Y.; Cheng, H.; Qu, L. A Graphene Fibriform Responsor for Sensing Heat, Humidity, and Mechanical Changes. *Angew. Chem.* **2015**, *127*, 15164–15168. [CrossRef]

145. Choi, S.-J.; Yu, H.; Jang, J.-S.; Kim, M.-H.; Kim, S.-J.; Jeong, H.S.; Kim, I.-D. Nitrogen-Doped Single Graphene Fiber with Platinum Water Dissociation Catalyst for Wearable Humidity Sensor. *Small* **2018**, *14*, 1703934. [CrossRef] [PubMed]

146. Li, B.; Xiao, G.; Liu, F.; Qiao, Y.; Li, C.M.; Lu, Z. A flexible humidity sensor based on silk fabrics for human respiration monitoring. *J. Mater. Chem. C* **2018**, *6*, 4549–4554. [CrossRef]

147. Li, X.; Zong, L.; Wu, X.; You, J.; Li, M.; Li, C. Biomimetic engineering of spider silk fibres with graphene for electric devices with humidity and motion sensitivity. *J. Mater. Chem. C* **2018**, *6*, 3212–3219. [CrossRef]

148. Ma, R.; Tsukruk, V.V. Seriography-Guided Reduction of Graphene Oxide Biopapers for Wearable Sensory Electronics. *Adv. Funct. Mater.* **2017**, *27*, 1604802. [CrossRef]

149. Wang, X.; Xiong, Z.; Liu, Z.; Zhang, T. Exfoliation at the Liquid/Air Interface to Assemble Reduced Graphene Oxide Ultrathin Films for a Flexible Noncontact Sensing Device. *Adv. Mater.* **2015**, *27*, 1370–1375. [CrossRef]

150. Pang, Y.; Jian, J.; Tu, T.; Yang, Z.; Ling, J.; Li, Y.; Wang, X.; Qiao, Y.; Tian, H.; Yang, Y.; et al. Wearable humidity sensor based on porous graphene network for respiration monitoring. *Biosens. Bioelectron.* **2018**, *116*, 123–129. [CrossRef]

151. Hosseini, Z.S.; Iraji zad, A.; Ghiass, M.A.; Fardindoost, S.; Hatamie, S. A New Approach to Flexible Humidity Sensors Using Graphene Quantum Dots. *J. Mater. Chem. C* **2017**, *5*, 8966–8973. [CrossRef]

152. Ruiz, V.; Fernández, I.; Carrasco, P.; Cabañero, G.; Grande, H.J.; Herrán, J. Graphene quantum dots as a novel sensing material for low-cost resistive and fast-response humidity sensors. *Sens. Actuat. B Chem.* **2015**, *218*, 73–77. [CrossRef]

153. Sreeprasad, T.S.; Rodriguez, A.A.; Colston, J.; Graham, A.; Shishkin, E.; Pallem, V.; Berry, V. Electron-Tunneling Modulation in Percolating Network of Graphene Quantum Dots: Fabrication, Phenomenological Understanding, and Humidity/Pressure Sensing Applications. *Nano Lett.* **2013**, *13*, 1757–1763. [CrossRef]

154. Huang, Q.; Zeng, D.; Tian, S.; Xie, C. Synthesis of defect graphene and its application for room temperature humidity sensing. *Mater. Lett.* **2012**, *83*, 76–79. [CrossRef]

155. Ben Aziza, Z.; Zhang, K.; Baillargeat, D.; Zhang, Q. Enhancement of Humidity Sensitivity of Graphene through Functionalization with Polyethylenimine. *Appl. Phys. Lett.* **2015**, *107*, 134102. [CrossRef]

156. Chen, L.; Li, Z.; Wu, G.; Wang, Y.; Wang, T.; Ma, Y.; Fei, B. Ultra-strong polyethyleneimine-graphene oxide nanocomposite film via synergistic interactions and its use for humidity sensing. *Compos. Part A Appl. Sci. Manuf.* **2018**, *115*, 341–347. [CrossRef]

157. Su, P.-G.; Lu, Z.-M. Flexibility and electrical and humidity-sensing properties of diamine-functionalized graphene oxide films. *Sens. Actuat. B Chem.* **2015**, *211*, 157–163. [CrossRef]

158. Lee, S.-W.; Choi, B.I.; Kim, J.C.; Woo, S.-B.; Kim, Y.-G.; Kwon, S.; Yoo, J.; Seo, Y.-S. Sorption/desorption hysteresis of thin-film humidity sensors based on graphene oxide and its derivative. *Sens. Actuat. B Chem.* **2016**, *237*, 575–580. [CrossRef]

159. Wang, S.; Chen, Z.; Umar, A.; Wang, Y.; Tian, T.; Shang, Y.; Fan, Y.; Qi, Q.; Xu, D. Supramolecularly Modified Graphene for Ultrafast Responsive and Highly Stable Humidity Sensor. *J. Phys. Chem. C* **2015**, *119*, 28640–28647. [CrossRef]

160. Teradal, N.L.; Marx, S.; Morag, A.; Jelinek, R. Porous graphene oxide chemi-capacitor vapor sensor array. *J. Mater. Chem. C* **2017**, *5*, 1128–1135. [CrossRef]

161. Ali, S.; Hassan, A.; Hassan, G.; Bae, J.; Lee, C.H. All-Printed Humidity Sensor Based on Graphene/Methyl-red Composite with High Sensitivity. *Carbon* **2016**, *105*, 23–32. [CrossRef]
162. Tao, J.; Wang, Y.; Xiao, Y.; Yao, P.; Chen, C.; Zhang, D.; Pang, W.; Yang, H.; Sun, D.; Wang, Z.; et al. One-Step Exfoliation and Functionalization of Graphene by Hydrophobin for High Performance Water Molecular Sensing. *Carbon* **2017**, *116*, 695–702. [CrossRef]
163. Chen, Z.; Wang, Y.; Shang, Y.; Umar, A.; Xie, P.; Qi, Q.; Zhou, G. One-Step Fabrication of Pyranine Modified-Reduced Graphene Oxide with Ultrafast and Ultrahigh Humidity Response. *Sci. Rep.* **2017**, *7*, 2713. [CrossRef] [PubMed]
164. Lim, M.-Y.; Shin, H.; Shin, D.M.; Lee, S.-S.; Lee, J.-C. Poly(vinyl alcohol) nanocomposites containing reduced graphene oxide coated with tannic acid for humidity sensor. *Polymer* **2016**, *84*, 89–98. [CrossRef]
165. Wen, Z.; Siyu, M.; Hui, W.; Yongning, H. Metal organic frameworks enhanced graphene oxide electrode for humidity sensor. *J. Phys. Conf. Ser.* **2018**, *986*, 012013.
166. Sadasivuni, K.K.; Kafy, A.; Zhai, L.; Ko, H.-U.; Mun, S.; Kim, J. Transparent and Flexible Cellulose Nanocrystal/Reduced Graphene Oxide Film for Proximity Sensing. *Small* **2014**, *11*, 994–1002. [CrossRef]
167. Kafy, A.; Akther, A.; Shishir, M.I.R.; Kim, H.C.; Yun, Y.; Kim, J. Cellulose nanocrystal/graphene oxide composite film as humidity sensor. *Sens. Actuat. A Phys* **2016**, *247*, 221–226. [CrossRef]
168. Xu, S.; Yu, W.; Yao, X.; Zhang, Q.; Fu, Q. Nanocellulose-assisted dispersion of graphene to fabricate poly(vinyl alcohol)/graphene nanocomposite for humidity sensing. *Compos. Sci. Technol.* **2016**, *131*, 67–76. [CrossRef]
169. Chen, Y.; Pötschke, P.; Pionteck, J.; Voit, B.; Qi, H. Smart cellulose/graphene composites fabricated by in situ chemical reduction of graphene oxide for multiple sensing applications. *J. Mater. Chem. A* **2018**, *6*, 7777–7785. [CrossRef]
170. Chen, C.; Wang, X.; Li, M.; Fan, Y.; Sun, R. Humidity Sensor Based on Reduced Graphene Oxide/Lignosulfonate Composite Thin-Film. *Sens. Actuat. B Chem.* **2018**, *255*, 1569–1576. [CrossRef]
171. Li, Y.; Deng, C.; Yang, M. Facilely prepared composites of polyelectrolytes and graphene as the sensing materials for the detection of very low humidity. *Sens. Actuat. B Chem.* **2014**, *194*, 51–58. [CrossRef]
172. Li, Y.; Fan, K.; Ban, H.; Yang, M. Detection of Very Low Humidity Using Polyelectrolyte/Graphene Bilayer Humidity Sensors. *Sens. Actuat. B Chem.* **2016**, *222*, 151–158. [CrossRef]
173. Zhang, D.; Tong, J.; Xia, B. Humidity-sensing properties of chemically reduced graphene oxide/polymer nanocomposite film sensor based on layer-by-layer nano self-assembly. *Sens. Actuat. B Chem.* **2014**, *197*, 66–72. [CrossRef]
174. Zhang, D.; Tong, J.; Xia, B.; Xue, Q. Ultrahigh performance humidity sensor based on layer-by-layer self-assembly of graphene oxide/polyelectrolyte nanocomposite film. *Sens. Actuat. B Chem.* **2014**, *203*, 263–270. [CrossRef]
175. Hwang, S.-H.; Kang, D.; Ruoff, R.S.; Shin, H.S.; Park, Y.-B. Poly(vinyl alcohol) Reinforced and Toughened with Poly(dopamine)-Treated Graphene Oxide, and Its Use for Humidity Sensing. *ACS Nano* **2014**, *8*, 6739–6747. [CrossRef]
176. He, J.; Xiao, P.; Shi, J.; Liang, Y.; Lu, W.; Chen, Y.; Wang, W.; Théato, P.; Kuo, S.-W.; Chen, T. High Performance Humidity Fluctuation Sensor for Wearable Devices via a Bioinspired Atomic-Precise Tunable Graphene-Polymer Heterogeneous Sensing Junction. *Chem. Mater.* **2018**, *30*, 4343–4354. [CrossRef]
177. Zhang, J.; Shen, G.; Wang, W.; Zhou, X.; Guo, S. Individual nanocomposite sheets of chemically reduced graphene oxide and poly(N-vinyl pyrrolidone): Preparation and humidity sensing characteristics. *J. Mater. Chem.* **2010**, *20*, 10824–10828. [CrossRef]
178. Santra, S.; Hu, G.; Howe, R.C.T.; De Luca, A.; Ali, S.Z.; Udrea, F.; Gardner, J.W.; Ray, S.K.; Guha, P.K.; Hasan, T. CMOS integration of inkjet-printed graphene for humidity sensing. *Sci. Rep.* **2015**, *5*, 17374. [CrossRef] [PubMed]
179. Su, Y.; Xie, G.; Wang, S.; Tai, H.; Zhang, Q.; Du, H.; Zhang, H.; Du, X.; Jiang, Y. Novel high-performance self-powered humidity detection enabled by triboelectric effect. *Sens. Actuat. B Chem.* **2017**, *251*, 144–152. [CrossRef]
180. Hernández-Rivera, D.; Rodríguez-Roldán, G.; Mora-Martínez, R.; Suaste-Gómez, E. A Capacitive Humidity Sensor Based on an Electrospun PVDF/Graphene Membrane. *Sensors* **2017**, *17*, 1009. [CrossRef]
181. Lin, W.-D.; Chang, H.-M.; Wu, R.-J. Applied Novel Sensing Material Graphene/Polypyrrole for Humidity Sensor. *Sens. Actuat. B Chem.* **2013**, *181*, 326–331. [CrossRef]
182. Trung, T.Q.; Duy, L.T.; Ramasundaram, S.; Lee, N.-E. Transparent, Stretchable, and Rapid-Response Humidity Sensor for Body-Attachable Wearable Electronics. *Nano Res.* **2017**, *10*, 2021–2033. [CrossRef]

183. Leng, X.; Luo, D.; Xu, Z.; Wang, F. Modified Graphene Oxide/Nafion Composite Humidity Sensor and Its Linear Response to the Relative Humidity. *Sens. Actuat. B Chem.* **2018**, *257*, 372–381. [CrossRef]

184. Li, X.; Chen, X.; Yu, X.; Chen, X.; Ding, X.; Zhao, X. A High-Sensitive Humidity Sensor Based on Water-Soluble Composite Material of Fullerene and Graphene Oxide. *IEEE Sens. J.* **2018**, *18*, 962–966. [CrossRef]

185. Li, X.; Chen, X.; Chen, X.; Ding, X.; Zhao, X. High-sensitive humidity sensor based on graphene oxide with evenly dispersed multiwalled carbon nanotubes. *Mater. Chem. Phys.* **2018**, *207*, 135–140. [CrossRef]

186. Fu, T.; Zhu, J.; Zhuo, M.; Guan, B.; Li, J.; Xu, Z.; Li, Q. Humidity sensors based on graphene/SnO$_x$/CF nanocomposites. *J. Mater. Chem. C* **2014**, *2*, 4861–4866. [CrossRef]

187. Xu, J.; Gu, S.; Lu, B. Graphene and graphene oxide double decorated SnO$_2$ nanofibers with enhanced humidity sensing performance. *RSC Adv.* **2015**, *5*, 72046–72050. [CrossRef]

188. Zhang, D.; Chang, H.; Liu, R. Humidity-Sensing Properties of One-Step Hydrothermally Synthesized Tin Dioxide-Decorated Graphene Nanocomposite on Polyimide Substrate. *J. Electron. Mater.* **2016**, *45*, 4275–4281. [CrossRef]

189. Toloman, D.; Popa, A.; Stan, M.; Socaci, C.; Biris, A.R.; Katona, G.; Tudorache, F.; Petrila, I.; Iacomi, F. Reduced graphene oxide decorated with Fe doped SnO$_2$ nanoparticles for humidity sensor. *Appl. Surf. Sci.* **2017**, *402*, 410–417. [CrossRef]

190. Wang, Z.; Xiao, Y.; Cui, X.; Cheng, P.; Wang, B.; Gao, Y.; Li, X.; Yang, T.; Zhang, T.; Lu, G. Humidity-Sensing Properties of Urchinlike CuO Nanostructures Modified by Reduced Graphene Oxide. *ACS Appl. Mater. Interfaces* **2014**, *6*, 3888–3895. [CrossRef]

191. Lin, W.-D.; Liao, C.-T.; Chang, T.-C.; Chen, S.-H.; Wu, R.-J. Humidity sensing properties of novel graphene/TiO$_2$ composites by sol–gel process. *Sens. Actuat. B Chem.* **2015**, *209*, 555–561. [CrossRef]

192. Zhang, D.; Liu, J.; Xia, B. Layer-by-Layer Self-Assembly of Zinc Oxide/Graphene Oxide Hybrid Toward Ultrasensitive Humidity Sensing. *IEEE Electron Device Lett.* **2016**, *37*, 916–919. [CrossRef]

193. Hassan, G.; Bae, J.; Lee, C.H.; Hassan, A. Wide range and stable ink-jet printed humidity sensor based on graphene and zinc oxide nanocomposite. *J. Mater. Sci. Mater. Electron.* **2018**, *29*, 5806–5813. [CrossRef]

194. Sun, L.; Haidry, A.A.; Fatima, Q.; Li, Z.; Yao, Z. Improving the humidity sensing below 30% RH of TiO$_2$ with GO modification. *Mater. Res. Bull.* **2018**, *99*, 124–131. [CrossRef]

195. Liu, Y.; Wang, L.; Zhang, H.; Ran, F.; Yang, P.; Li, H. Graphene oxide scroll meshes encapsulated Ag nanoparticles for humidity sensing. *RSC Adv.* **2017**, *7*, 40119–40123. [CrossRef]

196. Su, P.-G.; Shiu, W.-L.; Tsai, M.-S. Flexible humidity sensor based on Au nanoparticles/graphene oxide/thiolated silica sol–gel film. *Sens. Actuat. B Chem.* **2015**, *216*, 467–475. [CrossRef]

197. Yeo, C.S.; Kim, H.; Lim, T.; Kim, H.J.; Cho, S.; Cho, K.R.; Kim, Y.S.; Shin, M.K.; Yoo, J.; Ju, S.; et al. Copper-embedded reduced graphene oxide fibers for multi-sensors. *J. Mater. Chem. C* **2017**, *5*, 12825–12832. [CrossRef]

198. Burman, D.; Ghosh, R.; Santra, S.; Guha, P.K. Highly proton conducting MoS$_2$/graphene oxide nanocomposite based chemoresistive humidity sensor. *RSC Adv.* **2016**, *6*, 57424–57433. [CrossRef]

199. Park, S.Y.; Kim, Y.H.; Lee, S.Y.; Sohn, W.; Lee, J.E.; Kim, D.H.; Shim, Y.-S.; Kwon, K.C.; Choi, K.S.; Yoo, H.J.; et al. Highly selective and sensitive chemoresistive humidity sensors based on rGO/MoS$_2$ van der Waals composites. *J. Mater. Chem. A* **2018**, *6*, 5016–5024. [CrossRef]

200. Park, S.Y.; Lee, J.E.; Kim, Y.H.; Kim, J.J.; Shim, Y.-S.; Kim, S.Y.; Lee, M.H.; Jang, H.W. Room temperature humidity sensors based on rGO/MoS$_2$ hybrid composites synthesized by hydrothermal method. *Sens. Actuat. B Chem.* **2018**, *258*, 775–782. [CrossRef]

201. Jha, R.K.; Burman, D.; Santra, S.; Guha, P.K. WS$_2$/GO Nanohybrids for Enhanced Relative Humidity Sensing at Room Temperature. *IEEE Sens. J.* **2017**, *17*, 7340–7347. [CrossRef]

202. Phan, D.-T.; Park, I.; Park, A.-R.; Park, C.-M.; Jeon, K.-J. Black P/graphene hybrid: A fast response humidity sensor with good reversibility and stability. *Sci. Rep.* **2017**, *7*, 10561. [CrossRef]

203. Goldsmith, B.R.; Locascio, L.; Gao, Y.; Lerner, M.; Walker, A.; Lerner, J.; Kyaw, J.; Shue, A.; Afsahi, S.; Pan, D.; et al. Digital Biosensing by Foundry-Fabricated Graphene Sensors. *Sci. Rep.* **2019**, *9*, 434. [CrossRef]

Review

Electronic and Thermal Properties of Graphene and Recent Advances in Graphene Based Electronics Applications

Mingyu Sang †, Jongwoon Shin †, Kiho Kim and Ki Jun Yu *

School of Electrical & Electronic Engineering, Yonsei University, Seoul 03722, Korea;
sangmg315@yonsei.ac.kr (M.S.); shinjw6435@yonsei.ac.kr (J.S.); rlarlgh03@naver.com (K.K.)
* Correspondence: kijunyu@yonsei.ac.kr; Tel.: +82-2-2123-2769
† These authors contributed equally to this work.

Received: 1 February 2019; Accepted: 21 February 2019; Published: 5 March 2019

Abstract: Recently, graphene has been extensively researched in fundamental science and engineering fields and has been developed for various electronic applications in emerging technologies owing to its outstanding material properties, including superior electronic, thermal, optical and mechanical properties. Thus, graphene has enabled substantial progress in the development of the current electronic systems. Here, we introduce the most important electronic and thermal properties of graphene, including its high conductivity, quantum Hall effect, Dirac fermions, high Seebeck coefficient and thermoelectric effects. We also present up-to-date graphene-based applications: optical devices, electronic and thermal sensors, and energy management systems. These applications pave the way for advanced biomedical engineering, reliable human therapy, and environmental protection. In this review, we show that the development of graphene suggests substantial improvements in current electronic technologies and applications in healthcare systems.

Keywords: graphene; electronic and thermal properties; electronic and thermal conductivity; quantum Hall effect; Dirac fermions; Seebeck coefficient; thermoelectric effect; graphene-based applications

---

1. Introduction

Graphene is a recently discovered two-dimensional (2D) carbon allotrope that consists of only a single layer of carbon atoms arranged in a honeycomb lattice and is a base unit for other graphitic materials. Although graphene has been theoretically studied for several decades [1–3], freestanding graphene was first obtained in 2004 by using sticky tape and a pencil, and follow-up experiments opened 'the world of graphene' [4–7]. Over the past decades, various electronics, such as next-generation radio-frequency and high-speed devices, sensors, thermally and electrically conductive composites, and transparent electrodes for solar cells and displays, have been widely developed [8]. Additionally, novel materials such as quantum dots (QDs) and rare earth elements have been thoroughly studied to support the fast-growing technology and to continuously enhance the performance of electronics [9–11]. However, the difficulty of synthesis and high cost due to limited supply have made the broad exploitation of these materials difficult [12]. Graphene is an excellent candidate to replace these materials. Graphene exhibits remarkable electronic and thermal properties and shows unusual electronic properties, such as Dirac fermions [13], the quantum Hall effect (QHE) [7], and an ambipolar electric field effect [14]. Low-energy excitations of graphene are massless; Dirac fermions behave uniquely in magnetic fields and lead to the QHE. Additionally, graphene is highly thermally conductive, exhibiting a thermal conductivity of ~4000 Wm^{-1} K^{-1} [15–17]. Moreover, graphene has a high Seebeck coefficient and figure of merit, which make it easier to convert electrical current to heat [18]. Thus, graphene has the potential for use in energy-harvesting applications. Additionally, this unique material has impressive mechanical and optical properties, such as a fracture strength of 130 GPa [19], optical transparency [20], high room-temperature carrier mobility, and an ultrabroad optical absorption spectrum [21]. Graphene has attracted explosive interest in physics, material science and wide technological applications [22–27]. Here, we present a review of the electronic and thermal properties of graphene and its up-to-date applications, including high conductivity, the quantum Hall effect, Dirac fermions, a high Seebeck coefficient, thermoelectric effects, optical devices, electronic and thermal sensors, and energy management systems. In addition, the remarkable potential to extend the field of applications based on graphene is suggested.

2. Electronic Properties and Applications

2.1. Electronic Properties

The remarkable electronic and optical properties observed for graphene crystallites are the primary reasons for the exclusive focus of experimental and theoretical efforts on graphene while ignoring the existence of other 2D materials. The electrons in graphene have long mean free paths without disrupting the electron-electron interactions and disorder. Therefore, the properties of graphene differ from those of other common metals and semiconductors associated with the physical structures and electronic properties.

2.1.1. Honeycomb Lattice and Brillouin Zone

Figure 1a shows the hexagonal structure of carbon atoms in graphene [13]. A triangular lattice with a basis of two atoms forms the honeycomb lattice structure. The unit vectors of the lattice can be expressed as

$$a_1 = \frac{a}{2}\left(3, \sqrt{3}\right), \ a_2 = \frac{a}{2}\left(3, -\sqrt{3}\right), \tag{1}$$

where the distance between two carbon atoms is approximately 1.42 Å. The reciprocal-lattice vectors can be written as

$$b_1 = \frac{2\pi}{3a}\left(1, \sqrt{3}\right), \ b_2 = \frac{2\pi}{3a}\left(1, -\sqrt{3}\right). \tag{2}$$

The two points K and K' at the edge of the graphene Brillouin zone (BZ), called Dirac points, are essential for the physics of graphene. Their positions in momentum space can be expressed as

$$K = \left(\frac{2\pi}{3a}, \frac{2\pi}{3\sqrt{3}a}\right), \ K' = \left(\frac{2\pi}{3a}, -\frac{2\pi}{3\sqrt{3}a}\right). \tag{3}$$

The three nearest-neighbor vectors are expressed by

$$\delta_1 = \frac{a}{2}\left(1, \sqrt{3}\right), \delta_2 = \frac{a}{2}\left(1, -\sqrt{3}\right), \delta_3 = -a(1, 0) \tag{4}$$

while the six second-nearest neighbors are positioned at $\delta'_1 = \pm a_1$, $\delta'_2 = \pm a_2$, and $\delta'_3 = \pm(a_2 - a_1)$.

The graphene electronic structure can be predicted by graphene simulation, especially using density functional theory (DFT). The adsorption energy of graphene and other materials (such as lithium, sodium, hydrogen and potassium), density of states (DOS), geometry, work function and dipole moment of each adatom-graphene system are studied and calculated using DFT [28–34].

2.1.2. Ambipolar Electric Field Effect in Single-Layer Graphene

As shown in Figure 1b, the exceptional quality of graphene can be precisely seen in its pronounced ambipolar electric field effect [14]. Charge carriers can have high mobilities μ over 15,000 cm^2 V^{-1} s^{-1} at ambient conditions and can be continually switched between electrons and holes in concentrations n of 10^{13} cm^{-2} [4–7]. The mobility of graphene has a consistently high value despite the high n (10^{13} cm^{-2}) in electrically and chemically doped devices.

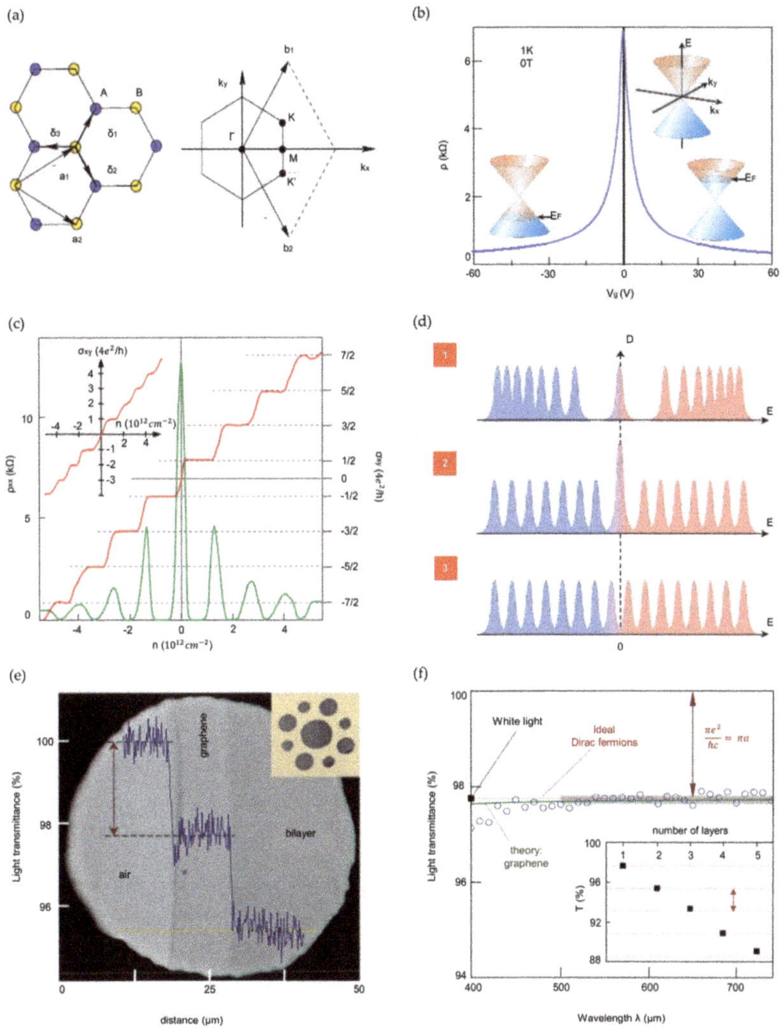

Figure 1. Electronic and optical properties of graphene, (**a**) Hexagonal lattice structure (left) and its Brillouin zone (right); (**b**) Graph of the ambipolar electric field effect of single-layer graphene. The inset indicates the change in Fermi energy E$_F$ in response to the changing gate voltage V_g in the conical low-energy spectrum E(k); (**c**) Hall conductivity σ_{xy} (red) and longitudinal resistivity ρ_{xx} (green) of graphene versus carrier concentration. The inset shows σ_{xy} in double-layer graphene; (**d**) Three examples of Landau quantization in graphene. *D* is density of states; (**e**) Image of a 50-μm diameter aperture covered partly by graphene and its bilayer. (Inset) The designed sample; (**f**) The open circles show the transmittance spectrum of single-layer graphene. (Inset) Black squares indicate the transmittance of white light versus the number of graphene layers; (a) Reproduced with permission from [13]. Copyright Reviews of Modern Physics, 2009. (b), (d) Reproduced with permission from [14]. Copyright Nature Materials, 2007. (c) Reproduced with permission from [7]. Copyright Nature, 2005. (e), (f) Reproduced with permission from [20]. Copyright Science, 2008.

2.1.3. Quantum Hall Effect

The QHE is another factor indicating the system's outstanding electronic quality. Because the temperature range of the QHE for graphene is 10 times broader than that of other 2D materials, the QHE in graphene can be seen at room temperature. The unique reaction of massless fermions in a magnetic field is made more evident by their behavior at the high-field limit, where Shubnikov-de Haas oscillations (SdHOs) progress to the QHE [7]. In Figure 1c, the Hall conductivity σ_{xy} and longitudinal resistivity ρ_{xx} of graphene are shown with electron or hole concentrations in a constant magnetic field B. Pronounced QHE plateaus are detectable but do not behave in the expected progression $\sigma_{xy} = \left(4e^2/h\right)N$, where N is an integer. The first plateau occurs at $\left(2e^2/h\right)$, and the progression is $\left(4e^2/h\right)(N+1/2)$ because the plateaus correspond to half-integer ν values. In graphene, the movement from the lowest hole (ν = -1/2) to the lowest electron (ν = +1/2) Landau level (LL) needs the same amount of carriers ($\Delta n = 4B/\varphi_0 \approx 1.2 \times 10^{12}\text{cm}^{-2}$), as does movement between other adjacent levels. To highlight this extraordinary property, Figure 1c shows the σ_{xy} of a graphite film comprising only two graphene layers. The existence of a quantized level at zero E, which is shared by electrons and holes, is the most important factor in realizing the unusual QHE sequence [14,35–39] (Figure 1d(1)). Another means of describing the half-integer QHE is to identify the coupling between pseudospin and orbital motion [5,7,40]. The pseudospin related to massive Dirac fermions induces a geometrical phase of 2π, which appears in the double degeneracy of the zero-E LL (Figure 1d(2)). Interestingly, the 'standard' QHE with all the plateaus present can be recovered in bilayer graphene by the electric field effect. A gate voltage causes an asymmetry between the two graphene layers, resulting in a semiconducting gap. The electric-field-induced gap results in a sustained QHE sequence by dividing the double step into two (Figure 1d(3)).

2.1.4. Visual Transparency of Graphene

The transparency or opacity of suspended graphene is dependent only on the fine-structure constant, $\alpha = e^2/hc \approx 7.299 \times 10^{-3}$ (where c is the speed of light) [14]. This parameter, conventionally related to quantum electrodynamics, defines the binding of light and relativistic electrons. Nair et al. reported that graphene absorbs a large portion ($\pi\alpha = 2.3\%$) of incident white light despite being only one atom thick [20]. The universal G (high-frequency conductivity) for Dirac fermions means that detectable quantities such as graphene's optical transmittance T and reflectance R are also universal (expressed by $T \equiv \left(1 + \frac{2\pi G}{c}\right)^{-2} = \left(1 + \frac{1}{2}\pi\alpha\right)^{-2}$ and $R \equiv \frac{1}{4}\pi^2\alpha^2 T$, respectively, for normally incident light) [41,42]. Additionally, the opacity of graphene is yielded by $(1 - T) \approx \pi\alpha$, which can be written by computing the absorption of light by two-dimensional Dirac fermions with Fermi's golden rule. Figure 1e displays a graphene crystal sample covering submillimeter apertures in a metal scaffold (Figure 1e inset) in transmitted white light. The opacities of different areas can be compared because the schematic in Figure 1e shows only an aperture partially covered by suspended graphene. Differences in the observed light intensity can be explained by the line scan traveling across the image. Figure 1f illustrates an opacity of graphene of $2.3 \pm 0.1\%$ with minor reflectance ($< 0.1\%$), in contrast to optical spectroscopy, which indicates that the opacity is nearly independent of the wavelength, λ. As shown in the inset of Figure 1f, the opacity increases by 2.3% for each graphene layer added to the membrane. The above results also demonstrate a universal dynamic conductivity G = (1.01 ± 0.04) $e^2/4\hbar$ (here e is the elementary charge and $\hbar$ is the Planck constant) in the visible-frequency region, which is the expected performance for ideal Dirac fermions.

2.2. Electronic Applications

2.2.1. Optical Devices

Photodetectors

The ability of photodetectors to detect broad spectral ranges of light with high responsivity and fast photodetection is critical in optoelectronic applications ranging from sensing, imaging, and spectroscopy to communications. Therefore, numerous studies have reported the use of traditional semiconductors for broader ranges of photodetection, even in the mid-infrared to far-infrared regime. As a result, narrow-bandgap semiconductor compounds such as indium antimonide (InSb), mercury cadmium telluride (HgCdTe), lead sulfide (PbS), and lead selenide (PbSe) are now widely applied in mid-infrared or far-infrared photodetectors [43–45]. However, the cryogenic operation, high cost, and complicated growth processes of these narrow-bandgap semiconductors has limited their practical usage as broad-range photodetectors [46]. Therefore, graphene has advantages over narrow-bandgap semiconductor compounds in terms of its ultrabroad optical absorption spectrum and high room-temperature electron and hole mobilities.

Liu et al. developed photodetectors consisting of a thin tunnel barrier sandwiched between two graphene layers [47]. The graphene/5-nm-thick Ta_2O_5/graphene heterostructure photodetector shows a remarkably high responsivity of ~1000 A W^{-1} under a low excitation power (~10^{-9} W) of 532 nm laser light at a 1 V source-drain bias voltage. Additionally, the authors fabricated a similar graphene/6-nm-thick intrinsic silicon/graphene heterostructure device with improved photoresponsivity in the infrared regime. The intrinsic-silicon-incorporated photodetector shows high responsivity in a broadband regime ranging from near-infrared (4 A W^{-1} at 1.3 μm light and 1.9 A W^{-1} at 2.1 μm light) to mid-infrared light (1.1 A W^{-1} at 3.2 μm). Yu et al. fabricated a mid-infrared photodetector by hybridizing graphene with an innovative narrow-bandgap semiconductor, titanium sesquioxide (Ti_2O_3) nanoparticles [48]. The schematic in Figure 2a shows the representative hybrid graphene/Ti_2O_3 nanoparticle photodetector, with a high responsivity of ~300 A W^{-1} and high detectivity of ~7×10^8 cm $Hz^{1/2}$ W^{-1} at room temperature. Furthermore, a high responsivity of ~120 A W^{-1} can be achieved from the photodetector over a wide range of wavelengths from 4.5 μm to 10 μm (Figure 2b). Figure 2c shows the superior performance of this material compared to those in other recent studies, such as hybrid ZnO/graphene [49], perovskite/graphene [50], QDs/graphene [51], etc. [52]. Wang et al. proposed seamless lateral graphene p-n junctions by performing selective-area ion implantation followed by the in situ chemical vapor deposition (CVD) of graphene [53]. The photodetector based on these seamless lateral graphene p-n junctions offers a high responsivity of 1.4~4.7 A W^{-1} and detectivity of ~10^{12} cm $Hz^{1/2}$ W^{-1} in the broad spectral range of visible light (532 nm) to near-infrared light (1550 nm) under an illuminated light intensity of 15 mW/ cm^2. Fang et al. proposed a graphene photodetector with a gold snowflake-like fractal metasurface design [54]. The authors measured the photovoltages generated when light was illuminated on the fractal metasurface and plain gold-graphene edge ($V_{fractal}$ and V_{edge}, respectively). The results showed that photovoltage enhancement factors ($V_{fractal}/V_{edge}$) of 8 to 13 over the visible-light range (476 to 647 nm) can be achieved. Cakmakyapan et al. fabricated a photodetector with gold-patched graphene nanostripes showing ultrafast photodetection and high responsivity [55]. The photodetector operates in a wide spectral range of visible light (800 nm) to infrared light (20 μm) with a high responsivity ranging from 0.6 A W^{-1} to 8 A W^{-1}. Additionally, the photodetector shows an ultrafast photodetection speed exceeding 50 GHz.

Light-Emitting Diodes

Solid-state light-emitting diodes (LEDs) are currently widely used in the semiconductor industry [56] due to their unique electronic properties. Recently, the improved performance of LEDs has enabled various electronic applications, such as smart displays, light communications, and optoelectronics [57–60]. The external quantum efficiency (EQE), current efficiency, and power efficiency are important factors that determine the performance of LEDs. Among these factors, the

EQE is known to be dominated by the light extraction efficiency and internal quantum efficiency [61]. To date, the internal quantum efficiency of LEDs has been remarkably improved and has already reached the theoretical limit, which is larger than 90% [62]. In contrast, the light extraction efficiency has not reached the theoretical limit. Therefore, improving the light extraction efficiency is key to obtaining a higher EQE. A high optical transparency, high reliability and low sheet resistance are required for better light extraction efficiency. Thus, the use of transparent conductive electrodes (TCEs) is necessary; indium tin oxide (ITO) is one of the most widely used materials because of its high optical transmittance (above 80% at 450 to 550 nm) and low sheet resistance [63,64]. However, ITO has considerable disadvantages, which are its mechanical brittleness and high cost [65]. Thus, metal nanowires (NWs) and carbon-based materials such as carbon nanotubes and graphene have been investigated to replace ITO [22,66–69].

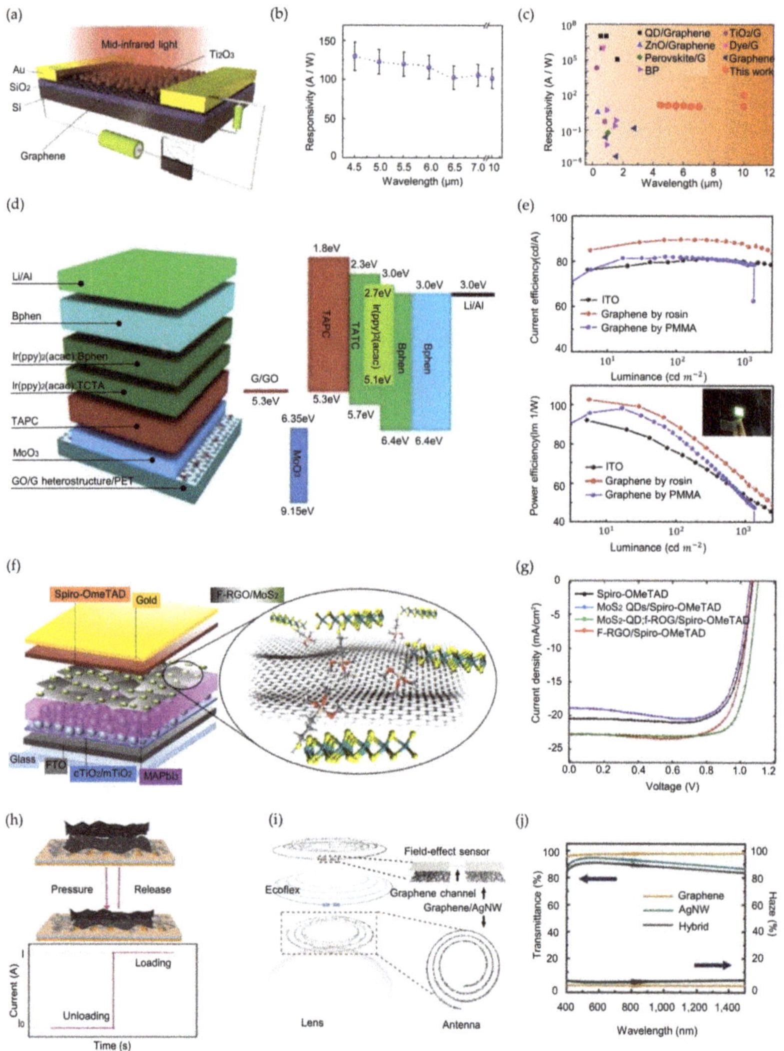

Figure 2. Electronic applications of graphene (**a**) Schematic showing a hybrid graphene/Ti2O3 photodetector for mid-infrared detection; (**b**) High photodetector responsivity when illuminated with

wavelengths of light from 4.5 to 10 µm; (**c**) Comparison of photodetector responsivity with the results of other recent studies; (**d**) Structure of a graphene OLED (left) and its energy-level diagram (right); (**e**) Current efficiency (top) and power efficiency (bottom) versus luminance characteristics of three types of OLEDs. The inset indicates a flexible green OLED with a rosin-transferred graphene anode; (**f**) Schematic of a mesoscopic MAPbI3-based perovskite solar cell; (**g**) I-V curves of measured perovskite solar cells exploiting MoS2 QDs, f-RGO, and MoS2 QDs:f-RGO as an ABL between spiro-OMeTAD and MAPbI3; (**h**) Sensing mechanism of a piezoresistive pressure sensor and current response to loading and unloading. (**i**) Image of multifunctional sensors on a soft contact lens; (**j**) Transmittance and haze of graphene, AgNW films, and graphene/AgNW hybrid structures. (a), (b), (c) Reproduced with permission from [48]. Copyright Nature Communications, 2018. (d), (e) Reproduced with permission from [21]. Copyright Nature Communications, 2017. (f), (g) Reproduced with permission from [70]. Copyright ACS Nano, 2018. (h) Reproduced with permission from [71]. Copyright Small, 2018. (i), (j) Reproduced with permission from [72]. Copyright Nature Communications, 2017.

Zhang et al. fabricated four-inch flexible organic light-emitting diodes (OLEDs) showing a high brightness of ~10,000 cd m^{-2} by transferring ultraclean and damage-free graphene supported by rosin ($C_{19}H_{29}COOH$); Figure 2d shows the device structure [21]. To date, various materials, such as small organic molecules and macromolecular polymers, have been used as supporting layers for transferring graphene [73–75]. However, low solubility in all solvents and strong interactions with graphene make the elimination of the supporting layers after the transfer difficult. These problems damage and leave polymer residues on the transferred graphene layer, degrading the electrical and optical properties. Rosin shows high solubility in solvents and weak interactions with graphene, thus enabling an ultraclean and damage-free transfer. As a result, OLEDs with rosin-transferred graphene showed a maximum current efficiency of 89.7 cd A^{-1} and power efficiency of 102.6 lm W^{-1}, exceeding the performance of ITO and poly(methyl) methacrylate (PMMA)-transferred graphene OLEDs (Figure 2e). In addition, the inset shows an optical image of a rosin-transferred graphene-based OLED showing uniformly bright green light. Lee et al. proposed flexible OLEDs with an electrode structure based on low-index hole injection layers (HILs) and high-index TiO$_2$ layers sandwiching graphene electrodes [76]. This structure enables losses in both the surface plasmon polariton (SPP) and cavity resonance enhancement to be controlled, thereby leading to a high EQE. The proposed OLEDs showed ultrahigh EQEs of 40.8% and 62.1% for single- and multijunction structures, respectively. Additionally, the OLEDs were bendable to a radius of 2.3 mm because the TiO$_2$ layer is capable of withstanding flexural strains up to 4%. Wang et al. developed a single tunable LED emitting light ranging from blue (~450 nm) to red (~750 nm) composed of semireduced graphene oxide sandwiched by graphene oxide (GO) and reduced graphene oxide (RGO) [77]. The LED produced red and green lights at 4.8 lm W^{-1} and 6000 cd m^{-2} under a 12 V bias voltage and a 0.1 A drive current and blue light at 0.67 lm W^{-1} and 800 cd m^{-2} under a 16.5 V bias voltage and a 0.1 A drive current. This experiment enabled in situ wavelength tuning by controlling the bias voltage and doping level. Huang et al. introduced transparent LEDs based on graphene-encapsulated Cu NWs [78]. Cu NWs are a strong candidate to replace ITO TCEs owing to their high transmittance (~93%), low sheet resistance (51 ohm/sq), and low cost [79,80]. However, the low stability against oxidation of Cu NWs limits their use as electrodes for LEDs. To solve this problem, gas-phase graphene can be used to encapsulate Cu NWs to enhance their stability: the material shows strong antioxidant stability while maintaining high optoelectronic properties (33 ohm/sq and 95% transmittance) over a broad transparency range (200~3000 nm). Therefore, the encapsulated NWs serve as electrodes by forming good ohmic contact with both n-GaN and p-GaN, thereby leading to the successful fabrication of blue-light LEDs. Seo et al. conducted similar research by integrating silver nanowires (AgNWs) with a graphene protecting layer to achieve high performance [81]. High-quality graphene was obtained by a two-step growth method, which results in reduced point defects and an increased graphene domain size. The measured sheet resistance of AgNWs with graphene produced via the two-step growth method (A-2GE) was 77.5 ±

10 ohm/sq, and that of AgNWs was 205.1 ± 40 ohm/sq. Additionally, the oxygen transmission rate (OTR) was measured after the graphene films had been transferred to a polyethylene terephthalate (PET) film. The OTR of 2-G/PET was 5.39 ± 0.9 cc/m^2-day, 74% less than that of bare PET. Thus, A-2GE showed high optoelectronic properties and strong antioxidant stability and can be used for TCEs for LEDs.

Solar Cells

Solar cells based on organic-inorganic perovskite materials have recently been widely studied due to their high power conversion efficiencies (PCEs), long-range carrier diffusion length, intense light absorption, and facile fabrication (low cost and low temperature) [82–89]. The active device layers of perovskite solar cells (PSCs) are sandwiched by electron and hole transport layers (ETLs and HTLs, respectively) [90–92]. Since the charge carrier transport properties can be affected by both the electronic structures of the various interfaces and the morphology of PSCs [93,94], graphene and related two-dimensional materials (GRMs), such as graphene quantum dots (GQDs) [95–97], RGO [97–99], and fullerene [100], have been widely used to tune the interfacial properties and morphologies of PSCs. GRMs have been exploited as the top electrodes of PSCs, as an interlayer between the ETL/HTL and the perovskite layer, and by integration with the ETL/HTL to provide efficient charge transfer [101,102].

You et al. proposed semitransparent PSCs with top graphene electrodes [103]. The fabricated PSCs showed maximum PCEs of 11.65 ± 0.35% and 12.02 ± 0.32% from the graphene and fluorine-doped tin oxide (FTO) sides, respectively, with optimized conditions of double-layer graphene and an ~70 mg mL^{-1} 2,2′,7,7′-tetrakis-(*N*,*N*-di-p-methoxyphenylamine)-9,9′-spirobifluorine (spiro-OMeTAD) solution due to the low surface roughness. Sung et al. first fabricated PSCs with over 17% efficiency by exploiting graphene as a transparent conducting anode [104]. The presence of molybdenum trioxide (MoO$_3$) controlled the contact angle between poly(3,4-ethylenedioxythiophene):poly(styrene sulfonate) (PEDOT:PSS) and the graphene surface and helped successfully form PEDOT:PSS/MAPbI$_3$ layers. The thickness of the MoO$_3$ layer affected the PCE of the PSCs. In the experiment, a 1 nm-thick MoO$_3$ layer with graphene-based devices showed a maximum PCEs of 17.1%, larger than 90% of those of ITO-based devices (18.8%). The flexibility and high PCE of this graphene-based device pave the way for the further development of flexible solar cells. Najafi et al. fabricated a PSC with a high PCE of over 20% using MoS$_2$ QDs and a functionalized reduced graphene oxide (f-RGO) hybrid [70]. MoS$_2$ QDs:f-RGO hybrids are used as both the HTL and active buffer layer (ABL), and a schematic of the PSC is shown in Figure 2f. The PSC structure consists of FTO/compact TiO$_2$ (cTiO$_2$)/mesoporous TiO$_2$ (mTiO$_2$)/MAPbI$_3$/f-RGO:MoS$_2$/spiro-OMeTAD/Au. The I-V characteristics of MoS$_2$ QDs, f-RGO, and MoS$_2$ QDs:f-RGO as ABLs are shown in Figure 2g. The PSC using MoS$_2$ QDs:f-RGO showed a further improved performance, with a maximum PCE of 20.12% and a fill factor (FF) of 79.75%, compared to the reference device showing a maximum PCE of 16.85% and an FF of 76.9%. Agresti et al. fabricated a large-area PSC of 50.56 cm^2 with a PCE of 12.6% by doping the mTiO$_2$ layer with graphene flakes and inserting lithium-neutralized graphene oxide flakes (GO-Li) between the interface of the mTiO$_2$ and perovskite [105]. The charge injection from perovskite to mTiO$_2$ is improved by the GO-Li layer, and the graphene flakes dispersed in the mTiO$_2$ layer help speed up the charge dynamic at the electrode. The interface engineering of PSCs by GRMs leads to an increased PCE, from 11.6% for the reference PSC to 12.6% for the PSC with GO-Li and mTiO$_2$. Li et al. proposed a PSC with an MAI perovskite layer incorporating graphene nanofibers [106]. Integrating graphene nanofibers with perovskite leads to improved charge injection and the formation of larger crystals. Therefore, the PSCs with graphene nanofibers showed an improved PCE of 19.83% compared to the PCE of the reference device of 17.51%.

2.2.2. Sensors

Electronic Sensors

Graphene can be utilized in various electronic sensors due to its unique device properties. Recently, novel integrated electronic circuitries with tactile pressure sensors based on piezoelectricity [107,108], piezoresistivity [109,110], capacitance [111,112], and field-effect transistors (FETs) [113–115] have been fabricated for healthcare and health-monitoring applications [107,108,115]. Haniff et al. successfully fabricated a flexible piezoresistive-type pressure sensor based on graphene synthesized at various substrate temperatures (750, 850 and 1000 °C) with hot-filament thermal chemical vapor deposition (HFTCVD) [116]. The authors found that the sensitivity of the flexible graphene-based piezoresistive-type pressure sensor could be tuned through this technique (graphene deposited at 750 °C has a four-fold higher sensitivity than graphene deposited at 1000 °C). A flexible and highly sensitive piezoresistive pressure sensor based on an ultrathin wrinkled graphene film (WGF), interdigital electrodes (IDEs), polyvinyl alcohol (PVA) NWs, and an interconnected isolation effect was reported by Liu et al. [71]. The sensing mechanism and current response to loading and unloading are described in Figure 2h. The WGF and interconnected PVA NWs produce a synergistic effect, resulting in a piezoresistive pressure sensor with an ultrasensitivity of 28.34 kPa^{-1} and mechanical durability and reliability (repeated tests of 6000 cycles). Shin et al. paved the way to use pressure sensors for diverse application areas, such as medical diagnosis, robotics and automatic electronics [117]. The authors developed transparent tactile pressure sensors covering a wide pressure range (250 Pa~3 MPa) by forming fully integrated active-matrix pressure-sensitive graphene FET arrays with air-dielectric layers between two folded opposing panels.

For the true integration of woven electronics and/or optoelectronics into textiles, the direct fabrication of the device on the fiber itself with high-performance materials allowing easy unification into fabrics is required [118,119]. Alonso et al. completely integrated flexible (up to 10 mm, the radius of a human finger), transparent and durable graphene-based LED and touch sensors on textile fibers [120]. Roll-to-roll and printing-compatible patterning techniques were adopted to fabricate the LED and the capacitive touch sensors. Transparent electrodes coated by monolayer graphene on a textile fiber can be utilized for wearable electronics with high conductivity and flexibility [121]. Graphene-based optical waveguide tactile sensors can overcome the disadvantage of optical sensors with a directional coupler [122]. Additionally, transparent and flexible UV sensors play an important role in the field of wearable and/or portable optoelectronic systems. Pyo et al. demonstrated a fully transparent, exceedingly sensitive, and flexible UV sensor with 1D hybrid carbon nanotubes (CNTs) combined with a 2D graphene electrode [123]. Because of the provided effective charge transfer and the minimized effect of contact resistance by the integration of the CNTs and graphene without a potential barrier, the CNT–graphene UV sensors have a 30 times higher photoresponse (45% under 254 nm UV illumination with a power intensity of 1.91 mW cm^{-2}) than that of the CNT-Au electrode sensors. Additionally, thanks to the outstanding properties of graphene and CNT, the device shows a high optical transparency (over 80% at 550 nm) and remarkable mechanical flexibility (bending radius of 5.5 mm) with high electrical reliability. Goossens et al. reported the monolithic integration of graphene with a complementary metal–oxide–semiconductor (CMOS)-integrated circuit operating as a high-mobility phototransistor [124]. Broadband high-resolution image sensing and a sensitive digital camera in the range of ultraviolet, visible and infrared light was demonstrated.

Research on the safe and accurate monitoring of human body signals, hazardous disease detection, and stimulation systems has been highly focused in areas ranging from biological research to clinical medicine [125–132]. Electrooculography (EOG) is a technique for recording the corneo-retinal standing potential between the retina and the cornea of human eyes. Ameri et al. described a noninvasive graphene electronic tattoo (GET)-based invisible EOG sensor [133]. Ultrathin (350 nm thickness), stretchable (up to 50%), soft, fully transparent (85% in the visible regime), and breathable EOG sensors based on GETs have a high resolution (approximately 4° to detect eye movement); wireless communication can be achieved by connecting the sensor to an OpenBCI Cyton board, and the sensors can be successfully applied to a human–robot interface (HRI). Kuzum et al. simultaneously recorded electrophysiological spikes in the brain with flexible, fully transparent graphene-based neural electrodes and neuro-optical images with confocal or multiphoton microscopy [134]. The graphene electrodes were highly transparent, such that both electrophysiological recording and optical imaging, leading to brain imaging with high spatiotemporal dynamics, could be achieved without perturbing either technique. Additionally, the simultaneous optical stimulation of neural tissue and electrical recording of the brain can be achieved by using µ-LEDs and transparent graphene electrodes, respectively. [135]. Such powerful graphene-based systems that resolve individual cells and their connections through optical imaging and electrical recording of brain activity are a breakthrough in neuroscience. Monitoring glucose [136,137] and the intraocular pressure [138–141] are particularly crucial for detecting and managing diabetes and glaucoma, respectively. Kim et al. fabricated a highly conductive, flexible and transparent wearable smart contact lens using a reliable and robust hybrid structure of 1D and 2D nanomaterials [72]. Wearable contact lenses can directly and noninvasively detect and wirelessly monitor biomarkers such as the glucose contained in tears and intraocular pressure by measuring the change in resistance and capacitance of the electronic device. Figure 2i shows a schematic of fully integrated multifunctional sensors on a soft contact lens. As shown in Figure 2j, a lower optical transmittance, haziness, and sheet resistance were obtained when graphene and AgNWs were integrated in a hybrid system than when the single materials of graphene or AgNWs were used.

Biomolecule Sensors

For early and reliable clinical cancer diagnosis and treatments, cost-efficient, dependable and sensitive monitoring and detecting systems are crucial. Wu et al. fabricated a functionalized graphene-based electrochemical sensor array system for cell sensing [142]. The graphene-based chemical nose/tongue approach system can discern 100 cell samples consisting of (i) artificial circulating tumor cells (CTCs); (ii) cancerous, multidrug-resistant cancerous and metastatic human breast cells; and (iii) different cell types with almost 100% classification accuracy.

Detecting and monitoring various electrolytes, such as biological, organic, and inorganic electrolytes, and electroactive materials in human body fluids are important for clinical diagnosis and analytical applications [61,143–145]. Wang et al. developed electrochemical biosensors to selectively and simultaneously detect five analytes in human serum: ascorbic acid (AA), dopamine (DA), nitrite (NO_2), tryptophan (Trp), and uric acid (UA) [146]. A CVD method was used to fabricate freestanding graphene nanosheets on tantalum (Ta) wire. Graphene-based biosensors provided highly selective, sensitive results (detection limits of AA, DA, NO_2, Trp, and UA of 1.58, 0.06, 6.45, 0.10, and 0.09 µM, respectively (S/N = 3)) with differential pulse voltammetry (DPV). As predicted by the World Health Organization (WHO), the number of diabetes patients has been gradually increasing [147], and such patients require the continuous monitoring of glucose levels in the interstitial fluid. Currently, only invasive methods are available for monitoring the glucose in the blood [148–150]. Lipani et al. demonstrated a noninvasive, transdermal, path-selective, and specific glucose monitoring and detection system based on a miniaturized device array platform fabricated by a graphene thin-film method or screen-printing technology [151]. The authors found via in vivo testing on healthy humans that the graphene-based glucose monitoring system can continuously record the blood sugar for 6 h.

The most important consideration for systems biology and personalized and precision medicine is an acceptable affinity and binding efficiency of deoxyribonucleic acid (DNA) hybridization and the ability to distinguish DNA sequences with single-nucleotide substitutions. Xu et al. suggested next-generation multichannel graphene-based DNA sensors for a cost-effective, fast, simple and label-free biosensing system to perform kinetic studies (e.g., detection limit of 10 pM for DNA) [152]. Reproducible and reliable G-FET DNA biosensors have been designed using a mature FET fabrication method characterized by low cost, low power, and simplicity of miniaturization. Diffusive transport is a general mechanism for biomolecules in viscous media, and a crucial prerequisite for biosensing is placing the target biomolecules at the most sensitive point. Because of the powerful dielectrophoresis (DEP) forces of monolayer graphene, the sharp edge of monolayer graphene can successfully produce singular electrical field gradients, accurately trapping and positioning biomolecules (nanoparticles and DNA molecules) [153]. Seo et al. fabricated a sensitive and selective electrochemical genosensor integrated with grown graphene films, realizing reliable biodetection and demonstrating the functionality of the graphene films [154].

A provisional WHO guideline has established a concentration limit of microcystin-LR (MC-LR) of 1 μg/L in drinking water [155] because the continuous monitoring of drinking water quality and appropriate treatment are extremely important for human healthcare and the quality of life worldwide [156,157]. Zhang et al. reported a novel graphene film-integrated biosensor for MC-LR detection that is time- and cost-effective, portable, scalable to large-scale manufacturing, and easy to handle. A graphene film/polyethylene terephthalate (GF/PET) composite grown through the CVD method was used to develop a graphene-based biosensor, resulting in a detection limit of 2.3 ng/L, which fully satisfies the safety guideline of the WHO.

Gas Molecule Sensors

With fast-growing industrial development, toxic gases endangering human beings have been extensively produced and emitted [158,159]. Therefore, developing gas sensors with a high detection sensitivity or control capacity for hazardous gases is extremely important for the protection of human health and the environment [160]. Wu et al. reported a chemiresistor-type sensor based on a 3D sulfonated reduced graphene oxide hydrogel (S-RGOH); the sensor is capable of detecting a variety of important gases with high sensitivity, boosted selectivity, fast response, and good reversibility [161]. The $NaHSO_3$-functionalized RGOH displays 118.6 and 58.9 times higher responses than its unmodified RGOH counterpart for NO_3 and NH_3, respectively. Moreover, the response increases monotonically from 6.1% at 200 ppb NO_2 to 22.5% at 2 ppm NO_2. Guo et al. fabricated a NO_2 gas sensor from the room-temperature reduction of GO via two-beam-laser interference (TBLI) [162]. The fabricated RGO sensor enhanced the sensing response for NO_2 and accelerated the response/recovery rates. For 20 ppm NO_2, the response (R_a/R_g) of the sensor based on RGO hierarchical nanostructures is 1.27, which is higher than those of GO (1.06) and thermally reduced RGO (1.04). The response time and recovery time of the sensor based on laser-reduced RGO are 10 s and 7 s, respectively, which are much shorter than those of GO (34 s and 45 s).

Because of the significant properties (high heat of combustion and low minimum ignition energy) of hydrogen (H_2), it can be exploited in various applications [163,164]. However, because of its delicate properties (low level of explosion limit ~4%), the real-time and long-distance monitoring of H_2 concentrations is essential. Sharma et al. produced an FET hydrogen sensor integrated with a graphene-Pd-Ag-gate FET (GPA-FET) [165]. The GPA-FET showed an excellent sensing response to hydrogen gas at 25~254.5 °C.

Many studies have reported the integration of graphene-based mobile gas sensors with modern technologies such as smart phones, cloud computing and the Internet of Things (IoT) with graphene nanoribbons due to the high electrical conductivity of the nanoribbons. The first monolithically integrated CMOS-monolayer graphene gas sensor was developed by Zanjani et al., with a minimal number of post-CMOS processing steps, to realize a gas sensor platform that integrates the higher gas sensitivity of monolayer graphene with the low power consumption and cost advantages of a silicon CMOS platform [166]. Furthermore, Pour et al. improved graphene nanoribbons' electrical conductivity by lateral extension [167]. Improved graphene nanoribbons were synthesized in solution; the lateral extension decreased their bandgap and improved their electrical conductivity.

3. Thermal Properties and Applications

3.1. Thermal Properties

3.1.1. Specific Heat of Graphene and Graphite

The specific heat, C, is a distinct characteristic of a material expressing the change in the energy density U with respect to the change in temperature (1 K or 1 °C), represented as $C = dU/dT$, with units of joules per kelvin per unit mass, per unit volume, or per mole. The thermal time constant of a material, which indicates how quickly the material heats or cools, is determined by the specific heat. The thermal time constant is expressed as $\tau \approx RCV$, where V is the volume of the material and R is the thermal resistance for heat dissipation [168]. As the specific heat of graphene has not been measured directly, the calculation was performed by referencing the data for graphite [169–171]. The specific heat of a material is contributed by two components, phonons (or lattice vibrations) and electrons: $C = C_{ph} + C_{el}$. However, the contribution of phonons dominates the specific heat of graphene at all temperatures [172]. In addition, as shown in Figure 3a, the phonon specific heat increases with the temperature [171,173]. The specific heat of graphite at room temperature is $C_{ph} \approx 0.7 \, \text{J g}^{-1} \, \text{K}^{-1}$, which is approximately 30% higher than that of diamond due to the higher density of states (DOS) at low phonon frequencies caused by the weak coupling of graphite layers. For a graphene sheet at room temperature, a similar result is expected, but the specific heat can be altered when graphene interfaces with a substrate (e.g., graphene on insulators) [174].

3.1.2. Thermal Conductivity

The thermal conductivity K is explained by Fourier's law, $q = -K\nabla T$ [175]. In this equation, the negative sign means heat flows from high to low temperature; q is the heat flux per unit area; and ∇T is the temperature gradient. The thermal conductivity is related to the specific heat by $K \approx \sum C v \lambda$, where v is the average phonon group velocity and λ is the mean free path. Therefore, as the specific heat of graphene is dominated by phonon transport, the thermal conductivity is also dominated by phonon transport [171].

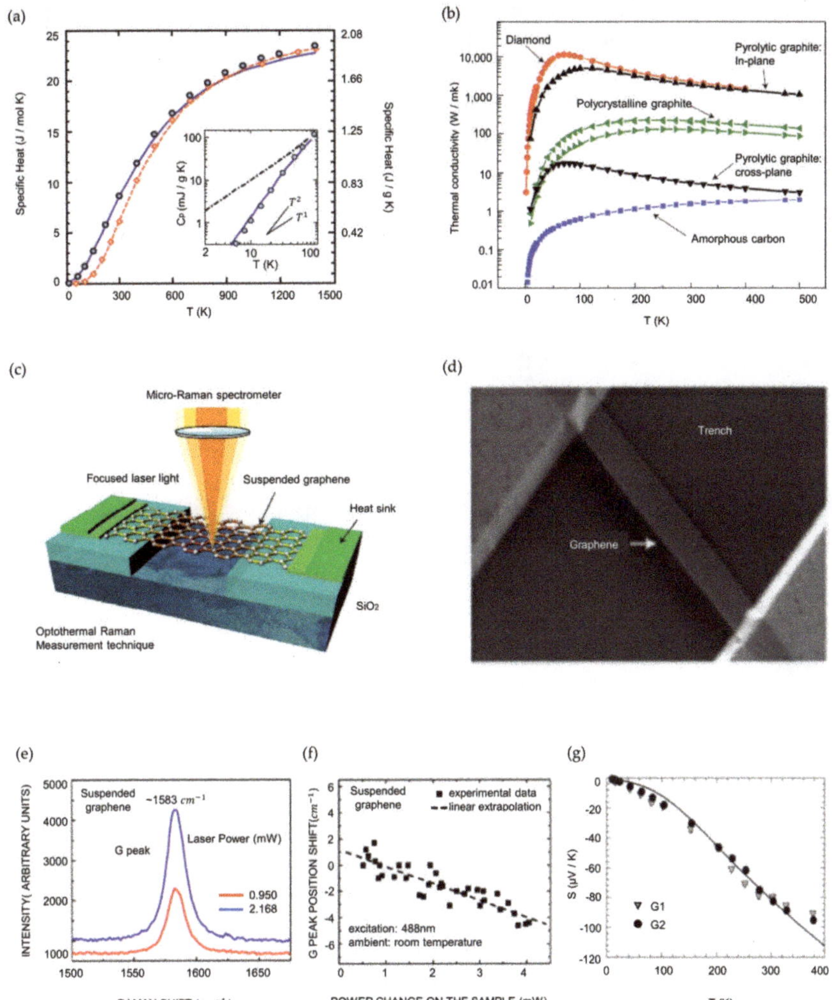

Figure 3. Thermal properties of graphene (**a**) Specific heats of graphite, diamond, and graphene. The inset compares the specific heats of graphene and graphite at low temperature; (**b**) Thermal conductivity of bulk carbon allotropes as a function of temperature; (**c**) Figure of a graphene layer suspended across a trench and the measurement of the thermal conductivity; (**d**) SEM image of a suspended graphene layer across a 3-μm-wide trench in a Si wafer; (**e**) Raman spectrum of graphene showing the G-peak region measured at two different power levels; (**f**) Graph showing the G-peak position shift versus power change; (**g**) Graph of measured Seebeck coefficients as a function of temperature for graphene samples. (a) Reproduced with permission from [171]. Copyright MRS Bulletin, 2012. (b), (c), (d) Reproduced with permission from [16]. Copyright Nature Materials, 2011. (e), (f) [176]. Copyright Nano Letters, 2008. (g) Reproduced with permission from [18]. Copyright Science, 2010.

Here, the thermal conductivity of various carbon allotropes, including two types of pyrolytic graphite (in-plane and cross-plane), diamond, and amorphous carbon, is presented in Figure 3b. Pyrolytic graphite is similar to highly oriented pyrolytic graphite (HOPG). The in-plane K of pyrolytic graphite is approximately 2000 W m^{-1} K^{-1} at room temperature, and the cross-plane K at room temperature is more than two orders of magnitude smaller. Because HOPG is composed of well-aligned large crystallites, the overall behavior is analogous to that of a single crystal; this fact explains the difference in K [177]. Heat is mainly transported by acoustic phonons in all bulk carbon allotropes. In HOPG and diamond, K reaches maximum values at approximately 100 K and 70 K, respectively. In amorphous carbon, K ranges from 0.01 W m^{-1} K^{-1} to 2 W m^{-1} K^{-1} at 4 K and 500 K, respectively. Similarly, carbon allotropes have different thermal conductivities depending on the temperature. The difference is attributed to the grain size and quality of graphite and the phonon DOS, as indicated by C_{ph}.

A method to measure the thermal conductivity of graphene by exploiting confocal micro-Raman spectroscopy is introduced in Figure 3c. The G peak in the graphene spectra depends strongly on the temperature [178]. This high temperature sensitivity of the G peak enables the monitoring of the local temperature change induced by laser light focused on a graphene layer. In this experiment, trenches were fabricated on Si/SiO$_2$ substrates by reactive ion etching (RIE), and graphene was suspended over the trenches. The depth of the trenches was 300 nm, and the width varied from 2 to 5 μm [176]. An optical image of the trenches taken by scanning electron microscopy (SEM) is shown in Figure 3d. The laser light focused on the middle of the suspended graphene layer generates heat in the graphene. The heat generated by laser excitation propagates laterally through the graphene due to the negligible thermal conductivity of air. Thus, even a small amount of heat propagated from the middle of the graphene can result in a detectable temperature increase. The heat front propagating through the graphene layers can be explained by two components: the plane-wave heat front and the radial heat wave [179]. The thermal conductivity from the plane-wave heat front can be expressed as $K = (L/2S)(\Delta P/\Delta T)$, where $\Delta P/\Delta T$ indicates the heating power change with respect to the temperature change, L is the distance from the center of the graphene to the heat sink, and S = h × W. For the radial heat wave case, $K = \chi_G(1/2h\pi)(\delta\omega/\delta P)^{-1}$, where $\delta\omega/\delta P$ indicates the G peak position shift due to the heating power change. Consequently, the thermal conductivity can be expressed as

$$K = \chi_G(L/2hW)(\delta\omega/\delta P)^{-1}. \tag{5}$$

As shown in Figure 3e, the excitation power dependence of the Raman G peak was measured for the suspended graphene layers. The increase in laser power induced an increase in the intensity and redshift of the G peak. The G peak position shift with respect to the power change in the suspended graphene layers is shown in Figure 3f. The slope from the measured data is $\delta\omega/\delta P_D \approx -1.29$ cm^{-1}mW^{-1}, where P_D is the total dissipated power and the temperature coefficient $\chi_G = -1.6 \times 10^{-2}$ cm^{-1}/K [180]. When these values are substituted into equation (5), a thermal conductivity of approximately 2000~4000 W m^{-1} K^{-1} is obtained for freely suspended graphenes; this value is the highest of any known material [15–17]. Raman spectroscopy is a basic technique for graphene analysis. In addition to thermal conductivity, Raman spectroscopy can help to characterize graphene flakes [181], detect traces of molecules [182] and achieve electronic properties [183].

3.1.3. Thermoelectric Effects of Graphene

The thermoelectric power (TEP) is the voltage induced by a temperature gradient. Experimental studies indicated that graphene has a TEP of ~ 50 to 100 μV K^{-1} [184]. In addition, other experiments showed that graphene has a maximum TEP value of ~80 μV K^{-1} at room temperature. It was theoretically verified that TEP behaves as $1/\sqrt{n_0}$ at a high carrier density (n_0) but is saturated at low densities. The values of the Seebeck coefficient range from ~100 to ~10 μV K^{-1} for temperatures ranging from ~100 to ~300 K. The Seebeck coefficient ($S = dV/dT$) shows lower values at high temperatures, as shown in Figure 3g [18]. Graphene has a higher S than those of semiconductors, and the polarity of S can be controlled by varying the gate voltage; these thermoelectric effects are intriguing [185]. The figure of merit (ZT) of the efficiency of thermoelectric energy conversion is defined as $ZT = S^2\sigma T/\left(K_{ph} + K_{el}\right)$, where σ is the electrical conductivity [186]. Graphene shows a high value of ZT due to suppressed phonon scattering, which leads to decreased K ($K_{ph} + K_{el}$) [187,188]. The high value of ZT for graphene suggests the possibility of its use in energy-harvesting applications. The ZT of various graphene nanostructures is very promising for the improvement of the thermoelectric energy conversion [189].

3.2. Thermal Applications

3.2.1. Sensors

Temperature Sensors

Electronic skins (E-skins) can generate electronic signals from external stimulation for use as human-machine interfaces (HMIs) and multifunctional smart skins [113,190–194]. In addition to pressure and strain sensing, achieving the simultaneous detection and monitoring of temperature with cost-efficient fabrication is important to make plausible mimics of multifunctional human skin. Ho et al. fabricated a fully transparent (over 90% transmittance in the range 400 ~ 1000 nm) and stretchable graphene-based multifunctional E-skin sensor matrix [195]. The matrix can detect and monitor humidity, temperature, and pressure through a simple lamination process. Figure 4a shows the fabrication technique of the sensor matrix. CVD-grown graphene interconnects electrodes with three sensors, and GO and rGO are the active sensing materials for humidity and temperature, respectively. Figure 4b shows the resistance change with respect to temperature, and real-time measurement results are shown. Each sensor was simultaneously and individually sensitive to only its relevant form of stimulation, and no response was obtained from other forms of stimulation. Vuorinen et al. reported a simple technique to fabricate graphene/PEDOT:PSS-based skin-conformable inkjet-printed temperature sensors [196]. A Phene Plus I3015 transparent graphene/PEDOT:PSS ink was printed onto polyurethane skin-conformable adhesive bandages to form transparent and flexible temperature sensors. The inkjet-printing method can significantly reduce the manufacturing costs and wasted materials and provides advantages for the production of disposable systems. The graphene/PEDOT:PSS temperature sensors present remarkable sensitivity to monitor temperature changes on the human skin with a temperature coefficient of resistance (TRC) higher than 0.06% per degree Celsius. Additionally, Trung et al. demonstrated an all-elastomeric transparent stretchable (TS) gated sensor with a simple method using graphene [197]. The schematic of the structure of the TS-gated sensor is shown in Figure 4c. The overall system integrated a graphene-based temperature sensor and strain sensor. A PEDOT:PSS/PU dispersion (PUD) composite elastomeric conductor serves as the source, drain, and gate contacts. A PU R-GO/PU nanocomposite and a composite of AgNWs/PEDOT:PSS/PUD as a gate dielectric are used to form a temperature-responsive channel layer and a strain sensing layer, respectively. The sensor shows a high stretchability of up to 70%, responsivity of up to 1.34% per 34 °C and durability (10,000 test cycles at a strain of 30%) and could be conformally attached to human body skin to monitor temperature changes. Figure 4d indicates the sensor's ability to monitor the surface temperature of targets: cold and hot water. The dashed line illustrates the device region. Such devices can be useful for applications including interactive remote healthcare systems, biomedical monitoring, and HMIs by integrating wireless communication units such as Bluetooth into the sensors.

Single-layer graphene (SLG) presents impressive thermal properties, so graphene can be used for thermal management and temperature sensing applications. Davaji et al. fabricated and compared micropatterned SLGs on a silicon dioxide (SiO_2)/Si substrate, a silicon nitride (SiNx) membrane, and a Si wafer with etched rectangular boxes (10×20 μm^2) [198]. These sensors show a quadratic dependence of the resistance versus temperature in the range of 283 ~ 303 K, as determined through analyzing the temperature-dependent carrier density, electron mobility relationship ($\sim T^{-4}$) and electron-phonon scattering. The graphene-based sensor fabricated on a SiNx membrane has a substantially faster response, better sensitivity because of the low thermal mass, and better mechanical stability than other suspended graphene sensors. As a result, a temperature sensor comprising the suggested SLG on a SiNx membrane can be exploited for various highly sensitive and fast applications.

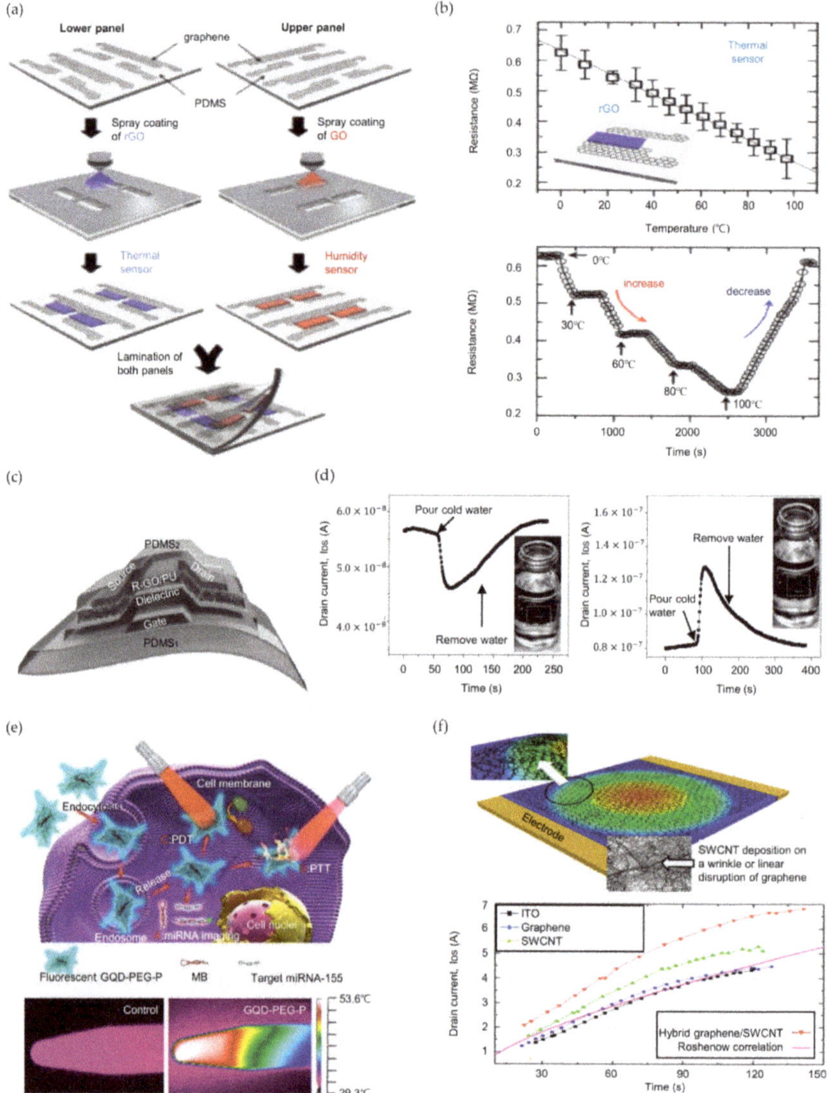

Figure 4. Thermal applications of graphene. (**a**) Schematic of the fabrication process of an all-graphene electronic skin sensing matrix; (**b**) Graph showing resistance vs temperature of an rGO-based thermal sensor (top) and real-time temperature sensing results (bottom); (**c**) Schematic of a device with a transparent and stretchable gate; (**d**) Drain current responses of a TS-gated sensor according to low-temperature water (left) and high-temperature water (right); (**e**) Schematic of a GO-PEG-P theragnostic platform (top) and thermal images of vials with water and GQD-PEG-P solution (100 µg/mL) (bottom); (**f**) Hybrid graphene/SWCNT film deposited on an ITO surface showing high effusivity (top) and heat transfer coefficients on various heating surfaces (bottom). (a), (b) Reproduced with permission from [195]. Copyright Advanced Materials, 2016. (c), (d) Reproduced with permission from [197]. Copyright Advanced Materials, 2016. (e) Reproduced with permission from [199]. Copyright ACS Appl. Mater. Interfaces, 2017. (f) Reproduced with permission from [200]. Copyright Nano Letters, 2016.

Thermoelectric Sensors

Graphene can be utilized as not only a temperature sensor but also a sensor of other crucial parameters. Though terahertz radiation has been used in various applications ranging from security to medicine [201], the sensitive room-temperature detection of terahertz radiation is very difficult [202]. Cai et al. successfully demonstrated a graphene thermoelectric terahertz photodetector with highly impressive sensing ability (over $10 \mid V \mid W^{-1}$ ($700 \mid V \mid W^{-1}$) at room temperature) and low noise-equivalent power (below $1100 \mid pW \mid Hz^{-1/2}$ ($20 \mid pW \mid Hz^{-1/2}$)) [203]. The compared reference was the incident (absorbed) power. The detection mechanism of the sensor is the hot-electron photothermoelectric effect in graphene. Because of strong electron-electron interactions, photoexcited carriers quickly gain heat [204,205], and the energy of the lattice decreases more slowly [204,206]. Electron diffusion is caused by the temperature gradient of electrons. As a results, a net current is generated by asymmetry due to local gating [207,208] or dissimilar contact metals [209] through a thermoelectric effect. The operation of graphene-based sensors is similar to that of state-of-the-art room-temperature terahertz detectors [210], and the ability of the former to obtain time-resolved measurements is eight to nine times faster than that of the latter [208,211]. Additionally, one of the important considerations in on-chip nano-optical processing is the detection, control and generation of propagating plasmons [212–214]. Graphene shows intensely confined and controllable (>0.5 ps) plasmons induced by electrostatic fields [215–218]. An all-graphene-based mid-infrared plasmon detector that precisely converts the natural decay product of the plasmon (electronic heat) to voltage via the thermoelectric effect [219,220] was presented by Lunderberg et al. [221]. The detection system, constructed with the plasmonic medium of graphene encapsulated in hexagonal boron nitride (hBN), operates at room temperature; a single graphene sheet simultaneously acts as the plasmonic medium and detector. The junction induced by two local gates was used to completely control the thermoelectric and plasmonic behavior of the graphene.

Thermal Biosensors

Recently, studies aiming to realize theranostic systems with simultaneous detection and therapeutic services in one single nanostructure have increased in the field of personalized nanomedicine and clinical biomedical applications [222–224]. Cao et al. reported a multifunctional theranostic system coupling diagnostic and therapeutic functions with a porphyrin derivative (P) and GQDs [199]. The P is capable of high singlet oxygen production, and GQDs have excellent fluorescence properties. A GQD- polyethylene glycol (PEG)-P system was fabricated by integrating the P into PEGylated and aptamer-functionalized GQDs; the system showed exceptional physiological stability, impressive biocompatibility and low cytotoxicity. The intrinsic fluorescence of the GQDs can distinguish cancer cells from somatic cells and helps to detect intracellular cancer-related microRNA (miRNA). The photothermal conversion properties of GQD-PEG-P are shown at the bottom of Figure 4e. The image results indicated that the temperature of 100 µg/mL GQD-PEG-P solution increased to 53.6 °C, while the temperature of control water remained at 33.2 °C. A cancer treatment system combining progressive photothermal therapy (PTT) and powerful photodynamic therapy (PDT) demonstrated outstanding efficiency for both in vitro cancer cells and in vivo multicellular tumor spheroids (MCTS). A total GQD-PEG-P theranostic system for intracellular miRNA detection and PTT and PDT therapy is presented at the top of Figure 4e (the molecular beacon (MB) is the detection probe for miRNA-155). After monitoring and detecting a cancer, highly effective and reliable treatment is crucial. Synergistic cancer cures such as drug delivery, magnetic hyperthermia, photothermal therapy, gene therapy, and radiotherapy have been spotlighted [225–230]. To accomplish safe and successful treatment, Yao et al. suggested a multifunctional platform with GQD-capped magnetic mesoporous silica nanoparticles (MMSNs); a composite of GQDs was used for caps and local photothermal generators, and MMSNs were used for controlled drug release and magnetic hyperthermia [231]. Drug release can be handled by changing the interaction between drug molecules and carriers and by controlling the state (open or closed) of the outlets of mesoporous channels [232,233]. Monodisperse MMSN/GQD nanoparticles (particle size of 100 nm) can carry doxorubicin (DOX), provoke DOX discharge and adequately generate heat in an alternating magnetic field (AMF) or by near-infrared irradiation. An integrated graphene-based chemo-magnetic hyperthermia therapy or chemo-photothermal therapy platform for destroying breast cancer 4T1 cells (the model cellular system) represents a remarkable synergistic effect. As a result, the MMSN/GQD nanoparticles have excellent potential for efficient and accurate cancer therapy.

3.2.2. Energy Management Systems

Efficient heat management systems have become extremely important in various fields, such as electronic, optoelectronic, and thermoelectric applications, with the fast-growing development of state-of-the-art technologies to transport heat to the field environment; a high power heating density is required, or excessive heat must be prevented from deteriorating the device reliability, lifetime, and performance [234,235]. In this regard, the development of materials with high thermal conductivity is urgently needed. Heat management systems are normally based on metallic materials such as Al and Cu because of their relatively high thermal conductivity, approximately 200 to 400 W m^{-1} K^{-1}, and low cost [236,237]. However, these materials show insufficient thermal conductivity when fabricated with nm thickness due to the linear relation between the thickness and thermal conductivity [238]. Therefore, graphene, which has superior thermal conductivity, is a promising alternative for use in practical heat management systems. Seo et al. studied the pool boiling of hybrid graphene/single-walled carbon nanotubes (SWCNTs), graphene, and SWCNT films deposited on ITO surfaces to prove that the hybrid graphene/SWCNT material is the layer with the most enhanced heat transfer coefficient [200]. The authors tested four types of heating surfaces: (1) a bare ITO surface, (2) SWCNTs, (3) graphene, and (4) hybrid graphene/SWCNT layers deposited on an ITO surface. CVD was used to deposit graphene on the ITO surface, and SWCNTs were spray-coated on the surface to fabricate hybrid graphene/SWCNTs. The critical heat flux (CHF) of the SWCNTs, graphene, and hybrid graphene/SWCNT heaters were measured to be 123.0, 130.5, and 141.6 kW/m^2, respectively. The CHF of the hybrid graphene/SWCNT heater showed the greatest improvement of 18.2% compared to that of the bare ITO heater. As shown in Figure 4f, the maximum heat transfer coefficient (HTC) values of the bare ITO, SWCNT, graphene, and hybrid graphene/SWCNT heaters were found to be 4.41, 5.31, 4.49, and 6.83 kW/m^2 K, respectively. The HTC of the hybrid graphene/SWCNT heater improved by 55% compared to that of the bare ITO heater. These improvements are attributed to the deposition of SWCNTs on the graphene, which compensates the disconnected areas and wrinkles of the graphene. Han et al. introduced a high-performance heat spreader to manage excessive heat [239]. Graphene-based films (GBFs) were used in this experiment because of their high thermal conductivity of 1600 Wm^{-1} K^{-1} [240]. With a heat flux of 1300 W cm^{-2}, the temperature of the hotspot decreased by 17 °C with a GBP deposited on nonfunctionalized GO. In addition, 3-amino-propyltriethoxysilane (APTES) functionalization resulted in an additional decrease in temperature of 11 °C, or a total decrease of 28 °C. Bernal et al. proposed edge-functionalized graphene nanoplatelets (GnPs) to improve the thermal conductivity [241]. The diazonium reaction was exploited to synthesize phenol-functionalized GnPs (GnP-OH) and dianiline-bridged GnPs (E-GnPs). The in-plane thermal conductivity of both GnP-OH and E-GnPs was enhanced by 20% compared to that of GnPs. Even the cross-plane thermal conductivity of E-GnPs was remarkably increased by 190% compared to that of GnPs. Ma et al. conducted research on the relation between the thermal conductivity and grain size of graphene [242]. Segregation adsorption CVD (SACVD) was exploited to grow high-quality graphene films with a controllable grain size ranging from ~200 nm to ~1 μm. The experimental results demonstrated that the thermal conductivity increased exponentially from ~610 to ~5230 W m^{-1} K^{-1}, with grain sizes ranging from ~200 nm to 10 μm. The thermal conductivity of graphene was found to be significantly decreased with the decreasing grain size by a small thermal boundary conductance of ~3.8 × 10^9 W m^{-2} K^{-1}. Ding et al. proposed graphene nanosheets (GNSs) exhibiting high dispersibility in water and thermal conductivity [243]. Graphite powder and a modifier of sodium lignosulfonate (LS) were used to fabricate the GNSs by a ball-milling method. The GNSs fabricated with this method exhibited a high thermal conductivity of 1324 W m^{-1} K^{-1}.

Currently, with the rapid development of industry, the energy consumption of mainly fossil fuels is causing dramatic depletion worldwide [244]. Additionally, various environmental problems, such as desertification, the greenhouse effect, and pollution, are accompanied by the excessive use of energy [245,246]. In this regard, the development of sustainable energy is urgently required, and heat, which is approximately 60% of all energy waste, is rising as a promising candidate [247]. Thus, thermoelectric materials are of high interest. As mentioned above, the figure of merit ZT is defined as ZT = $S^2\sigma T/(K_{ph} + K_{el})$; thus, enhancing the Seebeck coefficient and electrical conductivity and reducing the thermal conductivity are essential to achieving a higher efficiency. Cho et al. proposed multidimensional nanomaterials based on organic thin films exhibiting a high thermoelectric power factor at room temperature [248]. The thin film consists of the organic materials polyaniline (PANi), PEDOT:PSS-stabilized graphene, and PEDOT:PSS-stabilized double-walled nanotubes (DWNTs). These materials are deposited as a quad-layer of PANi/graphene-PEDOT:PSS/PANi/DWNT. The quad-layer film showed a gradual increase in the electrical conductivity and Seebeck coefficient with the increased number of layers. The electrical conductivity and Seebeck coefficient of the film were found to be 1.9 $\times$ 10^5 S m^{-1} and 120 μV K^{-1} at 80 QLs, respectively. These results led to a high value of the thermoelectric power factor (PF) (~2710 μW m^{-1} K^{-2}), where PF = $S^2\sigma$. Ma et al. doped bromine (Br) onto graphene fibers to enhance the thermoelectric properties [27]. The porous structure of the Br-doped graphene fibers enhances phonon scattering and leads to limited thermal conductivity compared to that of other graphene materials [249,250]. Additionally, doping with Br induces a downshift in the Fermi level, which causes an increased Seebeck coefficient and increased electrical conductivity [251,252]. As a result, the experiment exhibited a maximum figure of merit of 2.76 $\times$ 10^{-3} and room-temperature PF of 624 μW m^{-1} K^{-2}. These values are better than those of materials composed of solely graphene and CNTs. Lin et al. observed the enhanced thermoelectric effect of lanthanum strontium titanium oxide (LSTO) upon adding graphene [253]. The addition of graphene to LSTO led to enhancements in the electrical conductivity, the Seebeck coefficient and power factor. It was found that LSTO integrated with 0.6 wt % graphene exhibited the best figure of merit and power factor. The figure of merit was over 0.25 at room temperature.

4. Conclusions

In this paper, we presented a review of the electronic and thermal properties of graphene and its up-to-date applications. Graphene has exceptional physical characteristics compared to other usual metals and semiconductors and has the potential for extensive applications. The ambipolar electric field effect in SLG proves that charge carriers have higher mobilities than those of semiconductors. Additionally, the QHE, the unique phenomenon observed with LL, is explained by graphs and unique geometrical phases. Using the experimental results of graphene crystal samples, the visual transparency of graphene was characterized by the fine-structure constant α. In addition to its electronic features, the superior thermal properties of graphene were reviewed. The specific heats of graphene and graphite were derived, and the thermal conductivity of carbon allotropes containing graphene was explained. Graphene has high thermoelectric power, which suggests its potential for use in energy-harvesting applications. These remarkable properties of graphene can be incorporated into versatile systems in various application fields: optical devices, electronic and thermal sensors and energy management systems. Here, state-of-the-art examples based on graphene were presented. Despite the development of many graphene-based electronic systems with unique properties, challenges and problems related to the development of materials for commercial electronics still exist and need to be studied further. By resolving such challenges and problems, graphene will be in the limelight as an active material for future electronics.

Author Contributions: M.S., J.S., K.K., and K.J.Y. collected the data, contributed to the scientific discussions, and cowrote the manuscript.

Nanomaterials **2019**, *9*, 374

Funding: This research received no external funding.

Acknowledgments: M.S., J.S., K.K. and K.J.Y. acknowledge the support from the National Research Foundation of Korea (Grant No. NRF-2017M1A2A2048904) and the Yonsei University Future-Leading Research Initiative of 2017 (RMS22017-22-00).

Conflicts of Interest: The authors declare no competing financial interests.

Acronym

Acronym Word	Original Word
2D	two-dimensional
QDs	quantum dots
QHE	quantum Hall effect
D (DOS)	density of states
BZ	Brillouin zone
DFT	Density Functional Theory
SdHOs	Shubnikov-de Haas oscillations
B	magnetic field
σ	Hall conductivity
ρ	longitudinal resistivity
LL	Landau level
c	speed of light
G	high-frequency conductivity
T	transmittance
R	reflectance
InSb	indium antimonide
HgCdTe	mercury cadmium telluride
Pbs	lead sulfide
PbSe	lead selenide
CVD	chemical vapor deposition
LEDs	light-emitting diodes
EQE	external quantum efficiency
TCEs	transparent conductive electrodes
ITO	indium tin oxide
NWs	nanowires
OLEDs	organic light-emitting diodes
PMMA	poly(methyl) methacrylate
HILs	hole injection layers
SPP	surface plasmon polariton
GO	graphene oxide
RGO	reduced graphene oxide
AgNWs	silver nanowires
A-2GE	two-step growth method
OTR	oxygen transmission rate
PET	polyethylene terephthalate
PCEs	power conversion efficiencies
PSCs	perovskite solar cells
ETLs	electron transport layers
HTLs	hole transport layers
GRMs	graphene and related two-dimensional materials
GQDs	graphene quantum dots
PEDOT:PSS	poly(3,4-ethylenedioxythiophene):poly(styrene sulfonate)

f-RGO	functionalized reduced graphene oxide
ABL	active buffer layer
FF	fill factor
GO-Li	lithium-neutralized graphene oxide flakes
FETs	field-effect transistors
HFTCVD	hot-filament thermal chemical vapor deposition
WGF	wrinkled graphene film
IDEs	interdigital electrodes
PVA	polyvinyl alcohol
CNTs	carbon nanotubes
CMOS	complementary metal–oxide–semiconductor
EOG	Electrooculography
GET	graphene electronic tattoo
CTCs	circulating tumor cells
AA	ascorbic acid
DA	dopamine
NO2	nitrite
Trp	tryptophan
UA	uric acid
WHO	World Health Organization
DNA	deoxyribonucleic acid
DEP	dielectrophoresis
MC-LR	microcystin-LR
GF/PET	graphene film/polyethylene terephthalate
S-RGOH	sulfonated reduced graphene oxide hydrogel
TBLI	two-beam-laser interference
GPA-FET	graphene-Pd-Ag-gate FET
IoT	Internet of Things
C	Specific heat
K	thermal conductivity
HOPG	highly oriented pyrolytic graphite
RIE	reactive ion etching
SEM	scanning electron microscopy
TEP	thermoelectric power
ZT	the figure of merit
E-skins	Electronic skins
HMIs	human-machine interfaces
TRC	temperature coefficient of resistance
TS	transparent stretchable
PUD	PU dispersion
SLG	single-layer graphene
SiNx	silicon nitride
hBN	hexagonal boron nitride
P	porphyrin derivative
PEG	polyethylene glycol
miRNA	microRNA
PTT	photothermal therapy
PDT	photodynamic therapy
MCTS	multicellular tumor spheroids
MB	molecular beacon
MMSNs	magnetic mesoporous silica nanoparticles
DOX	doxorubicin
AMF	alternating magnetic field
SWCNTs	single-walled carbon nanotubes

CHF	critical heat flux
HTC	heat transfer coefficient
GBFs	Graphene-based films
GnPs	graphene nanoplatelets
SACVD	segregation adsorption CVD
GNSs	graphene nanosheets
LS	lignosulfonate
PANi	polyaniline
DWNTs	double-walled nanotubes
PF	power factor
LSTO	lanthanum strontium titanium oxide

References

1. Wallace, P.R. The band theory of graphite. *Phys. Rev.* **1947**, *71*, 622–634. [CrossRef]
2. McClure, J.W. Diamagnetism of graphite. *Phys. Rev.* **1956**, *104*, 666–671. [CrossRef]
3. Slonczewski, J.C.; Weiss, P.R. Band structure of graphite. *Phys. Rev.* **1958**, *109*, 272–279. [CrossRef]
4. Novoselov, K.S.; Geim, A.K.; Morozov, S.V.; Jiang, D.; Zhang, Y.; Dubonos, S.V.; Grigorieva, I.V.; Firsov, A.A. Electric field effect in atomically thin carbon films. *Science* **2004**, *306*, 666–669. [CrossRef] [PubMed]
5. Zhang, Y.; Tan, Y.-W.; Stormer, H.L.; Kim, P. Experimental observation of the quantum hall effect and berry's phase in graphene. *Nature* **2005**, *438*, 201. [CrossRef] [PubMed]
6. Novoselov, K.S.; Jiang, D.; Schedin, F.; Booth, T.J.; Khotkevich, V.V.; Morozov, S.V.; Geim, A.K. Two-dimensional atomic crystals. *Proc. Natl. Acad. Sci. USA* **2005**, *102*, 10451–10453. [CrossRef] [PubMed]
7. Novoselov, K.S.; Geim, A.K.; Morozov, S.V.; Jiang, D.; Katsnelson, M.I.; Grigorieva, I.V.; Dubonos, S.V.; Firsov, A.A. Two-dimensional gas of massless dirac fermions in graphene. *Nature* **2005**, *438*, 197. [CrossRef] [PubMed]
8. Kang, K.; Cho, Y.; Yu, K.J. Novel nano-materials and nano-fabrication techniques for flexible electronic systems. *Micromachines* **2018**, *9*, 263. [CrossRef] [PubMed]
9. Medintz, I.L.; Uyeda, H.T.; Goldman, E.R.; Mattoussi, H. Quantum dot bioconjugates for imaging, labelling and sensing. *Nat. Mater.* **2005**, *4*, 435. [PubMed]
10. Harman, T.C.; Taylor, P.J.; Walsh, M.P.; LaForge, B.E. Quantum dot superlattice thermoelectric materials and devices. *Science* **2002**, *297*, 2229–2232. [CrossRef] [PubMed]
11. Song, X.; Zhang, J.; Yue, M.; Li, E.; Zeng, H.; Lu, N.; Zhou, M.; Zuo, T. Technique for preparing ultrafine nanocrystalline bulk material of pure rare-earth metals. *Adv. Mater.* **2006**, *18*, 1210–1215. [CrossRef]
12. Massari, S.; Ruberti, M. Rare earth elements as critical raw materials: Focus on international markets and future strategies. *Resour. Policy* **2013**, *38*, 36–43. [CrossRef]
13. Castro Neto, A.H.; Guinea, F.; Peres, N.M.R.; Novoselov, K.S.; Geim, A.K. The electronic properties of graphene. *Rev. Mod. Phys.* **2009**, *81*, 109–162. [CrossRef]
14. Geim, A.K.; Novoselov, K.S. The rise of graphene. *Nat. Mater.* **2007**, *6*, 183. [CrossRef] [PubMed]
15. Chen, S.; Moore, A.L.; Cai, W.; Suk, J.W.; An, J.; Mishra, C.; Amos, C.; Magnuson, C.W.; Kang, J.; Shi, L.; et al. Raman measurements of thermal transport in suspended monolayer graphene of variable sizes in vacuum and gaseous environments. *ACS Nano* **2011**, *5*, 321–328. [CrossRef] [PubMed]
16. Balandin, A.A. Thermal properties of graphene and nanostructured carbon materials. *Nat. Mater.* **2011**, *10*, 569. [CrossRef] [PubMed]
17. Chen, S.; Wu, Q.; Mishra, C.; Kang, J.; Zhang, H.; Cho, K.; Cai, W.; Balandin, A.A.; Ruoff, R.S. Thermal conductivity of isotopically modified graphene. *Nat. Mater.* **2012**, *11*, 203. [CrossRef] [PubMed]
18. Seol, J.H.; Jo, I.; Moore, A.L.; Lindsay, L.; Aitken, Z.H.; Pettes, M.T.; Li, X.; Yao, Z.; Huang, R.; Broido, D.; et al. Two-dimensional phonon transport in supported graphene. *Science* **2010**, *328*, 213–216. [CrossRef] [PubMed]
19. Lee, C.; Wei, X.; Kysar, J.W.; Hone, J. Measurement of the elastic properties and intrinsic strength of monolayer graphene. *Science* **2008**, *321*, 385–388. [CrossRef] [PubMed]
20. Nair, R.R.; Blake, P.; Grigorenko, A.N.; Novoselov, K.S.; Booth, T.J.; Stauber, T.; Peres, N.M.R.; Geim, A.K. Fine structure constant defines visual transparency of graphene. *Science* **2008**, *320*, 1308. [CrossRef] [PubMed]

21. Zhang, Z.; Du, J.; Zhang, D.; Sun, H.; Yin, L.; Ma, L.; Chen, J.; Ma, D.; Cheng, H.-M.; Ren, W. Rosin-enabled ultraclean and damage-free transfer of graphene for large-area flexible organic light-emitting diodes. *Nat. Commun.* **2017**, *8*, 14560. [CrossRef] [PubMed]

22. Kim, K.S.; Zhao, Y.; Jang, H.; Lee, S.Y.; Kim, J.M.; Kim, K.S.; Ahn, J.-H.; Kim, P.; Choi, J.-Y.; Hong, B.H. Large-scale pattern growth of graphene films for stretchable transparent electrodes. *Nature* **2009**, *457*, 706. [CrossRef] [PubMed]

23. Radisavljevic, B.; Radenovic, A.; Brivio, J.; Giacometti, V.; Kis, A. Single-layer MoS$_2$ transistors. *Nat. Nanotechnol.* **2011**, *6*, 147. [CrossRef] [PubMed]

24. Schedin, F.; Geim, A.K.; Morozov, S.V.; Hill, E.W.; Blake, P.; Katsnelson, M.I.; Novoselov, K.S. Detection of individual gas molecules adsorbed on graphene. *Nat. Mater.* **2007**, *6*, 652. [CrossRef] [PubMed]

25. Bolotin, K.I.; Sikes, K.J.; Jiang, Z.; Klima, M.; Fudenberg, G.; Hone, J.; Kim, P.; Stormer, H.L. Ultrahigh electron mobility in suspended graphene. *Solid State Commun.* **2008**, *146*, 351–355. [CrossRef]

26. Zhu, Y.; Murali, S.; Cai, W.; Li, X.; Suk, J.W.; Potts, J.R.; Ruoff, R.S. Graphene and graphene oxide: Synthesis, properties, and applications. *Adv. Mater.* **2010**, *22*, 3906–3924. [CrossRef] [PubMed]

27. Ma, W.; Liu, Y.; Yan, S.; Miao, T.; Shi, S.; Xu, Z.; Zhang, X.; Gao, C.J.N.R. Chemically doped macroscopic graphene fibers with significantly enhanced thermoelectric properties. *Nano Res.* **2018**, *11*, 741–750. [CrossRef]

28. Chan, K.T.; Neaton, J.B.; Cohen, M.L. First-principles study of metal adatom adsorption on graphene. *Phys. Rev. B* **2008**, *77*, 235430. [CrossRef]

29. Dimakis, N.; Valdez, D.; Flor, F.A.; Salgado, A.; Adjibi, K.; Vargas, S.; Saenz, J. Density functional theory calculations on alkali and the alkaline ca atoms adsorbed on graphene monolayers. *Appl. Surf. Sci.* **2017**, *413*, 197–208. [CrossRef]

30. Medeiros, P.V.C.; de Brito Mota, F.; Mascarenhas, A.J.S.; de Castilho, C.M.C. Adsorption of monovalent metal atoms on graphene: A theoretical approach. *Nanotechnology* **2010**, *21*, 115701. [CrossRef] [PubMed]

31. Lee, E.; Persson, K.A. Li absorption and intercalation in single layer graphene and few layer graphene by first principles. *Nano Lett.* **2012**, *12*, 4624–4628. [CrossRef] [PubMed]

32. Okamoto, Y. Density functional theory calculations of lithium adsorption and insertion to defect-free and defective graphene. *J. Phys. Chem. C* **2016**, *120*, 14009–14014. [CrossRef]

33. Yildirim, H.; Kinaci, A.; Zhao, Z.-J.; Chan, M.K.Y.; Greeley, J.P. First-principles analysis of defect-mediated li adsorption on graphene. *ACS Appl. Mater. Interfaces* **2014**, *6*, 21141–21150. [CrossRef] [PubMed]

34. Shiota, K.; Kawai, T. Li atom adsorption on graphene with various defects for large-capacity li ion batteries: First-principles calculations. *Jpn. J. Appl. Phys.* **2017**, *56*, 06GE11. [CrossRef]

35. Zheng, Y.; Ando, T. Hall conductivity of a two-dimensional graphite system. *Phys. Rev. B* **2002**, *65*, 245420. [CrossRef]

36. Gusynin, V.P.; Sharapov, S.G. Unconventional integer quantum hall effect in graphene. *Phys. Rev. Lett.* **2005**, *95*, 146801. [CrossRef] [PubMed]

37. Peres, N.M.R.; Guinea, F.; Castro Neto, A.H. Electronic properties of disordered two-dimensional carbon. *Phys. Rev. B* **2006**, *73*, 125411. [CrossRef]

38. MacDonald, A.H. Quantized hall conductance in a relativistic two-dimensional electron gas. *Phys. Rev. B* **1983**, *28*, 2235–2236. [CrossRef]

39. Novoselov, K.S.; McCann, E.; Morozov, S.V.; Fal'ko, V.I.; Katsnelson, M.I.; Zeitler, U.; Jiang, D.; Schedin, F.; Geim, A.K. Unconventional quantum hall effect and berry's phase of 2π in bilayer graphene. *Nat. Phys.* **2006**, *2*, 177. [CrossRef]

40. Mikitik, G.P.; Sharlai, Y.V. Manifestation of berry's phase in metal physics. *Phys. Rev. Lett.* **1999**, *82*, 2147–2150. [CrossRef]

41. Gusynin, V.P.; Sharapov, S.G.; Carbotte, J.P. Unusual microwave response of dirac quasiparticles in graphene. *Phys. Rev. Lett.* **2006**, *96*, 256802. [CrossRef] [PubMed]

42. Ando, T.; Zheng, Y.; Suzuura, H. Dynamical conductivity and zero-mode anomaly in honeycomb lattices. *J. Phys. Soc. Jpn.* **2002**, *71*, 1318–1324. [CrossRef]

43. Kershaw, S.V.; Susha, A.S.; Rogach, A.L. Narrow bandgap colloidal metal chalcogenide quantum dots: Synthetic methods, heterostructures, assemblies, electronic and infrared optical properties. *Chem. Soc. Rev.* **2013**, *42*, 3033–3087. [CrossRef] [PubMed]

44. Musca, C.; Antoszewski, J.; Dell, J.; Faraone, L.; Piotrowski, J.; Nowak, Z. Multi-heterojunction large area hgcdte long wavelength infrared photovoltaic detector for operation at near room temperatures. *J. Electron. Mater.* **1998**, *27*, 740–746. [CrossRef]

45. Taojian, S.; Haijuan, C.; Chunjie, F.; Bo, H.; Weile, L.; Junfeng, X.; Yi, T.; Shengyi, Y.; Bingsuo, Z. Influence of the active layer nanomorphology on device performance for ternary PbS_xSe_{1-x} quantum dots based solution-processed infrared photodetector. *Nanotechnology* **2016**, *27*, 165202.

46. Lhuillier, E.; Keuleyan, S.; Zolotavin, P.; Guyot-Sionnest, P. Mid-infrared $HgTe/As_2S_3$ field effect transistors and photodetectors. *Adv. Mater.* **2013**, *25*, 137–141. [CrossRef] [PubMed]

47. Liu, C.-H.; Chang, Y.-C.; Norris, T.B.; Zhong, Z. Graphene photodetectors with ultra-broadband and high responsivity at room temperature. *Nat. Nanotechnol.* **2014**, *9*, 273. [CrossRef] [PubMed]

48. Yu, X.; Li, Y.; Hu, X.; Zhang, D.; Tao, Y.; Liu, Z.; He, Y.; Haque, M.A.; Liu, Z.; Wu, T.; et al. Narrow bandgap oxide nanoparticles coupled with graphene for high performance mid-infrared photodetection. *Nat. Commun.* **2018**, *9*, 4299. [CrossRef] [PubMed]

49. Nie, B.; Hu, J.-G.; Luo, L.-B.; Xie, C.; Zeng, L.-H.; Lv, P.; Li, F.-Z.; Jie, J.-S.; Feng, M.; Wu, C.-Y.; et al. Monolayer graphene film on zno nanorod array for high-performance schottky junction ultraviolet photodetectors. *Small* **2013**, *9*, 2872–2879. [CrossRef] [PubMed]

50. Lee, Y.; Kwon, J.; Hwang, E.; Ra, C.-H.; Yoo, W.J.; Ahn, J.-H.; Park, J.H.; Cho, J.H. High-performance perovskite–graphene hybrid photodetector. *Adv. Mater.* **2015**, *27*, 41–46. [CrossRef] [PubMed]

51. Konstantatos, G.; Badioli, M.; Gaudreau, L.; Osmond, J.; Bernechea, M.; de Arquer, F.P.G.; Gatti, F.; Koppens, F.H.L. Hybrid graphene–quantum dot phototransistors with ultrahigh gain. *Nat. Nanotechnol.* **2012**, *7*, 363. [CrossRef] [PubMed]

52. Manga, K.K.; Wang, J.; Lin, M.; Zhang, J.; Nesladek, M.; Nalla, V.; Ji, W.; Loh, K.P. High-performance broadband photodetector using solution-processible $PbSe–TiO_2$–graphene hybrids. *Adv. Mater.* **2012**, *24*, 1697–1702. [CrossRef] [PubMed]

53. Wang, G.; Zhang, M.; Chen, D.; Guo, Q.; Feng, X.; Niu, T.; Liu, X.; Li, A.; Lai, J.; Sun, D.; et al. Seamless lateral graphene p–n junctions formed by selective in situ doping for high-performance photodetectors. *Nat. Commun.* **2018**, *9*, 5168. [CrossRef] [PubMed]

54. Fang, J.; Wang, D.; DeVault, C.T.; Chung, T.-F.; Chen, Y.P.; Boltasseva, A.; Shalaev, V.M.; Kildishev, A.V. Enhanced graphene photodetector with fractal metasurface. *Nano Lett.* **2017**, *17*, 57–62. [CrossRef] [PubMed]

55. Cakmakyapan, S.; Lu, P.K.; Navabi, A.; Jarrahi, M. Gold-patched graphene nano-stripes for high-responsivity and ultrafast photodetection from the visible to infrared regime. *Light Sci. Appl.* **2018**, *7*, 20. [CrossRef]

56. Zheludev, N. The life and times of the led—A 100-year history. *Nat. Photonics* **2007**, *1*, 189. [CrossRef]

57. Ge, D.; Lee, E.; Yang, L.; Cho, Y.; Li, M.; Gianola, D.S.; Yang, S. A robust smart window: Reversibly switching from high transparency to angle-independent structural color display. *Adv. Mater.* **2015**, *27*, 2489–2495. [CrossRef] [PubMed]

58. Li, X.; Cho, W.K.; Hussain, B.; Kwok, H.S.; Yue, C.P. 66-3: Micro-led display with simultaneous visible light communication function. *SID Symp. Dig. Tech. Pap.* **2018**, *49*, 876–879. [CrossRef]

59. Dursun, I.; Shen, C.; Parida, M.R.; Pan, J.; Sarmah, S.P.; Priante, D.; Alyami, N.; Liu, J.; Saidaminov, M.I.; Alias, M.S.; et al. Perovskite nanocrystals as a color converter for visible light communication. *ACS Photonics* **2016**, *3*, 1150–1156. [CrossRef]

60. Wang, Q.H.; Kalantar-Zadeh, K.; Kis, A.; Coleman, J.N.; Strano, M.S. Electronics and optoelectronics of two-dimensional transition metal dichalcogenides. *Nat. Nanotechnol.* **2012**, *7*, 699. [CrossRef] [PubMed]

61. Pramanick, B.; Cadenas, L.B.; Kim, D.-M.; Lee, W.; Shim, Y.-B.; Martinez-Chapa, S.O.; Madou, M.J.; Hwang, H. Human hair-derived hollow carbon microfibers for electrochemical sensing. *Carbon* **2016**, *107*, 872–877. [CrossRef]

62. Park, S.H.; Roy, A.; Beaupré, S.; Cho, S.; Coates, N.; Moon, J.S.; Moses, D.; Leclerc, M.; Lee, K.; Heeger, A.J. Bulk heterojunction solar cells with internal quantum efficiency approaching 100%. *Nat. Photonics* **2009**, *3*, 297. [CrossRef]

63. Kulkarni, A.K.; Schulz, K.H.; Lim, T.S.; Khan, M. Dependence of the sheet resistance of indium-tin-oxide thin films on grain size and grain orientation determined from x-ray diffraction techniques. *Thin Solid Films* **1999**, *345*, 273–277. [CrossRef]

64. Margalith, T.; Buchinsky, O.; Cohen, D.A.; Abare, A.C.; Hansen, M.; DenBaars, S.P.; Coldren, L.A. Indium tin oxide contacts to gallium nitride optoelectronic devices. *Appl. Phys. Lett.* **1999**, *74*, 3930–3932. [CrossRef]

65. Hecht, D.S.; Hu, L.; Irvin, G. Emerging transparent electrodes based on thin films of carbon nanotubes, graphene, and metallic nanostructures. *Adv. Mater.* **2011**, *23*, 1482–1513. [CrossRef] [PubMed]

66. Zhou, L.; Xiang, H.-Y.; Shen, S.; Li, Y.-Q.; Chen, J.-D.; Xie, H.-J.; Goldthorpe, I.A.; Chen, L.-S.; Lee, S.-T.; Tang, J.-X. High-performance flexible organic light-emitting diodes using embedded silver network transparent electrodes. *ACS Nano* **2014**, *8*, 12796–12805. [CrossRef] [PubMed]

67. Li, N.; Oida, S.; Tulevski, G.S.; Han, S.-J.; Hannon, J.B.; Sadana, D.K.; Chen, T.-C. Efficient and bright organic light-emitting diodes on single-layer graphene electrodes. *Nat. Commun.* **2013**, *4*, 2294. [CrossRef] [PubMed]

68. Wu, Z.; Chen, Z.; Du, X.; Logan, J.M.; Sippel, J.; Nikolou, M.; Kamaras, K.; Reynolds, J.R.; Tanner, D.B.; Hebard, A.F.; et al. Transparent, conductive carbon nanotube films. *Science* **2004**, *305*, 1273–1276. [CrossRef] [PubMed]

69. Lee, J.-Y.; Connor, S.T.; Cui, Y.; Peumans, P. Solution-processed metal nanowire mesh transparent electrodes. *Nano Lett.* **2008**, *8*, 689–692. [CrossRef] [PubMed]

70. Najafi, L.; Taheri, B.; Martín-García, B.; Bellani, S.; Di Girolamo, D.; Agresti, A.; Oropesa-Nuñez, R.; Pescetelli, S.; Vesce, L.; Calabrò, E.; et al. Mos2 quantum dot/graphene hybrids for advanced interface engineering of a ch3nh3pbi3 perovskite solar cell with an efficiency of over 20%. *ACS Nano* **2018**, *12*, 10736–10754. [CrossRef] [PubMed]

71. Liu, W.; Liu, N.; Yue, Y.; Rao, J.; Cheng, F.; Su, J.; Liu, Z.; Gao, Y. Piezoresistive pressure sensor based on synergistical innerconnect polyvinyl alcohol nanowires/wrinkled graphene film. *Small* **2018**, *14*, 1704149. [CrossRef] [PubMed]

72. Kim, J.; Kim, M.; Lee, M.-S.; Kim, K.; Ji, S.; Kim, Y.-T.; Park, J.; Na, K.; Bae, K.-H.; Kyun Kim, H.; et al. Wearable smart sensor systems integrated on soft contact lenses for wireless ocular diagnostics. *Nat. Commun.* **2017**, *8*, 14997. [CrossRef] [PubMed]

73. Bae, S.; Kim, H.; Lee, Y.; Xu, X.; Park, J.-S.; Zheng, Y.; Balakrishnan, J.; Lei, T.; Ri Kim, H.; Song, Y.I.; et al. Roll-to-roll production of 30-inch graphene films for transparent electrodes. *Nat. Nanotechnol.* **2010**, *5*, 574. [CrossRef] [PubMed]

74. Kim, J.; Park, H.; Hannon, J.B.; Bedell, S.W.; Fogel, K.; Sadana, D.K.; Dimitrakopoulos, C. Layer-resolved graphene transfer via engineered strain layers. *Science* **2013**, *342*, 833–836. [CrossRef] [PubMed]

75. Lin, Y.-C.; Jin, C.; Lee, J.-C.; Jen, S.-F.; Suenaga, K.; Chiu, P.-W. Clean transfer of graphene for isolation and suspension. *ACS Nano* **2011**, *5*, 2362–2368. [CrossRef] [PubMed]

76. Lee, J.; Han, T.-H.; Park, M.-H.; Jung, D.Y.; Seo, J.; Seo, H.-K.; Cho, H.; Kim, E.; Chung, J.; Choi, S.-Y.; et al. Synergetic electrode architecture for efficient graphene-based flexible organic light-emitting diodes. *Nat. Commun.* **2016**, *7*, 11791. [CrossRef] [PubMed]

77. Wang, X.; Tian, H.; Mohammad, M.A.; Li, C.; Wu, C.; Yang, Y.; Ren, T.-L. A spectrally tunable all-graphene-based flexible field-effect light-emitting device. *Nat. Commun.* **2015**, *6*, 7767. [CrossRef] [PubMed]

78. Huang, Y.; Huang, Z.; Zhong, Z.; Yang, X.; Hong, Q.; Wang, H.; Huang, S.; Gao, N.; Chen, X.; Cai, D.; et al. Highly transparent light emitting diodes on graphene encapsulated cu nanowires network. *Sci. Rep.* **2018**, *8*, 13721. [CrossRef] [PubMed]

79. Guo, H.; Lin, N.; Chen, Y.; Wang, Z.; Xie, Q.; Zheng, T.; Gao, N.; Li, S.; Kang, J.; Cai, D.; et al. Copper nanowires as fully transparent conductive electrodes. *Sci. Rep.* **2013**, *3*, 2323. [CrossRef] [PubMed]

80. Song, M.; You, D.S.; Lim, K.; Park, S.; Jung, S.; Kim, C.S.; Kim, D.-H.; Kim, D.-G.; Kim, J.-K.; Park, J.; et al. Highly efficient and bendable organic solar cells with solution-processed silver nanowire electrodes. *Adv. Funct. Mater.* **2013**, *23*, 4177–4184. [CrossRef]

81. Seo, T.H.; Lee, S.; Min, K.H.; Chandramohan, S.; Park, A.H.; Lee, G.H.; Park, M.; Suh, E.-K.; Kim, M.J. The role of graphene formed on silver nanowire transparent conductive electrode in ultra-violet light emitting diodes. *Sci. Rep.* **2016**, *6*, 29464. [CrossRef] [PubMed]

82. Stranks, S.D.; Eperon, G.E.; Grancini, G.; Menelaou, C.; Alcocer, M.J.P.; Leijtens, T.; Herz, L.M.; Petrozza, A.; Snaith, H.J. Electron-hole diffusion lengths exceeding 1 micrometer in an organometal trihalide perovskite absorber. *Science* **2013**, *342*, 341–344. [CrossRef] [PubMed]

83. Green, M.A.; Ho-Baillie, A. Perovskite solar cells: The birth of a new era in photovoltaics. *ACS Energy Lett.* **2017**, *2*, 822–830. [CrossRef]

84. Zuo, C.; Bolink, H.J.; Han, H.; Huang, J.; Cahen, D.; Ding, L. Advances in perovskite solar cells. *Adv. Sci.* **2016**, *3*, 1500324. [CrossRef] [PubMed]

85. Kojima, A.; Teshima, K.; Shirai, Y.; Miyasaka, T. Organometal halide perovskites as visible-light sensitizers for photovoltaic cells. *J. Am. Chem. Soc.* **2009**, *131*, 6050–6051. [CrossRef] [PubMed]

86. Xing, G.; Mathews, N.; Sun, S.; Lim, S.S.; Lam, Y.M.; Grätzel, M.; Mhaisalkar, S.; Sum, T.C. Long-range balanced electron- and hole-transport lengths in organic-inorganic $CH_3NH_3PbI_3$. *Science* **2013**, *342*, 344–347. [CrossRef] [PubMed]

87. Ball, J.M.; Lee, M.M.; Hey, A.; Snaith, H.J. Low-temperature processed meso-superstructured to thin-film perovskite solar cells. *Energy Environ. Sci.* **2013**, *6*, 1739–1743. [CrossRef]

88. Snaith, H.J. Perovskites: The emergence of a new era for low-cost, high-efficiency solar cells. *J. Phys. Chem. Lett.* **2013**, *4*, 3623–3630. [CrossRef]

89. Ke, W.; Fang, G.; Liu, Q.; Xiong, L.; Qin, P.; Tao, H.; Wang, J.; Lei, H.; Li, B.; Wan, J.; et al. Low-temperature solution-processed tin oxide as an alternative electron transporting layer for efficient perovskite solar cells. *J. Am. Chem. Soc.* **2015**, *137*, 6730–6733. [CrossRef] [PubMed]

90. Zhou, Z.; Li, X.; Cai, M.; Xie, F.; Wu, Y.; Lan, Z.; Yang, X.; Qiang, Y.; Islam, A.; Han, L. Stable inverted planar perovskite solar cells with low-temperature-processed hole-transport bilayer. *Adv. Energy Mater.* **2017**, *7*, 1700763. [CrossRef]

91. Kakavelakis, G.; Maksudov, T.; Konios, D.; Paradisanos, I.; Kioseoglou, G.; Stratakis, E.; Kymakis, E. Efficient and highly air stable planar inverted perovskite solar cells with reduced graphene oxide doped pcbm electron transporting layer. *Adv. Energy Mater.* **2017**, *7*, 1602120. [CrossRef]

92. Malinkiewicz, O.; Yella, A.; Lee, Y.H.; Espallargas, G.M.; Graetzel, M.; Nazeeruddin, M.K.; Bolink, H.J. Perovskite solar cells employing organic charge-transport layers. *Nat. Photonics* **2013**, *8*, 128. [CrossRef]

93. Palma, A.L.; Cinà, L.; Pescetelli, S.; Agresti, A.; Raggio, M.; Paolesse, R.; Bonaccorso, F.; Di Carlo, A. Reduced graphene oxide as efficient and stable hole transporting material in mesoscopic perovskite solar cells. *Nano Energy* **2016**, *22*, 349–360. [CrossRef]

94. Wang, Z.-K.; Li, M.; Yuan, D.-X.; Shi, X.-B.; Ma, H.; Liao, L.-S. Improved hole interfacial layer for planar perovskite solar cells with efficiency exceeding 15%. *ACS Appl. Mater. Interfaces* **2015**, *7*, 9645–9651. [CrossRef] [PubMed]

95. Zhu, Z.; Ma, J.; Wang, Z.; Mu, C.; Fan, Z.; Du, L.; Bai, Y.; Fan, L.; Yan, H.; Phillips, D.L.; et al. Efficiency enhancement of perovskite solar cells through fast electron extraction: The role of graphene quantum dots. *J. Am. Chem. Soc.* **2014**, *136*, 3760–3763. [CrossRef] [PubMed]

96. Fang, X.; Ding, J.; Yuan, N.; Sun, P.; Lv, M.; Ding, G.; Zhu, C. Graphene quantum dot incorporated perovskite films: Passivating grain boundaries and facilitating electron extraction. *Phys. Chem. Chem. Phys.* **2017**, *19*, 6057–6063. [CrossRef] [PubMed]

97. Petridis, C.; Kakavelakis, G.; Kymakis, E. Renaissance of graphene-related materials in photovoltaics due to the emergence of metal halide perovskite solar cells. *Energy Environ. Sci.* **2018**, *11*, 1030–1061. [CrossRef]

98. Cho, K.T.; Grancini, G.; Lee, Y.; Konios, D.; Paek, S.; Kymakis, E.; Nazeeruddin, M.K. Beneficial role of reduced graphene oxide for electron extraction in highly efficient perovskite solar cells. *ChemSusChem* **2016**, *9*, 3040–3044. [CrossRef] [PubMed]

99. Chung, C.-C.; Narra, S.; Jokar, E.; Wu, H.-P.; Wei-Guang Diau, E. Inverted planar solar cells based on perovskite/graphene oxide hybrid composites. *J. Mater. Chem. A* **2017**, *5*, 13957–13965. [CrossRef]

100. Abrusci, A.; Stranks, S.D.; Docampo, P.; Yip, H.-L.; Jen, A.K.Y.; Snaith, H.J. High-performance perovskite-polymer hybrid solar cells via electronic coupling with fullerene monolayers. *Nano Lett.* **2013**, *13*, 3124–3128. [CrossRef] [PubMed]

101. Zhang, L.; Liu, T.; Liu, L.; Hu, M.; Yang, Y.; Mei, A.; Han, H. The effect of carbon counter electrodes on fully printable mesoscopic perovskite solar cells. *J. Mater. Chem. A* **2015**, *3*, 9165–9170. [CrossRef]

102. Yan, K.; Wei, Z.; Li, J.; Chen, H.; Yi, Y.; Zheng, X.; Long, X.; Wang, Z.; Wang, J.; Xu, J.; et al. High-performance graphene-based hole conductor-free perovskite solar cells: Schottky junction enhanced hole extraction and electron blocking. *Small* **2015**, *11*, 2269–2274. [CrossRef] [PubMed]

103. You, P.; Liu, Z.; Tai, Q.; Liu, S.; Yan, F. Efficient semitransparent perovskite solar cells with graphene electrodes. *Adv. Mater.* **2015**, *27*, 3632–3638. [CrossRef] [PubMed]

104. Sung, H.; Ahn, N.; Jang, M.S.; Lee, J.-K.; Yoon, H.; Park, N.-G.; Choi, M. Transparent conductive oxide-free graphene-based perovskite solar cells with over 17% efficiency. *Adv. Energy Mater.* **2016**, *6*, 1501873. [CrossRef]

105. Agresti, A.; Pescetelli, S.; Palma, A.L.; Del Rio Castillo, A.E.; Konios, D.; Kakavelakis, G.; Razza, S.; Cinà, L.; Kymakis, E.; Bonaccorso, F.; et al. Graphene interface engineering for perovskite solar modules: 12.6% power conversion efficiency over 50 cm2 active area. *ACS Energy Lett.* **2017**, *2*, 279–287. [CrossRef]

106. Li, Y.; Leung, W.W.-F. Introduction of graphene nanofibers into the perovskite layer of perovskite solar cells. *ChemSusChem* **2018**, *11*, 2921–2929. [CrossRef] [PubMed]

107. Zang, Y.; Zhang, F.; Di, C.-A.; Zhu, D.J.M.H. Advances of flexible pressure sensors toward artificial intelligence and health care applications. *Mater. Horiz.* **2015**, *2*, 140–156. [CrossRef]

108. Zang, Y.; Zhang, F.; Huang, D.; Gao, X.; Di, C.-A.; Zhu, D.J. Flexible suspended gate organic thin-film transistors for ultra-sensitive pressure detection. *Mater. Horiz.* **2015**, *2*, 140–156. [CrossRef]

109. Jung, S.; Kim, J.H.; Kim, J.; Choi, S.; Lee, J.; Park, I.; Hyeon, T.; Kim, D.H. Reverse-micelle-induced porous pressure-sensitive rubber for wearable human–machine interfaces. *Adv. Mater.* **2014**, *26*, 4825–4830. [CrossRef] [PubMed]

110. Tian, H.; Shu, Y.; Wang, X.-F.; Mohammad, M.A.; Bie, Z.; Xie, Q.-Y.; Li, C.; Mi, W.-T.; Yang, Y.; Ren, T.-L. A graphene-based resistive pressure sensor with record-high sensitivity in a wide pressure range. *Sci. Rep.* **2015**, *5*, 8603. [CrossRef] [PubMed]

111. Mannsfeld, S.C.; Tee, B.C.; Stoltenberg, R.M.; Chen, C.V.H.; Barman, S.; Muir, B.V.; Sokolov, A.N.; Reese, C.; Bao, Z. Highly sensitive flexible pressure sensors with microstructured rubber dielectric layers. *Nat. Mater.* **2010**, *9*, 859. [CrossRef] [PubMed]

112. Lipomi, D.J.; Vosgueritchian, M.; Tee, B.C.; Hellstrom, S.L.; Lee, J.A.; Fox, C.H.; Bao, Z. Skin-like pressure and strain sensors based on transparent elastic films of carbon nanotubes. *Nat. Nanotechnol.* **2011**, *6*, 788. [CrossRef] [PubMed]

113. Takei, K.; Takahashi, T.; Ho, J.C.; Ko, H.; Gillies, A.G.; Leu, P.W.; Fearing, R.S.; Javey, A. Nanowire active-matrix circuitry for low-voltage macroscale artificial skin. *Nat. Mater.* **2010**, *9*, 821. [CrossRef] [PubMed]

114. Sun, Q.; Kim, D.H.; Park, S.S.; Lee, N.Y.; Zhang, Y.; Lee, J.H.; Cho, K.; Cho, J.H. Transparent, low-power pressure sensor matrix based on coplanar-gate graphene transistors. *Adv. Mater.* **2014**, *26*, 4735–4740. [CrossRef] [PubMed]

115. Honda, W.; Harada, S.; Arie, T.; Akita, S.; Takei, K. Wearable, human-interactive, health-monitoring, wireless devices fabricated by macroscale printing techniques. *Adv. Funct. Mater.* **2014**, *24*, 3299–3304. [CrossRef]

116. Mohammad Haniff, M.A.S.; Muhammad Hafiz, S.; Wahid, K.A.A.; Endut, Z.; Wah Lee, H.; Bien, D.C.S.; Abdul Azid, I.; Abdullah, M.Z.; Ming Huang, N.; Abdul Rahman, S. Piezoresistive effects in controllable defective hftcvd graphene-based flexible pressure sensor. *Sci. Rep.* **2015**, *5*, 14751. [CrossRef] [PubMed]

117. Shin, S.-H.; Ji, S.; Choi, S.; Pyo, K.-H.; Wan An, B.; Park, J.; Kim, J.; Kim, J.-Y.; Lee, K.-S.; Kwon, S.-Y.; et al. Integrated arrays of air-dielectric graphene transistors as transparent active-matrix pressure sensors for wide pressure ranges. *Nat. Commun.* **2017**, *8*, 14950. [CrossRef] [PubMed]

118. Tao, X.; Koncar, V.; Huang, T.-H.; Shen, C.-L.; Ko, Y.-C.; Jou, G.-T. How to make reliable, washable, and wearable textronic devices. *Sensors* **2017**, *17*, 673. [CrossRef] [PubMed]

119. Zeng, W.; Shu, L.; Li, Q.; Chen, S.; Wang, F.; Tao, X.-M. Fiber-based wearable electronics: A review of materials, fabrication, devices, and applications. *Adv. Mater.* **2014**, *26*, 5310–5336. [CrossRef] [PubMed]

120. Torres Alonso, E.; Rodrigues, D.P.; Khetani, M.; Shin, D.-W.; De Sanctis, A.; Joulie, H.; de Schrijver, I.; Baldycheva, A.; Alves, H.; Neves, A.I.S.; et al. Graphene electronic fibres with touch-sensing and light-emitting functionalities for smart textiles. *NPJ Flex. Electron.* **2018**, *2*, 25. [CrossRef]

121. Neves, A.I.S.; Bointon, T.H.; Melo, L.V.; Russo, S.; de Schrijver, I.; Craciun, M.F.; Alves, H. Transparent conductive graphene textile fibers. *Sci. Rep.* **2015**, *5*, 9866. [CrossRef] [PubMed]

122. Kim, J.T.; Choi, H.; Shin, E.; Park, S.; Kim, I.G. Graphene-based optical waveguide tactile sensor for dynamic response. *Sci. Rep.* **2018**, *8*, 16118. [CrossRef] [PubMed]

123. Pyo, S.; Choi, J.; Kim, J. A fully transparent, flexible, sensitive, and visible-blind ultraviolet sensor based on carbon nanotube-graphene hybrid. *Adv. Electron. Mater.* **2018**. [CrossRef]

124. Goossens, S.; Navickaite, G.; Monasterio, C.; Gupta, S.; Piqueras, J.J.; Pérez, R.; Burwell, G.; Nikitskiy, I.; Lasanta, T.; Galán, T.; et al. Broadband image sensor array based on graphene–cmos integration. *Nat. Photonics* **2017**, *11*, 366. [CrossRef]

125. Fang, H.; Yu, K.J.; Gloschat, C.; Yang, Z.; Song, E.; Chiang, C.-H.; Zhao, J.; Won, S.M.; Xu, S.; Trumpis, M.; et al. Capacitively coupled arrays of multiplexed flexible silicon transistors for long-term cardiac electrophysiology. *Nat. Biomed. Eng.* **2017**, *1*, 0038. [CrossRef] [PubMed]

126. Li, J.; Song, E.; Chiang, C.-H.; Yu, K.J.; Koo, J.; Du, H.; Zhong, Y.; Hill, M.; Wang, C.; Zhang, J.; et al. Conductively coupled flexible silicon electronic systems for chronic neural electrophysiology. *Proc. Natl. Acad. Sci. USA* **2018**, *115*, E9542–E9549. [CrossRef] [PubMed]

127. Jeong, J.-W.; McCall, J.G.; Shin, G.; Zhang, Y.; Al-Hasani, R.; Kim, M.; Li, S.; Sim, J.Y.; Jang, K.-I.; Shi, Y.; et al. Wireless optofluidic systems for programmable in vivo pharmacology and optogenetics. *Cell* **2015**, *162*, 662–674. [CrossRef] [PubMed]

128. Montgomery, K.L.; Yeh, A.J.; Ho, J.S.; Tsao, V.; Mohan Iyer, S.; Grosenick, L.; Ferenczi, E.A.; Tanabe, Y.; Deisseroth, K.; Delp, S.L. Wirelessly powered, fully internal optogenetics for brain, spinal and peripheral circuits in mice. *Nat. Methods* **2015**, *12*, 969. [CrossRef] [PubMed]

129. Kim, T.-I.; McCall, J.G.; Jung, Y.H.; Huang, X.; Siuda, E.R.; Li, Y.; Song, J.; Song, Y.M.; Pao, H.A.; Kim, R.-H.; et al. Injectable, cellular-scale optoelectronics with applications for wireless optogenetics. *Science* **2013**, *340*, 211–216. [CrossRef] [PubMed]

130. Canales, A.; Jia, X.; Froriep, U.P.; Koppes, R.A.; Tringides, C.M.; Selvidge, J.; Lu, C.; Hou, C.; Wei, L.; Fink, Y.; et al. Multifunctional fibers for simultaneous optical, electrical and chemical interrogation of neural circuits in vivo. *Nat. Biotechnol.* **2015**, *33*, 277. [CrossRef] [PubMed]

131. Viventi, J.; Kim, D.-H.; Vigeland, L.; Frechette, E.S.; Blanco, J.A.; Kim, Y.-S.; Avrin, A.E.; Tiruvadi, V.R.; Hwang, S.-W.; Vanleer, A.C.; et al. Flexible, foldable, actively multiplexed, high-density electrode array for mapping brain activity in vivo. *Nat. Neurosci.* **2011**, *14*, 1599. [CrossRef] [PubMed]

132. Kim, T.; Cho, M.; Yu, K.J. Flexible and stretchable bio-integrated electronics based on carbon nanotube and graphene. *Materials* **2018**, *11*, 1163. [CrossRef] [PubMed]

133. Ameri, S.K.; Kim, M.; Kuang, I.A.; Perera, W.K.; Alshiekh, M.; Jeong, H.; Topcu, U.; Akinwande, D.; Lu, N. Imperceptible electrooculography graphene sensor system for human–robot interface. *NPJ 2D Mater. Appl.* **2018**, *2*, 19. [CrossRef]

134. Kuzum, D.; Takano, H.; Shim, E.; Reed, J.C.; Juul, H.; Richardson, A.G.; de Vries, J.; Bink, H.; Dichter, M.A.; Lucas, T.H.; et al. Transparent and flexible low noise graphene electrodes for simultaneous electrophysiology and neuroimaging. *Nat. Commun.* **2014**, *5*, 5259. [CrossRef] [PubMed]

135. Lee, J.; Song, Y.; Ozden, I.; Nurmikko, A.V. Transparent micro-optrode arrays for simultaneous multichannel optical stimulation and electrical recording. In Proceedings of the CLEO: 2013, San Jose, CA, USA, 9–14 June 2013.

136. Zhang, J.; Hodge, W.; Hutnick, C.; Wang, X. Noninvasive diagnostic devices for diabetes through measuring tear glucose. *J. Diabetes. Sci. Technol.* **2011**, *5*, 166–172. [CrossRef] [PubMed]

137. Ascaso, F.J.; Huerva, V. Noninvasive continuous monitoring of tear glucose using glucose-sensing contact lenses. *Optom. Vis. Sci.* **2016**, *93*, 426–434. [CrossRef] [PubMed]

138. Chen, G.-Z.; Chan, I.-S.; Lam, D.C.C. Capacitive contact lens sensor for continuous non-invasive intraocular pressure monitoring. *Sens. Actuators A Phys.* **2013**, *203*, 112–118. [CrossRef]

139. Leonardi, M.; Leuenberger, P.; Bertrand, D.; Bertsch, A.; Renaud, P. First steps toward noninvasive intraocular pressure monitoring with a sensing contact lens. *Investig. Ophthalmol. Vis. Sci.* **2004**, *45*, 3113–3117.

140. Mansouri, K.; Shaarawy, T. Continuous intraocular pressure monitoring with a wireless ocular telemetry sensor: Initial clinical experience in patients with open angle glaucoma. *Br. J. Ophthalmol.* **2011**, *95*, 627–629. [CrossRef] [PubMed]

141. Mansouri, K.; Medeiros, F.A.; Tafreshi, A.; Weinreb, R.N. Continuous 24-hour monitoring of intraocular pressure patterns with a contact lens sensor: Safety, tolerability, and reproducibility in patients with glaucoma. *Arch. Ophthalmol.* **2012**, *130*, 1534–1539. [CrossRef] [PubMed]

142. Wu, L.; Ji, H.; Guan, Y.; Ran, X.; Ren, J.; Qu, X. A graphene-based chemical nose/tongue approach for the identification of normal, cancerous and circulating tumor cells. *Npg Asia Mater.* **2017**, *9*, e356. [CrossRef]

143. Zhou, Y.; Dong, H.; Liu, L.; Hao, Y.; Chang, Z.; Xu, M. Fabrication of electrochemical interface based on boronic acid-modified pyrroloquinoline quinine/reduced graphene oxide composites for voltammetric determination of glycated hemoglobin. *Biosens. Bioelectron.* **2015**, *64*, 442–448. [CrossRef] [PubMed]

144. Viswambari Devi, R.; Doble, M.; Verma, R.S. Nanomaterials for early detection of cancer biomarker with special emphasis on gold nanoparticles in immunoassays/sensors. *Biosens. Bioelectron.* **2015**, *68*, 688–698. [CrossRef] [PubMed]

145. Arvand, M.; Dehsaraei, M. A simple and efficient electrochemical sensor for folic acid determination in human blood plasma based on gold nanoparticles–modified carbon paste electrode. *Mat. Sci. Eng. C* **2013**, *33*, 3474–3480. [CrossRef] [PubMed]

146. Wang, X.; Gao, D.; Li, M.; Li, H.; Li, C.; Wu, X.; Yang, B. Cvd graphene as an electrochemical sensing platform for simultaneous detection of biomolecules. *Sci. Rep.* **2017**, *7*, 7044. [CrossRef] [PubMed]

147. Wild, S.; Roglic, G.; Green, A.; Sicree, R.; King, H. Global prevalence of diabetes: Estimates for the year 2000 and projections for 2030. *Diabetes Care* **2004**, *27*, 1047–1053. [CrossRef] [PubMed]

148. McGarraugh, G.; Brazg, R.; Weinstein, R. Freestyle navigator continuous glucose monitoring system with trustart algorithm, a 1-hour warm-up time. *J. Diabetes. Sci. Techonol.* **2011**, *5*, 99–106. [CrossRef] [PubMed]

149. Mamkin, I.; Ten, S.; Bhandari, S.; Ramchandani, N. Real-time continuous glucose monitoring in the clinical setting: The good, the bad, and the practical. *J. Diabetes Sci. Technol.* **2008**, *2*, 882–889. [CrossRef] [PubMed]

150. Torjman, M.C.; Dalal, N.; Goldberg, M.E. Glucose monitoring in acute care: Technologies on the horizon. *J. Diabetes Sci. Technol.* **2008**, *2*, 178–181. [CrossRef] [PubMed]

151. Lipani, L.; Dupont, B.G.R.; Doungmene, F.; Marken, F.; Tyrrell, R.M.; Guy, R.H.; Ilie, A. Non-invasive, transdermal, path-selective and specific glucose monitoring via a graphene-based platform. *Nat. Nanotechnol.* **2018**, *13*, 504–511. [CrossRef] [PubMed]

152. Xu, S.; Zhan, J.; Man, B.; Jiang, S.; Yue, W.; Gao, S.; Guo, C.; Liu, H.; Li, Z.; Wang, J.; et al. Real-time reliable determination of binding kinetics of DNA hybridization using a multi-channel graphene biosensor. *Nat. Commun.* **2017**, *8*, 14902. [CrossRef] [PubMed]

153. Barik, A.; Zhang, Y.; Grassi, R.; Nadappuram, B.P.; Edel, J.B.; Low, T.; Koester, S.J.; Oh, S.-H. Graphene-edge dielectrophoretic tweezers for trapping of biomolecules. *Nat. Commun.* **2017**, *8*, 1867. [CrossRef] [PubMed]

154. Seo, D.H.; Pineda, S.; Fang, J.; Gozukara, Y.; Yick, S.; Bendavid, A.; Lam, S.K.H.; Murdock, A.T.; Murphy, A.B.; Han, Z.J.; et al. Single-step ambient-air synthesis of graphene from renewable precursors as electrochemical genosensor. *Nat. Commun.* **2017**, *8*, 14217. [CrossRef] [PubMed]

155. World Health Organization. *WHO Guidelines for Drinking Water Quality*, 2nd ed.; World Health Organization: Geneva, Switzerland, 1996.

156. Zhang, W.; Jia, B.; Wang, Q.; Dionysiou, D. Visible-light sensitization of tio2 photocatalysts via wet chemical n-doping for the degradation of dissolved organic compounds in wastewater treatment: A review. *J. Nanopart. Res.* **2015**, *17*, 221. [CrossRef]

157. Zhang, W.; Zou, L.; Wang, L. Visible-light assisted methylene blue (mb) removal by novel tio2/adsorbent nanocomposites. *Water Sci. Technol.* **2010**, *61*, 2863–2871. [CrossRef] [PubMed]

158. Zhang, J.; Liu, X.; Neri, G.; Pinna, N. Nanostructured materials for room-temperature gas sensors. *Adv. Mater.* **2016**, *28*, 795–831. [CrossRef] [PubMed]

159. Wu, J.; Tao, K.; Miao, J.; Norford, L.K. Improved selectivity and sensitivity of gas sensing using a 3d reduced graphene oxide hydrogel with an integrated microheater. *ACS Appl. Mater. Interfaces* **2015**, *7*, 27502–27510.

160. Davidovikj, D.; Bouwmeester, D.; van der Zant, H.S.J.; Steeneken, P.G. Graphene gas pumps. *2D Mater.* **2018**, *5*, 031009. [CrossRef]

161. Wu, J.; Tao, K.; Guo, Y.; Li, Z.; Wang, X.; Luo, Z.; Feng, S.; Du, C.; Chen, D.; Miao, J.; et al. A 3d chemically modified graphene hydrogel for fast, highly sensitive, and selective gas sensor. *Adv. Sci.* **2016**, *4*, 1600319–1600319. [CrossRef] [PubMed]

162. Guo, L.; Hao, Y.-W.; Li, P.-L.; Song, J.-F.; Yang, R.-Z.; Fu, X.-Y.; Xie, S.-Y.; Zhao, J.; Zhang, Y.-L. Improved no2 gas sensing properties of graphene oxide reduced by two-beam-laser interference. *Sci. Rep.* **2018**, *8*, 4918. [CrossRef] [PubMed]

163. Favier, F.; Walter, E.C.; Zach, M.P.; Benter, T.; Penner, R.M. Hydrogen sensors and switches from electrodeposited palladium mesowire arrays. *Science* **2001**, *293*, 2227–2231. [CrossRef] [PubMed]

164. Sharma, B.; Sharma, A.; Kim, J.-S. Recent advances on h2 sensor technologies based on mox and fet devices: A review. *Sens. Actuators B Chem.* **2018**, *262*, 758–770. [CrossRef]

165. Sharma, B.; Kim, J.-S. Mems based highly sensitive dual fet gas sensor using graphene decorated pd-ag alloy nanoparticles for h2 detection. *Sci. Rep.* **2018**, *8*, 5902. [CrossRef] [PubMed]

166. Mortazavi Zanjani, S.M.; Holt, M.; Sadeghi, M.M.; Rahimi, S.; Akinwande, D. 3d integrated monolayer graphene–si cmos rf gas sensor platform. *NPJ 2D Mater. Appl.* **2017**, *1*, 36. [CrossRef]

167. Mehdi Pour, M.; Lashkov, A.; Radocea, A.; Liu, X.; Sun, T.; Lipatov, A.; Korlacki, R.A.; Shekhirev, M.; Aluru, N.R.; Lyding, J.W.; et al. Laterally extended atomically precise graphene nanoribbons with improved electrical conductivity for efficient gas sensing. *Nat. Commun.* **2017**, *8*, 820. [CrossRef] [PubMed]

168. Armstrong, S.; Hurley, W.G. A thermal model for photovoltaic panels under varying atmospheric conditions. *Appl. Therm. Eng.* **2010**, *30*, 1488–1495. [CrossRef]

169. Nicklow, R.; Wakabayashi, N.; Smith, H.G. Lattice dynamics of pyrolytic graphite. *Phys. Rev. B* **1972**, *5*, 4951–4962. [CrossRef]

170. Nihira, T.; Iwata, T. Temperature dependence of lattice vibrations and analysis of the specific heat of graphite. *Phys. Rev. B* **2003**, *68*, 134305. [CrossRef]

171. Pop, E.; Varshney, V.; Roy, A.K. Thermal properties of graphene: Fundamentals and applications. *MRS Bull.* **2012**, *37*, 1273–1281. [CrossRef]

172. Benedict, L.X.; Louie, S.G.; Cohen, M.L. Heat capacity of carbon nanotubes. *Solid State Commun.* **1996**, *100*, 177–180. [CrossRef]

173. Yi, W.; Lu, L.; Dian-lin, Z.; Pan, Z.W.; Xie, S.S. Linear specific heat of carbon nanotubes. *Phys. Rev. B* **1999**, *59*, R9015–R9018. [CrossRef]

174. Ong, Z.-Y.; Pop, E. Effect of substrate modes on thermal transport in supported graphene. *Phys. Rev. B* **2011**, *84*, 075471. [CrossRef]

175. Bonetto, F.; Lebowitz, J.L.; Rey-Bellet, L. Fourier's law: A challenge to theorists. *Math. Phys.* **2000**, 128–150.

176. Balandin, A.A.; Ghosh, S.; Bao, W.; Calizo, I.; Teweldebrhan, D.; Miao, F.; Lau, C.N. Superior thermal conductivity of single-layer graphene. *Nano Lett.* **2008**, *8*, 902–907. [CrossRef] [PubMed]

177. Woodcraft, A.L.; Barucci, M.; Hastings, P.R.; Lolli, L.; Martelli, V.; Risegari, L.; Ventura, G. Thermal conductivity measurements of pitch-bonded graphites at millikelvin temperatures: Finding a replacement for agot graphite. *Cryogenics* **2009**, *49*, 159–164. [CrossRef]

178. Calizo, I.; Balandin, A.A.; Bao, W.; Miao, F.; Lau, C.N. Temperature dependence of the raman spectra of graphene and graphene multilayers. *Nano Lett.* **2007**, *7*, 2645–2649. [CrossRef] [PubMed]

179. Shiomi, J.; Maruyama, S. Non-fourier heat conduction in a single-walled carbon nanotube: Classical molecular dynamics simulations. *Phys. Rev. B* **2006**, *73*, 205420. [CrossRef]

180. Calizo, I.; Miao, F.; Bao, W.; Lau, C.N.; Balandin, A.A. Variable temperature raman microscopy as a nanometrology tool for graphene layers and graphene-based devices. *Appl. Phys. Lett.* **2007**, *91*, 071913. [CrossRef]

181. Beams, R.; Gustavo Cançado, L.; Novotny, L. Raman characterization of defects and dopants in graphene. *J. Phys. Condens. Matter* **2015**, *27*, 083002. [CrossRef] [PubMed]

182. Xu, W.; Mao, N.; Zhang, J. Graphene: A platform for surface-enhanced raman spectroscopy. *Small* **2013**, *9*, 1206–1224. [CrossRef] [PubMed]

183. Merlen, A.; Buijnsters, J.G.; Pardanaud, C. A guide to and review of the use of multiwavelength raman spectroscopy for characterizing defective aromatic carbon solids: From graphene to amorphous carbons. *Coatings* **2017**, *7*, 153. [CrossRef]

184. Wei, P.; Bao, W.; Pu, Y.; Lau, C.N.; Shi, J. Anomalous thermoelectric transport of dirac particles in graphene. *Phys. Rev. Lett.* **2009**, *102*, 166808. [CrossRef] [PubMed]

185. Small, J.P.; Perez, K.M.; Kim, P. Modulation of thermoelectric power of individual carbon nanotubes. *Phys. Rev. Lett.* **2003**, *91*, 256801. [CrossRef] [PubMed]

186. Kim, H.S.; Liu, W.; Chen, G.; Chu, C.-W.; Ren, Z. Relationship between thermoelectric figure of merit and energy conversion efficiency. *Proc. Natl. Acad. Sci. USA* **2015**, *112*, 8205–8210. [CrossRef] [PubMed]

187. Savin, A.V.; Kivshar, Y.S.; Hu, B. Suppression of thermal conductivity in graphene nanoribbons with rough edges. *Phys. Rev. B* **2010**, *82*, 195422. [CrossRef]

188. Hu, J.; Schiffli, S.; Vallabhaneni, A.; Ruan, X.; Chen, Y.P. Tuning the thermal conductivity of graphene nanoribbons by edge passivation and isotope engineering: A molecular dynamics study. *Appl. Phys. Lett.* **2010**, *97*, 133107. [CrossRef]

189. Dollfus, P.; Hung Nguyen, V.; Saint-Martin, J. Thermoelectric effects in graphene nanostructures. *J. Phys. Condens. Matter* **2015**, *27*, 133204. [CrossRef] [PubMed]

190. Webb, R.C.; Bonifas, A.P.; Behnaz, A.; Zhang, Y.; Yu, K.J.; Cheng, H.; Shi, M.; Bian, Z.; Liu, Z.; Kim, Y.-S.; et al. Ultrathin conformal devices for precise and continuous thermal characterization of human skin. *Nat. Mater.* **2013**, *12*, 938. [CrossRef] [PubMed]

191. Son, D.; Lee, J.; Qiao, S.; Ghaffari, R.; Kim, J.; Lee, J.E.; Song, C.; Kim, S.J.; Lee, D.J.; Jun, S.W.; et al. Multifunctional wearable devices for diagnosis and therapy of movement disorders. *Nat. Nanotechnol.* **2014**, *9*, 397. [CrossRef] [PubMed]

192. Tao, H.; Kainerstorfer, J.M.; Siebert, S.M.; Pritchard, E.M.; Sassaroli, A.; Panilaitis, B.J.B.; Brenckle, M.A.; Amsden, J.J.; Levitt, J.; Fantini, S.; et al. Implantable, multifunctional, bioresorbable optics. *Proc. Natl. Acad. Sci. USA* **2012**, *109*, 19584–19589. [CrossRef] [PubMed]

193. Kim, J.; Lee, M.; Shim, H.J.; Ghaffari, R.; Cho, H.R.; Son, D.; Jung, Y.H.; Soh, M.; Choi, C.; Jung, S.; et al. Stretchable silicon nanoribbon electronics for skin prosthesis. *Nat. Commun.* **2014**, *5*, 5747. [CrossRef] [PubMed]

194. Chen, L.Y.; Tee, B.C.K.; Chortos, A.L.; Schwartz, G.; Tse, V.; Lipomi, D.J.; Wong, H.S.P.; McConnell, M.V.; Bao, Z. Continuous wireless pressure monitoring and mapping with ultra-small passive sensors for health monitoring and critical care. *Nat. Commun.* **2014**, *5*, 5028. [CrossRef] [PubMed]

195. Ho, D.H.; Sun, Q.; Kim, S.Y.; Han, J.T.; Kim, D.H.; Cho, J.H. Stretchable and multimodal all graphene electronic skin. *Adv. Mater.* **2016**, *28*, 2601–2608. [CrossRef] [PubMed]

196. Vuorinen, T.; Niittynen, J.; Kankkunen, T.; Kraft, T.M.; Mäntysalo, M. Inkjet-printed graphene/pedot:Pss temperature sensors on a skin-conformable polyurethane substrate. *Sci. Rep.* **2016**, *6*, 35289. [CrossRef] [PubMed]

197. Trung, T.Q.; Ramasundaram, S.; Hwang, B.-U.; Lee, N.-E. An all-elastomeric transparent and stretchable temperature sensor for body-attachable wearable electronics. *Adv. Mater.* **2016**, *28*, 502–509. [CrossRef] [PubMed]

198. Davaji, B.; Cho, H.D.; Malakoutian, M.; Lee, J.-K.; Panin, G.; Kang, T.W.; Lee, C.H. A patterned single layer graphene resistance temperature sensor. *Sci. Rep.* **2017**, *7*, 8811. [CrossRef] [PubMed]

199. Cao, Y.; Dong, H.; Yang, Z.; Zhong, X.; Chen, Y.; Dai, W.; Zhang, X. Aptamer-conjugated graphene quantum dots/porphyrin derivative theranostic agent for intracellular cancer-related microrna detection and fluorescence-guided photothermal/photodynamic synergetic therapy. *ACS Appl. Mater. Interfaces* **2017**, *9*, 159–166. [CrossRef] [PubMed]

200. Seo, H.; Yun, H.D.; Kwon, S.-Y.; Bang, I.C. Hybrid graphene and single-walled carbon nanotube films for enhanced phase-change heat transfer. *Nano Lett.* **2016**, *16*, 932–938. [CrossRef] [PubMed]

201. Tonouchi, M. Cutting-edge terahertz technology. *Nat. Photonics* **2007**, *1*, 97. [CrossRef]

202. Wai Lam, C.; Jason, D.; Daniel, M.M. Imaging with terahertz radiation. *Rep. Prog. Phys.* **2007**, *70*, 1325.

203. Cai, X.; Sushkov, A.B.; Suess, R.J.; Jadidi, M.M.; Jenkins, G.S.; Nyakiti, L.O.; Myers-Ward, R.L.; Li, S.; Yan, J.; Gaskill, D.K.; et al. Sensitive room-temperature terahertz detection via the photothermoelectric effect in graphene. *Nat. Nanotechnol.* **2014**, *9*, 814. [CrossRef] [PubMed]

204. Breusing, M.; Ropers, C.; Elsaesser, T. Ultrafast carrier dynamics in graphite. *Phys. Rev. Lett.* **2009**, *102*, 086809. [CrossRef] [PubMed]

205. Tse, W.-K.; Hwang, E.H.; Sarma, S.D. Ballistic hot electron transport in graphene. *Appl. Phys. Lett.* **2008**, *93*,, 023128. [CrossRef]

206. Lui, C.H.; Mak, K.F.; Shan, J.; Heinz, T.F. Ultrafast photoluminescence from graphene. *Phys. Rev. Lett.* **2010**, *105*, 127404. [CrossRef] [PubMed]

207. Gabor, N.M.; Song, J.C.W.; Ma, Q.; Nair, N.L.; Taychatanapat, T.; Watanabe, K.; Taniguchi, T.; Levitov, L.S.; Jarillo-Herrero, P. Hot carrier–assisted intrinsic photoresponse in graphene. *Science* **2011**, *334*, 648–652. [CrossRef] [PubMed]

208. Graham, M.W.; Shi, S.-F.; Ralph, D.C.; Park, J.; McEuen, P.L. Photocurrent measurements of supercollision cooling in graphene. *Nat. Phys.* **2012**, *9*, 103. [CrossRef]

209. Xia, F.; Mueller, T.; Golizadeh-Mojarad, R.; Freitag, M.; Lin, Y.-m.; Tsang, J.; Perebeinos, V.; Avouris, P. Photocurrent imaging and efficient photon detection in a graphene transistor. *Nano Lett.* **2009**, *9*, 1039–1044. [CrossRef] [PubMed]

210. Sizov, F.; Rogalski, A. Thz detectors. *Prog. Quantum Electron.* **2010**, *34*, 278–347. [CrossRef]

211. Kim, M.H.; Yan, J.; Suess, R.J.; Murphy, T.E.; Fuhrer, M.S.; Drew, H.D. Photothermal response in dual-gated bilayer graphene. *Phys. Rev. Lett.* **2013**, *110*, 247402. [CrossRef] [PubMed]

212. Gramotnev, D.K.; Bozhevolnyi, S.I. Plasmonics beyond the diffraction limit. *Nat. Photonics* **2010**, *4*, 83. [CrossRef]

213. Vakil, A.; Engheta, N. Transformation optics using graphene. *Science* **2011**, *332*, 1291–1294. [CrossRef] [PubMed]

214. Dyakonov, M.; Shur, M. Detection, mixing, and frequency multiplication of terahertz radiation by two-dimensional electronic fluid. *IEEE Trans. Electron Dev.* **1996**, *43*, 380–387. [CrossRef]

215. Wunsch, B.; Stauber, T.; Sols, F.; Guinea, F. Dynamical polarization of graphene at finite doping. *New J. Phys.* **2006**, *8*, 318. [CrossRef]

216. Hwang, E.H.; Das Sarma, S. Dielectric function, screening, and plasmons in two-dimensional graphene. *Phys. Rev. B* **2007**, *75*, 205418. [CrossRef]

217. Jablan, M.; Buljan, H.; Soljačić, M. Plasmonics in graphene at infrared frequencies. *Phys. Rev. B* **2009**, *80*, 245435. [CrossRef]

218. Grigorenko, A.N.; Polini, M.; Novoselov, K.S. Graphene plasmonics. *Nat. Photonics* **2012**, *6*, 749. [CrossRef]

219. Innes, R.A.; Sambles, J.R. Simple thermal detection of surface plasmon-polaritons. *Solid State Commun.* **1985**, *56*, 493–496. [CrossRef]

220. Weeber, J.-C.; Hassan, K.; Bouhelier, A.; Colas-des-Francs, G.; Arocas, J.; Markey, L.; Dereux, A. Thermo-electric detection of waveguided surface plasmon propagation. *Appl. Phys. Lett.* **2011**, *99*, 031113. [CrossRef]

221. Lundeberg, M.B.; Gao, Y.; Woessner, A.; Tan, C.; Alonso-González, P.; Watanabe, K.; Taniguchi, T.; Hone, J.; Hillenbrand, R.; Koppens, F.H.L. Thermoelectric detection and imaging of propagating graphene plasmons. *Nat. Mater.* **2016**, *16*, 204. [CrossRef] [PubMed]

222. Chen, G.; Roy, I.; Yang, C.; Prasad, P.N. Nanochemistry and nanomedicine for nanoparticle-based diagnostics and therapy. *Chem. Rev.* **2016**, *116*, 2826–2885. [CrossRef] [PubMed]

223. Ma, Y.; Huang, J.; Song, S.; Chen, H.; Zhang, Z. Cancer-targeted nanotheranostics: Recent advances and perspectives. *Small* **2016**, *12*, 4936–4954. [CrossRef] [PubMed]

224. Li, L.; Chen, C.; Liu, H.; Fu, C.; Tan, L.; Wang, S.; Fu, S.; Liu, X.; Meng, X.; Liu, H. Multifunctional carbon–silica nanocapsules with gold core for synergistic photothermal and chemo-cancer therapy under the guidance of bimodal imaging. *Adv. Funct. Mater.* **2016**, *26*, 4252–4261. [CrossRef]

225. He, Q.; Shi, J. Msn anti-cancer nanomedicines: Chemotherapy enhancement, overcoming of drug resistance, and metastasis inhibition. *Adv. Mater.* **2014**, *26*, 391–411. [CrossRef] [PubMed]

226. Yang, P.; Gai, S.; Lin, J. Functionalized mesoporous silica materials for controlled drug delivery. *Chem. Soc. Rev.* **2012**, *41*, 3679–3698. [CrossRef] [PubMed]

227. Ramachandra Kurup Sasikala, A.; Thomas, R.G.; Unnithan, A.R.; Saravanakumar, B.; Jeong, Y.Y.; Park, C.H.; Kim, C.S. Multifunctional nanocarpets for cancer theranostics: Remotely controlled graphene nanoheaters for thermo-chemosensitisation and magnetic resonance imaging. *Sci. Rep.* **2016**, *6*, 20543. [CrossRef] [PubMed]

228. Liu, J.; Detrembleur, C.; De Pauw-Gillet, M.-C.; Mornet, S.; Jérôme, C.; Duguet, E. Gold nanorods coated with mesoporous silica shell as drug delivery system for remote near infrared light-activated release and potential phototherapy. *Small* **2015**, *11*, 2323–2332. [CrossRef] [PubMed]

229. Xu, Q.; Leong, J.; Chua, Q.Y.; Chi, Y.T.; Chow, P.K.-H.; Pack, D.W.; Wang, C.-H. Combined modality doxorubicin-based chemotherapy and chitosan-mediated p53 gene therapy using double-walled microspheres for treatment of human hepatocellular carcinoma. *Biomaterials* **2013**, *34*, 5149–5162. [CrossRef] [PubMed]

230. Im, J.H.; Seong, J.; Lee, I.J.; Park, J.S.; Yoon, D.S.; Kim, K.S.; Lee, W.J.; Park, K.R. Surgery alone versus surgery followed by chemotherapy and radiotherapy in resected extrahepatic bile duct cancer: Treatment outcome analysis of 336 patients. *Cancer Res. Treat* **2016**, *48*, 583–595. [CrossRef] [PubMed]

231. Yao, X.; Niu, X.; Ma, K.; Huang, P.; Grothe, J.; Kaskel, S.; Zhu, Y. Graphene quantum dots-capped magnetic mesoporous silica nanoparticles as a multifunctional platform for controlled drug delivery, magnetic hyperthermia, and photothermal therapy. *Small* **2017**, *13*, 1602225. [CrossRef] [PubMed]

232. Zhu, Y.-F.; Shi, J.-L.; Li, Y.-S.; Chen, H.-R.; Shen, W.-H.; Dong, X.-P. Storage and release of ibuprofen drug molecules in hollow mesoporous silica spheres with modified pore surface. *Microporous Mesoporous Mater.* **2005**, *85*, 75–81. [CrossRef]

233. Wang, Y.; Li, B.; Zhang, L.; Song, H.; Zhang, L. Targeted delivery system based on magnetic mesoporous silica nanocomposites with light-controlled release character. *ACS Appl. Mater. Interfaces* **2013**, *5*, 11–15. [CrossRef] [PubMed]

234. Yang, B.; Wang, P.; Bar-Cohen, A. Mini-contact enhanced thermoelectric cooling of hot spots in high power devices. *IEEE Trans. Compon. Packag. Technol.* **2007**, *30*, 432–438. [CrossRef]

235. Wachutka, G.K. Rigorous thermodynamic treatment of heat generation and conduction in semiconductor device modeling. *IEEE Trans. Comput.-Aided Des. Integr. Circuits Syst.* **1990**, *9*, 1141–1149. [CrossRef]

236. Patel, H.E.; Das, S.K.; Sundararajan, T.; Nair, A.S.; George, B.; Pradeep, T. Thermal conductivities of naked and monolayer protected metal nanoparticle based nanofluids: Manifestation of anomalous enhancement and chemical effects. *Appl. Phys. Lett.* **2003**, *83*, 2931–2933. [CrossRef]

237. Nath, P.; Chopra, K.L. Thermal conductivity of copper films. *Thin Solid Films* **1974**, *20*, 53–62. [CrossRef]

238. Langer, G.; Hartmann, J.; Reichling, M. Thermal conductivity of thin metallic films measured by photothermal profile analysis. *Rev. Sci. Instrum.* **1997**, *68*, 1510–1513. [CrossRef]

239. Han, H.; Zhang, Y.; Wang, N.; Samani, M.K.; Ni, Y.; Mijbil, Z.Y.; Edwards, M.; Xiong, S.; Sääskilahti, K.; Murugesan, M.; et al. Functionalization mediates heat transport in graphene nanoflakes. *Nat. Commun.* **2016**, *7*, 11281. [CrossRef] [PubMed]

240. Kumar, P.; Shahzad, F.; Yu, S.; Hong, S.M.; Kim, Y.-H.; Koo, C.M. Large-area reduced graphene oxide thin film with excellent thermal conductivity and electromagnetic interference shielding effectiveness. *Carbon* **2015**, *94*, 494–500. [CrossRef]

241. Bernal, M.M.; Di Pierro, A.; Novara, C.; Giorgis, F.; Mortazavi, B.; Saracco, G.; Fina, A. Edge-grafted molecular junctions between graphene nanoplatelets: Applied chemistry to enhance heat transfer in nanomaterials. *Adv. Funct. Mater.* **2018**, *28*, 1706954. [CrossRef]

242. Ma, T.; Liu, Z.; Wen, J.; Gao, Y.; Ren, X.; Chen, H.; Jin, C.; Ma, X.-L.; Xu, N.; Cheng, H.-M.; et al. Tailoring the thermal and electrical transport properties of graphene films by grain size engineering. *Nat. Commun.* **2017**, *8*, 14486. [CrossRef] [PubMed]

243. Ding, J.; Zhao, H.; Wang, Q.; Dou, H.; Chen, H.; Yu, H. An ultrahigh thermal conductive graphene flexible paper. *Nanoscale* **2017**, *9*, 16871–16878. [CrossRef] [PubMed]

244. Pérez-Lombard, L.; Ortiz, J.; Pout, C. A review on buildings energy consumption information. *Energy Build.* **2008**, *40*, 394–398. [CrossRef]

245. Schneider, S.H. The greenhouse effect: Science and policy. *Science* **1989**, *243*, 771–781. [CrossRef] [PubMed]

246. Wark, K.; Warner, C.F. *Air Pollution: Its Origin and Control*; Harper and Row Publishers: New York, NY, USA, 1981; 526p.

247. Worrell, E.; Bernstein, L.; Roy, J.; Price, L.; Harnisch, J. Industrial energy efficiency and climate change mitigation. *Energy Effic.* **2008**, *2*, 109. [CrossRef]

248. Cho, C.; Wallace, K.L.; Tzeng, P.; Hsu, J.-H.; Yu, C.; Grunlan, J.C. Outstanding low temperature thermoelectric power factor from completely organic thin films enabled by multidimensional conjugated nanomaterials. *Adv. Mater.* **2016**, *6*, 1502168. [CrossRef]

249. Cai, W.; Moore, A.L.; Zhu, Y.; Li, X.; Chen, S.; Shi, L.; Ruoff, R.S. Thermal transport in suspended and supported monolayer graphene grown by chemical vapor deposition. *Nano Lett.* **2010**, *10*, 1645–1651. [CrossRef] [PubMed]

250. Faugeras, C.; Faugeras, B.; Orlita, M.; Potemski, M.; Nair, R.R.; Geim, A.K. Thermal conductivity of graphene in corbino membrane geometry. *ACS Nano* **2010**, *4*, 1889–1892. [CrossRef] [PubMed]

251. Kim, K.K.; Bae, J.J.; Park, H.K.; Kim, S.M.; Geng, H.-Z.; Park, K.A.; Shin, H.-J.; Yoon, S.-M.; Benayad, A.; Choi, J.-Y.; et al. Fermi level engineering of single-walled carbon nanotubes by aucl3 doping. *J. Am. Chem. Soc.* **2008**, *130*, 12757–12761. [CrossRef] [PubMed]

252. Giovannetti, G.; Khomyakov, P.A.; Brocks, G.; Karpan, V.M.; van den Brink, J.; Kelly, P.J. Doping graphene with metal contacts. *Phys. Rev. Lett.* **2008**, *101*, 026803. [CrossRef] [PubMed]

253. Lin, Y.; Norman, C.; Srivastava, D.; Azough, F.; Wang, L.; Robbins, M.; Simpson, K.; Freer, R.; Kinloch, I.A. Thermoelectric power generation from lanthanum strontium titanium oxide at room temperature through the addition of graphene. *ACS Appl. Mater. Interfaces* **2015**, *7*, 15898–15908. [CrossRef] [PubMed]

 *nanomaterials* 

Article

Graphene Nanoplatelet-Reinforced Poly(vinylidene fluoride)/High Density Polyethylene Blend-Based Nanocomposites with Enhanced Thermal and Electrical Properties

**Kartik Behera [1], Mithilesh Yadav [1], Fang-Chyou Chiu [1,2,* and Kyong Yop Rhee [3,*

[1] Department of Chemical and Materials Engineering, Chang Gung University, Taoyuan 333, Taiwan; b.kartik1991@gmail.com (K.B.); dryadavin@gmail.com (M.Y.)

[2] Department of General Dentistry, Chang Gung Memorial Hospital, Taoyuan 333, Taiwan

[3] Department of Mechanical Engineering, College of Engineering, Kyung Hee University, Yongin 446-701, Korea

* Correspondence: maxson@mail.cgu.edu.tw (F.-C.C.); rheeky@khu.ac.kr (K.Y.R.); Tel.: +886-953678628 (F.-C.C.); +82-31-201-2565 (K.Y.R.)

Received: 29 January 2019; Accepted: 25 February 2019; Published: 4 March 2019

Abstract: In this study, a graphene nanoplatelet (GNP) was used as a reinforcing filler to prepare poly(vinylidene fluoride) (PVDF)/high density polyethylene (HDPE) blend-based nanocomposites through a melt mixing method. Scanning electron microscopy confirmed that the GNP was mainly distributed within the PVDF matrix phase. X-ray diffraction analysis showed that PVDF and HDPE retained their crystal structure in the blend and composites. Thermogravimetric analysis showed that the addition of GNP enhanced the thermal stability of the blend, which was more evident in a nitrogen environment than in an air environment. Differential scanning calorimetry results showed that GNP facilitated the nucleation of PVDF and HDPE in the composites upon crystallization. The activation energy for non-isothermal crystallization of PVDF increased with increasing GNP loading in the composites. The Avrami n values ranged from 1.9–3.8 for isothermal crystallization of PVDF in different samples. The Young's and flexural moduli of the blend improved by more than 20% at 2 phr GNP loading in the composites. The measured rheological properties confirmed the formation of a pseudo-network structure of GNP-PVDF in the composites. The electrical resistivity of the blend reduced by three orders at a 3-phr GNP loading. The PVDF/HDPE blend and composites showed interesting application prospects for electromechanical devices and capacitors.

Keywords: PVDF; HDPE; graphene nanoplatelet; nanocomposites; electrical properties; thermal properties

---

1. Introduction

In the past two decades, special attention has been paid to polymeric blend-based nanocomposites, because these systems benefit from the advantages of both blends and nanocomposites [1,2]. Various combinations of polymer matrices and a small amount of nanofillers have been designed to study their potential in improving the chemical and physical properties of neat polymers. The one-dimensional carbon nanotube (CNT) and two-dimensional nanoclays are two of the most studied nanofillers. Reviews on CNT-based polymer nanocomposites have been updated [3–6]. Additionally, graphene and its derivatives have also been recognized as appropriate nanofillers in fabricating polymer nanocomposites to improve the properties of parent polymers [7–10]. Two-dimensional graphene or graphene nanoplatelets reveal a 1 TPa Young's modulus, a 130 GPa ultimate strength, and a high electrical conductivity of 6000 S/cm [11]. Similar to CNT-based nanocomposites, the interfacial adhesion between graphene-polymers and the dispersion of graphene throughout the polymers are the key factors in achieving nanocomposites with an advanced performance [6]. Moreover, dispersion and distribution of the fillers are strongly affected by the melt-mixing methods.

Poly(vinylidene fluoride) (PVDF), a crystalline thermoplastic polymer, shows excellent properties, such as a high mechanical strength, worthy dielectric properties, and remarkable thermal/chemical stability. The disadvantages of PVDF include its high cost and low production volume. The five crystalline polymorphs (α, β, γ, δ, and ε) of PVDF have attracted much academic attention [12,13]. The non-polar α-form with trans-gauche-trans-gauche (TG^+TG^-) linkage conformation is the most stable and encountered polymorph. The polar β-form crystal with all-trans (TTTT) zig-zag conformation is the most attractive polymorph because of its unique piezo- and pyroelectric properties, which allow PVDF applications in sensor and actuators. PVDF-based blends and nanocomposites have been investigated for extending the versatility of PVDF [14–20]. Mago et al. [21] reported that the crystal size of PVDF decreases with increasing CNTs content, whereas its rate of crystallization increases with increasing β-form crystallization. Almasri et al. [20] investigated PVDF/double-walled carbon nanotube (DWNT) nanocomposite films fabricated by the solvent casting method. They found that the storage modulus of PVDF evidently increased by 48% below T_g and increased by up to 85% above it, and the percolation threshold at the addition of 0.23 vol.% DWNTs achieved electrical conductivity. Martins et al. [22] demonstrated that the electrical and rheological percolation threshold was achieved at 1.2 and 0.9 wt.% MWCNT loading, respectively. They showed that a 0.5 wt.% MWCNT nanocomposite revealed a uniform dispersion throughout the PVDF matrix, whereas a percolated network started to form at 1 wt.%. Wang et al. [15] used quaternary phosphorus salt to physically modify graphene and achieved an excellent dispersion of graphene within the PVDF matrix. Moreover, the electrical percolation threshold was achieved at 0.662 wt.%. The nanocomposite material films displayed tremendous electric and dielectric properties, and the salt modifier induced β- and γ-form crystals. Blend-based nanocomposites have received academic and industrial interest because of their potential to exhibit superior properties. Polyethylene (PE) is an important thermoplastic with versatile properties, which has been widely used in the agricultural, automotive, and packaging industry due to its low cost, easy processability, good mechanical/thermal properties, insulation capability, resistance to chemical solvents, and biological attack [23–26]. Its use has repetitively grown in the plastic industry. However, certain problems exist, caused by the inherent insulation of polymers, thus restricting all application fields, such as the material surface trend for the easy accumulation of charges trapping dust and the deterioration of product performance, possibly resulting in an explosion. Only a few studies have been reported on PVDF/PE blend-based nanocomposites [27–29]. Blending of PVDF with a suitable polymer such as PE can be an effective strategy to overcome its drawbacks. The blend of PVDF/HDPE and their conductive composites is especially cost effective and used in micro-electromechanical devices and high-charge storage capacitors. Chiu [30] used CNT, GNP, and organo-montmorillonite (Cloisite 15A) as reinforcing fillers to prepare PVDF/polycarbonate (PC) blend-based nanocomposites. He found that fillers were selectively located in the minor PC phase. The localization of some parts of the Cloisite 15A filler at the interface of the PVDF/PC blend facilitated the crystallization of PVDF and further induced the formation of β-form PVDF. Moreover, Chiu et al. [31] compared the thermal, mechanical, and electrical properties of PVDF/GNP nanocomposites and PVDF/PMMA/GNP blend-nanocomposites. They reported that GNP had a higher nucleation effect on crystallization of the PVDF in ternary composites compared with binary composites. The electrical percolation threshold was achieved at 1–2 phr GNP loading for the two composite systems, whereas ternary composites showed a lower electrical resistivity at identical GNP loadings. Rafei et al. [27] examined the morphology, rheological properties, and electrical conductivity of PVDF/PE/GNP ternary nanocomposites. They reported that double percolation was predicted for the PVDF/PE blend containing 0.9 wt.% GNP and was confirmed by direct electron microscopy and conductivity analysis.

From academic and industrial viewpoints, the influences of the incorporation of GNP on the various properties of PVDF should be thoroughly investigated. GNP-loaded PVDF/HDPE blend-based nanocomposites have been less reported. In the current study, commercialized GNP served as the nanofiller in preparing PVDF/HDPE blend-based nanocomposites through a melt-mixing method. It aimed to investigate the morphology and the resulting physical properties of the prepared samples. The thermal properties (including crystallization/melting behavior and thermal stability) of neat components, a blend, and composites were compared and reported. The mechanical and rheological properties and electrical resistivity of the samples were also characterized.

2. Materials and Methods

All materials used in this study are commercially available. PVDF with an average molecular weight of 1.8×10^5 g/mol in pellet form (Kynar 710, Arkema, Colombes, France) was used as the major component in the blend/composites. Commercial grade HDPE (Taisox 8050, MFR-6.0 g/10 min, density of 0.96 g/cm^3) was supplied by Formosa Plastic Corporation, Taoyuan, Taiwan. The graphene nanoplatelet (xGNP M-15, denoted as GNP), obtained from XG Sciences, Inc. (Lansing, MI, USA), was used as the nanofiller for composites fabrication. The components were pre-dried in a vacuum oven at 70 °C for 24 h to eliminate the absorbed moisture and then dry-mixed at given ratios before being fed into the mixer. All blend/composites samples were mixed using a Thermo Haake PolyDrive mixer (R600, Karlsruhe, Germany) at 190 °C for 10 min with a rotor speed of 60 rpm. The PVDF/HDPE blend was prepared with the weight ratio of 70:30. GNP was loaded at concentrations of 0.5–3 phr into the blend to prepare the composites. Neat PVDF and HDPE were also melt-treated for the purpose of comparison. The sample designation of F7E3 represents the PVDF/HDPE-70/30 blend. The codes of F7E3-# represent the # phr (1 phr = 0.99 wt.%, 2 phr = 1.96 wt.%, and 3 phr = 2.91 wt.%) of GNP loaded in the composites. To attain different shapes of specimens for further characterizations, the melt-mixed samples were consequently compression-molded at 190 °C for 3 min.

The fractured surface morphology of the blend and composites was examined using a field emission scanning electron microscope FESEM, Jeol JSM-7500F (Akishima, Tokyo, Japan). The cryo-fractured (in liquid nitrogen for 3 min) specimens were sputter-coated with gold prior to observation. Thermogravimetric analysis (TGA) was conducted by using a TA Q50 analyzer (TA instrument, New Castle, DE, USA) under air and nitrogen environments at a 10 °C/min heating rate. The crystallization kinetics and the resulting melting behavior of the samples were studied using a differential scanning calorimeter (DSC, TA instrument, New Castle, DE, USA) TA Q10. For non-isothermal crystallization experiments, the samples were first melted at 210 °C for 2 min to erase the thermal history. The samples were then cooled at various rates to 20 °C. The cooled samples were subsequently heated to 210 °C at 10 °C/min to evaluate their melting behavior. For isothermal crystallization kinetics study, samples were rapidly cooled at a rate of 100 °C/min (from 210 °C) to the pre-set crystallization temperatures (T_c), and the crystallized samples were then heated at 10 °C/min to evaluate the melting behavior. The crystal structures of PVDF and HDPE in the blend/composites were investigated using an X-ray diffractometer (AXS, D2, Bruker, Karlsruhe, Germany) operating at 40 kV and 40 mA. The X-ray source was CuK$_\alpha$ radiation with a wavelength of 1.54 Å.

Fourier-transform infrared spectroscopy FTIR spectra of the samples were recorded using a BRUKER, TENSOR 27 IR spectrometer (Karlsruhe, Germany). The samples were analyzed within the range of 700–2500 cm^{-1}. A Raman spectrometer (UniDRON, C. L. Technology Co., Ltd., New Taipei City, Taiwan) with a 532 nm laser excitation wavelength was used. All samples were compression molded into films at 200 °C and 100 bar. The samples were cooled to room temperature in the mold under pressure. The rheological properties of the samples were analyzed at 190 °C by using an Anton Paar Physica rheometer (MCR 101, Graz, Austria) in an oscillating mode. A parallel plate geometry with a 25 mm diameter and 1 mm gap was used, and the strain amplitude was set at 1%. The tensile properties of the compression-molded specimens (according to ASTM D638) were measured at a crosshead speed of 5 mm/min with a Gotech AI-3000 system (Taichung, Taiwan). The flexural modulus of the samples (according to ASTM D790) was measured using the same Gotech AI-3000 system at a cross-head speed of 5 mm/min. Izod impact tests (according to ASTM D256) were performed using a CEAST impact tester (Taichung, Taiwan). The reported data was expressed as the average of at least five specimens of the same formulation. The volume electrical resistivity of each specimen was measured using commercial resistivity meters (MCP-HT450 and MCP-T600, Mitsubishi Chemical Co., Yamato, Japan) at room temperature. The resistivity data was expressed as the average value of four repeated measurements at different positions.

3. Results

3.1. Phase Morphology and Selective Localization of GNP

The cryo-fractured surfaces of the prepared blend and blend-based composites were investigated using FESEM, as shown in Figure 1. Figure 1a depicts the F7E3 (PVDF70/HDPE30) blend, which evidently exhibited a bi-phasic (matrix-dispersed domain) morphology, indicating its immiscible character. The dispersed HDPE phase (minor component) exhibited a distribution of spherical domain size within the PVDF matrix phase, and the domain ranged from 8–30 μm. The size of domains was measured by using eight domains. The displayed piercing boundary clearly indicated no specific interaction between the PVDF and HDPE phases. Figure 1b–d depict the images of GNP-added representative composites. The domain size of HDPE decreased after the addition of GNP, and the size further decreased with increasing GNP loading. The added-GNP might have prevented the aggregation of HDPE domains forming large HDPE domains. Notably, the HDPE domains had a continuous-like morphology in the composites, which was more evident at high GNP loading. The GNP (arrowed) was mainly located within the PVDF matrix, which suggested that the interaction between PVDF and GNP was greater than that of HDPE and GNP. Thus, GNP thermodynamically prefers the PVDF phase. The inset Figure 1b clearly reveals the dispersion of GNP in the PVDF matrix. Additionally, the localization of GNP in the PVDF matrix caused the increased viscosity of the composites at the mixing temperature (see rheological data). According to the morphological observations, GNP was well-dispersed in the PVDF matrix phase, thereby confirming the achievement of blend-based nanocomposites. The thermal and mechanical properties of the F7E3 blend were improved after forming the nanocomposites. The results are discussed in the following sections.

The selective localization of filler in the immiscible polymer blend was executed by kinetic and thermodynamic factors. Sumita et al. [32] reported that through the wetting coefficient (ω) from Young's equation, the GNP localization could be predicted by the following equation in equilibrium:

$$\omega = \frac{\gamma_{\text{HDPE–GNP}} - \gamma_{\text{PVDF–GNP}}}{\gamma_{\text{PVDF–HDPE}}} \tag{1}$$

where $\gamma_{\text{PVDF-GNP}}$, $\gamma_{\text{HDPE-GNP}}$, and $\gamma_{\text{PVDF-HDPE}}$ are the interfacial tension between PVDF and GNP, HDPE and GNP, and PVDF and HDPE phases, respectively. If $\omega > 1$, the GNP is localized in the HDPE phase. If $-1 < \omega < 1$, GNP will be located at the interface of the two polymer phases. If $\omega < -1$, GNP will preferentially reside in the PVDF phase. The values of interfacial tension between different polymers can be evaluated from their surface energies based on the Harmonic-mean equation and the Geometric-mean equation:

$$\gamma_{12} = \gamma_1 + \gamma_2 - 4\left(\frac{\gamma_1^d \gamma_2^d}{\gamma_1^d + \gamma_2^d} + \frac{\gamma_1^p \gamma_2^p}{\gamma_1^p + \gamma_2^p}\right) \tag{2}$$

$$\gamma_{12} = \gamma_1 + \gamma_2 - 2\left(\sqrt{\gamma_1^d \gamma_2^d} + \sqrt{\gamma_1^p \gamma_2^p}\right) \tag{3}$$

where, γ^d and γ^p stand for the dispersive and polar parts of surface tension, respectively. The provided surface tension values of PVDF, HDPE, and GNP in Table 1 were followed by reported data [27,33,34]. The interfacial tension at 190 °C between the components was calculated using Equations (2) and (3) (see Table 2). According to the theoretical prediction by Equation (1), the ω values calculated by the harmonic-mean and geometric-mean equations were -1.37 and -1.17, respectively (Table 2), confirming the dispersion of GNP within the PVDF matrix during melt-mixing. These theoretical results agree with the SEM results, as discussed in the above section.

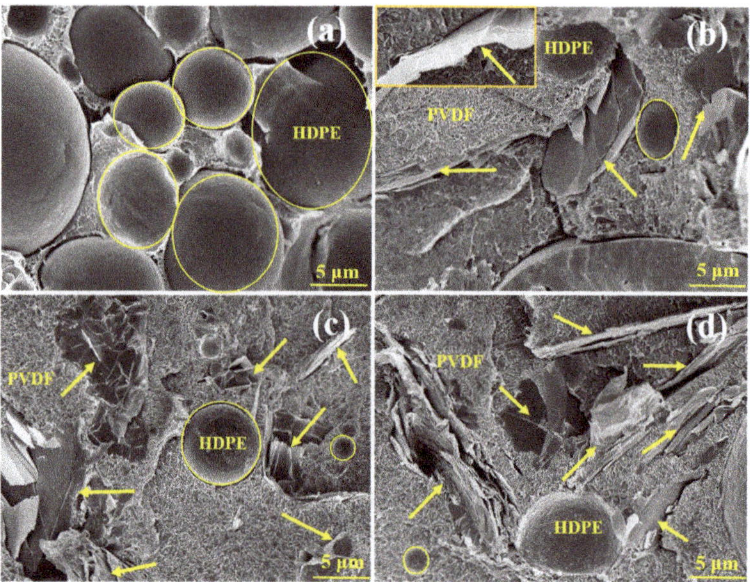

Figure 1. Scanning electron microscope (SEM) images of selected samples: (**a**) F7E3, (**b**) F7E3-1 (higher magnification image included in the inset), (**c**) F7E3-2, and (**d**) F7E3-3.

Table 1. Surface tension data of PVDF, HDPE, and GNP at 190 °C.

Samples	γ (mN m^{-1})	γ^d (mN m^{-1})	γ^p (mN m^{-1})	References
PVDF	38.0	32.6	5.4	[27]
HDPE	25.9	25.9	0	[33]
GNP	52.6	47.7	4.9	[34]

Table 2. Interfacial tension and calculated wetting coefficient of composites at 190 °C.

Samples	$\gamma_{PVDF\text{-}GNP}$	$\gamma_{HDPE\text{-}GNP}$	$\gamma_{PVDF\text{-}HDPE}$	ω (GNP)
Harmonic-mean Equation (2)	2.90	11.35	6.17	−1.37
Geometric-mean Equation (3)	1.45	8.20	5.78	−1.17

3.2. Thermal Stability

The TGA curves of the selected samples scanned under N_2 and air environments are shown in Figure 2. Figure 2a shows that neat HDPE displayed an evidently lower degradation temperature than PVDF under an N_2 environment. The temperature at 10% weight loss (T_{d10}) was 451 and 422 °C for PVDF and HDPE, respectively. The degradation marginally shifted to a higher temperature for the composites compared with that of the blend, and a higher GNP loading led to a higher degradation temperature. The thermal stability increased for the GNP-added composites, which was attributed to the high aspect ratio of GNP that served as a barrier and then prevented the emission of gaseous molecules during the thermal degradation. In addition, the radical scavenging function of GNP could inhibit the degradation process of the organic polymers. The GNP enhanced the thermal stability in some reported nanocomposite system [35]. The decomposition curves of the samples scanned under an air environment are compared in Figure 2b. The samples scanned in an N_2 environment exhibited higher degradation temperatures compared with those scanned in air because of the auto-oxidation reaction (due to the presence of oxygen). The neat PVDF again exhibited a higher degradation temperature than that of HDPE. The T_{d10} was 446 °C for PVDF and 376 °C for HDPE under an air environment. As anticipated, the F7E3 blend curve lied between those of PVDF and HDPE. After the formation of the blend and composites, two-step degradation corresponding to the individual degradation of PVDF and HDPE was observed. The composites displayed an improved degradation behavior compared to that of the F7E3 blend. The increase in thermal stability was demonstrated more evidently in the PVDF phase, because GNP was located in the PVDF phase (see above FESEM images). The temperatures at 10 and 50 wt.% loss (T_{d10} and T_{d50}) of the samples scanned under both air and N_2 environments are listed in Table 3. For the blend and composites, T_{d10} is mainly associated with the degradation of HDPE and T_{d50} corresponds to the degradation of PVDF.

Table 3. Thermogravimetric analysis (TGA) and differential scanning calorimeter (DSC) data of selected samples.

Samples	Properties						
	(T_{d10}) [a] (°C)	(T_{d50}) [a] (°C)	(T_{d10}) [b] (°C)	(T_{d50}) [b] (°C)	(T_p) [c] (°C)	(T_m) [d] (°C)	ΔE_c (kJ/mol)
PVDF	451	476	446	471	138.1	169.2	−303
HDPE	422	461	376	411	118.2	135.0	−153
F7E3	447	480	421	464	138.2	167.7	−326
F7E3-1	457	481	427	475	140.6	167.9	−330
F7E3-2	458	484	435	476	140.1	168.4	−363
F7E3-3	463	491	449	489	140.7	168.5	−374

[a] in N_2; [b] in air; [c] 5 °C/min cooling; [d] 5 °C/min pre-cooled.

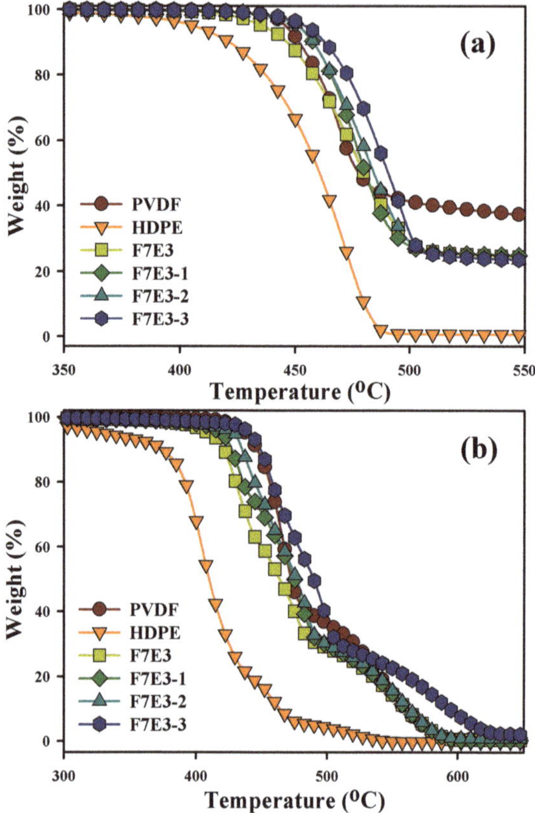

Figure 2. Thermogravimetric analysis (TGA)-scanned curves of selected samples under (**a**) N_2 environment and (**b**) air environment.

3.3. Crystallization Kinetics

Figure 3a shows the DSC cooling thermograms (from the melt state) of the prepared samples at 5 °C/min. Neat PVDF and HDPE showed their crystallization peak temperatures (T_ps, temperature at the exotherm minimum) at 138.1 and 118.2 °C, respectively. After blending with each other, the T_p value of PVDF remained unchanged and the crystallization temperature of HDPE slightly decreased. Two further observations can be made from this figure. First, the addition of GNP into the blend increased the T_p of individual PVDF and HDPE, and a higher GNP loading increased the T_p value. This observation suggests that the GNP acted as the nucleation agent for PVDF and HDPE crystallization. The determined T_p values of different samples are listed in Table 3. The formulation-dependent crystallization behavior of PVDF and HDPE at 5 °C/min-cooling was similarly observed in the samples cooled at 40 °C/min (Figure 3b). Lower T_p values were displayed for the samples cooled at a faster rate, mainly due to the thermal lag effect. For kinetic analysis, the crystallization of PVDF in the selected samples under different cooling rates was examined, as exhibited in Figure 3c–e. The T_p value decreased with the increased cooling rate for all samples, as anticipated. The GNP-added composites showed higher T_p values compared with the blend at each cooling rate. For example, F7E3 showed a T_p value ca. 141.6 °C at 2.5 °C/min cooling, whereas F7E3-3 had a T_p value ca. 142.7 °C at the same cooling rate. The activation energy of PVDF was studied due to its major component. The activation energy (ΔE_c) for the non-isothermal crystallization of PVDF in different samples (PVDF used as a major component in the whole experiment) was calculated using the Kissinger equation [36–38] (Equation (4)), as the typical plots shown in Figure 3f.

$$\left(\frac{\beta}{T_p{}^2}\right) = \text{Const} - \frac{\Delta E_c}{RT_p} \tag{4}$$

where T_p, R, and β stand for the crystallization peak temperature, universal gas constant, and cooling rate, respectively. The ΔE_c values of the representative samples are summarized in Table 3. Neat PVDF showed an absolute ΔE_c value of 303 kJ/mol, while the F7E3 blend showed ΔE_c values higher than that of neat PVDF. The composites showed that absolute ΔE_c increased with increasing GNP content in the blend. This result suggests that although GNP accelerated the nucleation of PVDF, it retarded the subsequent crystal growth of PVDF during the non-isothermal crystallization process. The isothermal crystallization kinetics of the selected samples was also analyzed. Figure 4a–d depict the DSC isothermal crystallization curves of the representative samples at different T_cs. According to the nucleation-controlled crystal growth theory, the shift in crystallization exothermic peak time from a lower to higher value with increasing T_c for individual samples was shown. The plots in Figure 4e show the reciprocal of the crystallization peak time (t_p) as a function of T_c. The reciprocal value ($t_p{}^{-1}$) was proportional to the overall isothermal crystallization rate. These results specified three pieces of information: First, the crystallization rate decreased with increasing T_c for the individual sample; second, the F7E3 blend revealed a lower overall crystallization rate than neat PVDF at identical T_c; and third, the composites (i.e., F7E3-1 and F7E3-3) crystallized at a faster rate than the blend at an identical T_c, due to the GNP nucleation effect. The Avrami equation (Equation (5)) [39] was used to evaluate the isothermal crystallization kinetics of PVDF in different samples based on the X_t (crystallinity at time t) versus t data.

$$X_t = 1 - \exp(-Kt^n) \tag{5}$$

where K is the crystallization rate constant and n is the Avrami exponent, depending on the type of nucleation and crystal growth geometry. Equation (5) can be transformed into the following equation:

$$\ln[-\ln(1 - X_t)] = \ln K + n \ln t \tag{6}$$

Figure 5a–d show the Avrami plots of the representative samples. The n and K values were determined from the linear region of the plots (slope and intercept). The n value provides qualitative information on the nature of nucleation and crystal growth geometry. The n value for PVDF ranged from 1.9–3.8 in different samples and decreased with increasing T_c, as shown in Figure 5. Furthermore, n decreased slightly after forming the composites. The decreased n values in the composites might indicate that nucleation changed from a thermal to athermal type, which was associated with the nucleation effect of GNP. The growth of PVDF crystals might also have changed from two-dimensional (at high T_cs) to three-dimensional (at low T_cs). The K value was noted to decrease with increasing T_c (not shown for brevity), corresponding to the $t_p{}^{-1}$ results.

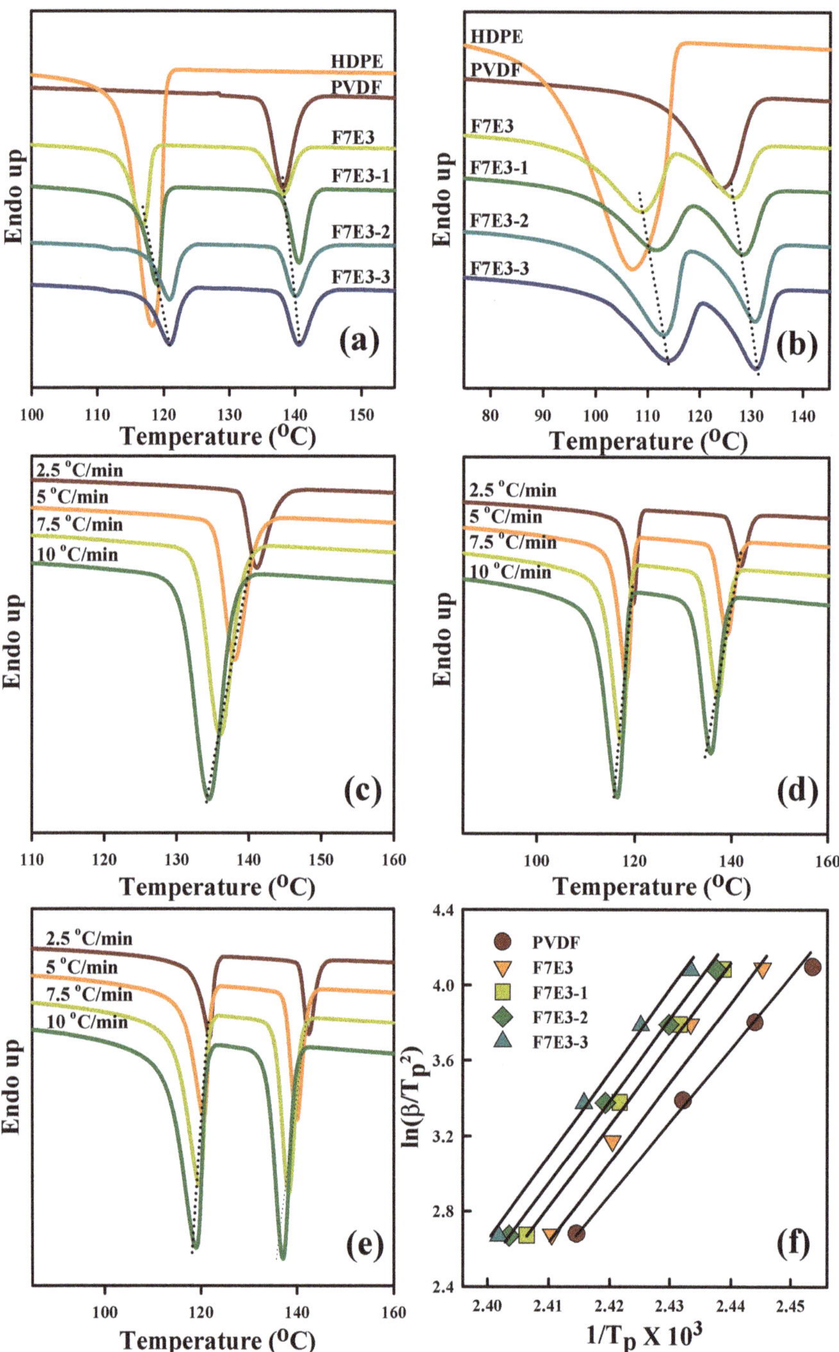

Figure 3. Differential scanning calorimeter (DSC) curves of the selected samples at (**a**) 5 °C/min cooling from the melt, (**b**) 40 °C/min cooling from the melt; DSC cooling curves of (**c**) PVDF, (**d**) F7E3, and (**e**) F7E3-3 at various rates; (**f**) Kissinger plots of selected samples for calculation of crystallization activation energy.

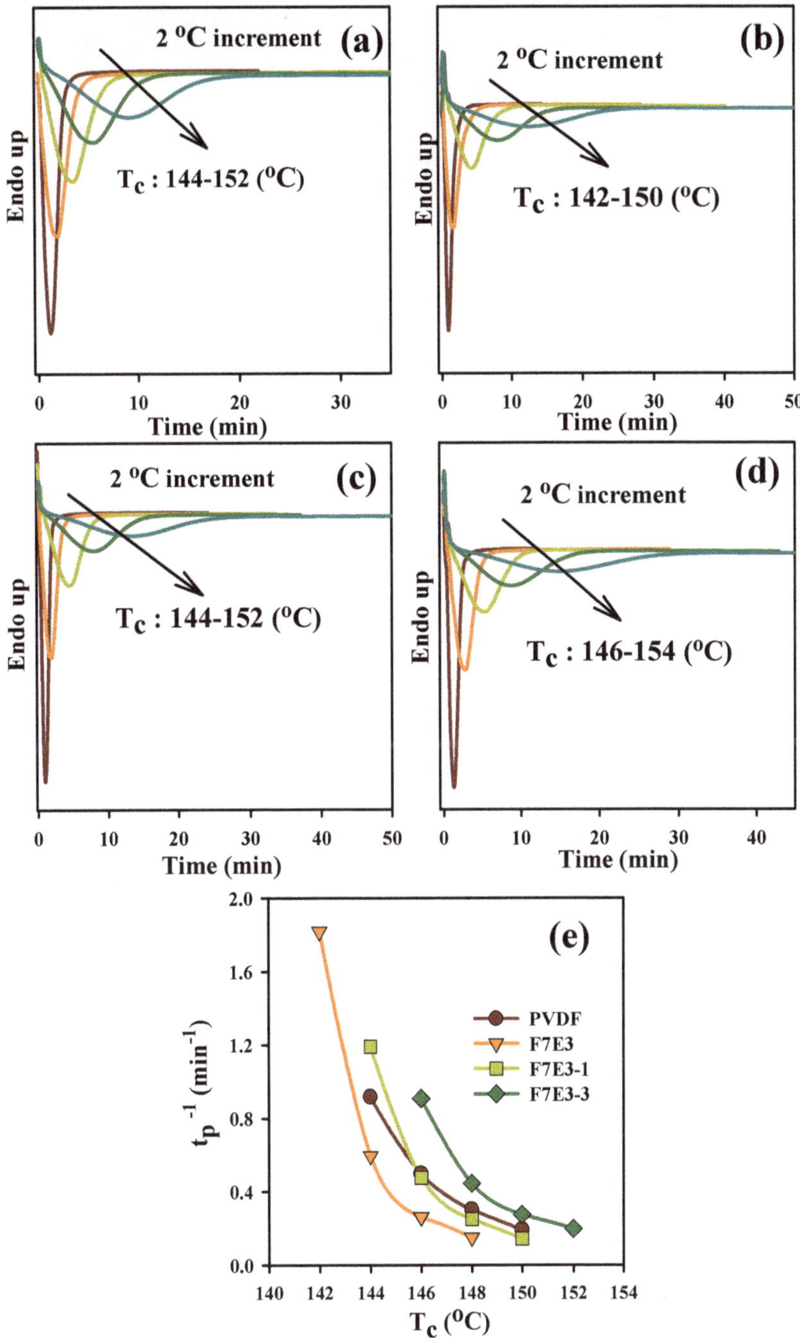

Figure 4. DSC traces of PVDF isothermally crystallized at indicated T_cs in different samples: (**a**) PVDF, (**b**) F7E3, (**c**) F7E3-1, (**d**) F7E3-3, and (**e**) t_p^{-1} vs. T_c of PVDF in representative samples.

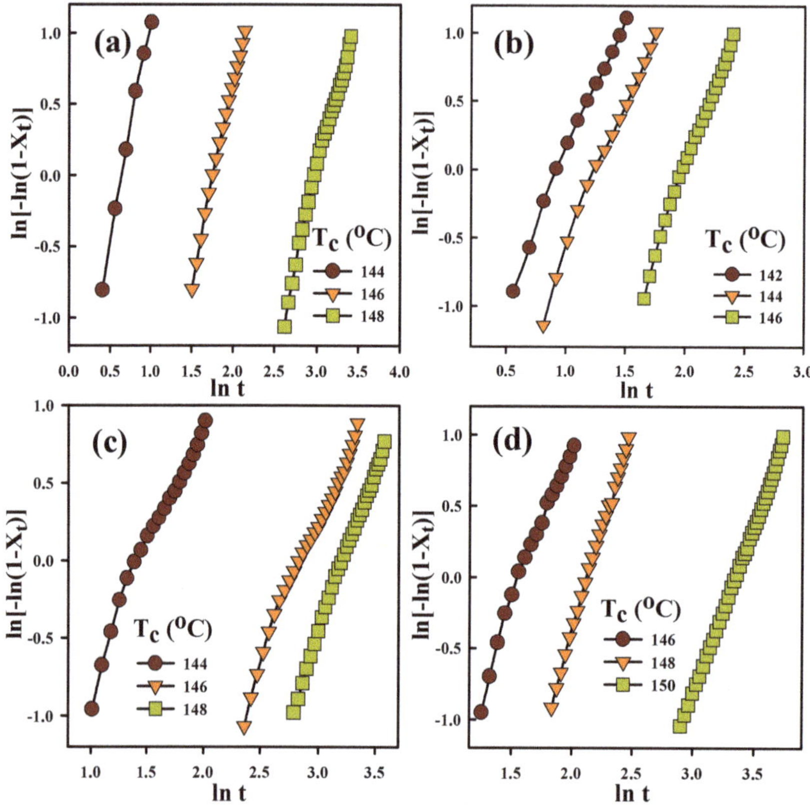

Figure 5. Avrami plots of PVDF isothermally crystallized at different T_cs in different samples: (**a**) PVDF, (**b**) F7E3, (**c**) F7E3-1, and (**d**) F7E3-3.

3.4. Melting Behavior

Figure 6a illustrates the DSC melting curves (10 °C/min heating) of the 5 °C/min pre-cooled samples. The melting temperatures (T_{ms}) of neat PVDF and HDPE were ca. 169.2 and 135.0 °C, respectively. The T_m of individual PVDF and HDPE slightly decreased in the blend. A higher loading of GNP slightly increased the T_m of PVDF and HDPE. This result suggests that PVDF crystals with a higher stability (more perfect) formed in the composite. The melting behavior of the 40 °C/min pre-cooled samples was also examined, as shown in Figure 6b. The neat PVDF and blend/composites showed multiple peaks from 165–170 °C. The multiple peaks were caused by the melting of original grown and heating-scan annealed α-form crystals (no β-form crystal was detected in the XRD data) [31]. The samples pre-cooled at 5 °C/min did not exhibit melting of the heating-annealed α-form crystals, because the slow cooling rate resulted in the growth of crystals with a sufficient stability. Accordingly, crystal annealing was prevented during heating scans. The T_m values of PVDF at different samples are summarized in Table 3. Figure 6c–f illustrate the melting curves of the representative samples that were isothermally crystallized at different T_cs. Some features were noticed in these figures. First, the T_m of PVDF increased with increasing T_c for the individual samples. Second, neat PVDF, the blend, and F7E3-1 showed similar melting behavior from 168–172 °C. Furthermore, the composites showed lower T_m values compared with the blend at each T_c. For example, F7E3 showed a T_m of 170.3 °C at T_c = 150 °C, whereas F7E3-3 had a T_m of 169.4 °C at an identical T_c. This is because the PVDF in composites displayed greater super cooling (due to higher T_m values) compared with in the blend, while they were crystallized at an identical T_c.

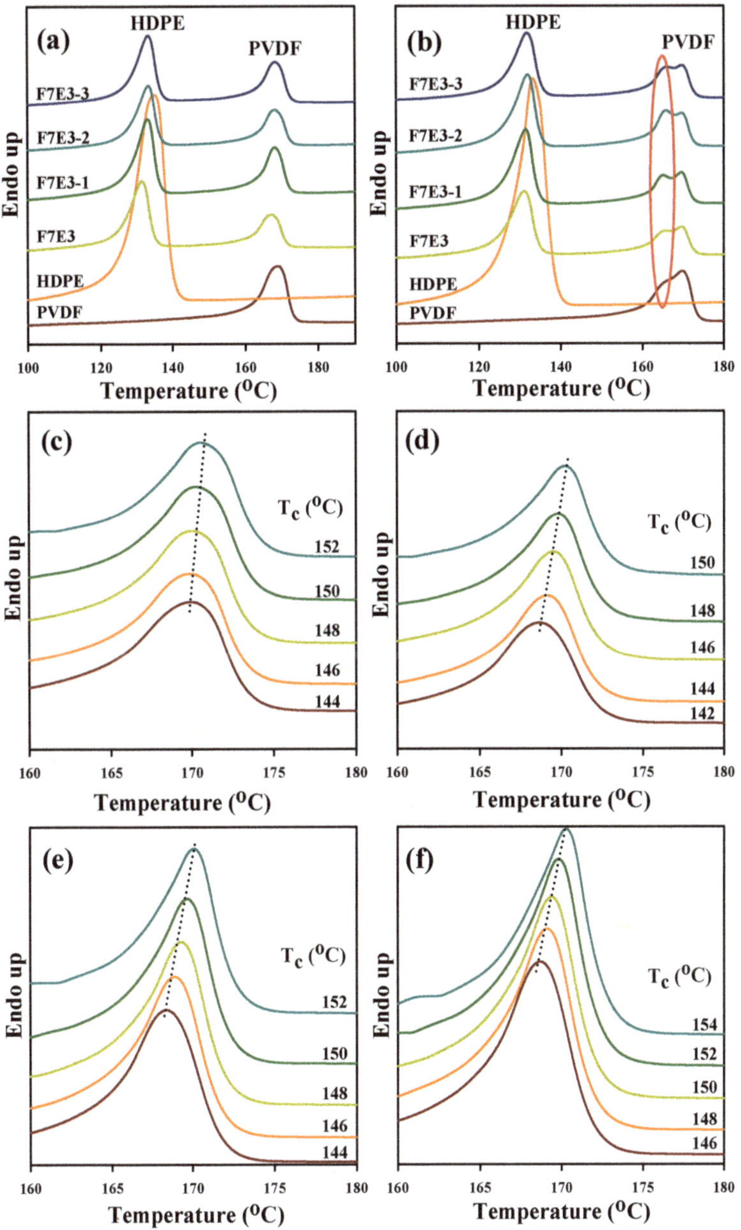

Figure 6. DSC heating curves of different samples: (**a**) 5 °C/min pre-cooled and (**b**) 40 °C/min pre-cooled; DSC melting curves of PVDF isothermally crystallized at indicated T_cs in different samples: (**c**) PVDF, (**d**) F7E3, (**e**) F7E3-1, and (**f**) F7E3-3.

3.5. Crystal Structure, FTIR, and Raman Spectra

Figure 7 shows the XRD patterns of the 5 °C/min pre-cooled samples. Neat PVDF revealed characteristic diffraction peaks at 2θ values of approximately 17.8° (100), 18.5° (020), 20.1° (110), and 26.8° (021) of the stable α-form [30,31], whereas neat HDPE exhibited characteristic peaks (orthorhombic structure) at 2θ values of approximately 21.6° (110) and 24.0° (200) [25]. The diffraction peaks of individual neat PVDF and HDPE with a lower intensity were observed in the blend and composites. The change in intensity ratio [(mainly between (020) and (110)] of PVDF in the blend/composites is because the growth of the (020) plane was somehow retarded compared to the growth of the (110) plane for PVDF crystals. Additionally, the composites exhibited a strong diffraction peak at ca. 2θ = 26.5, due to the layered structure of GNP. XRD results thus indicated that the crystal structure of PVDF and HDPE remained after forming the blend and composites. The presence of GNP did not affect the crystal structures of PVDF and HDPE.

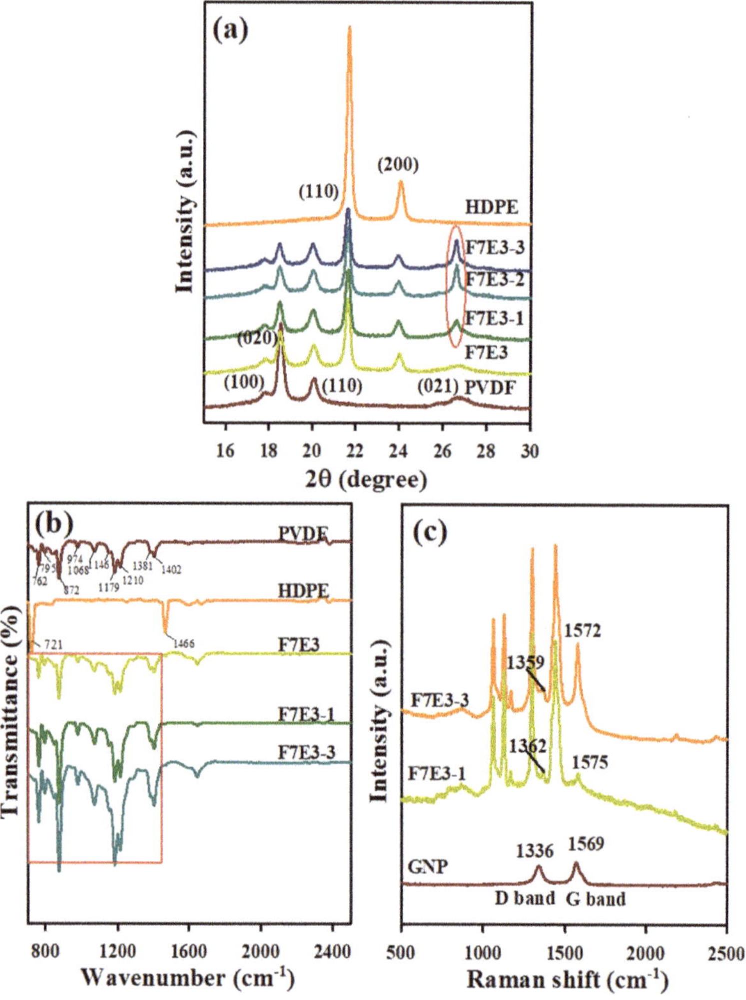

Figure 7. (**a**) X-ray diffraction (XRD) patterns of 5 °C/min-cooled representative samples; (**b**) Fourier-transform infrared spectroscopy (FTIR) spectra of selected samples; and (**c**) Raman spectra of GNP, F7E3-1, and F7E3-3.

Figure 7b shows the FTIR spectra of neat PVDF, HDPE, and selected composites. The bands located at 1381 and 1402 cm^{-1} correspond to the deformed vibration of the CH_2 groups in neat PVDF. The peaks at 1068, 1146, 1179, and 1210 cm^{-1} were related to the CF_2 stretching mode of PVDF chains [40,41]. The peaks at 762, 795, 839, 872, and 974 cm^{-1} indicated the high content α-form crystal in the neat PVDF sample [41]. F7E3 showed two extra peaks compared with the neat PVDF at the wave numbers of 721 and 1466 cm^{-1}, corresponding to the bending and rocking mode of the CH bonds of HDPE chains, respectively [42]. The peaks interrelated to the stretching mode of CH_2 in both PVDF and HDPE chains overlapped in the F7E3 blend, leading to broadened peaks (as shown in selected area). The intensity of the stretching of CH and CF_2 bonds of the composites increased compared with the F7E3 blend. The characteristic bands of composites were moderately broadened compared with the blend due to the charge-transfer-type interaction [43] between the electron pairs of fluorine atoms in PVDF and electrons in GNPs.

Figure 7c shows the Raman spectrum of the neat GNP and selected composites. Neat GNP showed a G band at 1569 cm^{-1}, which was attributed to the first-order scattering of the E_{2g} vibration mode, and the D band at 1336 cm^{-1} arose from the breathing mode of the j-point phonons of A_{1g} symmetry [44]. The location of the D band shifted from 1336 to 1362 cm^{-1} while GNP was loaded into the blend, suggesting a significant interaction between PVDF and GNP. Additionally, the G band of GNP shifted from 1569 to 1575 cm^{-1}, due to the charge-transfer interaction between the fluorine atoms of PVDF and the aromatic structure of GNP, noted as F–C bonding [42].

3.6. Melt Rheology

The rheological property measurements, including complex viscosity (η^*) and storage modulus (G'), can reveal the processability and change in the internal structure of the neat components after forming the blend and composites. Figure 8a provides a comparison of the η^* as a function of sweep frequency (ω) at 190 °C for the selected samples. The neat PVDF and HDPE showed similar Newtonian fluid behavior at the low-frequency region, whereas that in the high-frequency region exhibited non-Newtonian fluid (shear thinning) behavior. The F7E3 blend exhibited non-Newtonian fluid behavior of η^* values at all frequencies, and basically in-between that of individual PVDF and HDPE, except for displaying lower values than HDPE at high frequencies. The network structure of PVDF in the blend should have played a role for the observation. The η^* values evidently increased at all frequencies after the addition of GNP into the blend, and higher η^* values were observed at higher GNP contents. The selective localization of GNP within the PVDF matrix phase along with the formation of the PVDF-HDPE co-continuous morphology should be responsible for this observation. Moreover, η^* exhibited a non-Newtonian fluid behavior at all frequencies for the composites. The shear thinning behavior at low frequencies indicated the development of a GNP-in-PVDF pseudo-network structure (liquid-like to solid-like transition) in the composites. Figure 8b shows the G' versus ω plots of the representative samples. G' increased with increasing ω for all samples, and the values increased with the addition of GNP into the blend. The addition of 3 phr GNP resulted in the highest increase in G'. Additionally, the smaller slopes of the blend/composite plots compared with those of neat PVDF and HDPE at low frequencies suggest a solid-like behavior, which agrees with the η^* observation. The smaller slope values revealed more elastic characteristics of the blend and composites due to the pseudo-network structure (PVDF-GNP) development within the matrix.

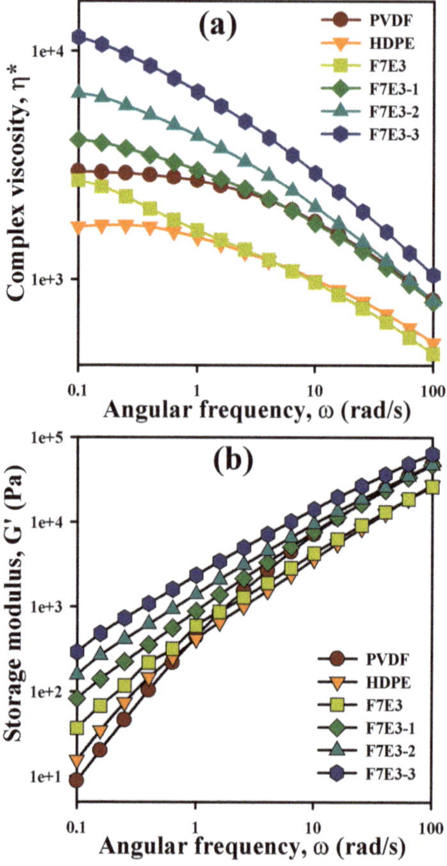

Figure 8. Rheological properties of the selected samples: (**a**) η^* vs. ω and (**b**) G' vs. ω.

3.7. Mechanical Properties

The mechanical properties of the prepared samples, including the Young's modulus (YM), flexural modulus (FM), and impact strength (IS), were measured. The neat PVDF had a YM value (1216 MPa) that was higher than that of the neat HDPE (665 MPa). The F7E3 blend showed a YM value (575 MPa) that was lower than those of its neat components, because of the immiscible character. The YM values and standard deviations (σ) of all samples are summarized in Table 4. The value increased after the GNP was loaded into the blend, and increased up to 2 phr GNP loading. F7E3-3 (654 MPa) showed a value lower than F7E3-2 (697 MPa, 21% increase at 2 phr GNP loading compared with the blend), which could be associated with the agglomeration of GNP at a higher loading. The noticeable improvement in YM after the GNP addition was mainly caused by the intrinsic high strength, high aspect ratio, and adequate interaction between the GNP and PVDF matrix phase [31]. The FM values of the individual samples were also compared and the values are listed in Table 4. PVDF possessed a much higher FM than that of HDPE, and the FM of HDPE significantly increased after blending with PVDF. The addition of GNP caused an evident increase in FM for the composites (24% increase at 2 phr GNP loading compared with the blend). The reasons behind this significant improvement are identical to those of the YM improvement after GNP addition. The notched IS values of the individual sample are listed in Table 4. The neat PVDF possessed an IS of 17.9 J/m, which evidently decreased after the formation of the blend because of the immiscibility. After the addition of 2 phr GNP, IS increased to approximately 1.54 times the original value of the F7E3 blend, and a higher GNP loading slightly reduced the IS value. The GNP was located in the PVDF matrix phase near the interface of PVDF-HDPE, leading to adhesion enhancement of the interface. The possibility of stress transfer between blend components through the potential bridging effect of the GNP at the blend interface caused the enhancement of toughness.

Table 4. Mechanical properties and electrical resistivity of the selected samples.

Samples	Properties			
	Young's Modulus (σ) (MPa)	Flexural Modulus (σ) (MPa)	Impact Strength (σ) (J/m)	Electrical Resistivity (Ω-cm)
PVDF	1216 (90)	927 (28)	17.9 (2.10)	1.03×10^{11}
HDPE	665 (41)	744 (36)	10.5 (0.98)	3.87×10^{11}
F7E3	575 (66)	815 (29)	6.4 (0.94)	1.95×10^{11}
F7E3-1	672 (34)	921 (18)	7.8 (0.73)	3.85×10^{10}
F7E3-2	697 (29)	1010 (25)	9.9 (0.45)	2.32×10^{9}
F7E3-3	654 (41)	937 (34)	8.5 (0.52)	1.79×10^{8}

3.8. Electrical Resistivity

The uniform distribution of electrically conductive fillers in the polymer matrices offers great potential for fabricating (semi)conductive polymers. The volume electrical resistivity values of the selected samples are summarized in Table 4. Neat PVDF, HDPE, and F7E3 exhibited resistivity values higher than 10^{11} Ω-cm, indicating their electrically insulating characteristic. PVDF shows an electrical resistivity value of 10^{11} Ω-cm, similar to a previously published paper [31]. However, the low loading of GNP (< 3 phr) could evidently reduce the resistivity for the composites. For example, the loading of 0.5 phr GNP decreased the resistivity value from 10^{11} to 10^{10} Ω-cm. The resistivity further decreased (down to 10^8 Ω-cm) with increasing GNP loading in the composites. The apparent decrease in resistivity was noticed at 1–1.5 phr GNP loading. The observations of a ca. three orders of magnitude drop resulted from the fine dispersion of GNP throughout the PVDF matrix phase. This result is consistent with the above SEM morphological observations.

4. Discussion

PVDF/HDPE blend and blend-based nanocomposites with PVDF as the major component were fabricated through the melt-mixing process. SEM results revealed that GNP was mainly located in the PVDF matrix phase. DSC data showed the nucleation effect of GNP on both PVDF and HDPE crystallization. The activation energy for the non-isothermal crystallization of PVDF increased from 303 kJ/mol in the neat state to 374 kJ/mol in the F7E3-3 composite. GNP increased the isothermal crystallization rate of PVDF in the composites. The Avrami n values ranged from 1.9–3.8 for PVDF in different samples, and the GNP caused an athermal nucleation mechanism for PVDF crystallization. TGA results confirmed the improvement in thermal stability of the blend after GNP addition under both air and nitrogen environments. XRD results revealed that the presence of GNP did not alter the crystalline polymorph of PVDF and HDPE. An increase in the rheological properties of η^* and G' was observed with increasing GNP loading. The pseudo-network structure was achieved in the composites at 3-phr loading GNP into the blend. The rigidity of the PVDF/HDPE blend increased with the incorporation of GNP. The toughness of the blend also increased with GNP inclusion, up to a 55% rise after 2 phr loading. The electrical resistivity of the blend dropped by more than three orders of magnitude with a GNP loading of 3 phr. The obtained results indicated that the prepared composites should have great potential applications in micro-electromechanical devices and storage capacitors.

Author Contributions: K.B. did the experimental work, analyzed the results, and wrote the whole manuscript. M.Y. reviewed and refined the manuscript. F.-C.C. supervised the whole experiments and revised the manuscript. K.Y.R. reviewed the manuscript. All authors have read and approved the final manuscript.

Funding: This research received no external funding.

Acknowledgments: The authors thankful to the Ministry of Science and Technology (MOST, Taiwan) for financially supporting this research under contracts MOST 106-2221-E-182-058-MY3 and MOST 107-2811-E-182-503.

Conflicts of Interest: The authors declare no conflict of interest.

References

1. Schaefer, D.W.; Justice, R.S. How nano are nanocomposites? *Macromolecules* **2007**, *40*, 8501–8517. [CrossRef]
2. Kotal, M.; Bhowmick, A.K. Polymer nanocomposites from modified clays: Recent advances and challenges. *Prog. Polym. Sci.* **2015**, *51*, 127–187. [CrossRef]
3. Sivanjineyulu, V.; Behera, K.; Chang, Y.H.; Chiu, F.C. Selective localization of carbon nanotube and organoclay in biodegradable poly(butylene succinate)/polylactide blend-based nanocomposites with enhanced rigidity, toughness and electrical conductivity. *Compos. Part A Appl. Sci. Manuf.* **2018**, *114*, 30–39. [CrossRef]
4. Kim, H.S.; Lee, S.K.; Wang, M.; Kang, J.; Sun, Y.; Jung, J.W.; Kim, K.; Kim, S.M.; Nam, J.D.; Suhr, J. Experimental investigation on 3D graphene-CNT hybrid foams with different interactions. *Nanomaterials* **2018**, *8*, 694–703. [CrossRef] [PubMed]
5. Punetha, V.D.; Rana, S.; Yoo, H.J.; Chaurasia, A.; McLeskey, J.T.; Ramasamy, M.S.; Sahoo, N.G.; Cho, J.W. Functionalization of carbon nanomaterials for advanced polymer nanocomposites: A comparison study between CNT and graphene. *Prog. Polym. Sci.* **2017**, *67*, 1–47. [CrossRef]
6. Zhou, S.; Hrymak, A.N.; Kamal, M.R. Effect of hybrid carbon fillers on the electrical and morphological properties of polystyrene nanocomposites in microinjection molding. *Nanomaterials* **2018**, *8*, 779–794. [CrossRef] [PubMed]
7. Kim, H.; Abdala, A.A.; Macosko, C.W. Graphene/polymer nanocomposites. *Macromolecules* **2010**, *43*, 6515–6530. [CrossRef]
8. Li, B.; Zhong, W.H. Review on polymer/graphite nanoplatelet nanocomposites. *J. Mater. Sci.* **2011**, *46*, 5595–5614. [CrossRef]
9. Huang, G.; Wang, S.; Song, P.a.; Wu, C.; Chen, S.; Wang, X. Combination effect of carbon nanotubes with graphene on intumescent flame-retardant polypropylene nanocomposites. *Compos. Part A Appl. Sci. Manuf.* **2014**, *59*, 18–25. [CrossRef]
10. Verdejo, R.; Bernal, M.M.; Romasanta, L.J.; Lopez-Manchado, M.A. Graphene filled polymer nanocomposites. *J. Mater. Chem.* **2011**, *21*, 3301–3310. [CrossRef]
11. Lee, C.; Wei, X.; Kysar, J.W.; Hone, J. Measurement of the elastic properties and intrinsic strength of monolayer graphene. *Science* **2008**, *321*, 385–388. [CrossRef] [PubMed]
12. El Mohajir, B.E.; Heymans, N. Changes in structural and mechanical behaviour of PVDF with processing and thermomechanical treatments. 1. Change in structure. *Polymer* **2001**, *42*, 5661–5667. [CrossRef]
13. Lovinger, A.J. Crystallization of the β phase of poly(vinylidene fluoride) from the melt. *Polymer* **1981**, *22*, 412–413. [CrossRef]
14. Dillon, D.R.; Tenneti, K.K.; Li, C.Y.; Ko, F.K.; Sics, I.; Hsiao, B.S. On the structure and morphology of polyvinylidene fluoride–nanoclay nanocomposites. *Polymer* **2006**, *47*, 1678–1688. [CrossRef]
15. Wang, J.; Wu, J.; Xu, W.; Zhang, Q.; Fu, Q. Preparation of poly(vinylidene fluoride) films with excellent electric property, improved dielectric property and dominant polar crystalline forms by adding a quaternary phosphorus salt functionalized graphene. *Compos. Sci. Technol.* **2014**, *91*, 1–7. [CrossRef]
16. Wu, C.M.; Chou, M.H.; Zeng, W.Y. Piezoelectric response of aligned electrospun polyvinylidene fluoride/carbon nanotube nanofibrous membranes. *Nanomaterials* **2018**, *8*, 420–453. [CrossRef] [PubMed]
17. Cendrowski, K.; Kukulka, W.; Kedzierski, T.; Zhang, S.; Mijowska, E. Poly(vinylidene fluoride) and carbon derivative structures from eco-friendly MOF-5 for supercapacitor electrode preparation with improved electrochemical performance. *Nanomaterials* **2018**, *8*, 890–910. [CrossRef] [PubMed]
18. Chiu, F.C. Comparisons of phase morphology and physical properties of PVDF nanocomposites filled with organoclay and/or multi-walled carbon nanotubes. *Mater. Chem. Phys.* **2014**, *143*, 681–692. [CrossRef]
19. Liu, Y.L.; Li, Y.; Xu, J.T.; Fan, Z.Q. Cooperative effect of electrospinning and nanoclay on formation of polar crystalline phases in poly(vinylidene fluoride). *ACS Appl. Mater. Inter.* **2010**, *2*, 1759–1768. [CrossRef] [PubMed]

20. Almasri, A.; Ounaies, Z.; Kim, Y.S.; Grunlan, J. Characterization of solution-processed double-walled carbon nanotube/poly(vinylidene fluoride) nanocomposites. *Macromol. Mater. Eng.* **2008**, *293*, 123–131. [CrossRef]

21. Mago, G.; Fisher, F.T.; Kalyon, D.M. Deformation-induced crystallization and associated morphology development of carbon nanotube-PVDF nanocomposites. *J. Nanosci. Nanotechnol.* **2009**, *9*, 3330–3340. [CrossRef] [PubMed]

22. Martins, J.N.; Bassani, T.S.; Barra, G.M.; Oliveira, R.V. Electrical and rheological percolation in poly(vinylidene fluoride)/multi-walled carbon nanotube nanocomposites. *Polym. Int.* **2011**, *60*, 430–435. [CrossRef]

23. Kanagaraj, S.; Varanda, F.R.; Zhil'tsova, T.V.; Oliveira, M.S.A.; Simões, J.A.O. Mechanical properties of high density polyethylene/carbon nanotube composites. *Compos. Sci. Technol.* **2007**, *67*, 3071–3077. [CrossRef]

24. Brandalise, R.N.; Zeni, M.; Martins, J.D.N.; Forte, M.M.C. Morphology, mechanical and dynamic mechanical properties of recycled high density polyethylene and poly(vinyl alcohol) blends. *Polym. Bull.* **2009**, *62*, 33–43. [CrossRef]

25. Wang, C.; Wang, S.; Cheng, H.; Xian, Y.; Zhang, S. Mechanical properties and prediction for nanocalcium carbonate-treated bamboo fiber/high-density polyethylene composites. *J. Mater. Sci.* **2017**, *52*, 11482–11495. [CrossRef]

26. Huang, C.Y.; Roan, M.L.; Kuo, M.C.; Lu, W.L. Effect of compatibiliser on the biodegradation and mechanical properties of high-content starch/low-density polyethylene blends. *Polym. Degrad. Stab.* **2005**, *90*, 95–105. [CrossRef]

27. Rafeie, O.; Razavi Aghjeh, M.K.; Tavakoli, A.; Salami Kalajahi, M.; Jameie Oskooie, A. Conductive poly(vinylidene fluoride)/polyethylene/graphene blend-nanocomposites: Relationship between rheology, morphology, and electrical conductivity. *J. Appl. Polym. Sci.* **2018**, *135*, 46333–46346. [CrossRef]

28. Deng, Y.; Song, X.; Ma, Z.; Zhang, X.; Shu, D.; Nan, J. Al$_2$O$_3$/PVDF-HFP-CMC/PE separator prepared using aqueous slurry and post-hot-pressing method for polymer lithium-ion batteries with enhanced safety. *Electrochim. Acta* **2016**, *212*, 416–425. [CrossRef]

29. Shao, Y.; Yang, Z.X.; Deng, B.W.; Yin, B.; Yang, M.B. Tuning PVDF/PS/HDPE polymer blends to tri-continuous morphology by grafted copolymers as the compatibilizers. *Polymer* **2018**, *140*, 188–197. [CrossRef]

30. Chiu, F.C. Poly(vinylidene fluoride)/polycarbonate blend-based nanocomposites with enhanced rigidity—Selective localization of carbon nanofillers and organoclay. *Polym. Test.* **2017**, *62*, 115–123. [CrossRef]

31. Chiu, F.C.; Chen, Y.J. Evaluation of thermal, mechanical, and electrical properties of PVDF/GNP binary and PVDF/PMMA/GNP ternary nanocomposites. *Compos. Part A Appl. Sci. Manuf.* **2015**, *68*, 62–71. [CrossRef]

32. Sumita, M.; Sakata, K.; Asai, S.; Miyasaka, K.; Nakagawa, H. Dispersion of fillers and the electrical conductivity of polymer blends filled with carbon black. *Polym. Bullet.* **1991**, *25*, 265–271. [CrossRef]

33. Wu, S. *Polymer Interface and Adhesion*; Marcel Dekker Inc.: New York, NY, USA, 1982.

34. Park, M.; Gu, G.Y.; Wang, Z.J.; Kwon, D.J.; Lawrence, K.; Vries, D. Interfacial durability and acoustical properties of transparent graphene nano platelets/poly(vinylidene fluoride) composite actuators. *Thin Solid Films* **2013**, *539*, 350–355.

35. Yu, J.; Huang, X.; Wu, C.; Jiang, P. Permittivity, thermal conductivity and thermal stability of poly(vinylidene fluoride)/graphene nanocomposites. *IEEE Trans. Dielectr. Electr. Insul.* **2011**, *18*, 478–484.

36. Kissinger, H.E. Variation of peak temperature with heating rate in differential thermal analysis. *J. Res. Natl. Bur. Stand.* **1956**, *17*, 217–221. [CrossRef]

37. Behera, K.; Sivanjineyulu, V.; Chang, Y.H.; Chiu, F.C. Thermal properties, phase morphology and stability of biodegradable PLA/PBSL/HAp composites. *Polym. Degrad. Stabil.* **2018**, *154*, 248–260. [CrossRef]

38. Behera, K.; Chang, Y.H.; Chiu, F.C.; Yang, J.C. Characterization of poly(lactic acid)s with reduced molecular weight fabricated through an autoclave process. *Polym. Test.* **2017**, *60*, 132–139. [CrossRef]

39. Avrami, M. Kinetics of phase change. I general theory. *J. Chem. Phys.* **1939**, *7*, 1103–1112. [CrossRef]

40. Martins, P.; Lopes, A.C.; Lanceros-Mendez, S. Electroactive phases of poly(vinylidene fluoride): Determination, processing and applications. *Prog. Polym. Sci.* **2014**, *39*, 683–706. [CrossRef]

41. Priya, L.; Jog, J.P. Polymorphism in intercalated poly(vinylidene fluoride)/clay nanocomposites. *J. Appl. Polym. Sci.* **2003**, *89*, 2036–2040. [CrossRef]

42. Kuila, T.; Bose, S.; Mishra, A.K.; Khanra, P.; Kim, N.H.; Lee, J.H. Effect of functionalized graphene on the physical properties of linear low density polyethylene nanocomposites. *Polym. Test.* **2012**, *31*, 31–38. [CrossRef]

43. Yang, J.; Feng, C.; Dai, J.; Zhang, N.; Huang, T.; Wang, Y. Compatibilization of immiscible nylon 6/poly(vinylidene fluoride) blends using graphene oxides. *Polym. Int.* **2013**, *62*, 1085–1093. [CrossRef]

44. Vasileiou, A.A.; Kontopoulou, M.; Docoslis, A. A noncovalent compatibilization approach to improve the filler dispersion and properties of polyethylene/graphene composites. *ACS Appl. Mater. Inter.* **2014**, *6*, 1916–1925. [CrossRef] [PubMed]

 *nanomaterials*

Review

Electrical Property of Graphene and Its Application to Electrochemical Biosensing

Jin-Ho Lee [1,2], Soo-Jeong Park [3] and Jeong-Woo Choi [1,3,*]

[1] Department of Chemical and Biomolecular Engineering, Sogang University, 35 Baekbeom-ro, Mapo-gu, Seoul 04107, Korea; jino@sogang.ac.kr

[2] Department of Chemistry and Chemical Biology, Rutgers, The State University of New Jersey, Piscataway, NJ 08854, USA

[3] Research Center for Disease Biophysics of Sogang-Harvard, Sogang University, 35 Baekbeom-ro, Mapo-gu, Seoul 04107, Korea; imation99@naver.com

* Correspondence: jwchoi@sogang.ac.kr; Tel.: +82-2-718-1976; Fax: +82-2-3273-0331

Received: 31 January 2019; Accepted: 18 February 2019; Published: 20 February 2019

Abstract: Graphene, a single atom thick layer of two-dimensional closely packed honeycomb carbon lattice, and its derivatives have attracted much attention in the field of biomedical, due to its unique physicochemical properties. The valuable physicochemical properties, such as high surface area, excellent electrical conductivity, remarkable biocompatibility and ease of surface functionalization have shown great potentials in the applications of graphene-based bioelectronics devices, including electrochemical biosensors for biomarker analysis. In this review, we will provide a selective overview of recent advances on synthesis methods of graphene and its derivatives, as well as its application to electrochemical biosensor development. We believe the topics discussed here are useful, and able to provide a guideline in the development of novel graphene and on graphene-like 2-dimensional (2D) materials based biosensors in the future.

Keywords: Graphene; Graphene Oxide; 2D materials; Electrochemical; Biosensor

---

1. Introduction

Graphene, a single 2-dimensional (2D) layer of a hexagonal structure consisting of sp^2 hybridized carbon atoms, and its derivatives have received increasing attention in biomedical fields, due to its unique physicochemical properties. This feature includes a high surface area, excellent electrical conductivity, strong mechanical strength, unparalleled thermal conductivity, and ease of surface functionalization (Table 1) [1–5].

Table 1. Physicochemical properties of graphene and its derivatives.

Physicochemical Property	Estimated Value	Ref.
High surface area	~2630 m^2g^{-1}	[1]
Excellent electrical conductivity	~1738 siemens/m	[2]
Strong mechanical strength	Young' Modulus ~1100 GPa, Fracture strength ~125 GPa	[3]
Thermal conductivity	5000 $Wm^{-1}K^{-1}$	[4]
Ease of functionalization	π–π stacking interaction Electrostatic interaction	[5]

Particularly, owing to the high surface area, excellent electrical conductivity, and capability to adsorb a variety of biomolecules, graphene has been considered as an ideal transducing material for constructing electrochemical biosensors [6,7]. It is well defined that the efficient electrochemical reaction takes place at the close distance between the electrode surface and the electroactive (reduction/oxidation, redox) site of a molecule. In detail, the electron transfer rate is inversely proportional to the exponential distance between the electrode surface and the electroactive redox site of the molecule [8,9]. Since the electron transfer between graphene and redox active molecule typically take place at either edge of the graphene layer or defects in the basal plane, the high surface area of 2D structure helps graphene to work as an excellent conducting material for electrical charge and heterogeneous electron transfer [10,11].

In addition, based on the unique atomic thin layer structure, the electrical properties of graphene are known to be highly sensitive to foreign atoms or absorbed molecules [12]. Favorably, graphene is known to be highly reliable for capturing aromatic molecules through a π–π stacking interaction. For example, single-stranded deoxyribonucleic acid (ssDNA) can bind to the graphene surface by π–π stacking interaction between deoxyribonucleic acid (DNA) and polyaromatic structures of graphene, and serve as a platform for various DNA based biosensing applications [13,14]. In addition, its derivatives graphene oxide (GO) are known to possess the oxygen-containing hydrophilic groups (hydroxyl and epoxy in the basal planes; carbonyl and carboxyl groups on the edges), that allows the electrostatic interaction [9,15]. Alternatively, graphene can be also functionalized by covalent bonding either through unsaturated p-bonds of graphene or oxygen-containing functional groups of GO [5,16]. As an example, various dienophiles, such as azomethine ylide, nitrene, and aryne, has been successfully generated a variety of terminal groups on the graphene surface for further modification [17–19]. The carboxyl groups of GO are used to link the amino groups of molecules by well-established carbodiimide chemistry [20]. According to the above mentioned unique physicochemical properties, the utilization of graphene as a functional component for an electrode has gained considerable interest in the field of electrochemical biosensors [19].

Although there exists an extensive collection of reviews for graphene synthesis methods and electrochemical sensing applications, the tremendous amount of recent activities and a new, live cell-based biosensing approach warrant a thorough review at this time. In this review, we will provide a selective overview of the recent advances on the synthesis methods of graphene and its derivatives, as well as its application in the electrochemical biosensor, which particularly covers small molecule, nucleic acid/protein, and live cell-based sensing. This review will provide an extensive analysis of the current state of the art and provide a perspective on key challenges that remain in the field. We hope that this review will inspire interest from various disciplines and highlight an important field wherein the advanced graphene-based electrochemical sensor is making great strides towards biomedical applications.

2. Synthesis of Graphene

As the outstanding physicochemical properties of graphene make this material promising candidate for electrochemical biosensor applications, the synthesis processes of the graphene which affect its properties also hold great influence on the proper development and performance of the biosensors. To this end, a number of a different synthesis method for graphene has been developed over the years. Particularly, in this review, most well-defined exfoliation phenomena or chemical vapor deposition (CVD) will be discussed as graphene and its derivatives synthesis methods (Figure 1) [9,21–23].

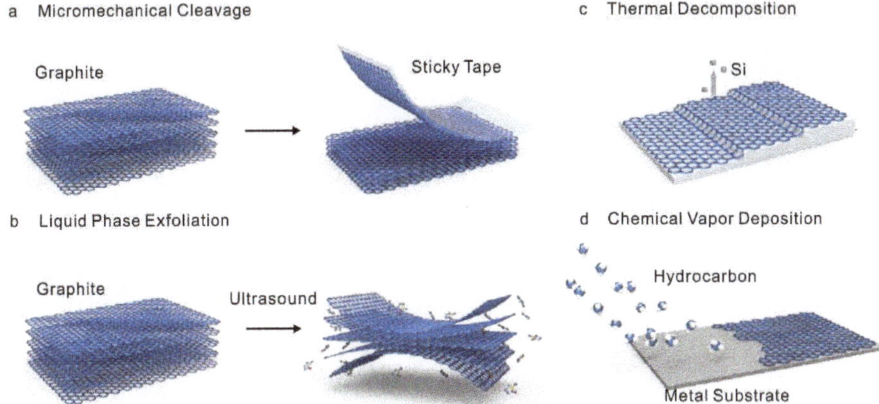

Figure 1. Schematic illustration of the graphene and its derivatives synthesis methods. (**a**) Mechanical exfoliation (**b**) liquid phase exfoliation, (**c**) thermal decomposition, and (**d**) chemical vapor deposition. Reproduced with permission from [23], Copyright Springer Nature, 2017.

2.1. Mechanical and Chemical Exfoliation Method

The phenomena of exfoliation are to separate a few layers from the bulk material by overcoming the strong van der Waals attractions between adjacent layers. Among the exfoliation based process, the mechanical approach, 'Scotch tape' based exfoliation was first developed to obtain graphene, where few layers of graphene are peeled off from a highly ordered pyrolytic graphite (HOPG) flakes by external mechanical force (adhesive tape) based on the relatively weak interaction between the thin layers and the bulk materials. After the mechanical exfoliation, micromechanical cleavage of graphite, by rubbing a bulk crystal flake against another flake, was also utilized to obtain individual crystal planes of graphene layers [24]. In principle, through these mechanical exfoliation methods, high structural and electronic quality graphene crystals can be obtained without hosting structural defects into 2D graphene layers repeatedly [25–27]. However, these mechanical exfoliation methods are often limited for biosensor application, due to the difficulties in the mechanical cleavage of graphite crystals in a controlled manner. Such as low yield in the production of single-layer or few-layer graphene and relatively large lateral dimensions of graphene (range in micrometer size) often restrict their application in biosensors. Apart from the mechanical exfoliation, the potential energy caused by van der Waals attractions could be overcome in the presence of solvents as well [28]. This process, which is also known as liquid phase exfoliation (LPE), generally requires dispersion of bulk materials in a solvent, exfoliation, and purification [29]. Thus, the selection of ideal solvents is critical for LPE process to obtain high yield and stability [30]. However, the commonly used solvents for LPE, such as N,N-dimethylformamide (DMF) and N-methyl pyrrolidinone (NMP) are usually known to have acute toxic effects. To overcome this limitation, numerous approaches, such as urea-based aqueous exfoliation, which even showed higher efficiency compare to conventional DMF based exfoliation, were recently developed [31]. In addition, graphene derivatives can be also synthesized by the chemical oxidation of graphite. For example, one of most well established Hummer's method for preparing graphite oxide includes the addition of potassium permanganate (KMnO4) to a mixture of graphite, sodium nitrate (NaNO3), and concentrated sulfuric acid (H_2SO_4) [32]. During the oxidation, small ions intercalate to bulk graphite oxide during oxidation and weaken the interlayer interactions. In detail, the sp^2 hybridized carbon bonding is disrupted during the oxidation process, yielding formation of sp^3 hybridized carbon bonding. Through this mechanism, sonication allows exfoliating GO layer from bulk graphite oxide flake and subsequent reduction of GO could also help to obtain reduce graphene oxide (rGO) [33–35]. Owing to ease of process, the exfoliation based on chemical method is known to be suitable for the synthesis of graphene at a large scale, which is important for the construction of bioelectronics devices, including electrochemical biosensors. In addition, the unique chemical structure of chemically derived GO and rGO, which differs from the pristine graphene or graphite, provide versatility in applications based on the chemical functionalization through the oxygen functional groups. However, large amounts of structural defects caused by inevitably introduced oxygen functional groups could affect the electrical properties of graphene and the performance of electrochemical biosensors.

2.2. Thermal Decomposition and Chemical Vapor Deposition Method

An alternative approach to synthesize a high quality graphene layer in a large scale comprises the self-organization of carbon atoms on the surface of the crystal by the thermal decomposition of hydrocarbon and segregation of the carbon monolayer on metal substrates through CVD [36–38]. For example, the evaporation of silicon from single-crystalline silicon carbide (SiC) substrates at a high annealing temperature results in the organized attachment of the remaining carbon atoms on the surface of lattice-matched SiC substrates. Though, this method directly provides graphene layers on insulating SiC substrates in a wafer-scale; however, the strong interaction of graphene with substrates limits doping property, as well as transfer efficiency to other substrates for biomedical applications. Among the numerous metal substrates, nickel (Ni) and (copper) Cu substrates demonstrated the potential to separate graphene layer from the substrates and transfer onto other substrates, including a solid substrate to flexible and bendable substrates [36]. In detail, the synthesis mechanism for the metals substrate with high solubility of carbon species, such as Ni substrate include catalytic decomposition of the precursor, dissolution of decomposed carbon species, segregation of dissolved carbon atoms onto the metal surface, and followed by nucleation and growth of graphene layer on the surface of the substrate [39]. Though, the several critical factors, such as the thickness of Ni films, growth time, and cooling rate has been already revealed to improve the quality of synthesized graphene layers; however, it is still difficult to obtain a single layer of graphene through the polycrystalline Ni substrate and the electrical property of the synthesized graphene layer were found to be not very satisfactory as well. Comparably, by using a Cu substrate, which has low solubility of carbon, highly uniformed single graphene layer with the excellent electrical property was synthesized [36,40]. Due to the low solubility of carbon, the formation of graphene happens through the self-limited nucleation and lateral growth by diffusion of carbon atoms on the surface directly after the decomposition of precursors [41]. In addition, free-floating graphene layers could be obtained by metal etching or electrochemical bubbling method for further applications [42]. Although, CVD method can synthesize graphene layer even in a large area (up to several inches) with high electronic quality (mobility up to 10^5 cm^{2}V^{-1}s^{-1}) comparable to the exfoliation methods [36,39,43], the structural defects or contamination which can be originated during the etching and transfer process are still limiting factor for obtaining high profile graphene layer and can also affect the performance of electrochemical biosensors as well.

3. Application to Electrochemical Sensing

The biosensor is the analytical device which consists of a biological component that recognizes the target analytes and an electrical component (transducer) which converts the recognition event into a measurable signal. To improve the performance of biosensors, tremendous efforts from multidiscipline fields has been established. Ever since the discovery of graphene by Geim and Novoselov in 2004, numerous approach has been conducted to utilize graphene as transducing material to improve the performance of electrochemical sensors [6,44]. Graphene and its derivatives modified electrodes have exhibited excellent electrochemical behavior in terms of their high surface area and active electron transfer sites [7], which makes graphene as a promising electrode material to improve the performance of graphene and its derivatives based electrochemical biosensors. In this review, the division of biomedical electrochemical sensors will be divided into three categories i) small molecules, ii) nucleic acids and proteins, and iii) Live cell-based sensing.

3.1. Small Molecule Sensing

There are many small molecules that are highly relevant to human health and disease. Even a subtle change of these biomolecules could cause a serious disease which threatens a patient's life [45–48]. For example, dopamine (DA) is one of the most important neurotransmitters that play a vital role in the central nervous system. Abnormal level of DA can cause severe neurological disorders, such as Parkinson's disease [49–51]. However, due to the complex matrices of nature, it is still challenging to develop a biosensor to distinguish the biomolecules which share a similar oxidation potential, such as ascorbic acid (AA), uric acid (UA) and other catecholamine molecules. To resolve this problem, graphene has been adopted as a transducing material in the development of the electrode. Through its phenyl structure, these molecules could adsorb on the graphene-modified electrode surface through the different π-π stacking interactions. Ping et al. introduced a graphene-based screen-printing ink which could selectively and sensitively analyze these molecules via differential pulse voltammetry (DPV) [52]. Even in the co-existence of these three molecules, the linear range and detection limit were found to be 0.5–2000 µM, 4.0–4500 µM, and 0.8–2500 µM and 0.12 µM, 0.95 µM, and 0.20 µM for DA, AA, and UA, respectively.

As an alternative approach, graphene was also utilized to provide large surface area, as well as a binding motif for the mediator (i.e., enzyme) to improve the electrocatalytic performance of electrochemical sensors to determine small molecules, such as hydrogen peroxide (H_2O_2), glucose, and nicotinamide adenine dinucleotide (NADH) sensitively. Besides its well-known cytotoxic effects, as an essential mediator in many biological processes, H_2O_2 detection has earned great attention. However, due to the co-existing other electro-active constituents, the detection of H_2O_2 is easily interfered [53,54]. To improve the H_2O_2 detection efficiency, graphene was also utilized to improve the performance of electrochemical sensors. Fan et al., designed a graphene capsule, which served as a carrier for horseradish peroxidase (HRP), to detect H_2O_2 in human serum [53]. Through the large surface area and high conductivity of graphene, the synergistic effect on catalytic activity was able to be obtained. Instead of using the enzyme, Wang et al., have grown Prussian blue nanocubes on the surface of nitrobenzene-functionalized reduced graphene oxide as an "artificial enzyme peroxidase" for constructing H_2O_2 electrochemical biosensors [55]. Another approach based on the incorporation of myoglobin (Mb) on graphene oxide encapsulated molybdenum disulfide (MoS_2) nanoparticle (Mb-GO@MoS_2) hybrid structure was also reported by Choi's group (Figure 2) [56]. Mb is a one of metalloprotein family with unique redox properties, due to the metal ion core integrated into the hemin group. Through this unique redox properties, Mb can be used to detect H_2O_2 through electrochemical reduction of H_2O_2 as well. The developed Mb-GO@MoS_2 structure and extended electroactive surface also affected the fast electron transfer and resulted in an enhanced amperometric response H_2O_2. Instead of using full protein, Song et al. utilized hemin porphyrin and functionalized on the graphene/GNP/glassy carbon electrode to avoid the possible insulating effect from protein structure [57]. Note that hemin porphyrin is a well-known natural metalloporphyrin which is the active site in heme-proteins, such as hemoglobins and myoglobins [58]. Alternatively, Shao et al. have shown that by just doping graphene nitrogen, the better electrocatalytic activity could be obtained for H_2O_2 detection as well [59]. The enhanced performance of nitrogen-doped graphene is expected, due to the existence of nitrogen functional groups in addition to oxygen-containing groups and structural defects. In a similar manner, graphene was also utilized to improve the electrocatalytic performance of electrochemical sensors to determine other small molecules, such as DA, glucose, and NADH as well. The recent researches on graphene-based electrochemical biosensors toward various small molecules are compared in Table 2.

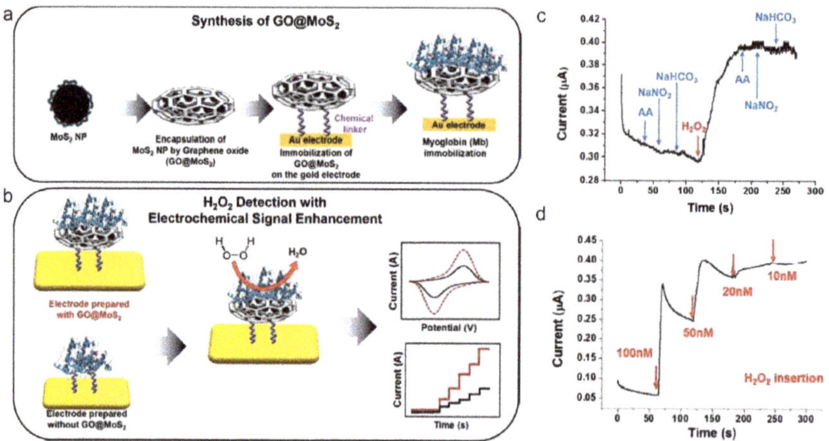

Figure 2. (**a**) Schematic of graphene oxide encapsulated molybdenum disulfide (MoS2) nanoparticle preparation for the fabrication of electrochemical biosensors composed of myoglobin (Mb) and (**b**) its application to H_2O_2 detection with improved electrochemical performance. (**c–d**) Amperometric response curves obtained from Mb/GO@MoS$_2$ upon successive (**c**) addition of 100 nM L-ascorbic acid (AA), 100 nM sodium nitrite (NaNO$_2$), 100 nM sodium bicarbonate (NaHCO$_3$), and 100 nM H_2O_2 solutions; and by (**d**) addition of 100, 50, 20, and 10 nM H_2O_2 solutions. Reproduced with permission from [56], Copyright Elsevier, 2017.

Table 2. Comparison of different graphene-based electrode for small molecule detection.

Electrode Materials	Target	Linear Range	Detection Limit	Ref.
Graphene capsule/horseradish peroxidase	H_2O_2	0.01–12 mM	3.3 µM	[53]
Prussian blue nanocubes/nitrobenzene/reduced graphene oxide	H_2O_2	1.2 µM–15.25 mM	0.4 µM	[55]
Myoglobin (Mb)/MoS2 nanoparticle/graphene oxide	H_2O_2	-	20 nM	[56]
Hemin porphyrin/graphene/gold nanoparticle	H_2O_2	0.3 µM–1.8 mM	0.11 µM	[57]
Cobalt ferrite nanoparticles decorated exfoliated graphene oxide	H_2O_2	0.9–900 µM	0.54 µM	[60]
	NADH	0.50–100 µM	0.38 µM	
Au-Ag nanoparticles/poly(L-Cysteine)/reduced graphene oxide	NADH	0.083 µM–1.05 mM	9.0 nM	[61]
	ethanol	0.017 µM–1.845 mM	5.0 µM	
Graphene-pyrroloquinoline quinone	NADH	0.32 µM–220 µM	0.16 µM	[62]
FeN nanoparticles/nitrogen-doped graphene core-shell	NADH	0.4 µM–718 µM	25 nM	[63]
Screen-printed graphene	Dopamine	0.5 µM–2000 µM	0.12 µM	[52]
	Ascorbic acid	4.0 µM–4500 µM	0.95 µM	
	Uric acid	0.8 µM–2500 µM	0.20 µM	
Nickel and copper oxides-decorated graphene	Dopamine	0.5 µM–20 µM	0.17 µM	[64]
Molecularly imprinted polymer modified graphene/carbon nanotube	Dopamine	2.0 fM–1.0 pM	667 aM	[65]
Gold nanoparticle-anchored nitrogen-doped graphene	Dopamine	30 nM–48 µM	10 nM	[66]
	glucose	40 µM–16.1 mM	12 µM	
Graphene-encapsulated gold nanoparticle	glucose	6 µM–28.5 mM	1 µM	[67]
Cobalt phthalocyanine–ionic liquid–graphene	glucose	0.01–1.3 mM and 1.3–5.0 mM	0.67 µM	[68]
Copper nanoparticle/graphene oxide/single wall carbon nanotube	glucose	1 µM–4.538 mM	0.34 µM	[69]

3.2. Nucleic Acid and Protein Sensing

A sensitive and selective nucleic acids (DNA/ribonucleic acid, RNA) sensor is in high demand for the diagnosing gene-related diseases. DNA sensors also referred to as geno-sensors, are an analytical system, which integrates a sequence-specific probe on a transducer. Thus, the immobilization of DNA strands greatly influences the performance of the electrochemical DNA sensor. In this manner, graphene and its derivatives provide an excellent avenue to develop electrochemical DNA sensor. Hu et al. utilized GO as a DNA probe immobilization layer for electrochemical detection of HIV-1 gene fragment [70]. First, GO is anchored on diazonium functionalized electrode surface via electrostatic attraction, hydrogen bonding or epoxy ring opening. The π-π stacking interaction between the aromatic ring of GO and DNA base ring facilitated DNA immobilization, and impedance measurement was used for the quantitative detection of HIV-1 gene fragment up to 0.11 pM. Moreover, Akhavan et al. developed a graphene nanowall structure and showed an extremely high response to single-strain DNA towards single-strain DNA electrochemical sensing. As a result, they have observed a unique response signal from each kind of basic group through DPV measurement (Figure 3) [71].

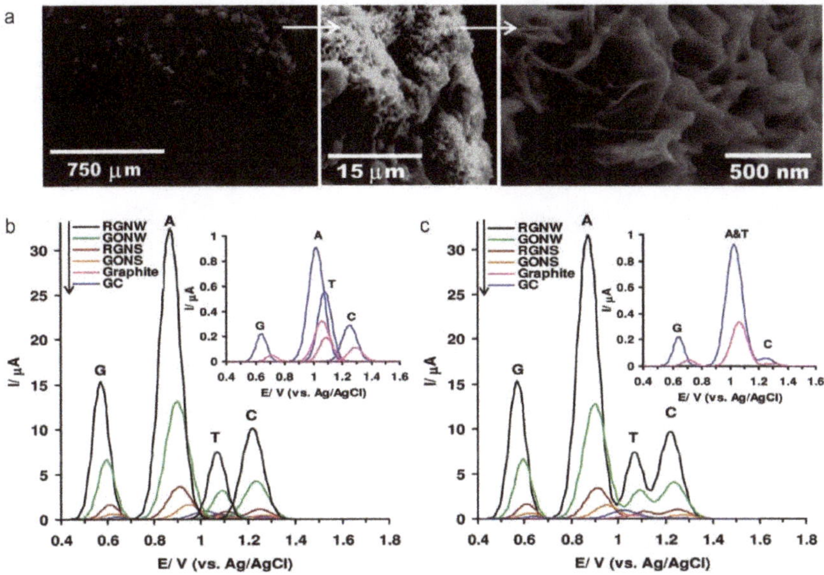

Figure 3. (a) Scanning electron microscopy images of the graphene oxide nanowalls deposited on a graphite rod by using electrophoretic deposition (b,c) Differential pulse voltammetric profiles of the reduced graphene nanowalls (RGNW), graphene oxide nanowalls (GONW), reduced graphene nanosheet (RGNS), and graphene oxide nanosheet (GONS) electrodes as compared to the graphite and glassy carbon (GC) electrodes for detection of (b) the four free bases of DNA (G, A, T, and C) separately, and (c) equimolar mixture of G, A, T, and C. Reproduced with permission from [71], Copyright American Chemical Society, 2012.

Although it is clear that nucleic acids can effectively immobilize on the graphene and its derivatives, many researchers have focused to modify the graphene and its derivatives electrode surface with various materials to achieve improved performance on electrochemical sensor. For example, Tiwari et al. electrophoretically deposited graphene oxide modified iron oxide-chitosan hybrid nanocomposite onto indium tin oxide (ITO) coated glass substrate and utilized for the detection of a pathogenic Escherichia coli DNA with a detection limit of 10 fM [72]. In addition to the surface modification, the enzyme was also utilized to improve electrochemical sensor. Esteban-Fernández de Ávila et al. designed a disposable electrochemical DNA sensor based on carboxymethyl-cellulose-rGO modified screen-printed carbon electrodes. And HRP was utilized to catalyze the redox mediator, tetramethylbenzidine (TMB), and the substrate (H_2O_2) for the detection of the p53 tumor suppressor gene [73]. Besides, graphene can be also utilized for DNA sequencing. With the presence of nanopore on the monolayer graphene, detailed electric signals can be sensed when DNA passes the pore by measuring transverse conductance of DNA. Through this mechanism, Freedman et al. have distinguished long and short DNA using nanopores with graphite polyhedral crystal (GPC)-edges (Figure 4) [74].

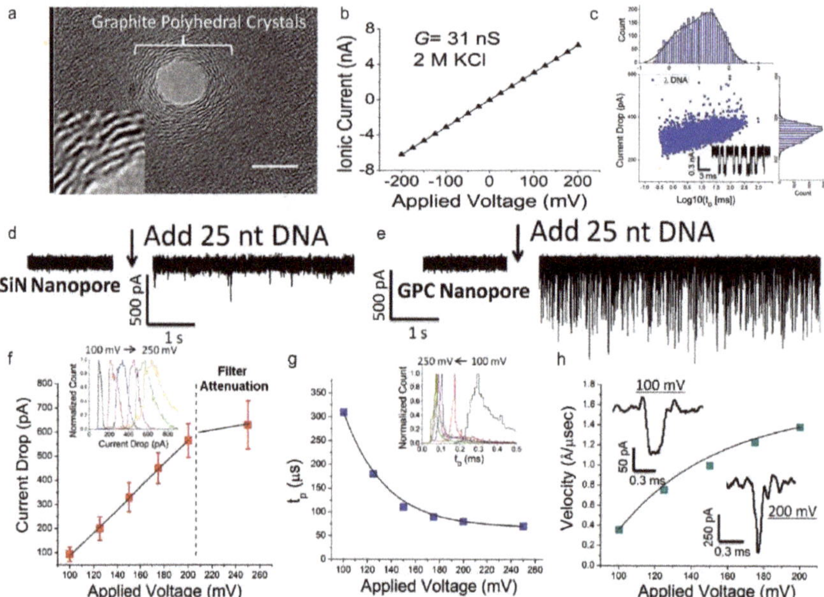

Figure 4. (**a**) Transmission electron microscopy image of the shrunked nanopore with graphite polyhedral crystal (GPC)-edges sculpted by the irradiation of electron beam (6.2 × 10^5 electrons/nm^2·s) on to the single layer graphene. Scale bar = 5 nm. (**b**) Current versus voltage plot for a 5 nm nanopore (in diameter) having GPC-edges in the 2M KCl condition. (**c**) Current drop translocation time scatter plot for double stranded λ-DNA (5 nM concentration, 48.5 kb long) using a 5 nm nanopore with a GPC-edges at 250 mV [1 M KCl, 10 mM Tris, 1 M ethylenediaminetetraacetic acid (EDTA)]. The mean current drop value was 332 ± 62 pA by the scatter plot. (**d–e**) Ionic current traces for single-stranded DNA (25 bases in length) in (**d**) a 5 nm silicon nitride nanopores (50 nm thick) and (**e**) 5 nm graphene nanopores with GPC-edges. (**f–h**) Detailed characterization of 5 nm graphene nanopores with GPC-edges for 25 nucleotide-long DNA fragment sensing. (**f**) The linear increase in current drop based on applied voltages, (**g**) The exponential decrease of peak translocation time based on applied voltages, and (**h**) Calculated translocation velocity from (**e**). The velocity of the 25 nucleotide-long DNA fragment in 5 nm graphene nanopores with GPC-edges was 0.35 Å/μs at 100 mV/room temperature, which demonstrates the slower velocity than the silicon nitride pores. Events were recorded at 100, 125, 150, 175, 200, and 250 mV in 2 M KCl, 10 mM Tris (pH 8), 1 mM EDTA and 10 nM DNA at room temperature. Reproduced from Reference [74] with permission from the American Chemical Society.

In parallel, many biological processes can be also monitored by quantification of specific proteins. Owing to the amphiphilic nature of graphene and its derivate, it provides sufficient active sites to immobilize these probes, including aptamer and antibody to detect specific proteins as well. Wen et al. explored hairpin-shaped DNA aptamer as a cognition element for carcinoembryonic antigen on the gold nanorods functionalized graphene electrode surface [75]. In addition, by targeting membrane protein of pathogenic microbes, Natarajan et al. successfully developed an immunoassay for white spot syndrome virus using a methylene blue dye (MB) immobilized graphene oxide modified glassy carbon electrode (GCE/GO@MB) (Figure 5) [76]. Here, graphene was also utilized to improve antibody immobilization efficiency and enhance electron transfer, the binding on the target produced an enhanced immune-recognition response by the sandwich assay with an enzyme reaction. As mentioned above, considerable approaches have been made to develop and improve graphene-based electrochemical DNA and protein sensors. However, the simultaneous detection of multiple targets in complex biological matrices still remains a major bottleneck for clinical analysis. The recent researches on graphene-based electrochemical biosensors toward various protein and nucleic acids are compared in Table 3.

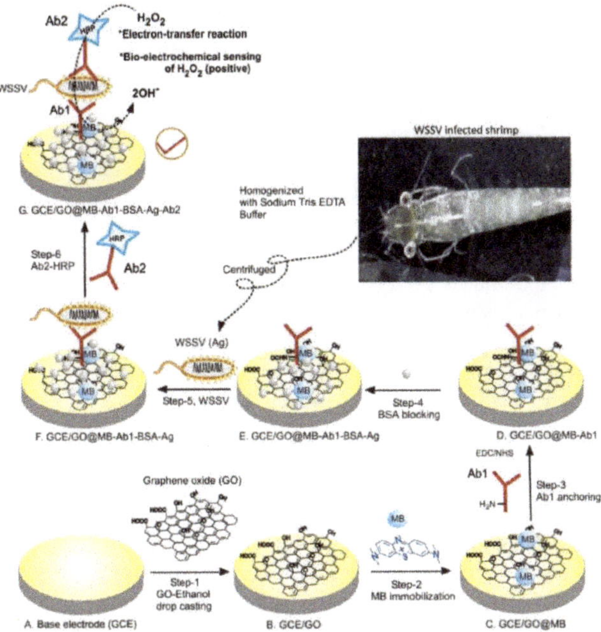

Figure 5. Schematic illustration for the development of electrochemical white spot syndrome virus immunosensor using a methylene blue dye (MB) immobilized graphene oxide modified glassy carbon electrode. Reproduced with permission from [76], Copyright Springer Nature, 2017.

Table 3. Comparison of different graphene-based electrode for protein/nucleic acid detection.

Electrode Materials	Target	Linear Range	Detection Limit	Ref.
Graphene Oxide/probe DNA	HIV-1 gene (cDNA)	1 pM–1 µM	0.11 pM	[70]
Reduced graphene nanowalls	dsDNA	0.1 fM–10 mM	9.4 zM	[71]
Graphene oxide modified iron Oxide/chitosan/probe DNA	*Escherichia coli* O157:¨H7 gene (cDNA)	10 fM–1 µM	10 fM	[72]
Screen-printed carbon/reduced graphene oxide/Carboxy-methyl-cellulose/probe DNA	p53 tumor suppressor gene (cDNA)	10 nM–0.1 µM	2.9 nM	[73]
Nitrogen-doped graphene/Au nanoparticles/probe DNA	multidrug resistance gene	10 fM–100 nM	3.12 fM	[77]
Fe3O4 Nanoparticles/reduced graphene oxide	HIV-1 gene (cDNA)	10 aM–100 pM	-	[78]
Glassy carbon/reduced graphene oxide/polypyrrole–3–carboxylic acid	Breast cancer 1 gene	1 pM–0.1 µM	0.3 pM	[79]
Gold nanorods/graphene/ hairpin-shaped DNA aptamer	Carcinoembryonic antigen	5 pg·mL^{-1}–50 ng·mL^{-1}	1.5 pg·mL^{-1}	[75]
Graphene quantum dot-ionic liquid-nafion/hairpin aptamer	Carcinoembryonic antigen	0.5 fg·mL^{-1}–0.5 ng mL^{-1}	0.34 fg·mL^{-1}	[80]
Graphene/glassy carbon/aptamer	Carcinoembryonic antigen	80 ag·mL^{-1}–950 fg·mL^{-1}	80 ag·mL^{-1}	[81]
Glassy carbon/graphene oxide methylene blue/Antibody	White spot syndrome virus	1.36×10^{-3}–10^7 copies·µL^{-1}	10^3 copies·µL^{-1}	[76]
Graphene-wrapped copper oxide/cysteine	*E. coli* O157:H7	10 CFU·mL^{-1}–10^8 CFU·mL^{-1}	3.8 CFU·mL^{-1}	[82]
Gold/reduced graphene oxide/polyethylenimine	*E. coli*	10 CFU·mL^{-1}–10^4 CFU·mL^{-1}	10 CFU·mL^{-1}	[83]

3.3. Live Cell-based Sensing

Owing to their excellent biocompatibility, solubility, and unique interactions with specific molecules, graphene, and its derivatives has been also utilized to detect a response from the biological process of living cells. For example, effective and accurate characterization of H_2O_2 concentration in a living cell is critical to achieving the normal physiological activities of cells. Wu et al. integrated nitrogen-doped graphene to monitor H_2O_2 release process from live cells through improved electrocatalytic activity [84]. After the injection of phorbol 12-myristate-13-acetate to induce H_2O_2 generation in the neutrophil cells, the rapid increase of amperometric response was able to be observed, which indicates a large amount of H_2O_2 release from the cells. Sun et al., also reported a graphene/Intermetallic platinum/lead (Pt/Pb) nanoplates composites for sensing H_2O_2 release from live macrophage cells (Raw 264.7) (Figure 6) [85]. Through the high-density of electrocatalytic active sites on the unique PtPb nanoplates and the synergistic effect with graphene contributed for outstanding electroanalytical performance. The proposed construct showed 12.7 times higher redox signals than that of commercial Pt/Carbon electrode and able to detect H_2O_2 with a wide linear detection range of 2 nM to 2.5 mM. Zhang et al. also proposed a way to monitor H_2O_2 secretion from viable cells with a freestanding nanohybrid paper electrode composed of 3D ionic liquid (IL) functionalized graphene framework (GF) decorated by gold nanoflowers [86]. The gold nanoflower modified IL–GF was synthesized by a dopamine-assisted one-pot self-assembly method. The resultant nanohybrid paper electrode exhibits good non-enzymatic electrochemical sensing performance toward H_2O_2. Through the real-time tracking of H_2O_2 release from different breast cells attached to the paper electrode allow to distinguish the normal breast cell line HBL-100 from the cancer breast cell line MDA-MB-231 and MCF-7 cells. Liu et al. utilized HRP on a porous graphene electrode to monitor the H_2O_2 release from living cells [87]. A simple method based on silver nanoparticles etching process was proposed to prepare porous graphene network. Owing to the versatile porous structure, the analysis performance was significantly improved by loading large amounts of enzyme and accelerating diffusion rate. A significant low detection limit of 0.0267 nM and wider linear range of 7 orders of magnitude were achieved. A rat adrenal medulla pheochromocytoma cell line PC12 was chosen as a model cancer cell, and H_2O_2 release was monitored within AA stimulation. In a similar manner, Li et al. monitored nitric oxide (NO) by developing a new 3D hydrogel composite via in situ reductions of Au^{3+} on three-dimensional graphene hydrogel [88]. The developed sensor showed improved electrochemical performance compare to pure gold nanoparticles, pure graphene, 3D graphene hydrogels, and gold nanoparticle-graphene hybrids. A linear relation was obtained for 0.05–0.4 mM of NO. Two different normal and cancer skin cell was stimulated with Ach, and concentration-dependent signal increments were analyzed.

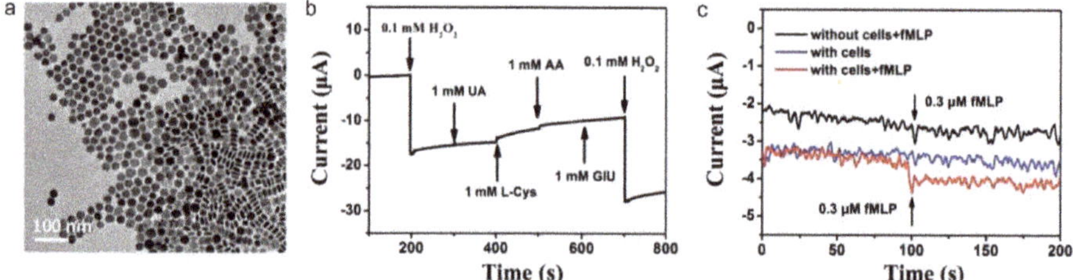

Figure 6. (**a**) Transmission electron microscopy image of PtPb nanoplates. (**b**) Chronoamperometric curves of the graphene/Intermetallic PtPb nanoplates composite (PtPb/G) electrode with the successive addition of 0.1 mM H_2O_2, 1 mM Uric Acid (UA), 1 mM L-Cysteine, 1 mM ascorbic acid (AA), and 1 mM glucose at a constant potential at −0.2 V. (**c**) Amperometric responses of the PtPb/G electrode to the addition of N-formyl methionyl-leucyl-phenylalanine (fMLP) with and without Raw 264.7 cells. Reproduced with permission from [85], Copyright American Chemical Society, 2017.

Differently, Lee et al., utilized graphene-Au hybrid nanoelectrode array (NEAs) to monitor stem differentiation in a non-destructive real-time manner (Figure 7) [89]. Typically, unique multifunctional graphene-Au hybrid NEAs were fabricated via laser interference lithography and physical vapor deposition methods. Followed by surface modification with reduced graphene oxide. The presence of reduced graphene oxide enhanced the cell adhesion and spreading without functionalization with any extracellular matrix proteins, which could work as an insulator and diminish ET between the electrode and electroactive molecules. Owing to the excellent biocompatibility and electrochemical performance of graphene-Au hybrid NEAs, the osteogenic differentiation of human mesenchymal stem cell was successfully monitored through an alkaline phosphatase (ALP)-based enzymatic reaction. During the osteogenesis, ALP expression level is known to be sequentially increased. P-aminophenyl phosphate (PAPP) were introduced to cell prior to electrochemical monitoring, the ALP expressed on the cell catalytically hydrolyzed the PAPP to produce electroactive p-aminophenol (PAP), and the redox reaction between PAP and Quinone imine (QI) was monitored by cyclic voltammogram. Through this mechanism, the osteogenic differentiation of human mesenchymal stem cell was successfully monitored in both non-destructive and real-time manner. Although, stem cell therapy has arisen as a promising method in the field of biomedicine owing to their unique ability to differentiate into multiple cell lineages [90], necessary required destructive analysis process, such as cell lysis and cell fixation were one of a critical limiting factor for further clinical applications. Such a novel electrochemical detection method proposed by graphene-Au hybrid NEAs could be a breakthrough in the preclinical investigation of differentiated stem cells. Consequently, this kind of work is expected to be highly potential to advance stem cell differentiation assays by providing a practical, non-destructive, real-time monitoring tool. The recent researches on graphene-based electrochemical biosensors toward various live cell-based sensing strategies are compared in Table 4.

Table 4. Comparison of different graphene-based electrode for live cell-based detection.

Electrode Materials	Target	Linear Range	Detection Limit	Ref.
Nitrogen doped graphene	H_2O_2	0.5 μM–1.2 mM	0.05 μM	[84]
Graphene/PtPb-nanoplate	H_2O_2	2 nM–2516 μM	2 nM	[85]
Gold nanoflowers modified ionic liquid functionalized graphene framework	H_2O_2	0.5 μM–2.3 mM	100 nM	[86]
HRP supported Porous graphene	H_2O_2	2.77 μM –835 μM	26.7 pM	[87]
Graphene-Pt nanocomposites	H_2O_2	0.5 μM–0.475 mM	0.2 μM	[91]
GNP deposited 3D graphene hydrogel	NO	200 nM –6 μM	9 nM	[88]
GNP/calf thymus DNA/nitrogen-doped graphene	NO	2 nM–500 nM	0.8 nM	[92]
Iron phthalocyanine decorated nitrogen-doped graphene on ITO	NO	0.18 μM–400 μM	0.18 μM	[93]
3-aminophenylboronic acid functionalized graphene foam network	H_2S	0.2 μM–10 μM	50 nM	[94]
Dendritic Pt nanoparticles decorated freestanding graphene paper	DA	87 nM–100 μM	5 nM	[95]
Zn-NiAl layered double hydroxide on reduced graphene oxide	DA	1 nM–1 μM	0.1 nM	[96]
Aryldiazonium Salts and GNP decorated reduced graphene oxide	TNF-α	0.1–150 pg·mL^{-1}	0.1 pg·mL^{-1}	[97]
Graphene-Au hybrid nanoelectrode array	ALP	0.1–10 unit·mL^{-1}	0.03 unit·mL^{-1}	[89]

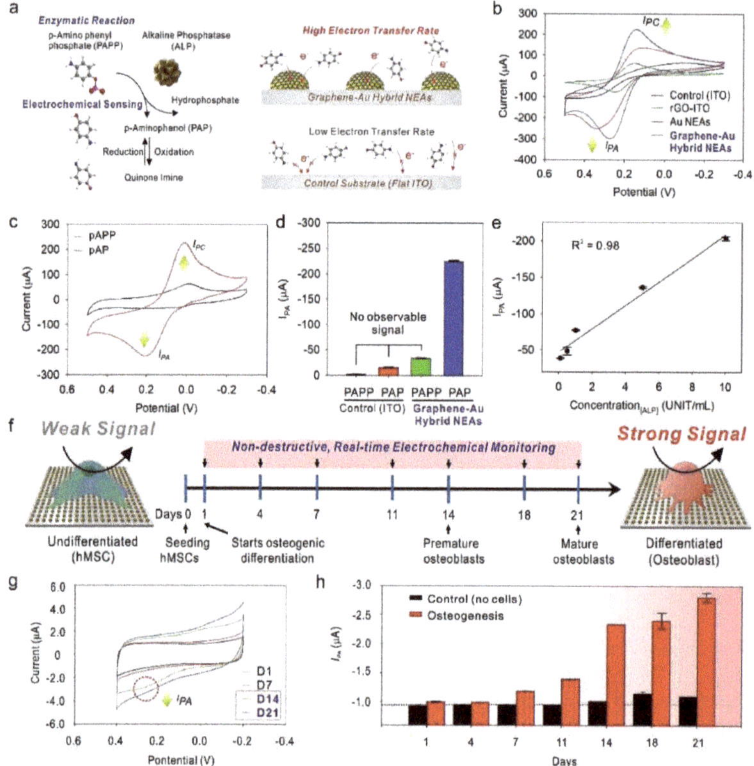

Figure 7. (**a**) Schematic illustration of alkaline phosphatase (ALP) based enzymatic reaction and electrochemical sensing mechanism on the 3D surface in graphene-Au hybrid nanoelectrode arrays (NEAs) compared to 2D flat ITO surface. (**b**) Improved voltammetric response of graphene-Au NEAs compares to bare ITO substrate, rGO-coated ITO substrate, and Au NEAs. (**c,d**) Cyclic voltammogram and Anodic peak (oxidation potential: I_{PA}) value of P-aminophenyl phosphate (PAPP) on graphene-Au NEAs before and after enzyme reaction with ALP. (**e**) The linear correlations between concentrations of ALP and the current signal at I_{PA}. (**f**) Schematic illustration of electrochemical signal change between undifferentiated hMSCs and differentiated osteocyte based on ALP generation. (**g**) Cyclic voltammetry, and (**h**) calculated I_{PC} values from time-dependent monitoring during osteogenesis of hMSCs (range from D1 to D21). Reproduced with permission from [89], Copyright John Wiley and Sons.

4. Conclusions and Future Outlook

It is evident that the exceptional physicochemical properties of graphene and its derivatives make them compelling for various electrochemical biosensor applications. The development and wide application of the electrochemical sensing based on these materials were hindered by the lack of facile and reproducible synthesis method of these materials with defined properties. In order to make it compatible to develop various electrochemical sensors, numerous efforts have been devoted to controllable and scalable production of graphene and its derivatives. Mechanical exfoliation method was first used to obtain graphene; however, the poor controllability and low production were the limiting factors. To achieve scalable production, LPE based methods were developed to obtain a dispersed solution of these materials and its products has been the most widely utilized for the electrochemical biosensors. More or less, recently CVD method was also adopted to synthesize graphene with controllable properties; however, limited scale production of graphene and its derivatives still hinders its applications. For example, various oxygenated surface functional groups provide a relatively high surface area, while it inhibits the performance of electrochemical biosensor as an insulator. Thus, a novel synthesis method which can consistently produce graphene and its derivatives with defined properties (high quality) in a large scale (high yield) and cost-effective manner is still required for future commercialization of graphene and its derivatives based electrochemical biosensors.

During the past years, many different strategies have been also explored to develop novel graphene and its derivatives based electrochemical biosensors for analyzing bio/chemical molecules. It is clear that the optimization through the selection of a suitable surface functionalization method still needs to be revealed, in order to develop a highly effective and reproducible electrochemical biosensors. In addition, the combination with functional materials, such as ionic liquids, nanomaterials and polymers, have provided numerous choices to take the synergistic effect to enhance the electroanalytical performances. However, designing and finding the appropriate combinatorial structure with functional materials must be addressed for each biomolecule to maximize the performance of electrochemical biosensors. Although extensive studies have been made to design and fabricate novel electrochemical biosensors, long-term stability of combinatorial structure and functionalized surfaces (with receptor), in complexed real sample matrices, should be considered. Real biological fluids, such as blood and plasma, always contain various molecules and ions which can cause the interference through nonspecific binding events and undermine the performance of electrochemical biosensors. Furthermore, to adopt graphene and its derivatives for recently arisen in situ live cells and in vivo sensing strategies, more effort, such as long-term toxicity of graphene and its derivatives, as well as incorporated functional materials, should be also discovered. Moreover, since graphene and its derivatives can serve as both the sensing component and transducer, utilization to flexible electrode and miniaturization in size by forming free-standing structures via self-assembly would be an excellent approach for developing an implantable (flexible) and portable (lightweight) sensors as well. However, the lack of a suitable power source is a limiting factor.

In addition, owing to the advantages of specific planar morphology, graphene-like 2D nanomaterials have attracted significant attention as emerging materials for electrochemical sensor approaches. Taking the advantages of diverse composition and structural effect, graphene-like 2D nanomaterials also have promoted great efforts to improve the performance of electrochemical biosensors. Similar to graphene and its derivatives, the defined synthesis method for high quality graphene-like 2D nanomaterials with controllable sizes and thickness, as well as tunable properties, and surface functionalization method is highly desirable for their practical application to electrochemical biosensor development as well. In witness of their current attention, we also endeavor that further improvement of graphene and its derivatives, as well as graphene-like 2D nanomaterial-based electrochemical biosensors, will lead to a significant advance in analytic applications for the highly effective and reliable detection of biomarker and open new avenues in biological and medical fields.

Funding: This research was supported by the Leading Foreign Research Institute Recruitment Program, through the National Research Foundation of Korea (NRF), funded by the Ministry of Science, ICT and Future Planning (MSIP) (2013K1A4A3055268) and Basic Science Research Program through the National Research Foundation of Korea (NRF) funded by the Ministry of Education (2016R1A6A1A03012845).

Conflicts of Interest: The authors declare no conflict of interest.

References

1. Stoller, M.D.; Park, S.; Zhu, Y.; An, J.; Ruoff, R.S. Graphene-based ultracapacitors. *Nano Lett.* **2008**, *8*, 3498–3502. [CrossRef] [PubMed]
2. Weiss, N.O.; Zhou, H.; Liao, L.; Liu, Y.; Jiang, S.; Huang, Y.; Duan, X. Graphene: An emerging electronic material. *Adv. Mater.* **2012**, *24*, 5782–5825. [CrossRef] [PubMed]
3. Lee, C.; Wei, X.; Kysar, J.W.; Hone, J. Measurement of the elastic properties and intrinsic strength of monolayer graphene. *Science* **2008**, *321*, 385–388. [CrossRef] [PubMed]

4. Balandin, A.A.; Ghosh, S.; Bao, W.; Calizo, I.; Teweldebrhan, D.; Miao, F.; Lau, C.N. Superior thermal conductivity of single-layer graphene. *Nano Lett.* **2008**, *8*, 902–907. [CrossRef] [PubMed]

5. Georgakilas, V.; Otyepka, M.; Bourlinos, A.B.; Chandra, V.; Kim, N.; Kemp, K.C.; Hobza, P.; Zboril, R.; Kim, K.S. Functionalization of graphene: Covalent and non-covalent approaches, derivatives and applications. *Chem. Rev.* **2012**, *112*, 6156–6214. [CrossRef] [PubMed]

6. Song, Y.; Luo, Y.; Zhu, C.; Li, H.; Du, D.; Lin, Y. Recent advances in electrochemical biosensors based on graphene two-dimensional nanomaterials. *Biosens. Bioelectron.* **2016**, *76*, 195–212. [CrossRef] [PubMed]

7. Bollella, P.; Fusco, G.; Tortolini, C.; Sanzo, G.; Favero, G.; Gorton, L.; Antiochia, R. Beyond graphene: Electrochemical sensors and biosensors for biomarkers detection. *Biosens. Bioelectron.* **2017**, *89 Pt 1*, 152–166. [CrossRef]

8. Rusling, J.F. Enzyme bioelectrochemistry in cast biomembrane-like films. *Acc. Chem. Res.* **1998**, *31*, 363–369. [CrossRef]

9. Lawal, A.T. Synthesis and utilisation of graphene for fabrication of electrochemical sensors. *Talanta* **2015**, *131*, 424–443. [CrossRef]

10. Pumera, M. Graphene in biosensing. *Mater. Today* **2011**, *14*, 308–315. [CrossRef]

11. Ostrovsky, P.M.; Gornyi, I.V.; Mirlin, A.D. Electron transport in disordered graphene. *Phys. Rev. B* **2006**, *74*, 235443. [CrossRef]

12. Schedin, F.; Geim, A.K.; Morozov, S.V.; Hill, E.W.; Blake, P.; Katsnelson, M.I.; Novoselov, K.S. Detection of individual gas molecules adsorbed on graphene. *Nat. Mater.* **2007**, *6*, 652–655. [CrossRef] [PubMed]

13. Lu, C.; Huang, P.J.; Liu, B.; Ying, Y.; Liu, J. Comparison of Graphene Oxide and Reduced Graphene Oxide for DNA Adsorption and Sensing. *Langmuir* **2016**, *32*, 10776–10783. [CrossRef] [PubMed]

14. Hwang, M.T.; Landon, P.B.; Lee, J.; Choi, D.; Mo, A.H.; Glinsky, G.; Lal, R. Highly specific SNP detection using 2D graphene electronics and DNA strand displacement. *Proc. Natl. Acad. Sci. USA* **2016**, *113*, 7088–7093. [CrossRef] [PubMed]

15. Liu, J.; Liu, Z.; Barrow, C.J.; Yang, W. Molecularly engineered graphene surfaces for sensing applications: A review. *Anal. Chim. Acta* **2015**, *859*, 1–19. [CrossRef] [PubMed]

16. Xu, J.H.; Wang, Y.Z.; Hu, S.S. Nanocomposites of graphene and graphene oxides: Synthesis, molecular functionalization and application in electrochemical sensors and biosensors. A review. *Microchim. Acta* **2017**, *184*, 1–44. [CrossRef]

17. Quintana, M.; Spyrou, K.; Grzelczak, M.; Browne, W.R.; Rudolf, P.; Prato, M. Functionalization of graphene via 1,3-dipolar cycloaddition. *ACS Nano* **2010**, *4*, 3527–3533. [CrossRef] [PubMed]

18. Fang, Y.; Wang, E. Electrochemical biosensors on platforms of graphene. *Chem. Commun. (Camb.)* **2013**, *49*, 9526–9539. [CrossRef]

19. Kuila, T.; Bose, S.; Mishra, A.K.; Khanra, P.; Kim, N.H.; Lee, J.H. Chemical functionalization of graphene and its applications. *Prog. Mater. Sci.* **2012**, *57*, 1061–1105. [CrossRef]

20. Srivastava, R.K.; Srivastava, S.; Narayanan, T.N.; Mahlotra, B.D.; Vajtai, R.; Ajayan, P.M.; Srivastava, A. Functionalized multilayered graphene platform for urea sensor. *ACS Nano* **2012**, *6*, 168–175. [CrossRef]

21. Novoselov, K.S.; Fal'ko, V.I.; Colombo, L.; Gellert, P.R.; Schwab, M.G.; Kim, K. A roadmap for graphene. *Nature* **2012**, *490*, 192–200. [CrossRef] [PubMed]

22. Bonaccorso, F.; Lombardo, A.; Hasan, T.; Sun, Z.P.; Colombo, L.; Ferrari, A.C. Production and processing of graphene and 2d crystals. *Mater. Today* **2012**, *15*, 564–589. [CrossRef]

23. Wang, X.-Y.; Narita, A.; Müllen, K. Precision synthesis versus bulk-scale fabrication of graphenes. *Nat. Rev. Chem.* **2017**, *2*, 0100. [CrossRef]

24. Novoselov, K.S.; Geim, A.K.; Morozov, S.V.; Jiang, D.; Katsnelson, M.I.; Grigorieva, I.V.; Dubonos, S.V.; Firsov, A.A. Two-dimensional gas of massless Dirac fermions in graphene. *Nature* **2005**, *438*, 197–200. [CrossRef] [PubMed]

25. Novoselov, K.S.; Geim, A.K.; Morozov, S.V.; Jiang, D.; Zhang, Y.; Dubonos, S.V.; Grigorieva, I.V.; Firsov, A.A. Electric field effect in atomically thin carbon films. *Science* **2004**, *306*, 666–669. [CrossRef] [PubMed]

26. Zhang, Y.; Tan, Y.W.; Stormer, H.L.; Kim, P. Experimental observation of the quantum Hall effect and Berry's phase in graphene. *Nature* **2005**, *438*, 201–204. [CrossRef] [PubMed]

27. Novoselov, K.S.; Jiang, Z.; Zhang, Y.; Morozov, S.V.; Stormer, H.L.; Zeitler, U.; Maan, J.C.; Boebinger, G.S.; Kim, P.; Geim, A.K. Room-temperature quantum Hall effect in graphene. *Science* **2007**, *315*, 1379. [CrossRef] [PubMed]

28. Cai, M.Z.; Thorpe, D.; Adamson, D.H.; Schniepp, H.C. Methods of graphite exfoliation. *J. Mater. Chem.* **2012**, *22*, 24992–25002. [CrossRef]

29. Ciesielski, A.; Samori, P. Graphene via sonication assisted liquid-phase exfoliation. *Chem. Soc. Rev.* **2014**, *43*, 381–398. [CrossRef]

30. Shen, J.; He, Y.; Wu, J.; Gao, C.; Keyshar, K.; Zhang, X.; Yang, Y.; Ye, M.; Vajtai, R.; Lou, J.; et al. Liquid Phase Exfoliation of Two-Dimensional Materials by Directly Probing and Matching Surface Tension Components. *Nano Lett.* **2015**, *15*, 5449–5454. [CrossRef]

31. He, P.; Zhou, C.; Tian, S.; Sun, J.; Yang, S.; Ding, G.; Xie, X.; Jiang, M. Urea-assisted aqueous exfoliation of graphite for obtaining high-quality graphene. *Chem. Commun. (Camb.)* **2015**, *51*, 4651–4654. [CrossRef] [PubMed]

32. Hummers Jr, W.S.; Offeman, R.E. Preparation of graphitic oxide. *J. Am. Chem. Soc.* **1958**, *80*, 1339. [CrossRef]

33. Ang, P.K.; Wang, S.; Bao, Q.; Thong, J.T.; Loh, K.P. High-throughput synthesis of graphene by intercalation-exfoliation of graphite oxide and study of ionic screening in graphene transistor. *ACS Nano* **2009**, *3*, 3587–3594. [CrossRef] [PubMed]

34. Stankovich, S.; Dikin, D.A.; Dommett, G.H.; Kohlhaas, K.M.; Zimney, E.J.; Stach, E.A.; Piner, R.D.; Nguyen, S.T.; Ruoff, R.S. Graphene-based composite materials. *Nature* **2006**, *442*, 282–286. [CrossRef] [PubMed]

35. Chen, D.; Feng, H.; Li, J. Graphene oxide: Preparation, functionalization, and electrochemical applications. *Chem. Rev.* **2012**, *112*, 6027–6053. [CrossRef] [PubMed]

36. Bae, S.; Kim, H.; Lee, Y.; Xu, X.; Park, J.S.; Zheng, Y.; Balakrishnan, J.; Lei, T.; Kim, H.R.; Song, Y.I.; et al. Roll-to-roll production of 30-inch graphene films for transparent electrodes. *Nat. Nanotechnol.* **2010**, *5*, 574–578. [CrossRef] [PubMed]

37. Forbeaux, I.; Themlin, J.M.; Debever, J.M. Heteroepitaxial graphite on6H−SiC(0001):Interface formation through conduction-band electronic structure. *Phys. Rev. B* **1998**, *58*, 16396–16406. [CrossRef]

38. Berger, C.; Song, Z.; Li, X.; Wu, X.; Brown, N.; Naud, C.; Mayou, D.; Li, T.; Hass, J.; Marchenkov, A.N.; et al. Electronic confinement and coherence in patterned epitaxial graphene. *Science* **2006**, *312*, 1191–1196. [CrossRef]

39. Kim, K.S.; Zhao, Y.; Jang, H.; Lee, S.Y.; Kim, J.M.; Kim, K.S.; Ahn, J.H.; Kim, P.; Choi, J.Y.; Hong, B.H. Large-scale pattern growth of graphene films for stretchable transparent electrodes. *Nature* **2009**, *457*, 706–710. [CrossRef]

40. Li, X.; Cai, W.; An, J.; Kim, S.; Nah, J.; Yang, D.; Piner, R.; Velamakanni, A.; Jung, I.; Tutuc, E.; et al. Large-area synthesis of high-quality and uniform graphene films on copper foils. *Science* **2009**, *324*, 1312–1314. [CrossRef]

41. Li, X.; Cai, W.; Colombo, L.; Ruoff, R.S. Evolution of graphene growth on Ni and Cu by carbon isotope labeling. *Nano Lett.* **2009**, *9*, 4268–4272. [CrossRef] [PubMed]

42. Gao, L.; Ren, W.; Xu, H.; Jin, L.; Wang, Z.; Ma, T.; Ma, L.P.; Zhang, Z.; Fu, Q.; Peng, L.M.; et al. Repeated growth and bubbling transfer of graphene with millimetre-size single-crystal grains using platinum. *Nat. Commun.* **2012**, *3*, 699. [CrossRef] [PubMed]

43. Banszerus, L.; Schmitz, M.; Engels, S.; Dauber, J.; Oellers, M.; Haupt, F.; Watanabe, K.; Taniguchi, T.; Beschoten, B.; Stampfer, C. Ultrahigh-mobility graphene devices from chemical vapor deposition on reusable copper. *Sci. Adv.* **2015**, *1*, e1500222. [CrossRef]

44. Geim, A.K.; Novoselov, K.S. The rise of graphene. *Nat. Mater.* **2007**, *6*, 183–191. [CrossRef] [PubMed]

45. Ascherio, A.; Schwarzschild, M.A. The epidemiology of Parkinson's disease: Risk factors and prevention. *Lancet Neurol.* **2016**, *15*, 1257–1272. [CrossRef]

46. Logroscino, G.; Piccininni, M.; Marin, B.; Nichols, E.; Abd-Allah, F.; Abdelalim, A.; Alahdab, F.; Asgedom, S.W.; Awasthi, A.; Chaiah, Y.; et al. Global, regional, and national burden of motor neuron diseases 1990-2016: A systematic analysis for the Global Burden of Disease Study 2016. *Lancet Neurol.* **2018**, *17*, 1083–1097. [CrossRef]

47. Trounson, A.; DeWitt, N.D. Pluripotent stem cells progressing to the clinic. *Nat. Rev. Mol. Cell Biol.* **2016**, *17*, 194–200. [CrossRef] [PubMed]

48. Fox, I.J.; Daley, G.Q.; Goldman, S.A.; Huard, J.; Kamp, T.J.; Trucco, M. Stem cell therapy. Use of differentiated pluripotent stem cells as replacement therapy for treating disease. *Science* **2014**, *345*, 1247391. [CrossRef]

49. Connolly, B.S.; Lang, A.E. Pharmacological treatment of Parkinson disease: A review. *JAMA* **2014**, *311*, 1670–1683. [CrossRef]

50. Choi, J.H.; Lee, J.H.; Oh, B.K.; Choi, J.W. Localized surface plasmon resonance-based label-free biosensor for highly sensitive detection of dopamine. *J. Nanosci. Nanotechnol.* **2014**, *14*, 5658–5661. [CrossRef]

51. Lee, J.H.; Lee, T.; Choi, J.W. Nano-Biosensor for Monitoring the Neural Differentiation of Stem Cells. *Nanomaterials (Basel)* **2016**, *6*, 224. [CrossRef] [PubMed]

52. Ping, J.; Wu, J.; Wang, Y.; Ying, Y. Simultaneous determination of ascorbic acid, dopamine and uric acid using high-performance screen-printed graphene electrode. *Biosens. Bioelectron.* **2012**, *34*, 70–76. [CrossRef] [PubMed]

53. Fan, Z.J.; Lin, Q.Q.; Gong, P.W.; Liu, B.; Wang, J.Q.; Yang, S.R. A new enzymatic immobilization carrier based on graphene capsule for hydrogen peroxide biosensors. *Electrochim. Acta* **2015**, *151*, 186–194. [CrossRef]

54. Panieri, E.; Gogvadze, V.; Norberg, E.; Venkatesh, R.; Orrenius, S.; Zhivotovsky, B. Reactive oxygen species generated in different compartments induce cell death, survival, or senescence. *Free Radic. Biol. Med.* **2013**, *57*, 176–187. [CrossRef] [PubMed]

55. Wang, L.; Ye, Y.J.; Lu, X.P.; Wu, Y.; Sun, L.L.; Tan, H.L.; Xu, F.G.; Song, Y.H. Prussian blue nanocubes on nitrobenzene-functionalized reduced graphene oxide and its application for H_2O_2 biosensing. *Electrochim. Acta* **2013**, *114*, 223–232. [CrossRef]

56. Yoon, J.; Lee, T.; Bapurao, G.B.; Jo, J.; Oh, B.K.; Choi, J.W. Electrochemical H_2O_2 biosensor composed of myoglobin on MoS_2 nanoparticle-graphene oxide hybrid structure. *Biosens. Bioelectron.* **2017**, *93*, 14–20. [CrossRef] [PubMed]

57. Song, H.; Ni, Y.; Kokot, S. A novel electrochemical biosensor based on the hemin-graphene nano-sheets and gold nano-particles hybrid film for the analysis of hydrogen peroxide. *Anal. Chim. Acta* **2013**, *788*, 24–31. [CrossRef] [PubMed]

58. Simplicio, J. Hemin monomers in micellar sodium lauryl sulfate. A spectral and equilibrium study with cyanide. *Biochemistry* **1972**, *11*, 2525–2528. [CrossRef]

59. Shao, Y.; Zhang, S.; Engelhard, M.H.; Li, G.; Shao, G.; Wang, Y.; Liu, J.; Aksay, I.A.; Lin, Y. Nitrogen-doped graphene and its electrochemical applications. *J. Mater. Chem.* **2010**, *20*, 7491. [CrossRef]

60. Ensafi, A.A.; Alinajafi, H.A.; Jafari-Asl, M.; Rezaei, B.; Ghazaei, F. Cobalt ferrite nanoparticles decorated on exfoliated graphene oxide, application for amperometric determination of NADH and H_2O_2. *Mater. Sci. Eng. C Mater. Biol. Appl.* **2016**, *60*, 276–284. [CrossRef]

61. Aydogdu Tig, G. Highly sensitive amperometric biosensor for determination of NADH and ethanol based on Au-Ag nanoparticles/poly(L-Cysteine)/reduced graphene oxide nanocomposite. *Talanta* **2017**, *175*, 382–389. [CrossRef] [PubMed]

62. Han, S.; Du, T.; Jiang, H.; Wang, X. Synergistic effect of pyrroloquinoline quinone and graphene nano-interface for facile fabrication of sensitive NADH biosensor. *Biosens. Bioelectron.* **2017**, *89 Pt 1*, 422–429. [CrossRef]

63. Balamurugan, J.; Thanh, T.D.; Kim, N.H.; Lee, J.H. Facile fabrication of FeN nanoparticles/nitrogen-doped graphene core-shell hybrid and its use as a platform for NADH detection in human blood serum. *Biosens. Bioelectron.* **2016**, *83*, 68–76. [CrossRef] [PubMed]

64. Liu, B.; Ouyang, X.; Ding, Y.; Luo, L.; Xu, D.; Ning, Y. Electrochemical preparation of nickel and copper oxides-decorated graphene composite for simultaneous determination of dopamine, acetaminophen and tryptophan. *Talanta* **2016**, *146*, 114–121. [CrossRef] [PubMed]

65. Li, Y.C.; Liu, J.; Liu, M.H.; Yu, F.; Zhang, L.; Tang, H.; Ye, B.C.; Lai, L.F. Fabrication of ultra-sensitive and selective dopamine electrochemical sensor based on molecularly imprinted polymer modified graphene@carbon nanotube foam. *Electrochem. Commun.* **2016**, *64*, 42–45. [CrossRef]

66. Thanh, T.D.; Balamurugan, J.; Lee, S.H.; Kim, N.H.; Lee, J.H. Effective seed-assisted synthesis of gold nanoparticles anchored nitrogen-doped graphene for electrochemical detection of glucose and dopamine. *Biosens. Bioelectron.* **2016**, *81*, 259–267. [CrossRef]

67. Thanh, T.D.; Balamurugan, J.; Hwang, J.Y.; Kim, N.H.; Lee, J.H. In situ synthesis of graphene-encapsulated gold nanoparticle hybrid electrodes for non-enzymatic glucose sensing. *Carbon* **2016**, *98*, 90–98. [CrossRef]

68. Chaiyo, S.; Mehmeti, E.; Siangproh, W.; Hoang, T.L.; Nguyen, H.P.; Chailapakul, O.; Kalcher, K. Non-enzymatic electrochemical detection of glucose with a disposable paper-based sensor using a cobalt phthalocyanine-ionic liquid-graphene composite. *Biosens. Bioelectron.* **2018**, *102*, 113–120. [CrossRef]

69. Yang, T.T.; Xu, J.K.; Lu, L.M.; Zhu, X.F.; Gao, Y.S.; Xing, H.K.; Yu, Y.F.; Ding, W.C.; Liu, Z. Copper nanoparticle/graphene oxide/single wall carbon nanotube hybrid materials as electrochemical sensing platform for nonenzymatic glucose detection. *J. Electroanal. Chem.* **2016**, *761*, 118–124. [CrossRef]

70. Hu, Y.; Li, F.; Han, D.; Wu, T.; Zhang, Q.; Niu, L.; Bao, Y. Simple and label-free electrochemical assay for signal-on DNA hybridization directly at undecorated graphene oxide. *Anal. Chim. Acta* **2012**, *753*, 82–89. [CrossRef]

71. Akhavan, O.; Ghaderi, E.; Rahighi, R. Toward single-DNA electrochemical biosensing by graphene nanowalls. *ACS Nano* **2012**, *6*, 2904–2916. [CrossRef] [PubMed]

72. Tiwari, I.; Singh, M.; Pandey, C.M.; Sumana, G. Electrochemical genosensor based on graphene oxide modified iron oxide–chitosan hybrid nanocomposite for pathogen detection. *Sens. Actuators B Chem.* **2015**, *206*, 276–283. [CrossRef]

73. Esteban-Fernandez de Avila, B.; Araque, E.; Campuzano, S.; Pedrero, M.; Dalkiran, B.; Barderas, R.; Villalonga, R.; Kilic, E.; Pingarron, J.M. Dual functional graphene derivative-based electrochemical platforms for detection of the TP53 gene with single nucleotide polymorphism selectivity in biological samples. *Anal. Chem.* **2015**, *87*, 2290–2298. [CrossRef] [PubMed]

74. Freedman, K.J.; Ahn, C.W.; Kim, M.J. Detection of long and short DNA using nanopores with graphitic polyhedral edges. *ACS Nano* **2013**, *7*, 5008–5016. [CrossRef] [PubMed]

75. Wen, W.; Huang, J.Y.; Bao, T.; Zhou, J.; Xia, H.X.; Zhang, X.H.; Wang, S.F.; Zhao, Y.D. Increased electrocatalyzed performance through hairpin oligonucleotide aptamer-functionalized gold nanorods labels and graphene-streptavidin nanomatrix: Highly selective and sensitive electrochemical biosensor of carcinoembryonic antigen. *Biosens. Bioelectron.* **2016**, *83*, 142–148. [CrossRef] [PubMed]

76. Natarajan, A.; Devi, K.S.; Raja, S.; Senthil Kumar, A. An Elegant Analysis of White Spot Syndrome Virus Using a Graphene Oxide/Methylene Blue based Electrochemical Immunosensor Platform. *Sci. Rep.* **2017**, *7*, 46169. [CrossRef] [PubMed]

77. Chen, M.; Hou, C.; Huo, D.; Bao, J.; Fa, H.; Shen, C. An electrochemical DNA biosensor based on nitrogen-doped graphene/Au nanoparticles for human multidrug resistance gene detection. *Biosens. Bioelectron.* **2016**, *85*, 684–691. [CrossRef] [PubMed]

78. Teymourian, H.; Salimi, A.; Khezrian, S. Development of a New Label-free, Indicator-free Strategy toward Ultrasensitive Electrochemical DNA Biosensing Based on Fe3O4 Nanoparticles/Reduced Graphene Oxide Composite. *Electroanalysis* **2017**, *29*, 409–414. [CrossRef]

79. Shahrokhian, S.; Salimian, R. Ultrasensitive detection of cancer biomarkers using conducting polymer/electrochemically reduced graphene oxide-based biosensor: Application toward BRCA1 sensing. *Sens. Actuators B-Chem.* **2018**, *266*, 160–169. [CrossRef]

80. Huang, J.Y.; Zhao, L.; Lei, W.; Wen, W.; Wang, Y.J.; Bao, T.; Xiong, H.Y.; Zhang, X.H.; Wang, S.F. A high-sensitivity electrochemical aptasensor of carcinoembryonic antigen based on graphene quantum dots-ionic liquid-nafion nanomatrix and DNAzyme-assisted signal amplification strategy. *Biosens. Bioelectron.* **2018**, *99*, 28–33. [CrossRef]

81. Ge, L.; Wang, W.; Sun, X.; Hou, T.; Li, F. Affinity-Mediated Homogeneous Electrochemical Aptasensor on a Graphene Platform for Ultrasensitive Biomolecule Detection via Exonuclease-Assisted Target-Analog Recycling Amplification. *Anal. Chem.* **2016**, *88*, 2212–2219. [CrossRef] [PubMed]

82. Pandey, C.M.; Tiwari, I.; Singh, V.N.; Sood, K.N.; Sumana, G.; Malhotra, B.D. Highly sensitive electrochemical immunosensor based on graphene-wrapped copper oxide-cysteine hierarchical structure for detection of pathogenic bacteria. *Sens. Actuators B-Chem.* **2017**, *238*, 1060–1069. [CrossRef]

83. Jijie, R.; Kahlouche, K.; Barras, A.; Yamakawa, N.; Bouckaert, J.; Gharbi, T.; Szunerits, S.; Boukherroub, R. Reduced graphene oxide/polyethylenimine based immunosensor for the selective and sensitive electrochemical detection of uropathogenic Escherichia coli. *Sens. Actuators B-Chem.* **2018**, *260*, 255–263. [CrossRef]

84. Wu, P.; Qian, Y.D.; Du, P.; Zhang, H.; Cai, C.X. Facile synthesis of nitrogen-doped graphene for measuring the releasing process of hydrogen peroxide from living cells. *J. Mater. Chem.* **2012**, *22*, 6402–6412. [CrossRef]

85. Sun, Y.; Luo, M.; Meng, X.; Xiang, J.; Wang, L.; Ren, Q.; Guo, S. Graphene/Intermetallic PtPb Nanoplates Composites for Boosting Electrochemical Detection of H_2O_2 Released from Cells. *Anal. Chem.* **2017**, *89*, 3761–3767. [CrossRef] [PubMed]

86. Zhang, Y.; Xiao, J.; Lv, Q.; Wang, L.; Dong, X.; Asif, M.; Ren, J.; He, W.; Sun, Y.; Xiao, F.; et al. In Situ Electrochemical Sensing and Real-Time Monitoring Live Cells Based on Freestanding Nanohybrid Paper Electrode Assembled from 3D Functionalized Graphene Framework. *ACS Appl. Mater. Interfaces* **2017**, *9*, 38201–38210. [CrossRef] [PubMed]

87. Liu, Y.; Liu, X.; Guo, Z.; Hu, Z.; Xue, Z.; Lu, X. Horseradish peroxidase supported on porous graphene as a novel sensing platform for detection of hydrogen peroxide in living cells sensitively. *Biosens. Bioelectron.* **2017**, *87*, 101–107. [CrossRef]

88. Li, J.; Xie, J.; Gao, L.; Li, C.M. Au nanoparticles-3D graphene hydrogel nanocomposite to boost synergistically in situ detection sensitivity toward cell-released nitric oxide. *ACS Appl. Mater. Interfaces* **2015**, *7*, 2726–2734. [CrossRef]

89. Lee, J.H.; Choi, H.K.; Yang, L.; Chueng, S.D.; Choi, J.W.; Lee, K.B. Nondestructive Real-Time Monitoring of Enhanced Stem Cell Differentiation Using a Graphene-Au Hybrid Nanoelectrode Array. *Adv. Mater.* **2018**, *30*, e1802762. [CrossRef]

90. Rathnam, C.; Chueng, S.D.; Yang, L.; Lee, K.B. Advanced Gene Manipulation Methods for Stem Cell Theranostics. *Theranostics* **2017**, *7*, 2775–2793. [CrossRef]

91. Zhang, Y.; Bai, X.; Wang, X.; Shiu, K.K.; Zhu, Y.; Jiang, H. Highly sensitive graphene-Pt nanocomposites amperometric biosensor and its application in living cell H_2O_2 detection. *Anal. Chem.* **2014**, *86*, 9459–9465. [CrossRef] [PubMed]

92. Dou, B.; Li, J.; Jiang, B.; Yuan, R.; Xiang, Y. DNA-Templated In Situ Synthesis of Highly Dispersed AuNPs on Nitrogen-Doped Graphene for Real-Time Electrochemical Monitoring of Nitric Oxide Released from Live Cancer Cells. *Anal. Chem.* **2019**, *91*, 2273–2278. [CrossRef] [PubMed]

93. Xu, H.; Liao, C.; Liu, Y.; Ye, B.C.; Liu, B. Iron Phthalocyanine Decorated Nitrogen-Doped Graphene Biosensing Platform for Real-Time Detection of Nitric Oxide Released from Living Cells. *Anal. Chem.* **2018**, *90*, 4438–4444. [CrossRef] [PubMed]

94. Hu, X.B.; Liu, Y.L.; Wang, W.J.; Zhang, H.W.; Qin, Y.; Guo, S.; Zhang, X.W.; Fu, L.; Huang, W.H. Biomimetic Graphene-Based 3D Scaffold for Long-Term Cell Culture and Real-Time Electrochemical Monitoring. *Anal. Chem.* **2018**, *90*, 1136–1141. [CrossRef] [PubMed]

95. Zan, X.; Bai, H.; Wang, C.; Zhao, F.; Duan, H. Graphene Paper Decorated with a 2D Array of Dendritic Platinum Nanoparticles for Ultrasensitive Electrochemical Detection of Dopamine Secreted by Live Cells. *Chemistry* **2016**, *22*, 5204–5210. [CrossRef] [PubMed]

96. Asif, M.; Aziz, A.; Wang, H.; Wang, Z.; Wang, W.; Ajmal, M.; Xiao, F.; Chen, X.; Liu, H. Superlattice stacking by hybridizing layered double hydroxide nanosheets with layers of reduced graphene oxide for electrochemical simultaneous determination of dopamine, uric acid and ascorbic acid. *Mikrochim. Acta* **2019**, *186*, 61. [CrossRef] [PubMed]

97. Qi, M.; Zhang, Y.; Cao, C.; Zhang, M.; Liu, S.; Liu, G. Decoration of Reduced Graphene Oxide Nanosheets with Aryldiazonium Salts and Gold Nanoparticles toward a Label-Free Amperometric Immunosensor for Detecting Cytokine Tumor Necrosis Factor-alpha in Live Cells. *Anal. Chem.* **2016**, *88*, 9614–9621. [CrossRef] [PubMed]

 *nanomaterials*

Review

The Thermal, Electrical and Thermoelectric Properties of Graphene Nanomaterials

Jingang Wang [1,†], Xijiao Mu [2,†] and Mengtao Sun [2,*]

[1] Computational Center for Property and Modification on Nanomaterials, College of Sciences, Liaoning Shihua University, Fushun 113001, China; jingang_wang@lnpu.edu.cn
[2] Center for Green Innovation, Beijing Key Laboratory for Magneto-Photoelectrical Composite and Interface Science, School of Mathematics and Physics, University of Science and Technology Beijing, Beijing 100083, China; shumuxijiao@163.com
* Correspondence: mengtaosun@ustb.edu.cn; Tel.: +86-18621012696
† These authors contributed equally to this work.

Received: 11 December 2018; Accepted: 30 January 2019; Published: 6 February 2019

Abstract: Graphene, as a typical two-dimensional nanometer material, has shown its unique application potential in electrical characteristics, thermal properties, and thermoelectric properties by virtue of its novel electronic structure. The field of traditional material modification mainly changes or enhances certain properties of materials by mixing a variety of materials (to form a heterostructure) and doping. For graphene as well, this paper specifically discusses the use of traditional modification methods to improve graphene's electrical and thermoelectrical properties. More deeply, since graphene is an atomic-level thin film material, its shape and edge conformation (zigzag boundary and armchair boundary) have a great impact on performance. Therefore, this paper reviews the graphene modification field in recent years. Through the change in the shape of graphene, the change in the boundary structure configuration, the doping of other atoms, and the formation of a heterostructure, the electrical, thermal, and thermoelectric properties of graphene change, resulting in broader applications in more fields. Through studies of graphene's electrical, thermal, and thermoelectric properties in recent years, progress has been made not only in experimental testing, but also in theoretical calculation. These aspects of graphene are reviewed in this paper.

Keywords: graphene; electrical; thermal; thermoelectric; applications

1. Introduction of Graphene

As an ideal two-dimensional (2D) material, graphene has become a popular topic of scientific research due to its distinctive physical and chemical properties since it was first proposed in 2004 [1–3]. Monolayer graphene has a hexagonal honeycomb structure comprised of the thinnest 2D crystal; graphene film is only one atom thick [2]. The sp^2 hybridization of C atoms also gives graphene its novel properties. As the basic unit of C-based materials, graphene can form zero-dimensional fullerenes, curl into one-dimensional C nanotubes, and form graphite by means of AA stacking or AB stacking; see Figure 1a [2].

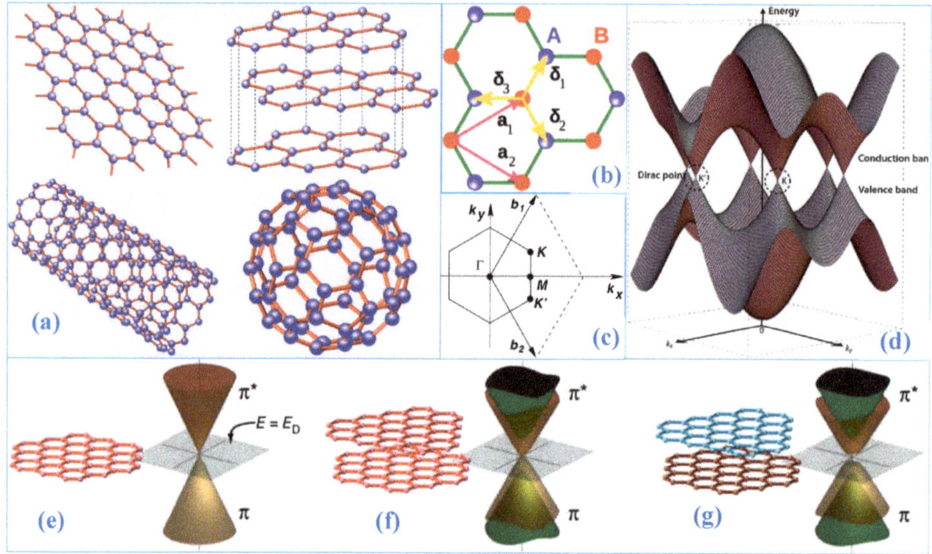

Figure 1. (**a**) Carbon-based nanomaterials [2]. Copyright 2008, Springer. (**b**) The planar crystal structure of graphene: a_1 and a_2 are the lattice vectors of the unit cell; δ_1, δ_2, and δ_3 are vectors mutually adjacent to each other [3]. (**c**) The Brillouin zone (BZ) corresponding to the graphene lattice (b_1 and b_2 are the reciprocal space bases; K, M, K′, etc. represent k-space high symmetry points. The coordinate axis is the k-space basis vector.) [3]. Copyright 2009, American Physical Society. (**d**) Energy band diagram of graphene's hexagonal lattice [4]. Copyright 2009, Springer. Electronic structure of a single layer (**e**), symmetric double layer (**f**), and asymmetric double layer (**g**) of graphene. The energy bands depend only on in-plane momentum because the electrons are restricted to motion in a two-dimensional plane. The Dirac crossing points (marked by a grey cross and square) are at energy E_D [5]. Copyright 2006, Science.

Graphene is a transparent conductor that can be used to replace the current liquid crystal display materials [5]. It is one of the smallest known inorganic nanomaterials, is very strong and hard, and is 100 times stronger than the best steel in the world [6]. The elastic modulus of the theoretical calculations and experimental measurements are 1.05–1.24 TPa and 1 TPa, respectively [7–9]. The thermal conductivity (TC) of graphene is up to 5300 $\text{Wm}^{-1}\,\text{K}^{-1}$ [10]. At room temperature, the carrier mobility of graphene is as high as 15,000 $\text{cm}^2\,\text{V}^{-1}\,\text{s}^{-1}$ [2], and its carrier mobility is not affected by temperatures in the range of 10–100 K, which proves that defect scattering is the main scattering mechanism of the electrons in graphene [11]. The light absorption of single-layer graphene is only 2.3% [12–14], and it has excellent nonlinear optical properties [15]. Graphene also has exceptional magnetic and spintronic properties [16].

2. The Crystal Structure and Electronic Structure of Graphene

The ideal graphene is the thinnest two-dimensional crystal (0.35 nm). The bond length between C and C is 0.142 nm (a_{c-c}). In the crystal structure of graphene, each cell contains two C atoms, A and B. Carbon atom A and atom B are not equivalent; see Figure 1b. The lattice constant is 2.460 Å [17]. Each C atom has four valence electrons, three of which form three σ bonds in sp^2 orbital hybridization with neighboring C atoms, respectively, in the graphene plane. This regular hexagonal crystal structure makes graphene's planar structure extremely stable. The tensile elastic modulus of graphene is as high as 1 TPa and the tensile strength is as high as 130 GPa [7–9]. The structure of graphene remains stable when external forces are applied to it; the other electron in the *P* orbital contributes to the nonlocalized

π and π^* bonds, which form the highest occupied molecular orbital (HOMO) and lowest unoccupied molecular orbital (LUMO). The π and π^* bonds degenerate at point K in the Brillouin zone (BZ) of graphene; see Figure 1c. The Fermi surface shrinks to a point, forming a bandwidth-free metal-like band structure [18–21], which gives graphene extremely high carrier mobility.

Researchers have described the interactions between graphene's π electrons using a tightly bound model [3,18,20]:

$$\varepsilon^\pm(k_x, k_y) = \pm\gamma_0\sqrt{1 + 4\cos\frac{\sqrt{3}k_x a}{2}\cos\frac{k_y a}{2} + 4\cos^2\frac{k_y a}{2}} \tag{1}$$

where the value of a is $\sqrt{3}a_{c-c}$ and γ_0 is the transfer integral that corresponds to the matrix element between π orbitals of adjacent carbon atoms, with the magnitude ranging from 2.8 eV to 3.1 eV. The plus energy of Equation (1) corresponds to the π^* band, and the minus energy corresponds to the π bond, respectively. Figure 1d illustrates the electronic dispersion in the honeycomb lattice, which is the electronic structure of monolayer graphene. Since the electronic states near the Dirac points (DPs) are composed of different sublattices, the dual component wave function corresponds to their relative contributions. The Schrödinger equation corresponding to these electrons is as follows [2]:

$$\hat{H} = \hbar v_F \begin{pmatrix} 0 & k_x - ik_y \\ k_x + ik_y & 0 \end{pmatrix} = \hbar v_F \vec{\sigma} \bullet \vec{k} \tag{2}$$

The Fermi group velocity corresponds to $v_F \approx 1 \times 10^6 ms^{-1}$, $\vec{\sigma}$ corresponds to the Pauli matrix, and $\vec{k}$ corresponds to the momentum of graphene's quasiparticles. Figure 1e–g shows the electronic structures of single-layer graphene and the electronic structure of double-layer graphene in the AA and AB stacking modes.

Graphene's unique physical and chemical properties are closely related to its electronic energy band structure. The energy level distribution of graphene can be calculated based on independent C atoms and the potential generated by the surrounding C atoms as perturbation. Spread out near the DPs, the energy is linearly related to the wave vector (similar to the dispersion of photons), and a singularity occurs at DPs. This means that the effective mass of the electrons in graphene is zero near the Fermi energy level, which also explains the material's exceptional electrical properties.

3. Graphene's Novel Electronic Properties

Graphene's exclusive crystal and electronic structures give it many novel and inimitable physical properties, including anomalous quantum hall effects (QHEs) [20,22,23], ambipolar electric field effects (AEFEs) [24], Klein tunneling (KT) [25,26], and ballistic transport (BT) [27,28]. With further research on graphene nanomaterials, more electrical characteristics have been discovered and recognized.

3.1. Current Vortices, Electron Viscosity, and Negative Nonlocal Resistance

Electron transport properties are similar to those of viscous fluid in strongly related electronics systems, known as the quantum critical effect, which is a typical collision-controlled mass transfer characteristic. However, the study of this phenomenon has been hindered by the lack of macroscopic features of electron viscosity. Levitov et al. determined the vortex characteristics associated with the easy verification of the macroscopic mass transfer performance [29–32]. The eddy current generated by viscous flow can be applied to the field of the opposite driving current, which results in a negative nonlocal voltage. This suggests that the latter plays an important role in the superconducting zero resistance viscosity system. In addition to providing a diagnosis of viscous mass transfer from the ohmic flow, the electrical response of the signal changes, providing a powerful way to directly measure the viscosity and resistivity ratios. The intense electron hole plasma interaction in high-mobility graphene provides an exclusive connection between the critical quantum electron transport and many hydrodynamic phenomena.

It has been a prominent problem that the zero resistance of the viscous flow is related to superconductivity. The viscous flow caused by the vortex leads to the unique macroscopic transport behavior. Levitov et al. predicted that the vorticity of the shear flow with viscosity may cause the return current [33], which is opposite to the applied field. Figure 2a,b shows a negative signature of the nonlocal voltage, which provides a clear signature of collective viscosity. The characteristic symbols associated with this behavior change the spatial pattern of the potential, as shown in Figure 2, which can be observed directly using modern scanning microscopy techniques [34].

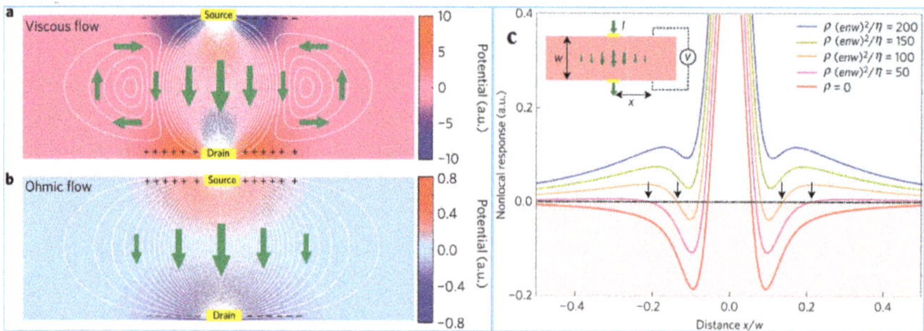

Figure 2. Current streamlines and potential map for viscous and ohmic flows. White lines show current streamlines, colors show electrical potential, and arrows show the direction of current. (**a**) Mechanism of a negative electrical response: viscous shear flow generates vorticity and a backflow on the side of the main current path, which leads to charge buildup of the sign opposing the flow and results in a negative nonlocal voltage. (**b**) In contrast, ohmic current flows down the potential gradient, producing a nonlocal voltage in the flow direction. (**c**) Nonlocal response for different resistivity-to-viscosity ratios, ρ/η [33]. Copyright 2016, Springer Nature.

Figure 2a,b shows the basic properties of the negative response from the shear flow. The collective behavior of the viscous system results from the momentum exchange of the fast exchange carrier while preserving the conservation of net momentum. The momentum is still referred to as a conserved quantity as it will produce a hydrodynamic momentum transfer mode; that is, the momentum flows in space, in a transverse diffusion to the source drain current and away from the nominal current path. As a result of this process, the shear flow is established to produce vorticity and the application field of the (noncompressible fluid) opposite reflux direction. The direct effect of such a complex and apparently noncurrent mode on the electrical response produces a reverse electric field acting on the opposite field driving the source drain current; see Figure 2c. This results in the presence of a negative nonlocal resistance, even in the presence of quite significant ohmic flow; see Figure 2c.

3.2. Transition Between Electrons and Photos

Graphene's inimitable electronic properties make it a new infrared frequency domain plasmon waveguide and terahertz metamaterial [35–49]. By excitation of photons or electrons, the collective oscillations of electrons on the surface of a conductor is the surface plasmon. When photons and graphene's surface plasmon are coupled, they form surface plasmon polaritons (SPPs). As an alternative to conventional metal plasmon, graphene surface plasmons (GSPs) can be used for optical metamaterials [40], optical absorption, and optical conversion [41,42].

Researchers have found that when light hits a single layer of graphene, it can slow down by hundreds of times [48,49]. Researchers have studied the hot carriers located inside graphene, which form GSPs on graphene's surface; see Figure 3a. Figure 3b–g shows the test results and analysis of the experimental results (The GP in the figure stands for graphene plasmon.). Slowing photons (particles of light) travel through a single layer of graphene much as electrons do through the same material. Electrons travel very fast in a single layer of graphene, at a million meters per second, about one-third the speed of light in a vacuum. The speed of the two particles is close enough to strongly interact. When the material is adjusted to match the speed of the two particles, graphene slows the speed of the photons and the rapid movement of the electrons across a graphene surface. This suggests that graphene can be used to create a possible intrinsic effect, producing light instead of capturing it. The theoretical research suggests this could generate light by a new way. This conversion is possible because electrons' drift velocity in graphene is close to light, breaking the "light barrier" just as shock waves produce sound when breaking the "sound barrier". In graphene, this will result in a shock wave of light on a two-dimensional level.

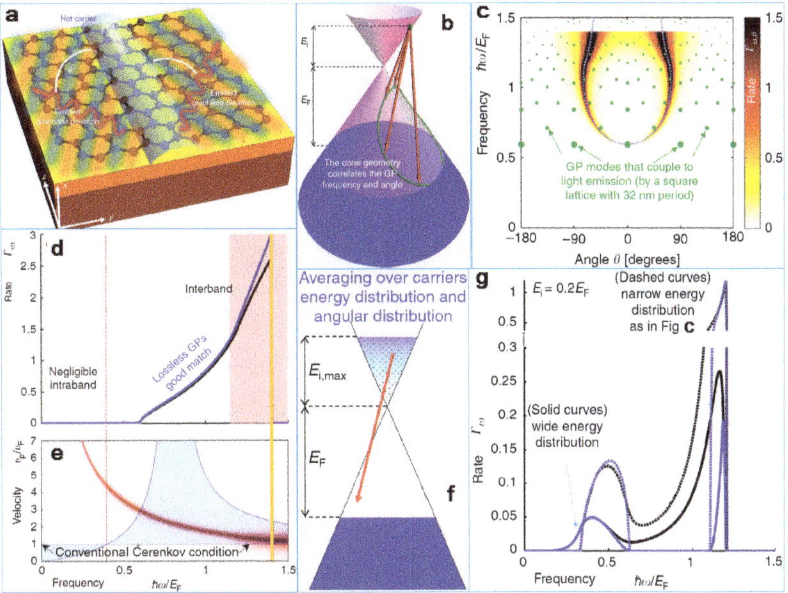

Figure 3. (**a**) Illustration of the plasmon emission from charge carriers in graphene via a 2D Čerenkov process. (**b**) Illustration of the possible transitions. The hot carrier (green dot) has a range of potential transitions (red arrows) with distinct final states (green curves and circles), emitting plasmons that satisfy conservation of momentum and energy (corresponding to the height and angle of the red arrows). In this way, the cone geometry correlates the graphene surface plasmon (GSP) frequency and angle. The projection of these arrows to a 2D plane predicts the in-plane angle y of the plasmonic emission, matching the (**c**) map of GSP emission rate as a function of frequency and angle. (**d**) Spectrum of the ČE (Čerenkov effects) GSP emission process of radiation rate. (**e**) Explaining the GSP emission with the quantum ČE. The GSP phase velocity is plotted as a red curve, with its thickness presenting the GSP loss. (**f**) Illustration of the distribution of hot carriers, which is taken to be an exponential multiplied by the linear electron density of states in graphene. The exponential decay is (**g**) $e^{E_1/0.01eV}$ with maximum hot carrier energy of $E_{i,max} = 0.2eV$ corresponding to (**d**) [48]. Copyright 2016, Springer Nature.

This new approach may eventually become more efficient and adjustable. Highly efficient and controllable plasmon generation properties that are compatible with current microchip technology are a new method of creating core chips for optical circuits, which is a new research direction in the evolution of computer technology toward smaller and more efficient devices.

3.3. Electron Transport Properties in Nitrogen-Doped Graphene

The presence of impurities, especially charged impurities in a lattice, can affect graphene's electrical properties [50–54]. When approaching the DPs, due to the loss of state density, graphene's transport properties are highly sensitive to the scattering of charged impurities. Therefore, physical and chemical doping of graphene can be used to improve its electrical properties. Among many dopants, both theories and experiments have shown that an n-doped graphene lattice can realize n-doping and the carrier can maintain high mobility [55].

High-quality nitrogen-doped graphene is prepared using chemical vapor deposition (CVD) methods. The spectral line in Figure 4a indicates that there is a characteristic peak at 0 eV, and it was confirmed that the doped nitrogen was graphitized nitrogen, in which the doping concentration was 2.0%. The doping of nitrogen atoms between graphene lattices was demonstrated using a scanning tunneling microscope (STM) graph. The white region in Figure 4b represents the location of nitrogen doping. In the figure, the color was caused by the intervalley scattering in the graphene lattice caused by single doping. The illustration is the fast Fourier transform (FTF) corresponding to the STM. In Figure 4c, the peak height of the D peak in the Raman spectrum increased with the increase in nitrogen doping. The 2D peak indicates that the nitrogen-doped graphene prepared via the experiment had good crystallization. The diagram is a test sample image. Figure 4d shows the relationship between the conductivity σ of nitrogen-doped graphene with different doping concentrations and the grid voltage V_g at temperature (T) = 9 K. Solid lines of different colors represent different doping concentrations. As shown in the figure, the negative location of the DPs is due to the sample's n-type doping characteristics, and as the nitrogen doping concentration increases, the DPs moves toward negative values. Figure 4e shows the evolution of the charge impurity density with the nitrogen doping concentration. Figure 4f shows that after the introduction of more scattering centers, carrier mobility in nitrogen-doped graphene decreases as the nitrogen doping concentration increases.

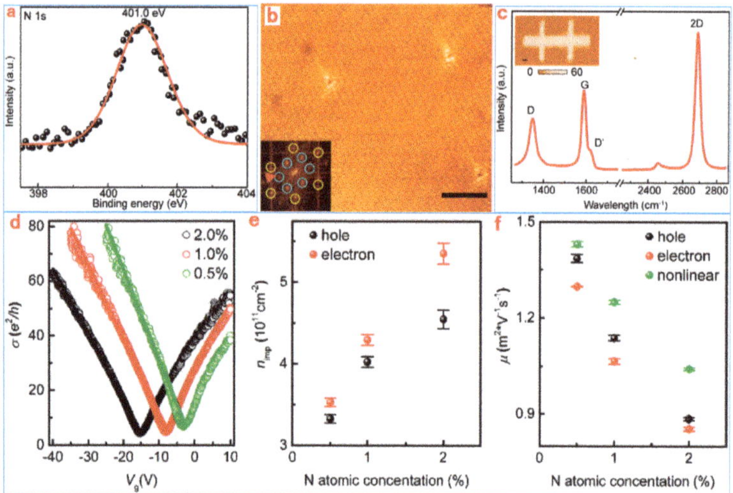

Figure 4. (**a**) N 1s core-level XPS (X-ray photoelectron spectroscopy) spectra of an as-formed graphitic N-doped (2% N atomic concentration) graphene film. (**b**) Scanning tunneling microscope (STM) image of an as-formed graphitic N-doped graphene sample ($V_{bias} = -10mV$ and $I_{set} = 100pA$). (**c**) Raman spectra of a graphitic N-doped (2% N atomic concentration) graphene film (The letters on each Raman peak in the figure indicate the vibrational symmetry of the Raman peak.). Transport properties (conductivity (**d**), carrier concentration (**e**) and mobility (**f**)) of as-formed N-doped graphene films [55]. Copyright 2017, American Chemical Society.

The test results confirmed the electron hole valley scattering asymmetry of the charge transport property of the nitrogen-doped graphene. Valley electron scattering is much stronger than hole scattering, and the scattering rate increases with the increase in the nitrogen doping concentration.

This is because the nitrogen-doped graphene forms a positive center, causing large-angle scattering for electron transport. Therefore, graphene carrier scattering can be effectively regulated by adding different amounts of graphene nitrogen to control graphene's electrical characteristics.

3.4. Strong Current Tolerance

Graphene can also withstand very high currents and quickly balance out missing charges. Researchers conducted an experiment to prove that the electrons in graphene are extremely mobile and react very quickly. Collisions of Xe^+ with the supercharged charges on graphene films can detach many electrons from a very precise point [56]. Within a few femtoseconds, graphene can quickly resupply the electrons. This will lead to the appearance of ultrastrong currents that cannot be sustained under normal circumstances. These unusual electrical properties make graphene a potential candidate for future electronic applications.

Researchers used a single xenon ion to travel through the graphene layer, so that each xenon ion carried approximately 20 electrons through the graphene region. At this point, the C atom with a missing electron is squeezed out of the graphene, and the location of the missing electron is immediately replenished by other electrons, which lasts only a matter of femtoseconds. This means there is a large amount of electron flow in a short period of time; that is, a high density of current is transmitted on graphene. Figure 5 shows a schematic diagram of the experimental equipment and test results. Strong currents from areas adjacent to the graphene membrane quickly replenish the electrons before the positive charges repel each other and cause an explosion. To accomplish this, graphene must carry a current density approximately 1000 times higher than any material would normally tolerate. The surrounding current density is 1000 times that of a normally destructible material, but graphene can withstand these extreme currents with no damage. The movement of extremely high charges on graphene is of great significance for a range of potential applications, and graphene may be used in superfast electronics in the future.

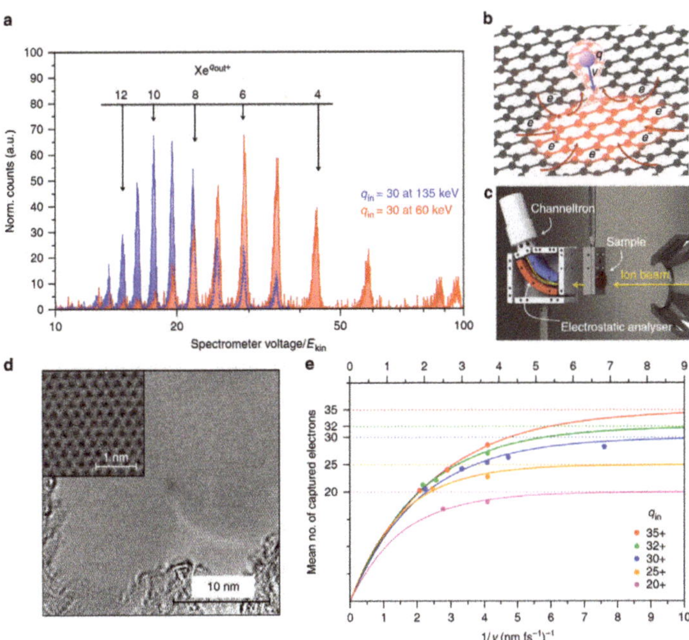

Figure 5. Experimental scheme and results. (**a**) Measured spectra of a Xe^{30+} beam at kinetic energies of

135 and 60 keV (blue and red, respectively) transmitted through a freestanding single-layer graphene (SLG) sheet with different amount of charge injected (q_{in}). Exit charge states q_{out} are calculated from the spectrometer voltage of the electrostatic analyzer. The exit charge state distribution shifts towards smaller average exit charge $\bar{q}_{out}$ for slower ions. (**b**) Schematic of the interaction process between freestanding SLG and an approaching highly charged ion (HCI). The HCI extracts a lot of charge from a very limited area on the femtosecond time scale, leading to a temporary charge-up of the impact region (q is the amount of electricity, and v is the speed.). (**c**) Sketch of the experimental setup with the target holder and electrostatic analyzer. (**d**) TEM image of a freestanding monolayer of graphene after irradiation with Xe^{40+} ions at 180 keV with an applied fluence of 10^{12} ions per cm^2 (about six impacts on the shown scale). No holes or nanosized topographic defects could be observed. (**e**) Average number of captured and stabilized electrons ($q_{in} - \bar{q}_{out}$) after transmission of $Xe^{q_{in}+}$ ions through a single layer of graphene as a function of the inverse projectile velocity for different incident charge states [56]. Copyright 2016, Springer Nature.

3.5. Novel Electrical Properties of Graphene/Graphene van der Waals Heterostructure

A van der Waals (VdW) heterostructure is a vertical stack of two-dimensional materials. Based on the rich functionality of two-dimensional materials, more engineering manipulation can be realized to obtain unexpected new characteristics. The graphene-based van der Waals heterostructure is introduced as a substrate from h-BN (hexagonal Boron Nitride). The peculiar physical and chemical characteristics of the heterogeneous structure have provoked much interest. Of the many VdW heterostructures based on graphene (graphene/graphene, graphene/h-BN, graphene/MoS_2, and graphene/Si, among others), VdW heterostructures produced by double layers of graphene are particularly interesting. Because the lattices of the two layers correspond precisely, the physical and chemical properties are equivalent [57–61]. Novel electrical properties are produced in the superlattices of graphene/graphene heterostructures [62–64].

Researchers from the Massachusetts Institute of Technology (MIT) have found that the electron properties of vertically stacked graphene heterostructures are closely related to the arrangement of C atoms in graphene [62]. The different arrangement of atoms also affects the movement of electrons between the layers. In general, electrical behavior is dominated by energy, and the energy of electrons involved in the movement of electrons between atoms within a single layer of graphene is on an order of magnitude of eV, while the energy involved in the movement of electrons within a double layer of graphene is on an order of magnitude of several hundred meV. Figure 6 shows a diagram of experimental equipment and calculation results [62].

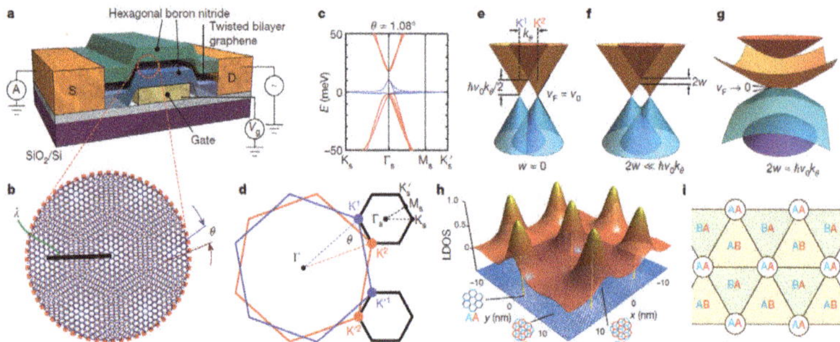

Figure 6. (**a**) Schematic diagram of twisted double graphene devices. (**b**) Moiré superlattice in graphene/graphene heterostructure. (**c**) The band structure of the double graphene at an angle of $\theta = 1.08°$. (**d**) Mini-BZ in graphene/graphene heterostructure. (**e–g**) A schematic diagram of overlapping bands in different energy regions. Localized density of states (LDOS) (**h**) of heterogeneous structural regions in different stacking modes(**i**). [62]. Copyright 2018, Springer Nature.

For graphene with a highly ordered structure, the electrical properties depend on symmetry. The MIT researchers created a rotating, twisted, two-layer graphene heterostructure that controls the electron state of the entire system through interactions between the electrons. The dislocation caused by rotation misaligns the electron band structure in the graphene layer and increases the single cell size. Twisted bilayer graphene (TwBLG) produces two new electron states [63,64]. One electronic state is the Mott insulated body that results from the strong repulsion between the electrons. The other is the superconducting state that results from the strong attraction between electrons to produce zero resistance. When the small rotation angle approaches the magic angle (<1.05°), distortion of the vertical stack of atoms in the double-layer graphene area will form a narrow electronic band, enhancing the electronic interaction effect and resulting in a nonconductive Mott insulation state. In a Mott insulation state, a small charge carrier can be successfully converted into a superconducting state. Figure 7 shows the test results of superconductivity in a graphene/graphene heterostructure [63].

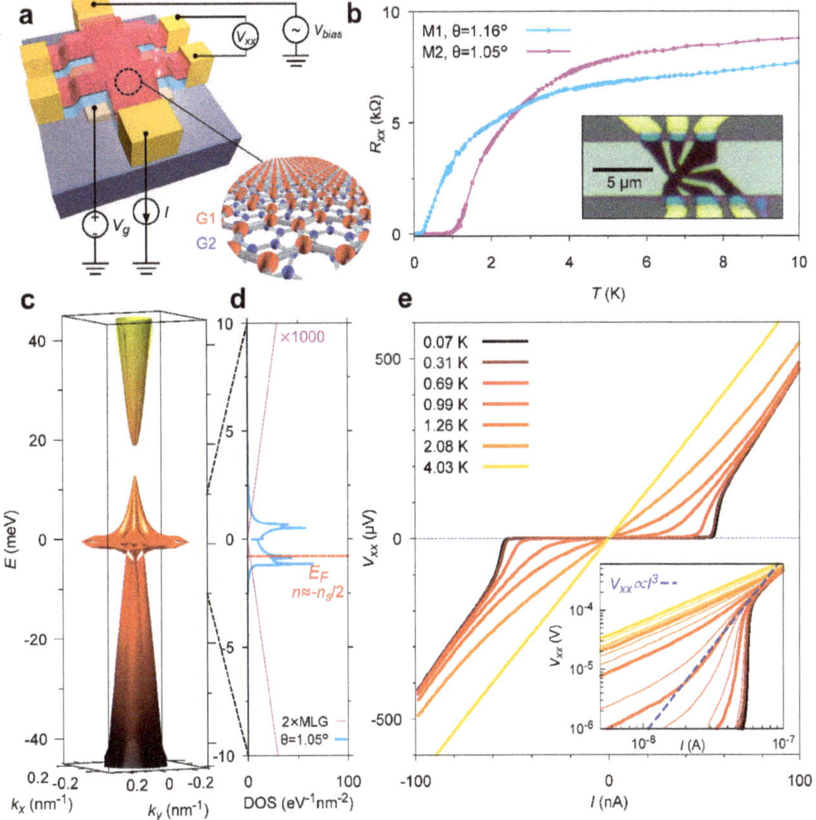

Figure 7. 2D superconductivity in a graphene superlattice. (**a**) Schematic of a typical twisted bilayer graphene (TwBLG) device and four-probe measurement scheme. The stack consists of top h-BN, rotated graphene bilayers (G1, G2), and bottom h-BN. (**b**) Measured four-probe resistance ($R_{xx} = V_{xx}/I$) in two devices M1 and M2, with twist angles $\theta = 1.16°$ and $\theta = 1.05°$, respectively. (**c**) The band structure of TBG at $\theta = 1.05°$ in the first mini-Brillouin zone (MBZ) of the superlattices. (**d**) The density of states (DOS) corresponding to the bands shown in (**c**), zoomed in to -10–10 meV. (**e**) I–V curves for device M2 measured at $n = -1.44 \times 10^{12}$ cm^{-2} and various temperatures. At the lowest temperature of 70 mK, the I–V curve shows a critical current of approximately 50 nA [63]. Copyright 2018, Springer Nature.

The MIT researchers created TwBLG that controls the electron states of the entire system through interactions between electrons. The dislocation caused by rotation misaligns the electron band structure in the graphene layer and increases the single cell size. The study found that in stacked graphene layers, electrical behavior is very sensitive to the arrangement of atoms, affecting the movement of electrons between layers.

3.6. The Interaction between Plasmons and Electrons in Graphene

The surface plasmons of 2D nanomaterials have distinctive electrical properties, and applications such as the interaction between plasmons and electrons and catalytic reactions are also well known, especially in surface plasmons based on grapheme [65–70].

Cao et al. reported on a plasmon–exciton coupling device [69]. They first prepared a complementary metal-oxide semiconductor (CMOS)-like device composed of gold dots and graphene. The device can easily be added to the gate voltage or can inject current into the device through the source and drain electrodes. They included a series of studies on the electrical properties of plasma–exciton-coupled devices under different gate voltages and source–drain voltages. The source–drain bias voltages at different gate voltages were measured before and after the catalyzed molecules were placed in the center of the device. The source–drain bias increased at the same gate voltage after the molecules were placed and showed varying bias change behaviors at different gate voltages; see Figure 8a,b. This occurred because the chemical potential of graphene is regulated by the gate voltage. Figure 8c,d shows the operation of the device current as the bias voltage changed. First, the volt–ampere characteristics of the bias voltage and current showed a perfect linear relationship. Second, the volt–ampere characteristic curve (slope) changed little at different gate voltages when no molecules were added to the device. When there was no molecule, the gate voltage had little ability to regulate the chemical potential of graphene. Third, after the device was added to the molecule, the slope of the volt–ampere characteristic curve was very different, indicating that the gate voltage had an increased regulation range for the device's chemical potential. These phenomena indicate that charge transfer occurs between the molecule and graphene and that the gate voltage also regulates the charge transfer. This can also be concluded by the relationship between the gate voltage and conductivity, as shown in Figure 8e,f. Focusing on the relationship between the volt–ampere characteristic curve and the gate voltage, the gate voltage can be regulated in both positive and negative directions for charge transfer. This is a very good property for photooxidative reduction catalysis. Since the oxidation reaction requires holes, the reduction reaction requires electrons. Such devices provide free holes or electrons for the reaction through different gate voltage controls.

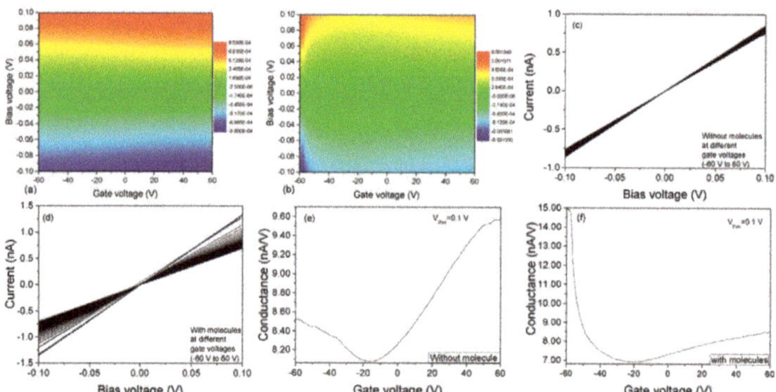

Figure 8. Gate- and bias-voltage-dependent electrical current for a device without (**a**) and with (**b**) addition of molecules; (**c**,**d**) the current varies with bias voltage at different gate voltages; (**e**,**f**) gate-voltage-dependent conductance for a device without and with molecules, where $V_{Bias} = 0.1\ V$ [69]. Copyright 2017, Wiley.

4. The Thermal and Thermoelectric Properties of Graphene

4.1. The Thermal Conductivity (TC) Measurement of Graphene

In 2008, researchers used confocal micro-Raman spectroscopy to measure the heat transport of suspended graphene nanoribbons [71,72]. They found that the thermal conductivity (κ, TC) of SLG reached 3500–5300 W/mK, and the mean free path of the phonons was 775 nm at room temperature [71]. The device is shown in Figure 9a. The experimental results confirmed that the heat transfer of graphene mainly comes from the phonon contribution. The same study was conducted again in 2011. To be immune to air heat transfer, researchers measured suspended graphene at the pore size of 2.9–9.7 µm in a vacuum environment to eliminate the loss of ambient air heat. Figure 9b shows the device. The TC of graphene was measured up to (2.6 ± 0.9)–$(3.1 \pm 1.0) \times 103$ Wm^{-1} K^{-1} at a temperature of 350 K. The experimental device configuration is shown in Figure 9b near the TC of graphene up to (2.6 ± 0.9)–$(3.1 \pm 1.0) \times 10^3$ Wm^{-1} K^{-1} [72].

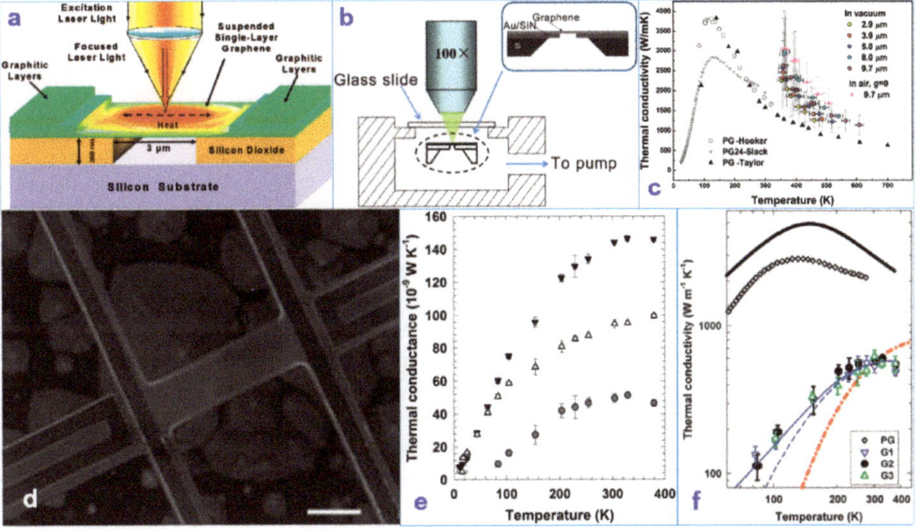

Figure 9. The thermal conductivity (TC) of graphene. (**a**) Schematic of the experiment showing the excitation laser light focused on a graphene layer suspended across a trench [71]. Copyright 2008, American Chemical Society. (**b**) Schematic of the experimental setup for thermal transport measurement of suspended graphene [72]. (**c**) Thermal conductivity of the suspended chemical vapor-deposited (CVD) graphene as a function of the measured temperature of the graphene monolayer suspended in a vacuum over holes of various diameters [72]. Copyright 2011, American Chemical Society. (**d**) SEM image of the suspended device, the central beam, and the folded edge of the SLG ribbon near the right electrode. (**e**) Measured thermal conductance of G2 before (solid downward triangles) and after (unfilled upward triangles) the SLG was etched, with the difference being the contribution from the SLG (circles). (**f**) The relation diagram of thermal conductivity of different samples with temperature change [73]. Copyright 2010, Science.

The phonon transmission mode and scattering mechanism of graphene's thermal conductivity have a significant impact. Numerous studies have shown that the substrate of graphene has an unavoidable effect on its thermal conductivity. After the two-dimensional material contacts the substrate, the thermal conductivity normally decreases significantly after the two-dimensional material contacts the substrate because the thermal conduction is mainly determined by the phonon conduction. When the graphene comes into contact with the substrate [73–75], the surface or edge of the graphene is very sensitive. In 2010, researchers measured the TC of graphene deposited on silica substrates and found that it had been reduced due to scattering at the substrate interface, but it still reached 1000 Wm^{-1} K^{-1} [73]. Figure 9c shows the thermal conductivity of the suspended CVD graphene as a function of the measured temperature of the graphene monolayer suspended in a vacuum over holes of various diameters. Figure 9d shows a STM representation of the graphene sample. The thermal properties of graphene nanostructures may play a very important role in future nanodevices. Figure 9e shows a comparison between the thermal conductivity of graphene before and after etching and that of single-layer graphene. Figure 9f shows the relation diagram of thermal conductivity of different samples with temperature changes.

It is apparent from the measurement of the thermal conductivity of graphene that all factors, such as the shape and thickness, affect graphene's thermal conductivity. In general, the size of the graphene in the heat transfer direction is positively correlated with the thermal conductivity. In general, the larger the linear size of graphene, the higher the thermal conductivity; when graphene is on the substrate, the substrate has a great influence on the thermal conductivity. This is because the interaction between the graphene and the substrate leads to changes in the graphene's lattice constant. The temperature also affects the magnitude of the TC.

4.2. Length-Dependent and Temperature-Dependent TC of Graphene

Based on the relationship between the TC and length, researchers also specifically analyzed the relationship between the TC and the temperature [75]. The results indicate that all of the graphene nanoribbons exhibit the same properties as the temperature. Thermal conductance (σ) adds to the platform as the temperature increases. The TC (κ) increases with the length of the graphene nanoribbons over the entire temperature range.

Earlier sections of this article cited examples of theoretical calculations for the TC of graphene [76]. However, the role of graphene in the ZA mode during heat transfer remains controversial. In particular, the solution to the Peierls–Boltzmann equation shows that the ZA mode is the main heat carrier for graphene transport and has a strong dimensional dependence [77–79]. Research conducted through experimental data and nonequilibrium molecular dynamics (NEMD) simulation resulted in the perfect combination [80,81]. Figure 10c shows the relationship between the TC (κ) and the length (L) [75]. The value of κ increased with the increase in L at a temperature of 300 K. The temperature did not saturate in the longest $L = 9$ μm of graphene. k was proportional to logL at L > 700 nm. This is because the length of graphene is comparable to the phonon mean free path. The researchers also observed that the size far exceeded the ballistic length at 1000 K and at 300 K; see Figure 10d. After comparing the different aspect ratios of the graphene nanoribbons, their simulated results agree with Figure 10d. The relationship between the TC and the length of the single-layer graphene nanoribbons has been revealed experimentally.

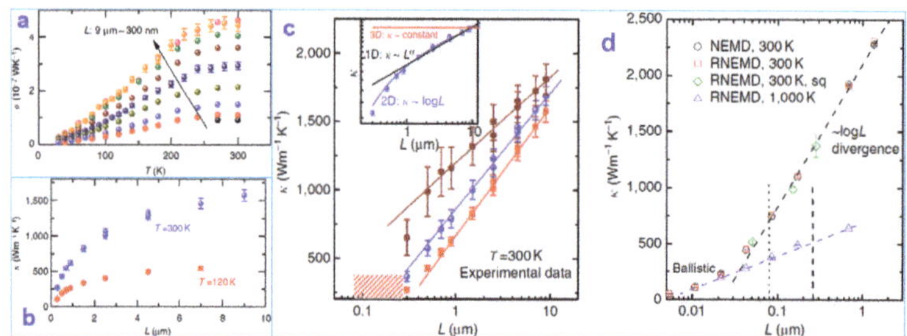

Figure 10. (**a**) The variation in the thermal conductance (σ) of a graphene nanoribbon with temperature (T) over a length of 9 μm to 300 nm. (**b**) Image of the TC (κ) with length (L) at temperatures of 120 K and 300 K. R_{total} is the thermal contact thermal resistance. Experimental (**c**) and various simulation methods (**d**) results on length-dependent thermal conductivity [76]. Copyright 2014, Springer Nature.

The researchers used experimental measurements and nonequilibrium molecular dynamics to simulate the thermal conduction of a single layer of suspended graphene as a function of the temperature and sample length. Different from bulk materials, the thermal conductivity continues to increase at 300 K. Although the length of the sample is much larger than the mean free path of the phonons, the thermal conductivity remains logarithmically discrete with the length of the sample. This is the result of the two-dimensional properties of phonons in graphene, which provides a basis for understanding heat transfer in two-dimensional materials.

4.3. Influence of Boundary or Configuration on Thermal Property and Thermal Rectification Effect

Many theoretical studies and experiments have shown that the different boundaries and configurations of graphene nanoribbons have a certain impact on the TC [82–93].

The TC of zigzag-type graphene nanoribbons is 30% higher than that of armchair-type graphene nanoribbons at room temperature, indicating obvious heat transport anisotropy [82]. Figure 11a,b demonstrates that the heat flux of Y-type graphene nanoribbons studied via molecular dynamic (MD) simulation shows a significant thermal rectification effect. Moreover, the bilayer Y-type nanoribbons can achieve a larger rectification ratio due to the interaction between the layers than the single-layered Y-type nanoribbons, thus providing a theoretical basis for the thermal management of nanoelectronics [84]. Figure 11c shows the phonon properties of three-terminal graphene nanojunctions (GNJs) [85]. This study demonstrates that the transport direction of the heat flux moves along the narrow end to the wide end and produces an obvious ballistic thermal rectification effect. The thermal rectification ratio of 200% is dependent on the asymmetry of the nanojunction, indicating excellent ballistic heat transport properties, making the preparation of nanodevices possible.

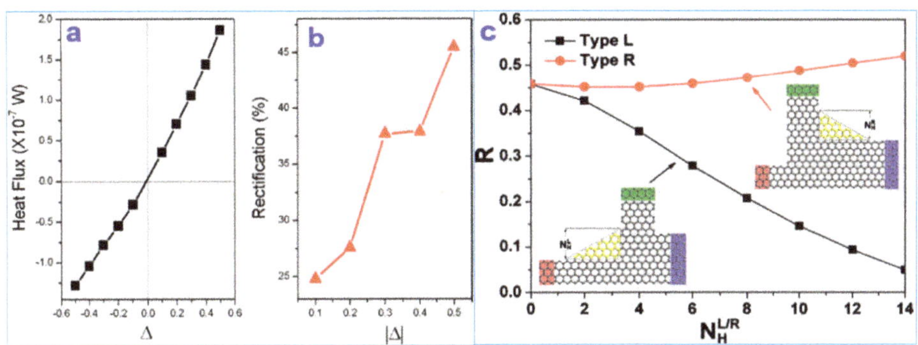

Figure 11. (**a**) Heat flux (*J*) versus Δ; (**b**) Rectifications versus Δ for the single-layer graphene Y junctions [84]. Copyright 2011, Royal Society of Chemistry. (**c**) The rectification ratio R versus the side height of corner $N_H^{L/R}$ for the two types of asymmetric three-terminal graphene nanojunctions (TGNJs) with armchair-edged corners [85]. Copyright 2011, American Physical Society.

4.4. The Effect of Atomic Edge Variation and Size Change on TC

Using the nonequilibrium green function (NEGF) method combined with first-principles calculations, researchers revealed the influence of the edge atom position on the TC of graphene nanoribbons [94]. Figure 12a,b presents the results of this study. When the width of the nanoribbons is $n > 12$, the TC possesses obvious quantization characteristics and is independent of the nanoribbon width. When the nanoribbon width is $2 < n < 12$, due to the obvious boundary effect of the nanoribbons, the quantum heat transport is destroyed.

Another theoretical simulation revealed the dependence of the TC on the length of graphene [95]. Researchers simulated the relationship between the TC and the lateral dimension of graphene nanoribbons and again revealed the mean free path of acoustic phonons and the nonmonotonic dependence of the nanoribbon length L. Moreover, past studies reported that the bulk three-dimensional phonons in graphene greatly contribute to the TC in the graphene plane. Figure 12c shows the results.

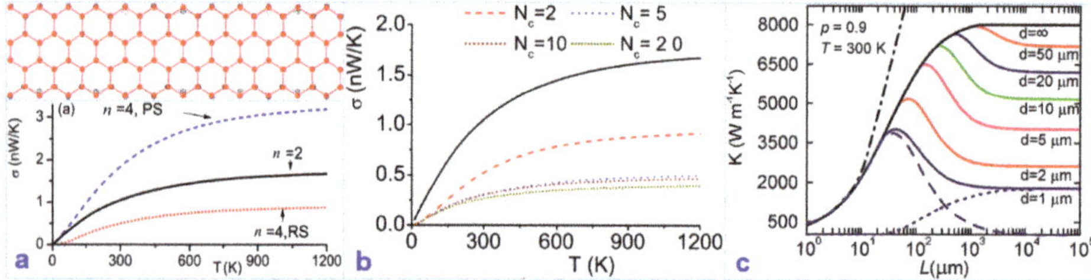

Figure 12. (**a**) Transport properties of graphene nanoribbons (GNRs) with width $n = 2$ and $n = 4$. Dotted curve corresponds to the case for the relaxed structure (RS) of GNR-4 and the dashed curve corresponds to the case for the perfect structure (PS) of GNR-4. Full circles represent the configuration of the perfect GNR-4. Circles with and without "+" represent two configurations after relaxation. (**b**) Thermal conductance of the sandwiched device versus temperature for different numbers of cells N_c in the central region. For comparison, the solid line corresponds to the conductance of a GNR-2 of infinite length [94]. Copyright 2009, American Physical Society. (**c**) Dependence of the thermal conductivity of the rectangular graphene ribbon on the ribbon length L shown for different ribbon width d. The specular parameter is fixed at $p = 0.9$ [95]. Copyright 2012, American Chemical Society.

Many theoretical calculations and experiments have revealed that graphene's TC can be changed through the hybridization of graphene [96], adsorption of metal atoms [96,97], gradient surface hydrogenation [76,98], grain size engineering [99], fluorination [100], carbon isotope doping [101], vacancies and defects engineering [102,103], and so on.

4.5. The Thermoelectric Properties of Graphene

The performance of thermoelectric materials is generally measured by the thermoelectric optimal value ZT (the thermoelectric figure of merit) [58,104]:

$$ZT = \sigma S^2 T / (\kappa_e, \kappa_{ph}) \tag{3}$$

where σ is the electronic conductivity, S is the Seebeck coefficient, κ_e and κ_{ph} are the electronic thermal conductance and phonon thermal conductance, respectively, and T is the average temperature of the device. The formula indicates that the higher the ZT, the higher the conversion efficiency between the device's thermal energy and electric energy. Outstanding thermoelectric materials must have higher Seebeck coefficients, good electrical conductivity, and very low thermal conductivity. Therefore, it has become a focus of research to identify thermoelectric materials with higher Seebeck coefficient, improve their electronic conductivity, and reduce their thermal conductivity [105–107].

4.5.1. Thermoelectric Properties in Graphene Nanoribbons (GNRs)

Thermoelectric conversion is currently a popular field of research. Researchers studied the ballistic thermoelectric properties of graphene nanoribbon-based heterojunctions [106]. The results reveal that the binding structure affects the transport of electrons, whereas the fluctuations in electrons are strongly enhanced by the thermoelectric power. The first four panels in Figure 13 show the ZT at $T = 300$ K and the ZT at $T = 100$ K for two different graphene edge heterojunctions. The calculation of the ballistic thermoelectric properties based on graphene nanoribbons heterostructures using the nonequilibrium green function and the Landau transport theory provides a method for understanding the thermoelectric properties of graphene. The phonon heat conduction under the influence of different heterostructures was basically similar, but the influence of the heterostructure geometry and geometric details on electron transport was substantial, and the change in the electronic transport greatly enhanced the thermoelectric properties. These parameters further improve the thermoelectric properties of nanomaterials.

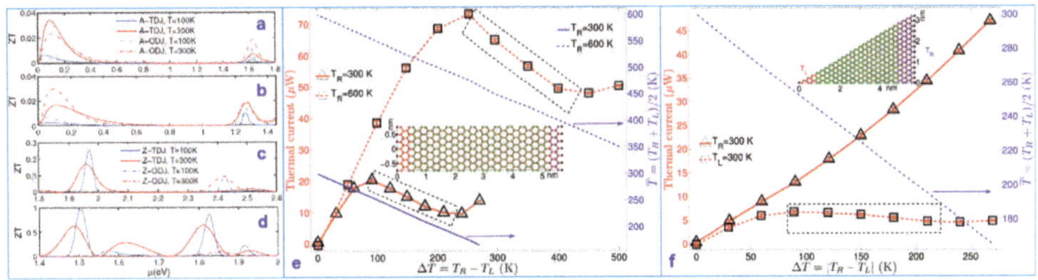

Figure 13. The figure of merit *ZT* at the two different temperatures of 100 K and 300 K in (**a**,**b**) for armchair edge junctions and (**c**,**d**) for zigzag edge junctions. Thermal current (left vertical axis) and average temperature (right vertical axis) vs. temperature difference ΔT [106]. Copyright 2012, American Physical Society. The dashed boxes highlight negative differential thermal conductance (NDTC) in (**e**). The inset shows the structure of the GNR (1.5 nm × 6 nm). (**f**) Thermal current (left vertical axis) and average temperature (right vertical axis) vs. temperature difference ΔT in triangular GNRs is shown in the inset. The dashed box highlights NDTC [107]. Copyright 2011, American Physical Society.

The nonlinear thermoelectric properties of triangular graphene nanoribbons at the armchair boundaries of armchair-GNR (AGNR) and zigzag-GNR (ZGNR) were also modeled by MD [107]. Researchers observed that negative differential thermal conductance (NDTC) appeared in the system when GNRs had large temperature deviations and beyond the range of the linear influence, and the NDTC could be controlled by the temperature. For rectangular GNRs, as the GNR length increases, the NDTC gradually decreases and eventually disappears; see Figure 13e. For triangular GNRs, the NDTC only exists when heat flows from the narrow end to the wide end; see Figure 13f. These results provide theoretical basis for the thermal management and thermal signal processing of nanomaterials.

4.5.2. Thermoelectric Spin Voltage (TSV) in Graphene

Spin-dependent thermal effects, or the interaction between spin and thermal current, have been demonstrated in ferromagnetic materials, and the spin Seebeck effect is one of the most interesting phenomena [16,108–112].

The spin current caused by the thermal gradient has been detected by the inverse spin Hall effect [113–115]. Graphene, by virtue of its highly efficient spin transmission [116–118], energy-dependent carrier mobility, and novel density of states [3,119], has become the focus material in this direction.

$$S_{Mott} = \frac{\pi^2 k_B^2 T}{3e} \bullet \frac{d \ln R}{d\mu}\bigg|_{\mu=\mu_F} \tag{4}$$

where k_B is the Boltzmann constant, R is the resistance and μ_F is Fermi energy.

Researchers prepared graphene-based detection devices and measured their properties and spin thermoelectric parameters. The results show that the thermal gradient of the carriers in a graphene-based transverse spin valve can lead to increased spin voltage in areas near the graphene charge neutrality point (CNP). Similar to the thermal voltage in a thermocouple, the effect caused by the thermoelectric spin voltage can be enhanced by the thermal carrier generated by applying the current [120–123]. These results and research methods such as maintaining the purity of the spin signals through the thermal gradient and adjusting the remote spin accumulation by changing the spin injection bias voltage are very important for driving graphene-based spin electronic devices through thermal spin.

Figure 14a shows the experimental device, which consists of two graphene nanosheets (GNs) [124]. Different carrier concentrations on the GNs lead to varying Seebeck coefficients S_1 and S_2. The temperature difference ΔT on both sides of the nanosheets leads to thermoelectric voltage versus ($V_s = V_s^+ - V_s^- = -(S_2 - S_1)\Delta T$), which is caused by the temperature gradient ΔT. When the carrier concentrations $n_1 = -n_2$, $V_S = \delta\mu/e$. Figure 14b demonstrates that nonlocal spin resistance R_{NL} changes slightly with the magnetic induction curve when the carrier concentration is $n = -2 \times 10^{11}$ cm^{-1} and the electrode current is $I_{\mathrm{dc}} = 0$ (black line) and $I_{\mathrm{dc}} = 50$ µA (red line), respectively. Figure 14b clearly indicates that the ferromagnetic property of the system switches (the blue arrow is the relative direction of the ferromagnetic magnetization) when the magnetic induction intensity is at $B_1 = 30$ and $B_2 = 40$. Figure 14c,d demonstrates the characteristics of the device. Figure 14e,f shows the qualitative representation of S (a) and its derivative dS/dn (b) about the CNP.

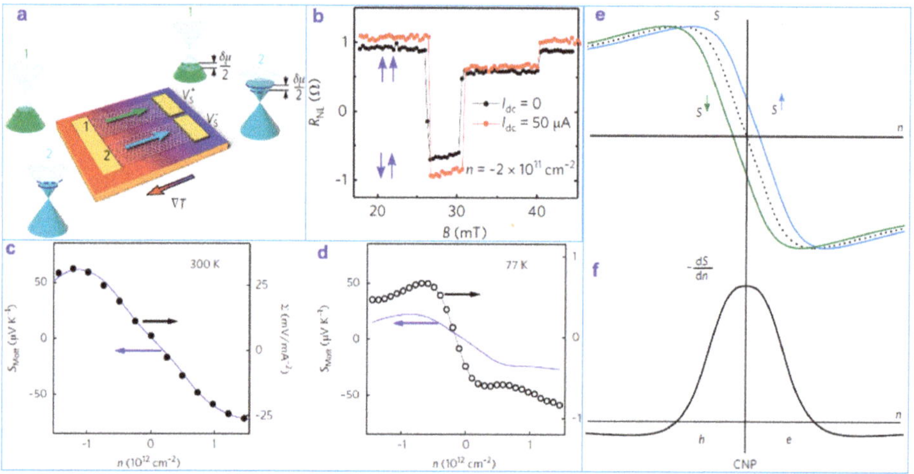

Figure 14. (**a**) Schematic of the test device. (**b**) Nonlocal spin resistance R_{NL} versus magnetic field B along the magnetization of the electrodes for $I_{\mathrm{dc}} = 0$ (black line) and $I_{\mathrm{dc}} = 50$ µA (red line). Blue arrows show the relative direction of the ferromagnetic magnetization. (**c**) Comparison between the Mott Seebeck coefficient S_{Mott} versus carrier density n (blue line and left axis) obtained from the graphene square resistance R at room temperature and the quadratic fitting coefficient Σ versus n. Modelling and roles of the Seebeck coefficient and the spin accumulation in (**e**,**f**), with qualitative representation of S (**a**) and its derivative dS/dn (**b**) about the CNP [124]. Copyright 2018, Springer Nature.

5. The Recent Applications in Electronic and Thermal Properties of Graphene

The novel electrical and thermal properties of graphene have been gradually recognized, and more applications are being widely used in photoelectric and thermoelectric devices.

5.1. Highly Efficient Thermal Conductivity Composite Film and Flexible Lateral Heat Spreaders

Polymer composite materials are ideal for horizontal heat dissipation in electronic equipment. Researchers have prepared polymer composites with highly efficient TC [125–129].

Ding et al. produced a graphene–nanocellulose composite film using vacuum-assisted self-assembly (VASA). The highly crystalline nanofibers driven by natural forces are conducive to the formation of thermal conduction paths; see Figure 15a [125]. Graphene's orientation was analyzed using effective medium approximation (EMA) to improve the TC of the composite. They changed the TC of the film by changing the defects of graphene; see Figure 15b. Through the qualitative and quantitative characterization of graphene's defects, the increase in the defects makes the TC decrease [126]. They fabricated composite films with high in-plane TC and thermal anisotropy using layer-by-layer assembly (LBL). The results show that when the content of graphene is reduced to 1.9 wt %, the in-plane TC of the composite film reaches 12.48 Wm^{-1} K^{-1} and the anisotropy coefficient is 279; see Figure 15c [127]. Composite films have great research and application potential in various fields due to their excellent TC preparation and adjustment methods. Figure 15d shows the TC of a composite film prepared by another group of researchers [128].

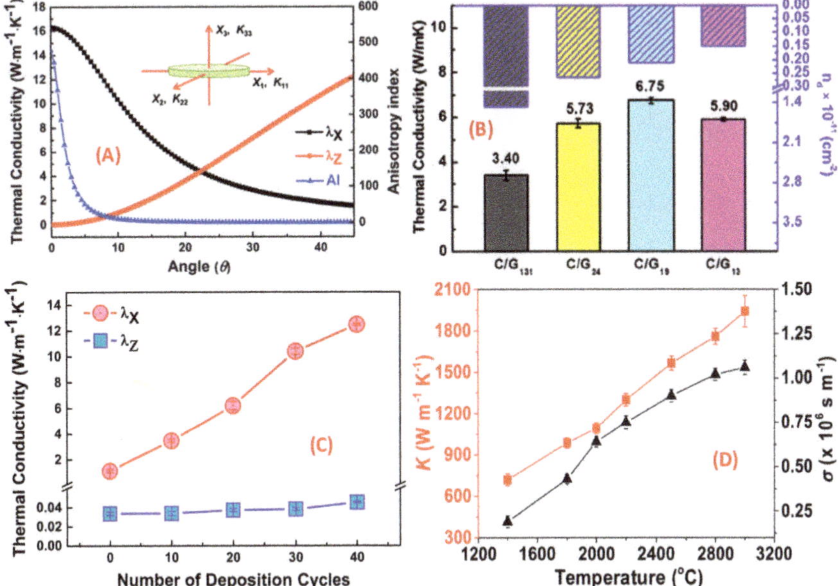

Figure 15. (**A**) Effect of graphene sheet orientation on theoretical heat transfer calculated by effective medium approximation (EMA) [125]. Copyright 2015, Royal Society of Chemistry. (**B**) The relationship between defects and thermal conductivity of graphene [126]. Copyright 2017, Elsevier. (**C**) The test results of thermal conductivity of a composite film [127]. Copyright 2017, American Chemical Society. (**D**) The thermal and electrical conductivity of debris-free graphene films (df-GFs) annealed at different temperatures [128]. Copyright 2018, Wiley.

5.2. Thermal Conductance Modulator

Based on graphene's robust TC and strength, a graphene nanoribbon-based TC modulator was developed in 2011 [130]. By changing the geometry of the graphene nanoribbons, the researchers were able to control and reverse the thermal conductance. This regulation can alter the conductance of unfolded graphene nanoribbons by up to 40%, as shown in Figure 16a,b. At this point, the folding angle of the GNRs exhibits a linear relationship with the conductivity and changes with the distance between the graphene layers. This thermal device has potential for use in phonon circuits and nanoscale thermal management.

Another interesting study was based on the 3D graphene structure of a curved fold that was transformed by planar graphene [131]. The distinctive properties of self-folding 3D graphene using multiscale molecular dynamics models have been revealed, making it possible to encapsulate cells or construct folded transistors; Figure 16c,d shows the results of this experiment.

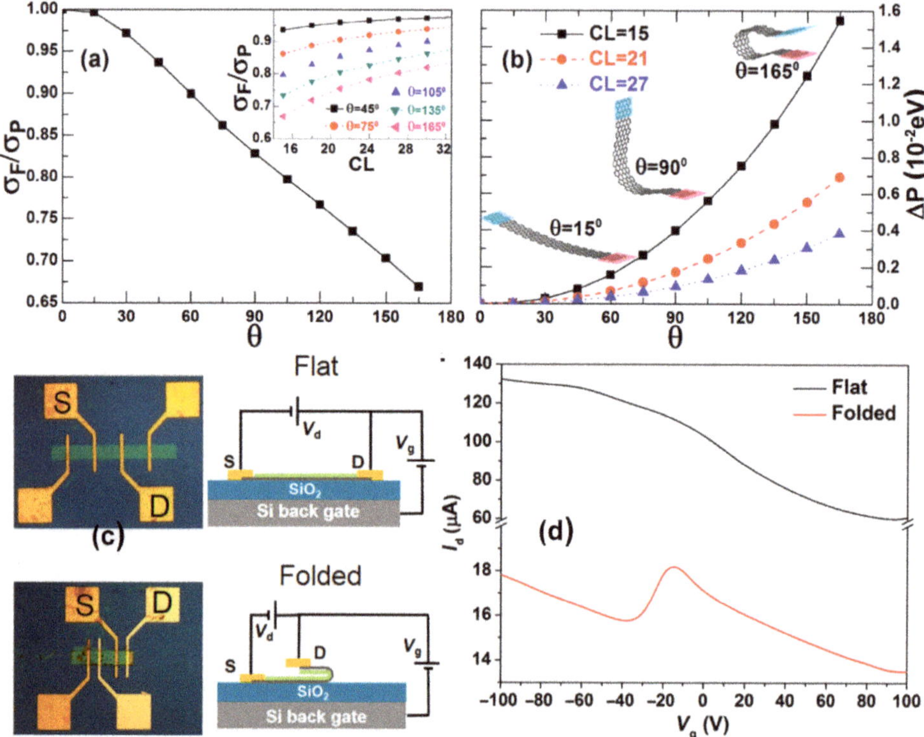

Figure 16. (**a**) σ_F/σ_P versus θ for the A-FGNR-α with CL(center region length) = 15 at T = 300 K. The inset shows the different θ values belonging to σ_F/σ_P versus CL [130]. Copyright 2011, American Physical Society. (**b**) The ΔP versus θ for the A-FGNR-α at different values of CL. (**c**) The optical images and circuit diagrams of graphene field-effect transistors (FETs); flat (top) and folded (bottom). (**d**) The transfer curves of the functionalized graphene FETs as a function of back-gate voltage in the flat (black line) and folded (red line) states [131]. Copyright 2017, Science.

5.3. Graphene Microheaters Based on Slow-Light-Enhanced Energy Efficiency

With high TC, graphene absorbs only 2.3% of light, which indicates that it is almost transparent. Based on these two properties, graphene is the best alternative to traditional metal thermometers. Graphene as a thermal microheater can be closely attached to the surface of an optical waveguide without considering the loss of graphene due to light absorption [132], while graphene's high TC can quickly transfer heat to the optical waveguide, thereby enhancing the speed regulation.

Figure 17a,b demonstrates graphene microheaters based on slow light enhancement by placing graphene on a photonic crystal waveguide with the light propagation speed reduced to 1/30 of the vacuum. The effective heating length of the optical signal is greatly increased, thus reducing the energy loss of the optical signal. Figure 17c shows the results of a graphene thermal microheater. The thermal regulation efficiency of the device is as high as 1.07 nm·mW^{-1}, which is nearly double that of traditional devices. The energy consumption of the optical signal reaching 2P phase shift is 3.99 mW, which is lower than that of traditional metal thermal heaters. The optical signal switching speed is 550 ns, which is three orders of magnitude faster than traditional metal thermal microheaters and is far from that of the fastest regulated nanothermal microheater; see Figure 17d,e. The comprehensive evaluation index of the device is 2.5413 mW, which is 30 times higher than the comprehensive evaluation index of the best nanothermal microheater. This study is expected to be widely used in integrated phased array radar systems and optical arbitrary waveform generators.

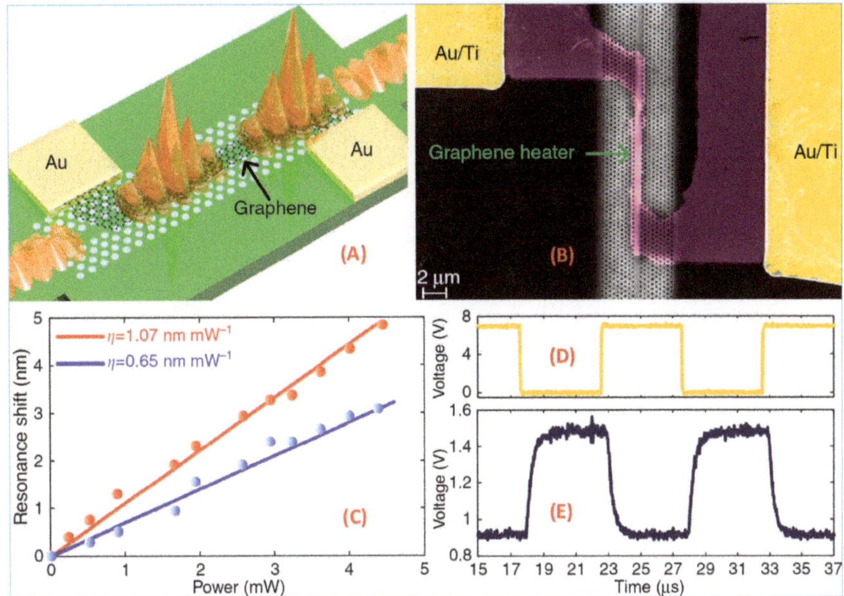

Figure 17. (**A**) Schematic of the slow-light-enhanced graphene heater. (**B**) False-color scanning electron microscope image of the slow-light-enhanced graphene heater. (**C**) Measured resonance shifts for the interference dips at 1525.12 nm (blue) and 1533.71 nm (red) as functions of the applied heating power. (**D**) Driving electrical signal and (**E**) corresponding temporal response signal [132]. Copyright 2017, Springer Nature.

5.4. Hybrid Graphene Tunneling Photoconductor

Composite photodetectors formed using highly efficient optical materials (such as quantum dots, carbon nanotubes, etc.) and graphene have attracted extensive attention. The photogenerated carriers in the absorbent materials can be effectively transferred to the graphene channel with high mobility to achieve superhigh light response gain [133–138]. However, due to a large number of trap states at the interface between the absorbent materials and the graphene, this type of photodetector is usually slow in response, which restricts its use in high-frequency applications.

Based on graphene/Si hybrid photodetectors, researchers inserted single-layer MoS_2 between graphene and silicon to improve the performance of the photoconductor [139]. The experimental results indicate that the photogenerated carriers flow out of the silicon and enter the graphene channel through the potential barrier of MoS_2 through the quantum tunneling effect under the condition of illumination. There are no suspension bonds on the surface of the molybdenum disulfide, which reduces the trap state and effectively passivates the interface. Comparing the detection performance of devices before and after MoS_2 insertion, the response speed of the latter is three orders of magnitude higher than the former; that is, the response time is 17 ns and the response degree is 3.4×10^4 A/W. Figure 18a presents a schematic diagram of the device. Figure 18b,c,d shows the device's photoelectric response characteristics. Figure 18e,f demonstrates the experimental results of the device. This type of nanodevice graphene-based heterostructure shows excellent performance, which provides new applications for graphene-based electrical characteristics.

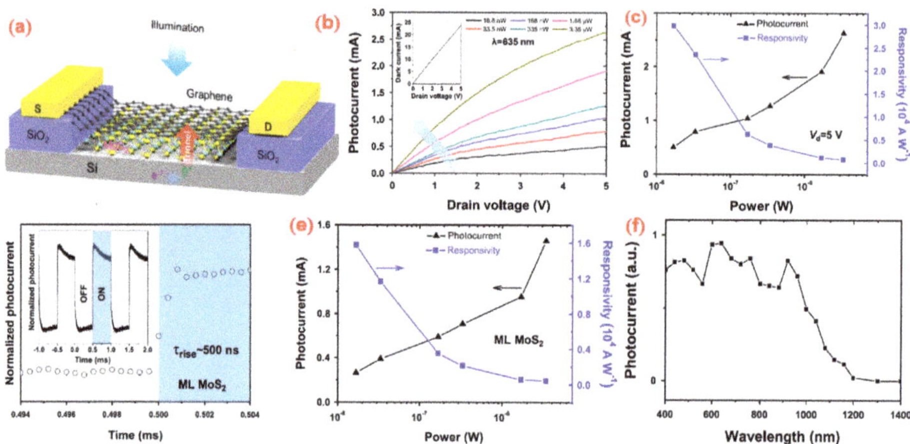

Figure 18. (**a**) Schematic diagram of a hybrid graphene photoconductor. (**b**) Photocurrent vs. drain voltage under various light powers at 635 nm wavelength. The arrow indicates the direction of light power increase. The inset shows the dark current of the device. (**c**) Power-dependent photocurrent and photoresponsivity at 5 V drain voltage calculated from the data in (**b**). (**d**) Normalized photocurrent vs. illumination wavelength. (**e**) Transient characteristics of the hybrid graphene photoconductor with MoS_2 under 635 nm illumination, showing a rising time of ~500 ns. Inset is the switching performance over three periods of square-wave modulation. (**f**) Photocurrent and responsivity as functions of the illumination power of the device with MoS_2 [139]. Copyright 2017, Springer Nature.

5.5. Graphene Electrode

P-doped graphene-based electrodes improve the performance of the device by reducing its resistance. However, the resistance of the electrode will gradually increase in the environment, which will affect the actual use of the graphene electrode [140–146].

Researchers recently used perfluorinated polymeric sulfonic acid (PFSA) as a dopant molecule to conduct p-doped graphene to build a PFSA-based p-doped graphene electrode [146]. The electric dipole of the sulfonate group proton in the PFSA molecule strongly attracts electrons, which leads to high ionization potential of the perfluorocarbon skeleton. Doped with PFSA graphene electrodes, the device's surface resistance (R_{sh}) decreased by 56% and its surface potential increased by 0.8 V. Moreover, the graphene electrode of this configuration, although treated with a chemical agent, was stable under high temperatures and long-term exposure to air. This graphene-based electrode can be used to produce phosphorescent organic light-emitting diodes with high hole injection and substantial luminescence efficiency.

Figure 19a presents a structural diagram of the PFSA-doped graphene electrode. Figure 19b shows the calculated results of the device. Figure 19c–e demonstrates the performance examination results of the electrode. Figure 19f shows the performance test results of organic light-emitting diodes (OLEDs) based on the graphene electrode. The construction of a doped graphene electrode has consistently been a popular topic of research, but many researchers focus on the regulation of its electrical properties and ignore the problem of its stability. The study used organic macromolecules as dopants to improve the electrical properties of graphene and its stability. This work can promote the construction of a stable graphene electrode and its application.

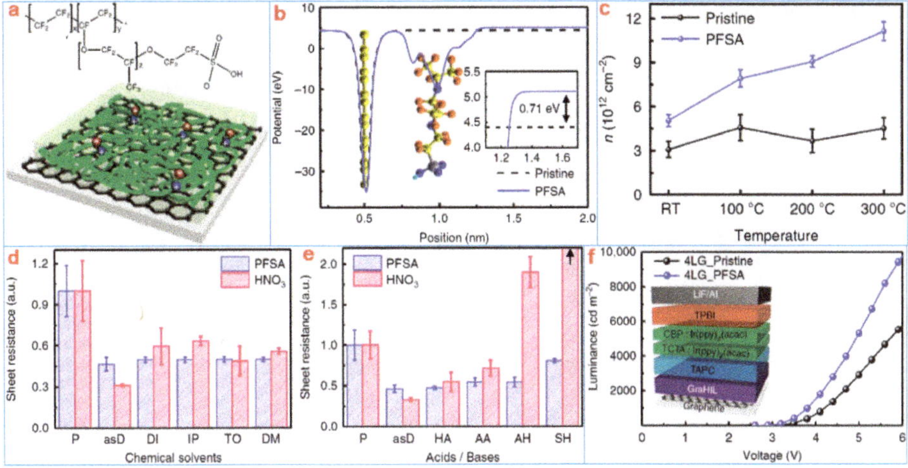

Figure 19. (**a**) Chemical structure of perfluorinated polymeric sulfonic acid (PFSA) and schematic drawings of graphene doped using PFSA (+: hole, −: electron). (**b**) Calculated electrostatic potential of the most stable configuration of PFSA-doped graphene (inset: difference in work function between pristine and PFSA-doped graphene). (**c**) Averaged n of thermally annealed pristine and PFSA-doped graphene with various T_a calculated from Raman spectroscopy results. (**d**) Various solvent treatments and (**e**) acid and base treatments as a function of exposure time. (**f**) Luminance vs. voltage of green phosphorescent organic light-emitting diodes (OLEDs) with pristine and PFSA-doped graphene anodes (inset: schematic device structure of OLEDs) [146]. Copyright 2018, Springer Nature.

5.6. Dirac-Source Field-Effect Transistors (DS-FETs)

The development trend of integrated circuits has changed from the pursuit of performance and integration to the most effective way to reduce power consumption, which is to reduce the working voltage. Currently, the working voltage of the integrated circuit (14/10 nm technical node) of a complementary metal-oxide semiconductor (CMOS) is reduced to 0.7 V, while the thermal excitation limit (60 mV/decade (Dec)) of the MOS transistor's subthreshold swing (*SS*) makes it impossible to reduce the working voltage of the integrated circuit to below 0.64 V. Existing transistors that are tunneling FET and negative capacitance FET can realize *SS* < 60 mV/Dec, but they have a low speed or important defects such as poor stability, unfavorable integration, and thus lack of practical value. The ultralow power consumption transistor to be used in future integrated circuits not only must obtain *SS* < 60 mV/Dec, ensuring the open state current is sufficiently large, but also requires stable performance and simple preparation [147–157].

Researchers in Beijing recently reexamined the MOS transistor and the physical limits of its threshold swing [158]. They proposed a new type of ultralow power field effect transistor and adopted doped graphene as a "cold" electronic source with carbon nanotubes as the active channel. Semiconductor sources with high-efficiency top grid structures have been built as DS-FETs. The threshold value of swing experiments has been implemented at 40 mV/Dec at room temperature (Figure 20). The results of variable temperature measurement indicate that there is an obvious linear relationship between the DS-FET subthreshold amplitude and temperature. This indicates that the carrier transport of transistors is a traditional thermal emission mechanism rather than a tunneling mechanism. The DS-FET has excellent scalability. When the channel length of the device decreases to 15 nm, it can still achieve a subthreshold swing of 60 mV/Dec.

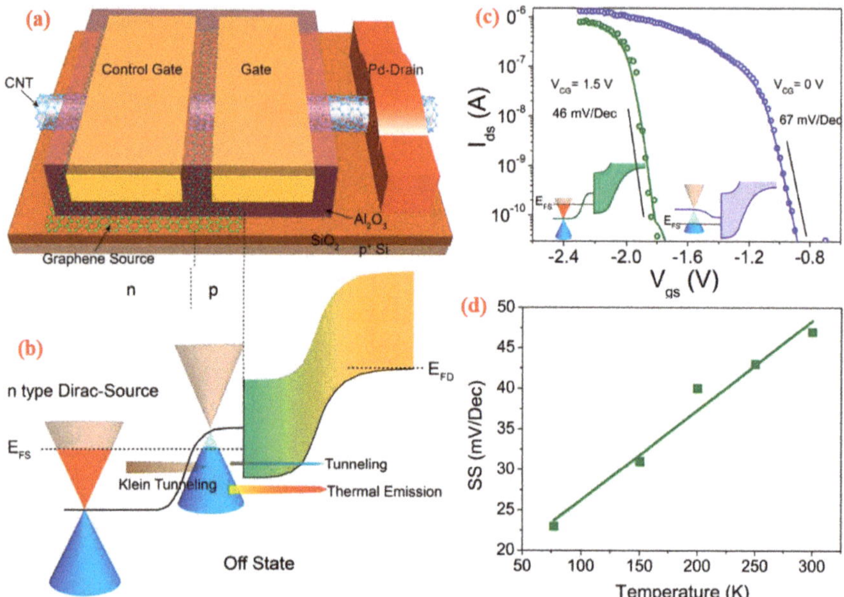

Figure 20. (**a**) Schematic diagram showing a Dirac-source field-effect transistor (DS-FET) with a control gate in addition to the normal gate. (**b**) Schematic diagrams illustrating the off-state of the DS-FET. (**c**) Transfer characteristics of a typical DS-FET at different V_{CG}. Circles and lines represent experimental and simulated results, respectively. Green color represents results obtained at V_{CG} = 1.5 V, and blue represents those obtained at V_{CG} = 0 V. Inset figures are schematic band edge profiles for fitted data situations. (**d**) Temperature-dependent SS of a typical DS-FET measured at temperatures between 77 K and 300 K; V_{CG} was set at 2 V to keep the device in Dirac-source mode. SS varied by more than 100% from 77 to 300 K. In all measurements, the substrate was biased with -20 V to keep the ungated region near the drain open [158]. Copyright 2018, Science.

Most importantly, the DS-FET has a proposed driving current much higher than that of tunneling transistors compared to metal-oxide semiconductor field-effect transistors, and its $SS < 60$ mV/Dec spans a larger range of currents. As the key parameter of the comprehensive index of open and closed state characteristics of sub-60 mV/Dec (that is, the current at $SS = 60$ mV/Dec), $I_{60} = 40$ μA/μm, which is 2000 times the published best tunneling transistor and fully meets the standards of the international semiconductor development roadmap (ITRS) for the practical application of sub-60 mV/Dec devices. The open and closed current of a typical DS transistor at a working voltage of 0.5 V is equivalent to that of a CMOS device at 14 nm (at a working voltage of 0.7 V). This indicates that the DS transistor can meet the requirements of future ultralow power consumption ($V_{dd} < 0.5$ V) integrated circuits. Moreover, the device structure of the DS does not rely on semiconductor materials and may be used in conventional CMOS transistors and field effect transistors in two-dimensional materials.

6. Conclusions and Prospects

Graphene, with its exceptional physical and chemical properties, has been increasingly applied in various fields of scientific research. When graphene's nanostructure changes (such as in terms of boundary configuration, shape, own defects, chemical doping, and the formation of heterogeneous structures, etc.), its physical and chemical properties show novel properties. With the improvements in the preparation of graphene nanomaterials and the enhancement of their measurement and regulation, more graphene nanomaterials and their hybrid structures have been applied in electronic, photothermal, thermoelectric, and photoelectric fields.

However, graphene's band gap characteristics limit its application. Graphene's physical and chemical properties are both closely related to its electronic properties. Graphene's electronic properties change its physical and chemical properties. There are two kinds of methods. One is a physical approach (for example, the change of the graphene nanostructures, the applied electric field or magnetic field, vertical configuration heterostructure or plane heterostructure, substrate, etc.). The other is a chemical method (chemical doping, other atoms or groups of adsorption, the use of chemical reagents, etc.). The fabrication level of graphene-based microscale or nanoscale devices also determines their application and development. Researchers are actively working in correlated fields, and more high-performance graphene-based devices will be prepared and used in the future.

Author Contributions: M.S. is responsible for the overall context of the article, the theme of the article and the design of the structure of the article. J.W. and M.X. are responsible for the writing of the article and the collection of the literature.

Funding: This work was supported by the National Nature Science Foundation of China (Grant No. 11374353, 91436102, 11474141). Fundamental Research Funds for the Central Universities and talent scientific research fund of LSHU (No. 2018XJJ-007).

Conflicts of Interest: The authors declare no conflict of interest.

References

1. Novoselov, K.S.; Geim, A.K.; Morozov, S.V.; Jiang, D.; Zhang, Y.; Dubonos, S.V.; Grigorieva, I.V.; Firsov, A.A. Electric Field Effect in Atomically Thin Carbon Films. *Science* **2004**, *306*, 666–669. [CrossRef] [PubMed]
2. Geim, A.K.; Novoselov, K.S. The rise of graphene. *Nat. Mater.* **2007**, *6*, 183–191. [CrossRef] [PubMed]
3. Castro Neto, A.H.; Guinea, F.; Peres, N.M.R.; Novoselov, K.S.; Geim, A.K. The electronic properties of graphene. *Rev. Mood Phys.* **2009**, *81*, 109–162. [CrossRef]
4. Ando, T. The electronic properties of graphene and carbon nanotubes. *Npg Asia Mater.* **2009**, *1*, 17–21. [CrossRef]
5. Ohta, T.; Bostwick, A.; McChesney, J.; Seyller, T.; Horn, K.; Rotenberg, E. Controlling the Electronic Structure of Bilayer Graphene. *Science* **2006**, *313*, 951–954. [CrossRef] [PubMed]
6. Lee, C.; Wei, X.; Kysar, J.W.; Hone, J. Measurement of the elastic properties and intrinsic strength of monolayer graphene. *Science* **2008**, *321*, 385–388. [CrossRef] [PubMed]
7. Liu, F.; Ming, P.; Li, J. Ab initio calculation of ideal strength and phonon instability of graphene under tension. *Phys. Rev. B* **2007**, *76*, 064120. [CrossRef]
8. Zakharchenko, K.V.; Katsnelson, M.I.; Fasolino, A. Finite temperature lattice properties of graphene beyond the quasiharmonic approximation. *Phys. Rev. Lett.* **2009**, *102*, 046808. [CrossRef]
9. Sahin, H.; Cahangirov, S.; Topsakal, M.; Bekaroglu, E.; Aktrk, E.; Senger, R.T.; Ciraci, S. Monolayer honeycomb structures of group IV elements and III-V binary compounds. *Phys. Rev. B* **2009**, *80*, 155453. [CrossRef]
10. Balandin, A.A.; Ghosh, S.; Bao, W.; Calizo, I.; Teweldebrhan, D.; Miao, F.; Lau, C.N. Superior thermal conductivity of single-layer graphene. *Nano Letters* **2008**, *8*, 902–907. [CrossRef]
11. Chen, J.; Jang, C.; Xiao, S.; Ishigami, M.; Fuhrer, M. Intrinsic and Extrinsic performance limits of graphene device on SiO$_2$. *Nat. Nanotechnol.* **2008**, *3*, 206–209. [CrossRef] [PubMed]
12. Avouris, P.; Chen, Z.H.; Perebeinos, V. Fine structure constant defines visual transparency of graphene. *Nat. Nanotechnol.* **2007**, *2*, 605–615. [CrossRef]
13. Wang, J.; Ma, F.; Sun, M. Graphene, hexagonal boron nitride, and their heterostructures: Properties and applications. *Rsc Adv.* **2017**, *7*, 16801. [CrossRef]
14. Wang, J.; Ma, F.; Liang, W.; Wang, R.; Sun, M. Optical, photonic and optoelectronic properties of graphene, h-NB and their hybrid materials. *Nanophotonics* **2017**, *6*, 943–976. [CrossRef]
15. Li, R.; Zhang, Y.; Xu, X.; Zhou, Y.; Chen, M.; Sun, M. Optical characterizations of two-dimensional materials using nonlinear optical microscopies of CARS, TPEF, and SHG. *Nanophotonics* **2018**, *7*, 873–881. [CrossRef]
16. Wang, J.; Xu, X.; Ma, F.; Sun, M. Magnetics and spintronics on two-dimensional composite materials of graphene/hexagonal boron nitride. *Mater. Today Phys.* **2017**, *3*, 93–117. [CrossRef]
17. Al-Jishi, R.; Elman, B.S.; Dresselhaus, G. Lattice dynamical model for graphite. *Carbon* **1982**, *20*, 4514–4552. [CrossRef]

18. Geim, A.K.; Philip, K. Carbon wonderland. *Sci. Am.* **2008**, *298*, 90–97. [CrossRef] [PubMed]
19. Nair, R.R.; Blake, P.; Grigorenko, A.N.; Novoselov, K.S.; Booth, T.J.; Stauber, T.; Peres, N.M.; Geim, A.K. Fine structure constant defines visual transparency of graphene. *Science* **2008**, *320*, 1308. [CrossRef] [PubMed]
20. Novoselov, K.S.; Geim, A.K.; Morozov, S.V.; Jiang, D.; Katsnelson, M.I.; Grigorieva, I.V.; Dubonos, S.V.; Firsov, A.A. Two-dimensional gas of massless Dirac fermions in graphene. *Nature* **2005**, *438*, 197–200. [CrossRef]
21. Elias, D.C.; Gorbachev, R.V.; Mayorov, A.S.; Morozov, S.V.; Zhukov, A.A.; Blake, P.; Ponomarenko, L.A.; Grigorieva, I.V.; Novoselov, K.S.; Guinea, F.; Geim, A.K. Dirac cones reshaped by interaction effects in suspended graphene. *Nat. Phys.* **2011**, *7*, 701–704. [CrossRef]
22. Kim, P. Experimental Observation of Quantum Hall Effect and Berry's Phase in Graphene. *Nature* **2005**, *438*, 201–204.
23. Novoselov, K.S.; Jiang, Z.; Zhang, Y.; Morozov, S.V.; Stormer, H.L.; Zeitler, U.; Maan, J.C.; Boebinger, G.S.; Kim, P.; Geim, A.K. Room-temperature quantum Hall effect in graphene. *Science* **2007**, *315*, 1379. [CrossRef] [PubMed]
24. Berger, C.; Song, Z.; Li, X.; Wu, X.; Brown, N.; Naud, C.; Mayou, D.; Li, T.; Hass, J.; Marchenkov, A.N. Electronic confinement and coherence in patterned epitaxial graphene. *Science* **2006**, *312*, 1191–1196. [CrossRef] [PubMed]
25. Katsnelson, M.I.; Novoselov, K.S.; Geim, A.K. Chiral tunneling and the Klein paradox in graphene. *Nat. Phys.* **2006**, *2*, 620–625. [CrossRef]
26. Allain, P.E.; Fuchs, J.N. Klein tunneling in graphene: Optics with massless electrons. *Eur. Phys. J. B* **2011**, *83*, 301–317. [CrossRef]
27. Du, X.; Skachko, I.; Barker, A.; Andrei, E.Y. Approaching ballistic transport in suspended graphene. *Nat. Nanotechnol.* **2008**, *3*, 491–495. [CrossRef] [PubMed]
28. Miao, F.; Wijeratne, S.; Zhang, Y.; Coskun, U.C.; Bao, W.; Lau, C.N. Phase-coherent transport in graphene quantum billiards. *Science* **2007**, *317*, 1530–1533. [CrossRef]
29. Damle, K.; Sachdev, S. Non-zero temperature transport near quantum critical points. *Phys. Rev. B* **1997**, *56*, 8714–8733. [CrossRef]
30. Kovtun, P.K.; Son, D.T.; Starinets, A.O. Viscosity in strongly interacting quantum field theories from black hole physics. *Phys. Rev. Lett.* **2005**, *94*, 111601. [CrossRef]
31. Son, D.T. Vanishing bulk viscosities and conformal invariance of the unitary fermi gas. *Phys. Rev. Lett.* **2007**, *98*, 020604. [CrossRef] [PubMed]
32. Karsch, F.; Kharzeev, D.; Tuchin, K. Universal properties of bulk viscosity near the QCD phase transition. *Phys. Lett. B* **2008**, *663*, 217–221. [CrossRef]
33. Levitov, L.; Falkovich, G. Electron viscosity, current vortices and negative nonlocal resistance in graphene. *Nat. Phys.* **2016**, *12*, 672–676. [CrossRef]
34. Yoo, M.J.; Fulton, T.A.; Hess, H.F.; Willett, R.L.; Dunkleberger, L.N.; Chichester, R.J.; Pfeiffer, L.N.; West, K.W. Scanning single-electron transistor microscopy: Imaging individual charges. *Science* **1997**, *276*, 579–582. [CrossRef] [PubMed]
35. Mikhailov, S.A.; Ziegler, K. New Electromagnetic Mode in Graphene. *Phys. Rev. Lett.* **2007**, *99*, 016803. [CrossRef] [PubMed]
36. Jablan, M.; Buljan, H.; Soljačić, M. Plasmonics in graphene at infrared frequencies. *Phys. Rev. B* **2009**, *80*, 245435. [CrossRef]
37. Hanson, G.W. Dyadic Green's functions and guided surface waves for a surface conductivity model of graphene. *J. Appl. Phys.* **2008**, *103*, 064302. [CrossRef]
38. Ju, L.; Geng, B.; Horng, J.; Girit, C.; Martin, M.; Hao, Z.; Bechtel, H.A.; Liang, X.; Zettl, A.; Shen, Y.R. Graphene plasmonics for tunable terahertz metamaterials. *Nat. Nanotechnol.* **2011**, *6*, 630–634. [CrossRef]
39. Thongrattanasiri, S.; Koppens, F.H.L.; Abajo, F.J.G.D. Total light absorption in graphene. *Phys. Rev. Lett.* **2011**, *108*, 047401. [CrossRef]
40. Vakil, A.; Engheta, N. Transformation optics using graphene. *Science* **2011**, *332*, 1291–1294. [CrossRef]
41. Koppens, F.H.L.; Chang, D.E.; Garcia de Abajo, F.J. Graphene plasmonics: A platform for strong light–matter interactions. *Nano Lett.* **2011**, *11*, 3370–3377. [CrossRef] [PubMed]
42. Chen, J.; Badioli, M.; Alonsogonzález, P.; Thongrattanasiri, S.; Huth, F.; Osmond, J.; Spasenović, M.; Centeno, A.; Pesquera, A.; Godignon, P. Optical nano-imaging of gate-tunable graphene plasmons. *Nature* **2012**, *487*, 77–81. [CrossRef]

43. Novoselov, K.S.; Mishchenko, A.; Carvalho, A.; Castro Neto, A.H. 2D materials and van der Waals heterostructures. *Science* **2016**, *353*, aac9439. [CrossRef] [PubMed]

44. Low, T.; Chaves, A.; Caldwell, J.D.; Kumar, A.; Fang, N.X.; Avouris, P.; Heinz, T.F.; Guinea, F.; Martinmoreno, L.; Koppens, F. Polaritons in layered two-dimensional materials. *Nat. Mater.* **2016**, *16*, 182. [CrossRef] [PubMed]

45. Basov, D.N.; Fogler, M.M.; Garcia de Abajo, F.J. Polaritons in van der Waals materials. *Science* **2016**, *354*, aag1992. [CrossRef] [PubMed]

46. Alonsogonzález, P.; Nikitin, A.Y.; Gao, Y.; Woessner, A.; Lundeberg, M.B.; Principi, A.; Forcellini, N.; Yan, W.; Vélez, S.; Huber, A.J. Acoustic terahertz graphene plasmons revealed by photocurrent nanoscopy. *Nat. Nanotechnol.* **2017**, *12*, 31–35. [CrossRef] [PubMed]

47. Lundeberg, M.B.; Gao, Y.; Asgari, R.; Tan, C.; Van, D.B.; Autore, M.; Alonsogonzález, P.; Woessner, A.; Watanabe, K.; Taniguchi, T. Tuning quantum nonlocal effects in graphene plasmonics. *Science* **2017**, *357*, 187–191. [CrossRef] [PubMed]

48. Ido, K.; Tenenbaum, K.Y.; Hrvoje, B.; Shen, Y.; Ognjen, I.; López, J.J.; Jie, W.L.; Joannopoulos, J.D.; Marin, S. Efficient plasmonic emission by the quantum Čerenkov effect from hot carriers in graphene. *Nat. Commun.* **2016**, *7*, 11880.

49. Alcaraz, D.I.; Nanot, S.; Dias, E.; Epstein, I.; Peng, C.; Efetov, D.K.; Lundeberg, M.B.; Parret, R.; Osmond, J.; Hong, J.Y. Probing the ultimate plasmon confinement limits with a van der Waals heterostructure. *Science* **2018**, *360*, 291–295. [CrossRef]

50. Tan, Y.W.; Zhang, Y.; Bolotin, K.; Zhao, Y.; Adam, S.; Hwang, E.H.; Das Sarma, S.; Stormer, H.L.; Kim, P. Measurement of Scattering Rate and Minimum Conductivity in Graphene. *Phys. Rev. Lett.* **2007**, *99*, 246803. [CrossRef]

51. Chen, J.H.; Jang, C.; Adam, S.; Fuhrer, M.S.; Williams, E.D.; Ishigami, M. Charged-Impurity Scattering in Graphene. *Nat. Phys.* **2008**, *4*, 377–381. [CrossRef]

52. Chen, J.H.; Cullen, W.G.; Jang, C.; Fuhrer, M.S.; Williams, E.D. Defect Scattering in Graphene. *Phys. Rev. Lett.* **2009**, *102*, 236805. [CrossRef] [PubMed]

53. Schedin, F.; Geim, A.K.; Morozov, S.V.; Hill, E.W.; Blake, P.; Katsnelson, M.I.; Novoselov, K.S. Detection of Individual Gas Molecules Adsorbed on Graphene. *Nat. Mater.* **2007**, *6*, 652–655. [CrossRef] [PubMed]

54. Jia, Z.; Yan, B.; Niu, J.; Han, Q.; Zhu, R.; Yu, D.; Wu, X. Transport Study of Graphene Adsorbed with Indium Adatoms. *Phys. Rev. B* **2015**, *91*, 085411. [CrossRef]

55. Li, J.; Lin, L.; Rui, D.; Li, Q.; Zhang, J.; Kang, N.; Zhang, Y.; Peng, H.; Liu, Z.; Xu, H.Q. Electron-Hole Symmetry Breaking in Charge Transport in Nitrogen-Doped Graphene. *ACS Nano* **2017**, *11*, 4641–4650. [CrossRef] [PubMed]

56. Gruber, E.; Wilhelm, R.A.; Pétuya, R.; Smejkal, V.; Kozubek, R.; Hierzenberger, A.; Bayer, B.C.; Aldazabal, I.; Kazansky, A.K.; Libisch, F. Ultrafast electronic response of graphene to a strong and localized electric field. *Nat. Commun.* **2016**, *7*, 13948. [CrossRef] [PubMed]

57. Wang, J.; Ma, F.; Liang, W.; Sun, M. Electrical properties and applications of graphene, hexagonal boron nitride (h-BN), and graphene/h-BN heterostructures. *Mater. Today Phys.* **2017**, *2*, 6–34. [CrossRef]

58. Wang, J.; Mu, X.; Wang, X.; Wang, N.; Ma, F.; Liang, W.; Sun, M. The thermal and thermoelectric properties of in-plane C-BN hybrid structures and graphene/h-BN van der Waals heterostructures. *Mater. Today Phys.* **2018**, *5*, 29–57. [CrossRef]

59. Yu, L.; Lee, Y.H.; Ling, X.; Santos, E.J.; Shin, Y.C.; Lin, Y.; Dubey, M.; Kaxiras, E.; Kong, J.; Wang, H. Graphene/MoS2 hybrid technology for large-scale two-dimensional electronics. *Nano Lett.* **2014**, *14*, 3055–3063. [CrossRef]

60. Fazio, D.D.; Goykhman, I.; Yoon, D.; Bruna, M.; Eiden, A.; Milana, S.; Sassi, U.; Barbone, M.; Dumcenco, D.; Marinov, K. High Responsivity, Large-Area Graphene/MoS2 Flexible Photodetectors. *ACS Nano* **2016**, *10*, 8252–8262. [CrossRef]

61. Bartolomeo, A.D.; Giubileo, F.; Luongo, G.; Iemmo, L.; Martucciello, N.; Niu, G.; Fraschke, M.; Skibitzki, O.; Schroeder, T.; Lupina, G. Tunable Schottky barrier and high responsivity in graphene/Si-nanotip optoelectronic device. *2d Mater.* **2016**, *4*, 015024. [CrossRef]

62. Cao, Y.; Fatemi, V.; Demir, A.; Fang, S.; Tomarken, S.L.; Luo, J.Y.; Sanchezyamagishi, J.D.; Watanabe, K.; Taniguchi, T.; Kaxiras, E. Correlated insulator behaviour at half-filling in magic-angle graphene superlattices. *Nature* **2018**, *556*, 80–84. [CrossRef]

63. Cao, Y.; Fatemi, V.; Fang, S.; Watanabe, K.; Taniguchi, T.; Kaxiras, E.; Jarillo-Herrero, P. Unconventional superconductivity in magic-angle graphene superlattices. *Nature* **2018**, *556*, 43–50. [CrossRef] [PubMed]

64. Mele, E.J. Novel electronic states seen in graphene. *Nature* **2018**, *556*, 37–38. [CrossRef] [PubMed]

65. Lin, W.; Shi, Y.; Yang, X.; Li, J.; Cao, E.; Xu, X.; Pullerits, T.; Liang, W.; Sun, M. Physical mechanism on exciton-plasmon coupling revealed by femtosecond pump-probe transient absorption spectroscopy. *Mater. Today Phys.* **2017**, *3*, 33–40. [CrossRef]

66. Yang, X.; Yu, H.; Guo, X.; Ding, Q.; Pullerits, T.; Wang, R.; Zhang, G.; Liang, W.; Sun, M. Plasmon-exciton coupling of monolayer MoS_2 -Ag nanoparticles hybrids for surface catalytic reaction. *Mater. Today Phys.* **2017**, *5*, 72–78.

67. Lin, W.; Cao, Y.; Wang, P.; Sun, M. Unified treatment for plasmon-exciton Co-driven reduction and oxidation reactions. *Langmuir* **2017**, *33*, 12102–12107. [CrossRef] [PubMed]

68. Lin, W.; Cao, E.; Zhang, L.; Xu, X.; Song, Y.; Liang, W.; Sun, M. Electrically enhanced hot hole driver oxidation catalysis at the interface of a plasmon-exciton hybrid. *Nanoscale* **2018**, *10*, 5482–5488. [CrossRef]

69. Cao, E.; Guo, X.; Zhang, L.; Shi, Y.; Lin, W.; Liu, X.; Fang, Y.; Zhou, L.; Sun, Y.; Song, Y. Electrooptical Synergy on Plasmon–Exciton-Codriven Surface Reduction Reactions. *Adv. Mater. Interfaces* **2017**, *4*, 1700869. [CrossRef]

70. Wang, J.; Lin, W.; Xu, X.; Ma, F.; Sun, M. Plasmon-Excition Coupling interaction for Surface Catalytic Reactions. *Chem. Rec.* **2018**, *18*, 481–490. [CrossRef]

71. Grigorenko, A.N.; Polini, M.; Novoselov, K.S. Graphene plasmonics. *Nat. Photonics* **2012**, *6*, 749–758. [CrossRef]

72. Chen, S.; Moore, A.L.; Cai, W.; Suk, J.W.; An, J.; Mishra, C.; Amos, C.; Magnuson, C.W.; Kang, J.; Shi, L.; et al. Raman measurements of thermal transport in suspended monolayer graphene of variable sizes in vacuum and gaseous environments. *ACS Nano* **2011**, *5*, 321–328. [CrossRef] [PubMed]

73. Seol, J.H.; Jo, I.; Moore, A.L.; Lindsay, L.; Aitken, Z.H.; Pettes, M.T.; Li, X.; Yao, Z.; Huang, R.; Broido, D. Two-Dimensional Phonon Transport in Supported Graphene. *Science* **2010**, *328*, 213–216. [CrossRef] [PubMed]

74. Ni, Z.; Wang, Y.; Yu, T.; Shen, Z. Raman spectroscopy and imaging of graphene. *Nano Res.* **2008**, *1*, 273–291. [CrossRef]

75. Wang, Y.Y.; Ni, Z.H.; Yu, T.; Shen, Z.X.; Wang, H.M.; Wu, Y.H.; Chen, W.; Wee, A.T.S. Raman Studies of Monolayer Graphene: The Substrate Effect. *J. Phys. Chem. C* **2008**, *112*, 10637–10640. [CrossRef]

76. Xu, X.; Pereira, L.F.; Wang, Y.; Wu, J.; Zhang, K.; Zhao, X.; Bae, S.; Tinh, B.C.; Xie, R.; Thong, J.T. Length-dependent thermal conductivity in suspended single-layer graphene. *Nat. Commun.* **2014**, *5*, 3689. [CrossRef]

77. Liu, T.H.; Lee, S.C.; Pao, C.W.; Chang, C.C. Anomalous thermal transport along the grain boundaries of bicrystalline graphene nanoribbons from atomistic simulations. *Carbon* **2014**, *73*, 432–442. [CrossRef]

78. Bonini, N.; Garg, J.; Marzari, N. Acoustic phonon lifetimes and thermal transport in free-standing and strained graphene. *Nano Lett.* **2012**, *12*, 2673–2678. [CrossRef]

79. Pereira, L.F.C.; Donadio, D. Divergence of the Thermal Conductivity in Uniaxially Strained Graphene. *Phys. Rev. B* **2013**, *87*, 125424. [CrossRef]

80. Lindsay, L.; Broido, D.A.; Mingo, N. Flexural phonons and thermal transport in graphene. *Phys. Rev. B* **2010**, *82*, 115427. [CrossRef]

81. Jund, P.; Jullien, R. Molecular dynamics calculation of the thermal conductivity of vitreous silica. *Phys. Rev. B* **2007**, *59*, 13707–13711. [CrossRef]

82. Xu, Y.; Chen, X.; Gu, B.L.; Duan, W. Intrinsic anisotropy of thermal conductance in graphene nanoribbons. *Appl. Phys. Lett.* **2009**, *95*, 233116. [CrossRef]

83. Khan, A.I.; Navid, I.A.; Hossain, F.F.; Noshin, M.; Subrina, S. A molecular dynamics study on thermal conductivity of armchair graphene nanoribbon. *TENCON* **2016**, 2775–2778.

84. Zhang, G.; Zhang, H. Thermal conduction and rectification in few-layer graphene Y junctions. *Nanoscale* **2011**, *3*, 4604–4607. [CrossRef] [PubMed]

85. Ouyang, T.; Chen, Y.; Xie, Y.; Wei, X.L.; Yang, K.; Yang, P.; Zhong, J. Ballistic thermal rectification in asymmetric three-terminal graphene nanojunctions. *Phys. Rev. B* **2010**, *82*, 245403. [CrossRef]

86. Nissimagoudar, A.S.; Sankeshwar, N.S. Electronic thermal conductivity and thermopower of armchair graphene nanoribbons. *Carbon* **2013**, *52*, 201–208. [CrossRef]

87. Barnard, A.S.; Snook, I.K. Thermal stability of graphene edge structure and graphene nanoflakes. *J. Chem. Phys.* **2008**, *128*, 094707. [CrossRef]

88. Ouyang, T.; Chen, Y.P.; Yang, K.K.; Zhong, J.X. Thermal transport of isotopic-superlattice graphene nanoribbons with zigzag edge. *Europhys. Lett.* **2009**, *88*, 28002. [CrossRef]

89. Yang, K.; Chen, Y.; Xie, Y.; Ouyang, T.; Zhong, J. Resonant splitting of phonon transport in periodic T-shaped graphene nanoribbons. *Europhys. Lett.* **2010**, *91*, 46006–46010. [CrossRef]

90. Hu, J.; Mazzamuto, F.; Hung Nguyen, V.; Apertet, Y.; Caër, C.; Chassat, C.; Saint-Martin, J.; Dollfus, P. Enhanced thermoelectric properties in graphene nanoribbons by resonant tunneling of electrons. *Phys. Rev. B* **2011**, *83*, 235426.

91. Yang, N.; Zhang, G.; Li, B. Thermal rectification in asymmetric graphene ribbons. *Appl. Phys. Lett.* **2009**, *95*, 033107. [CrossRef]

92. Zhong, W.R.; Huang, W.H.; Deng, X.R.; Ai, B.Q. Thermal rectification in thickness-asymmetric graphene nanoribbons. *Appl. Phys. Lett.* **2011**, *99*, 143501. [CrossRef]

93. Pei, Q.X.; Zhang, Y.W.; Sha, Z.D.; Shenoy, V.B. Carbon isotope doping induced interfacial thermal resistance and thermal rectification in graphene. *Appl. Phys. Lett.* **2012**, *100*, 101901. [CrossRef]

94. Lan, J.; Wang, J.S.; Gan, C.K.; Chin, S.K. Edge effects on quantum thermal transport in graphene nanoribbons: Tight-binding calculations. *Phys. Rev. B* **2009**, *79*, 115401. [CrossRef]

95. Nika, D.L.; Askerov, A.S.; Balandin, A.A. Anomalous size dependence of the thermal conductivity of graphene ribbons. *Nano Lett.* **2012**, *12*, 3238–3244. [CrossRef] [PubMed]

96. Goyal, V.; Balandin, A.A. Thermal properties of the hybrid graphene-metal nano-micro-composites: Applications in thermal interface materials. *Appl. Phys. Lett.* **2012**, *100*, 073113. [CrossRef]

97. Si, S.; Li, W.; Zhao, X.; Meng, H.; Yue, Y.; Wei, W.; Guo, S.; Zhang, X.; Dai, Z.; Wang, X. Significant Radiation Tolerance and Moderate Reduction in Thermal Transport of a Tungsten Nanofilm by Inserting Monolayer Graphene. *Adv. Mater.* **2016**, *29*, 1604623. [CrossRef]

98. Li, Y.; Wei, A.; Datta, D. Thermal characteristics of graphene nanoribbons endorsed by surface functionalization. *Carbon* **2017**, *113*, 274–282. [CrossRef]

99. Li, M.; Zhou, H.; Zhang, Y.; Liao, Y.; Zhou, H. Effect of defects on thermal conductivity of graphene/epoxy nanocomposites. *Carbon* **2018**, *130*, 295–303. [CrossRef]

100. Hu, J.; Ruan, X.; Jiang, Z.; Chen, Y. Thermal conductivity and thermal rectification in graphene nanoribbons: A molecular dynamics study. *Nano Lett.* **2009**, *9*, 2730–2735. [CrossRef]

101. Singh, S.K.; Srinivasan, S.G.; Neek-Amal, M.; Costamagna, S.; Duin, A.C.T.V.; Peeters, F.M. Thermal properties of fluorinated graphene. *Phys. Rev. B* **2013**, *87*, 104114. [CrossRef]

102. Hao, F.; Fang, D.; Xu, Z. Mechanical and thermal transport properties of graphene with defects. *Appl. Phys. Lett.* **2011**, *99*, 223115. [CrossRef]

103. Haskins, J.; Kınacı, A.; Sevik, C.; Sevinçli, H.; Cuniberti, G.; Çağın, T. Control of Thermal and Electronic Transport in Defect-Engineered Graphene Nanoribbons. *ACS Nano* **2011**, *5*, 3779–3787. [CrossRef] [PubMed]

104. Minnich, A.J.; Dresselhaus, M.S.; Ren, Z.F.; Chen, G. Bulk nanostructured thermoelectric materials: Current research and future prospects. *Energy Environ. Sci.* **2009**, *2*, 466–479. [CrossRef]

105. Zhang, H.; Lee, G.; Cho, K. Thermal transport in graphene and effects of vacancy defects. *Phys. Rev. B* **2011**, *84*, 115460. [CrossRef]

106. Pan, C.N.; Xie, Z.X.; Tang, L.M.; Chen, K.Q. Ballistic thermoelectric properties in graphene-nanoribbon-based heterojunctions. *Appl. Phys. Lett.* **2012**, *101*, 103115. [CrossRef]

107. Hu, J.; Wang, Y.; Vallabhaneni, A.; Ruan, X. Nonlinear thermal transport and negative differential thermal conductance in graphene nanoribbons. *Appl. Phys. Lett.* **2011**, *99*, 113101. [CrossRef]

108. Johnson, M.; Silsbee, R.H. Thermodynamic analysis of interfacial transport and of the thermomagnetoelectric system. *Phys. Rev. B* **1987**, *12*, 4959–4972. [CrossRef]

109. Bauer, G.E.W.; Saitoh, E.; Wees, B.J. Spin caloritronics. *Nat. Mater.* **2012**, *11*, 391–399. [CrossRef]

110. Uchida, K.; Takahashi, S.; Harii, K.; Ieda, J.; Koshibae, W.; Ando, K.; Maekawa, S.; Saitoh, E. Observation of the spin Seebeck effect. *Nature* **2008**, *455*, 778–781. [CrossRef]

111. Adachi, H. Spin Seebeck insulator. *Nat. Mater.* **2012**, *9*, 894–897.

112. Jaworski, C.M.; Yang, J.; Mack, S.; Awschalom, D.D.; Heremans, J.P.; Myers, R.C. Observation of the spin-Seebeck effect in a ferromagnetic semiconductor. *Nat. Mater.* **2010**, *9*, 898–903. [CrossRef] [PubMed]

113. Valenzuela, S.O.; Tinkham, M. Direct electronic measurement of the spin Hall effect. *Nature* **2006**, *442*, 176–179. [CrossRef] [PubMed]

114. Saitoh, E.; Ueda, M.; Miyajima, H.; Tatara, G. Conversion of spin current into charge current at room temperature: Inverse spin-Hall effect. *Appl. Phys. Lett.* **2006**, *88*, 182509. [CrossRef]

115. Sinova, J.; Valenzuela, S.O.; Wunderlich, J.; Back, C.H.; Jungwirth, T. Spin Hall effects. *Rev. Mod. Phys.* **2015**, *87*, 1213–1259. [CrossRef]

116. Van Wees, B. Electronic spin transport and spin precession in single graphene layers at room temperature. *Nature* **2007**, *448*, 571–574.

117. Han, W.; Kawakami, R.K.; Gmitra, M.; Fabian, J. Graphene spintronics. *Nat. Nanotechnol.* **2015**, *9*, 794–807. [CrossRef]

118. Roche, S.; Valenzuela, S.O. Graphene spintronics: Puzzling controversies and challenges for spin manipulation. *J. Phys. D Appl. Phys.* **2014**, *47*, 094011. [CrossRef]

119. Veramarun, I.J.; Ranjan, V.; Wees, B.J.V. Nonlinear detection of spin currents in graphene with non-magnetic electrodes. *Nat. Phys.* **2012**, *8*, 313–316. [CrossRef]

120. Berciaud, S.; Han, M.Y.; Mak, K.F.; Brus, L.E.; Kim, P.; Heinz, T.F. Electron and optical phonon temperatures in electrically biased graphene. *Phys. Rev. Lett.* **2010**, *104*, 227401. [CrossRef]

121. Betz, A.C.; Vialla, F.; Brunel, D.; Voisin, C.; Picher, M.; Cavanna, A.; Madouri, A.; Fève, G.; Berroir, J.M.; Plaçais, B. Hot electron cooling by acoustic phonons in graphene. *Phys. Rev. Lett.* **2012**, *109*, 056805. [CrossRef] [PubMed]

122. Betz, A.C.; Jhang, S.H.; Pallecchi, E.; Ferreira, R.; Fève, G.; Berroir, J.; Plaçais, B. Supercollision cooling in undoped graphene. *Nat. Phys.* **2012**, *9*, 109–112. [CrossRef]

123. Sierra, J.F.; Neumann, I.; Costache, M.V.; Valenzuela, S.O. Hot-Carrier Seebeck Effect: Diffusion and Remote Detection of Hot Carriers in Graphene. *Nano Lett.* **2015**, *15*, 4000–4005. [CrossRef] [PubMed]

124. Sierra, J.F.; Neumann, I.; Cuppens, J.; Raes, B.; Costache, M.V.; Valenzuela, S.O. Thermoelectric spin voltage in graphene. *Nat. Nanotechnol.* **2018**, *13*, 107–111. [CrossRef] [PubMed]

125. Song, N.; Jiao, D.; Ding, P.; Cui, S.; Tang, S.; Shi, L.Y. Anisotropic Thermally Conductive Flexible Films based on Nanofibrillated Cellulose and Aligned Graphene Nanosheets. *J. Mater. Chem. C* **2015**, *4*, 305–314. [CrossRef]

126. Song, N.; Cui, S.; Jiao, D.; Hou, X.; Ding, P.; Shi, L. Layered nanofibrillated cellulose hybrid films as flexible lateral heat spreaders: The effect of graphene defect. *Carbon* **2017**, *115*, 338–346. [CrossRef]

127. Song, N.; Jiao, J.; Cui, S.; Hou, X.; Ding, P.; Shi, L. Highly anisotropic thermal conductivity of layer-by-layer assembled nanofibrillated cellulose/graphene nanosheets hybrid films for thermal management. *ACS Appl. Mater. Interfaces* **2017**, *9*, 2924–2932. [CrossRef] [PubMed]

128. Peng, L.; Xu, Z.; Liu, Z.; Guo, Y.; Li, P.; Gao, C. Ultrahigh Thermal Conductive yet Superflexible Graphene Films. *Adv. Mater.* **2017**, *29*, 1700589. [CrossRef]

129. Song, N.; Hou, X.; Chen, L.; Cui, S.; Shi, L.; Ding, P. A green plastic constructed from cellulose and functionalized graphene with high thermal conductivity. *ACS Appl. Mater. Interfaces* **2017**, *9*, 17914–17922. [CrossRef]

130. Ouyang, T.; Chen, Y.; Xie, Y.; Stocks, G.M. Thermal conductance modulator based on folded graphene nanoribbons. *Appl. Phys. Lett.* **2011**, *99*, 233101. [CrossRef]

131. Xu, W.; Qin, Z.; Chen, C.; Kwag, H.R.; Ma, Q.; Sarkar, A.; Buehler, M.J.; Gracias, D.H. Ultrathin thermoresponsive self-folding 3D graphene. *Sci. Adv.* **2017**, *3*, e1701084. [CrossRef] [PubMed]

132. Yan, S.; Zhu, X.; Frandsen, L.H.; Xiao, S.; Mortensen, N.A.; Dong, J.; Ding, Y. Slow-light-enhanced energy efficiency for graphene microheaters on silicon photonic crystal waveguides. *Nat. Commun.* **2016**, *8*, 14411. [CrossRef] [PubMed]

133. Xia, F.; Mueller, T.; Lin, Y.; Valdesgarcia, A.; Avouris, P. Ultrafast graphene photodetector. *Nat. Nanotechnol.* **2009**, *4*, 839–843. [CrossRef] [PubMed]

134. Li, X.; Zhu, H.; Wang, K.; Cao, A.; Wei, J.; Li, C.; Jia, Y.; Li, Z.; Li, X.; Wu, D. Graphene-on-silicon Schottky junction solar cells. *Adv. Mater.* **2010**, *22*, 2743–2748. [CrossRef]

135. Urich, A.; Unterrainer, K.; Mueller, T. Intrinsic response time of graphene photodetectors. *Nano Lett.* **2011**, *11*, 2804–2808. [CrossRef]

136. Limmer, T.; Feldmann, J.; Da Como, E. Carrier lifetime in exfoliated few-layer graphene determined from intersubband optical transitions. *Phys. Rev. Lett.* **2013**, *110*, 217406. [CrossRef]

137. Chen, Z.; Li, X.; Wang, J.; Tao, L.; Long, M.; Liang, S.; Ang, L.K.; Shu, C.C.T.; Tsang, H.K.; Xu, J.B. Synergistic effects of plasmonics and electron trapping in graphene short-wave infrared photodetectors with ultrahigh responsivity. *ACS Nano* **2017**, *11*, 430–437. [CrossRef]

138. An, X.; Liu, F.; Jung, Y.J.; Kar, S. Tunable Graphene-Silicon Heterojunctions for Ultrasensitive Photodetection. *Nano Lett.* **2013**, *13*, 909–916. [CrossRef]

139. Tao, L.; Chen, Z.; Li, X.; Yan, K.; Xu, J. Hybrid graphene tunneling photoconductor with interface engineering towards fast photoresponse and high responsivity. *Npj 2d Mater. Appl.* **2017**, *1*, 19. [CrossRef]

140. Bae, S.; Kim, H.; Lee, Y.; Xu, X.; Park, J.S.; Zheng, Y.; Balakrishnan, J.; Lei, T.; Kim, H.R.; Song, Y.I.; et al. Roll-to-roll production of 30-inch graphene films for transparent electrodes. *Nat. Nanotechnol.* **2010**, *5*, 574–578. [CrossRef]

141. Han, T.H.; Lee, Y.; Choi, M.R.; Woo, S.H.; Bae, S.H.; Hong, B.H.; Ahn, J.H.; Lee, T.W. Extremely efficient flexible organic light-emitting diodes with modified graphene anode. *Nat. Photonics* **2012**, *6*, 105–110. [CrossRef]

142. Hwang, J.O.; Park, J.S.; Choi, D.S.; Kim, J.Y.; Lee, S.H.; Lee, K.E.; Kim, Y.H.; Song, M.H.; Yoo, S.; Kim, S.O. N-doped reduced graphene transparent electrodes for high-performance polymer light-emitting diodes. *ACS Nano* **2012**, *6*, 159–167. [CrossRef] [PubMed]

143. Xu, W.; Wang, L.; Liu, Y.; Thomas, S.; Seo, H.K.; Kim, K.I.; Kim, K.S.; Lee, T.W. Controllable n-type doping on CVD-grown single- and doublelayer graphene mixture. *Adv. Mater.* **2015**, *27*, 1029–1034. [CrossRef] [PubMed]

144. Kim, Y.; Ryu, J.; Park, M.; Kim, E.S.; Yoo, J.M.; Park, J.; Kang, J.H.; Hong, B.H. Vapor-phase molecular doping of graphene for highperformance transparent electrodes. *ACS Nano* **2014**, *8*, 868–874. [CrossRef]

145. Han, T.H.; Kwon, S.J.; Li, N.; Seo, H.K.; Xu, W.; Kim, K.S.; Lee, T.W. Versatile p-type chemical doping to achieve ideal flexible graphene electrodes. *Angew. Chem. Int. Ed. Engl.* **2016**, *55*, 6197–6210. [CrossRef]

146. Kwon, S.; Han, T.; Ko, T.; Li, N.; Kim, Y.; Kim, D.; Bae, S.; Yang, Y.; Hong, B.; Kim, K.; et al. Extremely stable graphene electrodes doped with macromolecular acid. *Nat. Commun.* **2018**, *9*, 2037. [CrossRef]

147. Chang, L.; Frank, D.J.; Montoye, R.K.; Koester, S.J.; Ji, B.L.; Coteus, P.W.; Dennard, R.H.; Haensch, W. Practical strategies for power-efficient computing technologies. *Proc. IEEE* **2010**, *98*, 215–236. [CrossRef]

148. Ionescu, A.M.; Heike, R. Tunnel field-effect transistors as energy-efficient electronic switches. *Nature* **2011**, *479*, 329–337. [CrossRef]

149. Li, X.; Zhu, M.; Du, M.; Lv, Z.; Zhang, L.; Li, Y.; Yang, Y.; Yang, T.; Li, X.; et al. High Detectivity Graphene-Silicon Heterojunction Photodetector. *Small* **2016**, *12*, 595–601. [CrossRef]

150. Kim, J.; Joo, S.S.; Lee, K.W.; Kim, J.H.; Shin, D.H.; Kim, S.; Choi, S.H. Near-ultraviolet-sensitive graphene/porous silicon photodetectors. *ACS Appl. Mater. Interfaces* **2014**, *6*, 20880–20886. [CrossRef]

151. Ni, Z.; Ma, L.; Du, S.; Xu, Y.; Yuan, M.; Fang, H.; Wang, Z.; Xu, M.; Li, D.; Yang, J.; et al. Plasmonic Silicon Quantum Dots Enabled High-Sensitivity Ultrabroadband Photodetection of Graphene-Based Hybrid Phototransistors. *ACS Nano* **2017**, *11*, 9854–9862. [CrossRef] [PubMed]

152. Koppens, F.H.L.; Mueller, T.; Avouris, P.; Ferrari, A.C.; Vitiello, M.S.; Polini, M. Photodetectors based on graphene, other two-dimensional materials and hybrid systems. *Nat Nanotechnol.* **2014**, *9*, 780–793. [CrossRef] [PubMed]

153. Seabaugh, A.C.; Zhang, Q. Low-voltage tunnel transistors for beyond CMOS logic. *Proc. IEEE* **2010**, *98*, 2095–2110. [CrossRef]

154. Gopalakrishnan, K.; Griffin, P.B.; Plummer, J.D. Impact ionization MOS (I-MOS)-Part I: Device and circuit simulations. *IEEE Trans. Electron. Devices* **2004**, *52*, 69–76. [CrossRef]

155. Salahuddin, S.; Datta, S. Use of negative capacitance to provide voltage amplification for low power nanoscale devices. *Nano Lett.* **2008**, *8*, 405–410. [CrossRef] [PubMed]

156. Jo, J.; Choi, W.Y.; Park, J.D.; Shim, J.W.; Yu, H.Y.; Shin, C. Negative capacitance in organic/ferroelectric capacitor to implement steep switching MOS devices. *Nano Lett.* **2015**, *15*, 4553–4556. [CrossRef]

157. Gnani, E.; Reggiani, S.; Gnudi, A.; Baccarani, G. Steep-slope nanowire FET with a superlattice in the source extension. *Solid State Electron.* **2011**, *65–66*, 108–113. [CrossRef]

158. Qiu, C.G.; Liu, F.; Xu, L.; Deng, B.; Xiao, M.M.; Si, J.; Lin, L.; Zhang, Z.Y.; Wang, J.; Guo, H.; et al. Dirac-source field-effect transistors as energy-efficient high-performance electronic switches. *Science* **2018**, *361*, 387–392. [CrossRef]

Article

Thermo-Responsive Graphene Oxide/Poly(Ethyl Ethylene Phosphate) Nanocomposite via Ring Opening Polymerization

Xue Jiang [1,2], Guolin Lu [2], Xiaoyu Huang [2,*], Yu Li [1], Fangqi Cao [1], Hong Chen [1] and Wenbin Liu [1,*]

[1] Shanghai Key Laboratory of Crime Scene Evidence, Shanghai Research Institute of Criminal Science and Technology, Zhongshan North No 1 Road, Shanghai 200083, China; 13795305289@163.com (X.J.); 13675113399@163.com (Y.L.); frankie-cao@163.com (F.C.); chenhong2898@163.com (H.C.)

[2] Key Laboratory of Synthetic and Self-Assembly Chemistry for Organic Functional Molecules, Shanghai Institute of Organic Chemistry, Chinese Academy of Sciences, 345 Lingling Road, Shanghai 200032, China; luguolin@mail.sioc.ac.cn

* Correspondence: wbliu1981@163.com (W.L.); xyhuang@sioc.ac.cn (X.H.); Tel.: +86-21-22028361 (W.L.); +86-21-54925310 (X.H.)

Received: 28 December 2018; Accepted: 29 January 2019; Published: 5 February 2019

Abstract: An efficient strategy for growing thermo-sensitive polymers from the surface of exfoliated graphene oxide (GO) is reported in this article. GO sheets with hydroxyls and epoxy groups on the surface were first prepared by modified Hummer's method. Epoxy groups on GO sheets can be easily modified through ring-opening reactions, involving nucleophilic attack by tris(hydroxymethyl) aminomethane (TRIS). The resulting GO-TRIS sheets became a more versatile precursor for next ring opening polymerization (ROP) of ethyl ethylene phosphate (EEP), leading to GO-TRIS/poly(ethyl ethylene phosphate) (GO-TRIS-PEEP) nanocomposite. The nanocomposite was characterized by ^{1}H NMR, Fourier transform infrared spectroscopy (FT-IR), X-ray photoelectron spectroscopy (XPS), thermogravimetric analysis (TGA), differential thermal gravity (DTG), transmission electron microscopy (TEM) and atomic force microscopy (AFM). Since hydrophilic PEEP chains make the composite separate into single layers through hydrogen bonding interaction, the dispersity of the functionalized GO sheets in water is significantly improved. Meanwhile, the aqueous dispersion of GO-TRIS-PEEP nanocomposite shows reversible temperature switching self-assembly and disassembly behavior. Such a smart graphene oxide-based hybrid material is promising for applications in the biomedical field.

Keywords: graphene oxide; PEEP; ROP; grafting-from

---

1. Introduction

Graphene oxide (GO), a single layer of carbon atoms in a closely packed honeycomb two-dimensional lattice with carboxylic acid, epoxide, and hydroxyl groups, has attracted considerable attention in recent years [1–9]. GO, which is highly hydrophilic, can be prepared using cheap graphite as raw material by cost-effective chemical methods with a high yield. The biocompatibility of GO renders it a good candidate for application in the biomedical field [10–13]. The presence of abundant functional groups at the surface of GO may be very interesting since that they provide enough reactive sites for the subsequent chemical modification using known carbon surface chemistry [14]. Recently, a considerable number of works have been performed on enhancing the properties of GO by adding additional functionality to the groups already present on the surface of GO [15–28]. In particular, "intelligentizing" graphene can been obtained by attaching stimuli-responsive polymers to the backbone of GO [21–28]. These attachments are typically made by either grafting-from or -onto approaches. The vast majority of the triggers reported so far are confined to temperature [21–24], pH [24,25], light [26,27] and redox [28].

Polyphosphoesters (PPEs) are known to be a class of biocompatible polymers with repeated phosphoester attachments in the backbone [29–33]. As the phosphorus atom has multivalent property, the physicochemical properties can be easily modified by means of introduction into such polymers with different functional groups. According to the type of side groups connected to the phosphorous atom, PPEs are classified as polyphosphate, polyphosphonate, polyphosphite or polyphosphoramidate [33]. Among them, polyphosphates are an important family of PPEs, which have attracted great attention for biological and phamaceutical applications such as drug release, gene transfer, and tissue engineering [34–37]. Under usual physiological conditions, polyphosphates can readily degrade through hydrolysis or enzymatic cleavage of the phosphoester bonds. In addition to the excellent biocompatibility and biodegradability, it has been demonstrated that polyphosphates also exhibit thermo-responsibility in aqueous solution so as to make them possess great potential as novel smart biomaterials [32,38]. However, no one has reported on a PPE-based GO/polymer nanocomposite until now.

Herein, we have successfully designed and prepared a kind of PPE-based temperature- responsive GO/polymer nanocomposite (Scheme 1), in which a typical hydrophilic polyphosphoester, i.e., poly(ethyl ethylene phosphate) (PEEP), was covalently connected on the surface of GO by ring opening polymerization (ROP) of ethyl ethylene phosphate (EEP) initiated by the hydroxyls on GO surface introduced by tris(hydroxymethyl) aminomethane (TRIS). The obtained GO-TRIS-PEEP nanocomposite has robust temperature-responsive properties resulting from a change in the PEEP conformation on GO surface. When raising the temperature, PEEP chains may become hydrophobic and collapse so that the nanocomposite precipitated from the solution. Owing to the excellent biocompatibility, biodegradability and thermo-responsibility of PEEP chains, the prepared nanocomposite shows a bright prospect in the field of smart nanocarrier for controlled release under temperature stimuli.

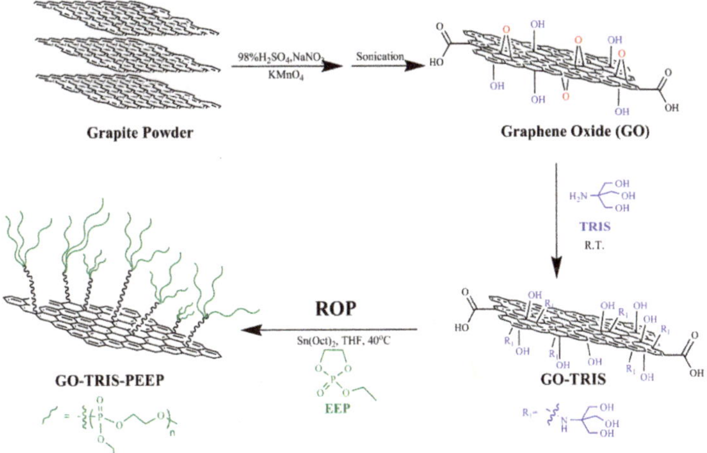

Scheme 1. Preparation of graphene oxide-tris(hydroxymethyl) aminomethane-poly(ethyl ethylene phosphate) (GO-TRIS-PEEP) nanocomposite.

2. Experimental

2.1.Characterization

^{1}H NMR measurements were performed on a JEOL resonance ECZ 400S (400 MHz) spectrophotometer (Tokyo, Japan) in CDCl$_3$ or D$_2$O, TMS was used as internal standard. Fourier transform infrared (FT-IR) spectra were recorded on a Nicolet AVATAR-360 FT-IR spectrophotometer (Waltham, MA, USA) with a resolution of 4 cm^{-1}. Elemental analysis was carried out on a Carlo-Erba 1108 system (Milan, Italy). X-ray photoelectron spectroscopy (XPS) was recorded on a Perkin Elmer PHI 5000c ESCA photoelectron spectrometer (Waltham, IN, USA). Thermogravimetric analysis (TGA) was carried out with a TA Q500 thermal analyzer (New Castle, IN, USA) from 50 °C to 600 °C at a heating rate of 10 °C/min in N$_2$. Transmission electron microscopy (TEM) images were obtained by a JEOL JEM 1230 instrument (Tokyo, Japan) operated at 80 kV. Atomic force microscopy (AFM) images were taken by a Veeco DI MultiMode SPM (Plainview, TX, USA) in the tapping mode of dropping the sample solution onto the freshly exfoliated mica substrate. Differential scanning calorimetry (DSC) measurement was run on a TA Q200 (New Castle, IN, USA) system under N$_2$ purge with a heating rate of 10 °C min^{-1}.

2.1. Preparation of GO Sheets

Exfoliated GO sheets were prepared by a modified Hummer's method using graphite powder as starting material. Graphite powder (99.99+%, Aldrich) was firstly oxidized by sulfuric acid (H$_2$SO$_4$, Aldrich, 95–98%) and potassium permanganate (KMnO$_4$, Aldrich, 99%) followed by filtration and subsequent dialysis or by several runs of centrifugation/washing to completely remove residual salts and acids. Finally, graphite oxide dispersion (0.1 mg mL^{-1}) was exfoliated by waterbath ultrasonication for 3 h. GO sheets were recovered by filtration and vacuum drying.

2.2. Preparation of TRIS-Bonded Graphene Sheets (GO-TRIS)

GO (2.3 g) was treated with 12.2 g of 1,1,1-tris(hydroxymethyl) methanamine (TRIS, Aldrich, 99%) in 500 mL of anhydrous *N,N*-dimethylformamide (DMF, Aldrich, 99.8%) at room temperature for 36 h to transform the epoxy groups on the basal plane of GO into hydroxyls. The crude product was filtered through a 0.22 μm filter and washed exhaustively with deionized water. The obtained GO-TRIS (2.4 g) was dried overnight at 40 °C in vacuo.

2.3. Preparation of Ethyl Ethylene Phosphate (EEP)

Tetrahydrofuran (THF, Aldrich, 99%) was dried over CaH$_2$ and distilled from sodium and benzophenone under N$_2$ prior to use. Triethylamine (Et$_3$N, Aldrich, 99.5%) was dried over KOH and distilled over CaH$_2$ under N$_2$ prior to use. A mixture of ethanol (EtOH, Aldrich, 99.8%, 8.8 mL, 0.15 mol) and Et$_3$N(20.9 mL, 0.15 mol) was added dropwise to a solution of 2-chloro-1,3,2-dioxaphospholane-2-oxide (COP, Aldrich, 90%–94%, 21.4 g, 0.15 mol) in anhydrous THF (250 mL) at −5 °C. The reaction mixture was stirred at −5 °C for 30 min and then at room temperature for additional 12 h. The resulting mixture was firstly filtered to remove the insoluble salt. The filtrate was concentrated and distilled under reduced pressure (107 Pa/95~97 °C) to give 15.5 g of yellowish liquid, ethyl ethylene phosphate (EEP), with a yield of 68%. ^{1}H NMR (400 MHz, CDCl$_3$): δ (ppm): 1.38 (3H, POCH$_2$CH$_3$), 4.22 (2H, POCH$_2$CH$_3$), 4.42 (4H, OCH$_2$CH$_2$).

2.4. Preparation of PEEP-Grafted Graphene Sheets (GO-TRIS-PEEP)

Tin-2-ethylhexanoate (Sn(Oct)$_2$, Aldrich, 98%) was purified by distilling under reduced pressure (20~40 Pa/152 °C) after co-boiling with *p*-xylene twice prior to use [39]. A 25 mL Schlenk flask was first dried by sonication in acetone. It was then immersed in the diethyl ether solution of trimethylchlorosilane (5% TMSCl) overnight to remove the hydroxyls on the surface of glass. GO-TRIS

(235 mg) was firstly added to the flask (flame-dried under vacuum prior to use) sealed with a rubber septum for degassing and kept under Ar. After three cycles of evacuation purging with purified Ar, anhydrous THF (15 mL) and DMF (2 mL) were added via a gastight syringe followed by water-bath ultra-sonication for 30 minutes. The flask was then degassed by three cycles of freezing-pumping-thawing after EEP (1.98 g, 13 mmol) was added. THF solution (0.9 mL) of $Sn(Oct)_2$ (0.18 mmol) was added via a gastight syringe. The flask was again degassed by three cycles of freezing-pumping-thawing, followed by immersing the flask into an oil bath set at 40 °C. The polymerization lasted 24 h and it was terminated by adding excess acetic acid. The crude product was filtered through a 0.22 μm filter followed by washing with THF and deionized water. The precipitate was dried *in vacuo* at 45 °C for 48 h to give the desired product of GO-TRIS-PEEP (156 mg). XPS (molecular molar ratio): C, 65.66%; N, 0.98%; O, 30.44%, P, 1.89%. ^{1}H NMR (400 MHz, D_2O): δ (ppm): 1.12 (3H, $POCH_2CH_3$), 3.81 (2H, $POCH_2CH_3$), 4.16 (4H, OCH_2CH_2). FT-IR: ν (cm^{-1}): 1727, 1590, 1226, 1030, 977, 800, 640, 518.

3. Results and discussion

3.1. Preparation of GO-TRIS-PEEP

The fabrication process of GO-TRIS-PEEP polymer-modified nanocomposite via ring opening polymerization of EEP monomer is outlined in Scheme 1. Commercial graphite powder was first oxidized to GO with the help of $KMnO_4$ under strong acidic environment using the modified Hummer's method [40], as a result, new epoxy and hydroxyl functionalities were introduced to the surface of a graphene oxide sheet. The structure of the obtained GO was confirmed by FT-IR (Figure S1), XRD (Figure S2) and elementary analysis (Table S1). The obtained epoxy groups can be easily modified through ring-opening reactions, on the basis of the mechanism of nucleophilic attack by amine groups [17]. 1,1,1-Tris(hydroxymethyl) methanamine (TRIS) is a kind of primary amine with three hydroxyls, which has been extensively used in chemical modification of macroporous materials [41,42]. In this work, TRIS is required to improve the opportunity of grafting from the plane of GO sheet, rendering GO a more versatile precursor for a wide range of applications. XPS measurement was used to confirm the introduction of TRIS moiety. XPS spectrum of GO-TRIS shown in Figure 1 exhibits a significant nitrogen peak around the binding energy of 400 eV, which is absent in that of GO (Figure S3). Zooming in further revealed that the N1s band appeared at 402.2 eV accompanied with a lower binding energy shoulder located at 399.8 eV, which is consistent with a previous report [43]. The appearance of N1s band suggested the successful covalent functionalization of TRIS units onto the surface of modified GO sheets; i.e., the formation of TRIS-modified GO (GO-TRIS). Moreover, Figure 1 also displays the elemental analysis result of GO-TRIS, which indicated that the nitrogen content (N%) increased from 0 to 1.40% after the functionalization with TRIS. The C/N ratio was estimated to be 38 = (46.18wt %/12)/(1.40 wt %/14) (1 TRIS group per 6.3 aromatic rings), which agreed with the value (45) derived from XPS measurement. Additionally, the dispersity of GO-TRIS kept as good as GO in water or polar organic solvents as shown in the inset of Figure 1. The yellow color of GO sheets remained after the incorporation of TRIS moieties, which also suggested that GO-TRIS retained the oxygen-containing groups so as to stabilize the corresponding aqueous solution.

Thermo-responsive PEEP chains then grew from the fabricated hydroxyls on the surface through ROP. To ensure that there was no polymer chains adsorbed on the surface of GO, the obtained product was filtered through a 0.22 μm filter followed by washing with THF constantly, until the P content determined by elementary analysis was detected to be zero in the filtrate. ^{1}H NMR and FT-IR analyses were employed to characterize the obtained product after ROP of EEP monomer for confirming GO functionalization and the presence of PEEP chains on the surface of GO. ^{1}H NMR spectrum of the resulting GO-TRIS-PEEP in D_2O is shown in Figure 2A and the peaks located at 1.12 (3H) and 3.81 (2H) ppm are attributed to the $POCH_2CH_3$ moiety in the side group of PEEP, respectively. On the other hand, the proton resonance signal of OCH_2CH_2 moiety in the backbone of PEEP is also seen at 4.16 ppm. Owing to the existence of the hydrophilic PEEP chains, GO-TRIS-PEEP might be well dispersed in D_2O. GO-TRIS-PEEP showed relative strong bands at 3217, 1726, 1577 and 1037 cm^{-1} in FT-IR spectrum (Figure 2B), corresponding to stretching vibrations of O-H, C=O, C=C and C-O groups, respectively. The stretching vibration peaks attributed to P=O and P-O-C appear at 1226 and 977 cm^{-1} after ROP. These characteristic peaks are originated from polyphosphoester chains attached to the surface through ROP. These data clearly illustrate the successful ROP of EEP initiated by GO-TRIS.

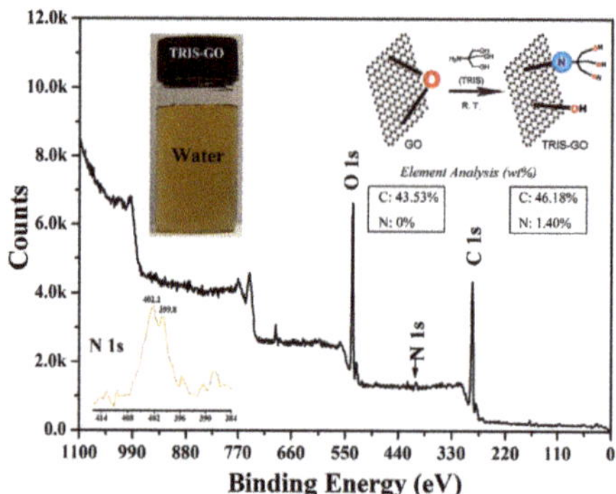

Figure 1. Survey X-ray photoelectron spectroscopy (XPS) data for GO-TRIS and photograph of well-dispersed GO-TRIS solution in water (inset).

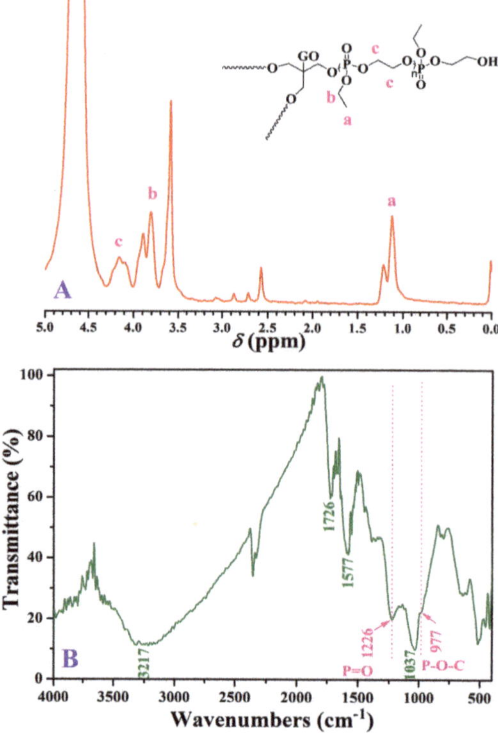

Figure 2. ^{1}H NMR (**A**) and Fourier transform infrared (FT-IR) (**B**) spectra of GO-TRIS-PEEP.

To further understand the effective functionalization of PEEP chains on GO sheets, XPS measurement was employed to provide qualitative analysis. GO-TRIS-PEEP shows a strong peak of C1s at a binding energy around 285 eV, a weak peak of N1s at around 401.0 eV and a peak at around 533 eV originating from O1s (Figure 3). Furthermore, a peak at around 134 eV is assigned to P2p3, which is resulted from PEEP chains. As the coordination between Sn of the catalyst and hydroxyls in the polymer chains [44], we also can observe the peaks of Sn4d around 24 eV, SnO2 around 487 eV and Sn3d around 497 eV. The molar contents of C, O, N and P are 65.66%, 30.44%, 0.98% and 1.89%, respectively. The weight percentage of grafted PEEP in GO-TRIS-PEEP is estimated to be 21.5 wt % from Equation (1).

$$\text{PEEP\%} = [c_P \times A_P / \sum (c_i \times A_i)] \times (152/31) \times 100\% \tag{1}$$

TGA analysis of GO-TRIS-PEEP was performed to elucidate its thermal behavior and chemical composition. In comparison with the initial GO (Figure S4), the thermograph of the obtained GO-TRIS-PEEP (Figure 4) exhibits three major stages of decomposition in the temperature range of 50–600 °C. The first stage shows a low weight loss until 125 °C, corresponding to the loss of adsorbed water into its structure. The second stage from 125 °C to 250 °C can be assigned to the decomposition of oxygen-containing functionalities by evaporation of CO and CO_2. The last weight loss occurs between 250 °C and 450 °C, indicating the degradation of PEEP segment. To further investigate the enhancement of thermal properties, differential thermal gravity (DTG) curve of GO-TRIS-PEEP is also plotted in Figure 4. The peaks in DTG curve correspond to the temperatures at maximum rate of weight loss (T_{max}), and the outcome is entirely consistent with the TGA data. The quantity of PEEP chains attached to the surface of GO sheets is also determined from TGA. Figure 4 shows a 14.6 wt % (= 79.6 wt % − 65.0 wt %) weight loss at the temperature range from 250 °C to 450 °C, and thus, the amount of PEEP chains grafted onto GO sheets is about 14.6 wt %. Compared to GO, most of PEEP chains were exposed on the surface of GO-TRIS-PEEP, the measured phosphorus content by XPS might be higher than actual data. As a result, the XPS data (21.5%) is a little higher than that obtained from TGA (14.6%).

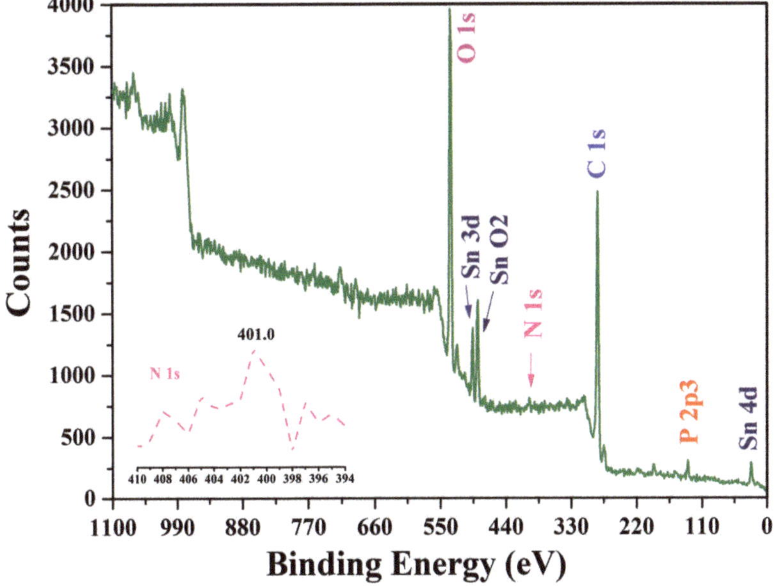

Figure 3. Survey XPS data for GO-TRIS-PEEP.

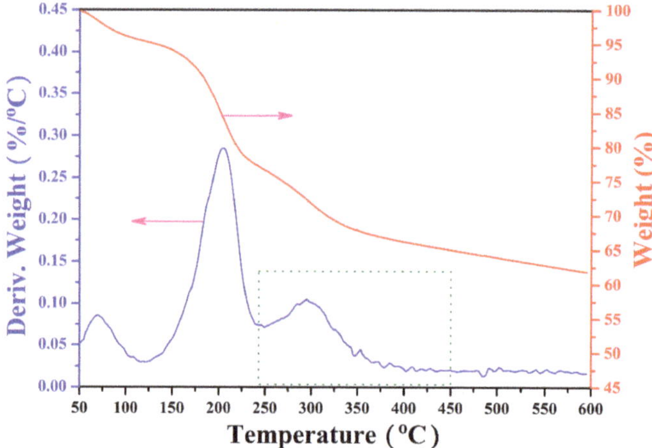

Figure 4. Thermogravimetric analysis (TGA)(red line) and differential thermal gravity (DTG) (blue line) curves of GO-TRIS-PEEP in N_2.

3.2.Surface Morphologies of GO-TRIS-PEEP Sheets

A TEM micrograph of the fabricated GO-TRIS-PEEP is presented in Figure 5. It can be observed that the lamellar morphology of GO is preserved after ROP of EEP monomer, and the individual sheets have sizes of about 1 μm. Compared to GO (Figure S5), new dark dots appear because of the tangled mess formed by the grafted PEEP chains. Previous studies on GO have identified that carboxyls are at the periphery of sheets, while the oxygen-containing functionalities such as hydroxyl and epoxy group are on the basal plane [45,46]. As hydroxyls were used as initiating sites to trigger the polymerization of EEP monomer via surface-initiated ROP, the blackish polymeric chains spread evenly on the surface of the GO sheets, which is consistent with previous reports [47,48]. The grafted polymeric chains also destroyed the ordered π-π stacking of graphitic basal planes so that functionalized GO sheets appeared to have better dispersion and exfoliation in methanol solution (Figure S6).

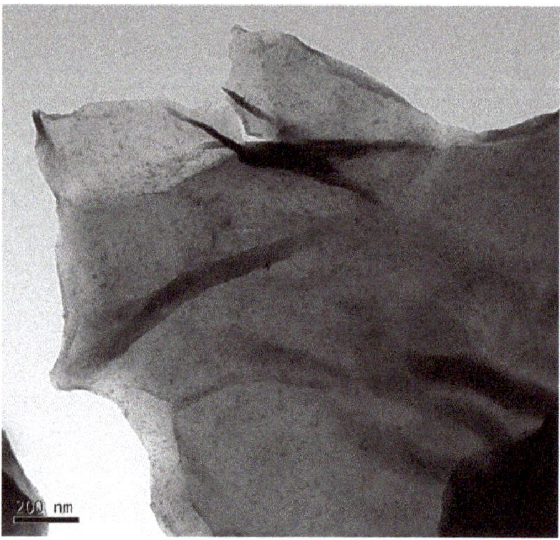

Figure 5. Transmission electron microscope (TEM) image of GO-TRIS-PEEP aqueous solutions with a concentration of 0.01 mg/mL using drop-casting method.

AFM provides a direct method to characterize the surface morphology and thickness of GO-TRIS-PEEP sheets. As shown in Figure 6, GO-TRIS-PEEP sheets have an average in-plane size with several hundreds of nanometers. The surfaces of GO-TRIS-PEEP sheets are very rough compared to those of pristine GO sheets (Figure S7) and the thickness of GO-TRIS-PEEP sheets is measured to be approximately 10 nm from the average height profiles, which is much higher than that of pristine GO sheets. Both the rough surfaces and higher thickness of GO-TRIS-PEEP sheets further demonstrate the successful polymeric grafting [49–51].

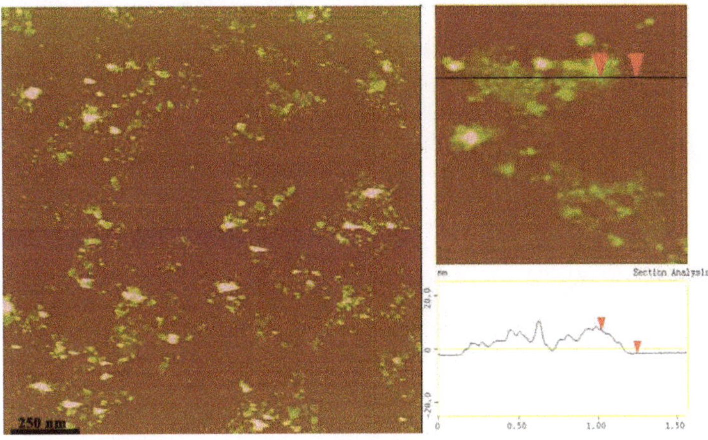

Figure 6. Atomic force microscopy (AFM) images of GO-TRIS-PEEP.

3.2. Thermo-Responsive Behavior of GO-TRIS-PEEP

As it is well known, PEEP has a phase transition behavior in water and form aggregates by making the solution turbid when the temperature is above its low critical solution temperature (LCST) [32,38]. The LCST behavior of PEEP is attributed to a balance between hydrophilic and hydrophobic interactions and the resulting hydrogen-bonding interactions between water molecules and the polymeric chains. It is one of the basic physical properties of thermo-responsive water-soluble polymers [52]. As expected, this thermal behavior in aqueous solution was well retained after PEEP was covalently bound to GO sheets (see the inset in Figure 7). In a typical experiment, we found that GO-TRIS-PEEP with a concentration of 0.5 mg/mL at 25 °C formed a stable homogeneous solution. If not heated, the homogeneous solution could last more than one month. When the solution was heated to 40 °C, GO-TRIS-PEEP began to aggregate. Further raising the temperature, GO-TRIS-PEEP would precipitate completely from the solution in a few minutes. When the mixture was cooled down to 25 °C, and then mildly shaken for several minutes, a homogeneous solution recovered again. This precipitation phenomenon was not observed for GO aqueous dispersion even if the temperature rose to 90 °C (Figure S8), which demonstrated that PEEP segments on the surface of GO-TRIS-PEEP take account for the above- mentioned thermal responsibility. The hydrophilic/hydrophobic phase transition of PEEP-grafted GO sheets should show exothermic or endothermic effects. The change of enthalpy during phase transition can be measured by DSC. The DSC curve (Figure 7) exhibits a small endothermic peak at ~74 °C, which indicates a phase transition process. Considering that GO-TRIS-PEEP nanocomposite begins to aggregate at 40 °C, the phase transition process should be gradual in a wide temperature range. It might be because the abundant oxygen-containing functionalities on the surface increased the hydrophilicity of GO-TRIS-PEEP. Therefore, it is not appropriate to provide a certain temperature as the LCST.

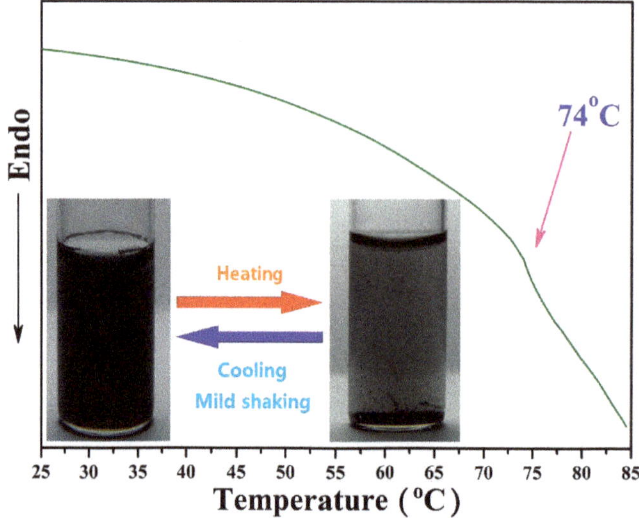

Figure 7. DSC curve of GO-TRIS-PEEP with a heating rate of 10 °C/min in N_2, the inset shows the thermo-stimuli reversible phase behavior of GO-TRIS-PEEP.

4. Conclusions

In summary, we have successfully modified GO with the thermo-responsive PEEP chains via the grafting-from approach. To the best knowledge, this is the first example of PPE-Based GO/polymer nanocomposite via covalent modification. The structure of the resulting GO-TRIS-PEEP nanocomposite was well studied by various characterizations to prove the change on functionalization. Moreover, GO-TRIS-PEEP nanocomposite shows good temperature responsive behavior as a result of a conformation change in PEEP chains grafted on the surface of GO. Our work provides a potent strategy for the design and synthesis of novel GO/stimuli- responsive polymer nanocomposites, which exhibit great application potential in biological and medical as well as interdisciplinary fields.

Supplementary Materials: The following are available online at http://www.mdpi.com/2079-4991/9/2/207/s1, Figure S1: FT-IR spectra of natural graphite powder and graphene oxide, Figure S2: X-ray diffraction patterns of natural graphite powder and graphene oxide, Table S1: Elementary analysis of natural graphite and graphene oxide, Figure S3: Survey XPS data for graphene oxide, the inset curve indicates that there is no N1s peak in XPS of graphene oxide, Figure S4: TGA (black) and DTG (blue) curves of graphene oxide, Figure S5: TEM images of graphene oxide with different scale bars in length, (A) 1 μm and (B) 200 nm, Figure S6: AFM images of graphene oxide, Figure S7: AFM images of graphene oxide, Figure S8: Photographs of graphene oxide aqueous solutions in different temperature with a concentration of ~0.5 mg/mL.

Author Contributions: Conceptualization, X.J.; methodology, X.J.; software, G.L.; validation, Y.L., F.C. and H.C.; formal analysis, X.J.; investigation, X.J.; resources, X.H. and W.L.; data curation, X.J.; writing—original draft preparation, X.J.; writing—review and editing, X.J.; visualization, X.J.; supervision, X.H. and W.L.; project administration, X.H. and W.L.; funding acquisition, X.H. and W.L.

Funding: The authors thank the financial supports from National Key Research & Development Program of China (213 project), National Natural Science Foundation of China (21704110, 21674124 and 51773222), Strategic Priority Research Program of Chinese Academy of Sciences (XDB20000000), Shanghai Scientific and Technological Innovation Project (16DZ1205600, 16JC1402500, 16520710300, 17DZ1205400 and 18ZR1449900), Opening Project of Shanghai Key Laboratory of Crime Scene Evidence (2015XCWZK02, 2017XCWZK24 and 2015XCWZK15) and Science and Technology Development Fund of Shanghai Municipal Public Security Bureau (2018002 and 2018003).

Conflicts of Interest: The authors declare no conflict of interest.

References

1. Dikin, D.A.; Stankovich, S.; Zimney, E.J.; Piner, R.D.; Dommett, G.H.B.; Evmenenko, G.; Nguyen, S.T.; Ruoff, R.S. Preparation and characterization of graphene oxide paper. *Nature* **2007**, *448*, 457–460. [CrossRef] [PubMed]

2. Becerril, H.A.; Mao, J.; Liu, Z.; Stoltenberg, R.M.; Bao, Z.; Chen, Y. Evaluation of solution-processed reduced graphene oxide films as transparent conductors. *ACS Nano* **2008**, *2*, 463–470. [CrossRef]

3. Georgakilas, V.; Tiwari, J.N.; Kemp, K.C.; Perman, J.A.; Bourlinos, A.B.; Kim, K.S.; Zboril, R. Noncovalent functionalization of graphene and graphene oxide for energy materials, biosensing, catalytic, and biomedical applications. *Chem. Rev.* **2016**, *116*, 5464–5519. [CrossRef] [PubMed]

4. Marcano, D.C.; Kosynkin, D.V.; Berlin, J.M.; Sinitskii, A.; Sun, Z.; Slesarev, A.; Alemany, L.B.; Lu, W.; Tour, J.M. Improved synthesis of graphene oxide. *ACS Nano* **2010**, *4*, 4806–4814. [CrossRef] [PubMed]

5. Dreyer, D.R.; Park, S.; Bielawski, C.W.; Ruoff, R.S. The chemistry of graphene oxide. *Chem. Soc. Rev.* **2010**, *39*, 228–240. [CrossRef] [PubMed]

6. Zhu, Y.W.; Murali, S.; Cai, W.; Li, X.; Suk, J.W.; Potts, J.R.; Ruoff, R.S. Graphene and graphene oxide: synthesis, properties, and applications. *Adv. Mater.* **2010**, *22*, 3906–3924. [CrossRef] [PubMed]

7. Narayanan, T.N.; Gupta, B.K.; Vithayathil, S.A.; Aburto, R.R.; Mani, S.A.; Tijerina, J.T.; Xie, B.; Kaipparettu, B.A.; Torti, S.V.; Ajayan, P.M. Hybrid 2D nanomaterials as dual-mode contrast agents in cellular imaging. *Adv. Mater.* **2012**, *24*, 2992–2998. [CrossRef]

8. Robinson, J.T.; Tabakman, S.M.; Liang, Y.; Wang, H.; Casalongue, H.S.; Vinh, D.; Dai, H.J. Ultrasmall reduced graphene oxide with high near-infrared absorbance for photothermal therapy. *J. Am. Chem. Soc.* **2011**, *133*, 6825–6831. [CrossRef]

9. Guan, F.L.; An, F.; Yang, J.; Li, X.; Li, X.H.; Yu, Z.Z. Fiber reinforced three-dimensional graphene aerogels for electrically conductive epoxy composites with enhanced mechanical properties. *Chin. J. Polym. Sci.* **2017**, *35*, 1381–1390. [CrossRef]

10. Liu, Z.; Robinson, J.T.; Sun, X.; Dai, H.J. PEGylated nano-graphene oxide for delivery of water insoluble cancer drugs. *J. Am. Chem. Soc.* **2008**, *130*, 10876–10877. [CrossRef]

11. Zhang, X.; Yin, J.; Peng, C.; Hu, W.; Zhu, Z.; Li, W.; Fan, C.H.; Huang, Q. Distribution and biocompatibility studies of graphene oxide in mice after intravenous administration. *Carbon* **2011**, *49*, 986–995. [CrossRef]

12. Bitounis, D.; Ali-Boucetta, H.; Hong, B.H.; Min, D.H.; Kostarelos, K. Prospects and challenges of graphene in biomedical applications. *Adv. Mater.* **2013**, *25*, 2258–2268. [CrossRef] [PubMed]

13. Liu, J.; Cui, L.; Losic, D. Graphene and graphene oxide as new nanocarriers for drug delivery applications. *Acta Biomater.* **2013**, *9*, 9243–9257. [CrossRef] [PubMed]

14. Loh, K.P.; Bao, Q.; Ang, P.K.; Yang, J. The chemistry of graphene. *J. Mater. Chem.* **2010**, *20*, 2277–2289. [CrossRef]

15. Chen, Y.; Li, D.X.; Yang, W.Y.; Xiao, C.G.; Wei, M.L. Effects of different amine-functionalized graphene on the mechanical, thermal, and tribological properties of polyimide nanocomposites synthesized by in situ polymerization. *Polymer* **2018**, *140*, 56–72. [CrossRef]

16. Nia, A.S.; Binder, W.H. Graphene as initiator/catalyst in polymerization chemistry. *Prog. Polym. Sci.* **2017**, *67*, 48–76.

17. Kuila, T.; Bose, S.; Mishra, A.K.; Khanra, P.; Kim, N.H.; Lee, J.H. Chemical functionalization of graphene and its applications. *Prog. Mater. Sci.* **2012**, *57*, 1061–1105. [CrossRef]

18. Goncalves, G.; Marques, P.A.A.P.; Granadeiro, C.M.; Nogueira, H.I.S.; Singh, M.K.; Gracio, J. Surface modification of graphene nanosheets with gold nanoparticles: The role of oxygen moieties at graphene surface on gold nucleation and growth. *Chem. Mater.* **2009**, *21*, 4796–4802. [CrossRef]

19. Luan, Y.G.; Zhang, X.A.; Jiang, S.L.; Chen, J.H.; Lyu, Y.F. Self-healing supramolecular polymer composites by hydrogen bonding interactions between hyperbranched polymer and graphene oxide. *Chin. J. Polym. Sci.* **2018**, *36*, 584–591. [CrossRef]

20. Wang, N.; Tian, H.; Zhu, S.Y.; Yan, D.Y.; Mai, Y.Y. Two-dimensional nitrogen-doped mesoporous carbon/graphene nanocomposites from the self-assembly of block copolymer micelles in solution. *Chin. J. Polym. Sci.* **2018**, *36*, 266–272. [CrossRef]

21. Balcioglu, M.; Buyukbekar, B.Z.; Yavuz, M.S.; Yigit, M.V. Smart-polymer-functionalized graphene nanodevices for thermo-switch-controlled biodetection. *ACS Biomater. Sci. Eng.* **2015**, *1*, 27–36. [CrossRef]

22. Pan, Y.Z.; Bao, H.Q.; Sahoo, N.G.; Wu, T.F.; Li, L. Water-soluble poly (*N*-isopropylacrylamide)-graphene sheets synthesized via click chemistry for drug delivery. *Adv. Funct. Mater.* **2011**, *21*, 2754–2763. [CrossRef]

23. Peng, X.; Liu, T.Q.; Shang, C.; Jiao, C.; Wang, H.L. Mechanically strong Janus poly(*N*-isopropylacrylamide)/graphene oxide hydrogels as thermo-responsive soft robots. *Chin. J. Polym. Sci.* **2017**, *35*, 1268–1275. [CrossRef]

24. Sun, S.; Wu, P. A One-Step Strategy for Thermal- and pH-responsive graphene oxide interpenetrating polymer hydrogel networks. *J. Mater. Chem.* **2011**, *21*, 4095–4097. [CrossRef]

25. Yang, Y.; Wang, J.; Zhang, J.; Liu, J.; Yang, X.; Zhao, H. Exfoliated graphite oxide decorated by PDMAEMA chains and polymer particles. *Langmuir* **2009**, *25*, 11808–11814. [CrossRef] [PubMed]

26. Wang, D.R.; Wang, X.G. Self-assembled graphene/azo polyelectrolyte multilayer film and its application in electrochemical energy storage device. *Langmuir* **2011**, *27*, 2007–2013. [CrossRef]

27. Tran, T.H.; Nguyen, H.T.; Pham, T.T.; Choi, J.Y.; Choi, H.G.; Yong, C.S.; Kim, J.O. Development of a graphene oxide nanocarrier for dual-drug chemo-phototherapy to overcome drug resistance in cancer. *ACS Appl. Mater. Interfaces* **2015**, *7*, 28647–28655. [CrossRef]

28. Jiang, X.; Deng, Y.; Liu, W.B.; Li, Y.J.; Huang, X.Y. Preparation of graphene/poly(2-acryloxyethyl ferrocenecarboxylate) nanocomposite via a "grafting-onto" strategy. *Polym. Chem.* **2018**, *9*, 184–192. [CrossRef]

29. Wang, Y.C.; Liu, X.C.; Sun, T.M.; Xiong, M.H.; Wang, J. Functionalized micelles from block copolymer of polyphosphoester and poly(ε-caprolactone) for receptor-mediated drug delivery. *J. Control. Release* **2008**, *128*, 32–40. [CrossRef]

30. Appukutti, N.; Serpell, C.J. High definition polyphosphoesters: between nucleic acids and plastics. *Polym. Chem.* **2018**, *9*, 2210–2226. [CrossRef]

31. Yuan, Y.Y.; Du, Q.; Wang, Y.C.; Wang, J. One-pot syntheses of amphiphilic centipede-like brush copolymers via combination of ring-opening polymerization and "click" chemistry. *Macromolecules* **2010**, *43*, 1739–1746. [CrossRef]

32. Iwasaki, Y.; Wachiralarpphaithoon, C.; Akiyoshi, K. Novel Thermoresponsive polymers having biodegradable phosphoester backbones. *Macromolecules* **2007**, *40*, 8136–8138. [CrossRef]

33. Liu, J.Y.; Huang, W.; Pang, Y.; Yan, D.Y. Hyperbranched polyphosphates: synthesis, functionalization and biomedical applications. *Chem. Soc. Rev.* **2015**, *44*, 3942–3952. [CrossRef] [PubMed]

34. Du, X.; Sun, Y.; Zhang, M.; He, J.L.; Ni, P.H. Polyphosphoester-camptothecin prodrug with reduction-response prepared via Michael addition polymerization and click reaction. *ACS Appl. Mater. Interfaces* **2017**, *9*, 13939–13949. [CrossRef] [PubMed]

35. Hindi, K.M.; Ditto, A.J.; Panzner, M.J.; Medvetz, D.A.; Han, D.S.; Hovis, C.E.; Hilliard, J.K.; Taylor, J.B.; Yun, Y.H.; Cannon, C.L.; et al. The antimicrobial efficacy of sustained release silver-carbine complex-loaded L-tyrosine polyphosphate nanoparticles: characterization, *in vitro* and *in vivo* Studies. *Biomaterials* **2009**, *30*, 3771–3779. [CrossRef] [PubMed]

36. Zhao, Z.; Wang, J.; Mao, H.Q.; Leong, K.W. Polyphosphoesters in drug and gene delivery. *Adv. Drug Deliv. Rev.* **2003**, *55*, 483–499. [CrossRef]

37. Gu, Z.; Xie, H.; Li, L.; Zhang, X.; Liu, F.; Yu, X. Application of strontium-doped calcium polyphosphate scaffold on angiogenesis for bone tissue engineering. *J. Mater. Sci. Mater. Med.* **2013**, *24*, 1251–1260. [CrossRef] [PubMed]

38. Wang, Y.C.; Tang, L.Y.; Li, Y.; Wang, J. Thermoresponsive block copolymers of poly (ethylene glycol) and polyphosphoester: thermo-induced self-assembly, biocompatibility, and hydrolytic degradation. *Biomacromolecules* **2009**, *10*, 66–73. [CrossRef]

39. Kricheldorf, H.R.; Kreiser-Saunders, I.; Stricker, A. Polylactones 48. SnOct$_2$-initiated polymerizations of lactide: A mechanistic study. *Macromolecules* **2000**, *33*, 702–709. [CrossRef]

40. Shen, J.F.; Hu, Y.Z.; Shi, M.; Li, N.; Ma, H.W.; Ye, M.X. One step synthesis of graphene oxide-magnetic nanoparticle composite. *J. Phys. Chem. C* **2010**, *114*, 1498–1503. [CrossRef]

41. Bui, T.N.H.; Verhage, J.J.; Irgum, K. Tris(hydroxymethyl)aminomethane-functionalized silica particles and their application for hydrophilic interaction chromatography. *J. Sep. Sci.* **2010**, *33*, 2965–2976. [CrossRef] [PubMed]

42. Mallakpour, S.; Abdolmaleki, A.; Khalesi, Z.; Borandeh, S. Surface functionalization of GO, preparation and characterization of PVA/TRIS-GO nanocomposites. *Polymer* **2015**, *81*, 140–150. [CrossRef]

43. Yang, H.F.; Shan, C.S.; Li, F.H.; Han, D.X.; Zhang, Q.X.; Niu, L. Covalent functionalization of polydisperse chemically-converted graphene sheets with amine-terminated ionic liquid. *Chem. Commun.* **2009**, 3880–3882. [CrossRef] [PubMed]

44. Ryner, M.; Stridsberg, K.; Albertsson, A.C. Mechanism of ring-opening polymerization of 1,5-dioxepan-2-oneandL-lactide with stannous 2-ethylhexanoate. A theoretical study. *Macromolecules* **2001**, *34*, 3877–3881. [CrossRef]

45. Nakajima, T.; Mabuchi, A.; Hagiwara, R. A new structure model of graphite oxide. *Carbon* **1988**, *26*, 357–361. [CrossRef]

46. Teng, C.Y.; Yeh, T.F.; Lin, K.I.; Chen, S.J.; Yoshimura, M.; Teng, H. Synthesis of graphene oxide dots for excitation-wavelength independent photoluminescence at high quantum yields. *J. Mater. Chem. C* **2015**, *3*, 4553–4562. [CrossRef]

47. Deng, Y.; Zhang, J.Z.; Li, Y.J.; Hu, J.H.; Yang, D.; Huang, X.Y. Thermoresponsive graphene oxide-PNIPAM nanocomposites with controllable grafting polymer chains via moderate *in situ* SET-LRP. *J. Polym. Sci. Polym. Chem.* **2012**, *50*, 4451–4458. [CrossRef]

48. Deng, Y.; Li, Y.J.; Dai, J.; Lang, M.D.; Huang, X.Y. Functionalization of graphene oxide towards thermo-sensitive nanocomposites via moderate *in situ* SET-LRP. *J. Polym. Sci. Polym. Chem.* **2011**, *49*, 4747–4755. [CrossRef]

49. Fang, M.; Wang, K.G.; Lu, H.B.; Yang, Y.L.; Nutt, S. Covalent polymer functionalization of graphene nanosheets and mechanical properties of composites. *J. Mater. Chem.* **2009**, *19*, 7098–7105. [CrossRef]

50. Xu, Z.; Gao, C. *In situ* Polymerization approach to graphene-reinforced nylon-6 composites. *Macromolecules* **2010**, *43*, 6716–6723. [CrossRef]

51. Fang, M.; Wang, K.G.; Lu, H.B.; Yang, Y.L.; Nutt, S. Single-layer graphene nanosheets with controlled grafting of polymer chains. *J. Mater. Chem.* **2010**, *20*, 1982–1992. [CrossRef]

52. Sánchez-Moreno, P.; Vicente, J.D.; Nardecchia, S.; Marchal, J.A.; Boulaiz, H. Thermo-sensitive nanomaterials: recent advance in synthesis and biomedical applications. *Nanomaterials* **2018**, *8*, 935. [CrossRef] [PubMed]

 *nanomaterials*

Article

Latent Heat Storage and Thermal Efficacy of Carboxymethyl Cellulose Carbon Foams Containing Ag, Al, Carbon Nanotubes, and Graphene in a Phase Change Material

Hong Gun Kim [†]**, Yong-Sun Kim** [†]**, Lee Ku Kwac, Hee Jae Shin, Sang Ok Lee, U Sang Lee and Hye Kyoung Shin** [*,†]

Institute of Carbon Technology, Jeonju University, 303 Cheonjam-ro, Wansan-gu, Jeonju-si, Jeollabuk-do 55069, Korea; hgkim@jj.ac.kr (H.G.K.); wva223g6@naver.com (Y.-S.K.); kwack29@jj.ac.kr (L.K.K.); ostrichs@jj.ac.kr (H.J.S.); lso0594@naver.com (S.O.L.); 2dbtkd@naver.com (U.S.L.)
* Correspondence: jokwanwoo@jj.ac.kr; Fax: +82-63220-3161
† These authors contributed equally to this work.

Received: 3 January 2019; Accepted: 23 January 2019; Published: 28 January 2019

Abstract: Carbon foam was prepared from carboxymethyl cellulose (CMC) and Ag, Al and carbon nanotubes (CNTs), and graphene was added to the foam individually, to investigate the enhancement effects on the thermal conductivity. In addition, we used the vacuum method to impregnate erythritol of the phase change material (PCM) into the carbon foam samples to maximize the latent heat and minimize the latent heat loss during thermal cycling. Carbon foams containing Ag (CF-Ag), Al (CF-Al), CNT (CF-CNT) and graphene (CF-G) showed higher thermal conductivity than the carbon foam without any nano thermal conducting materials (CF). From the variations in temperature with time, erythritol added to CF, CF-Ag, CF-Al, CF-CNT, and CF-G was observed to decrease the time required to reach the phase change temperature when compared with pure erythritol. Among them, erythritol added to CF-G had the fastest phase change temperature, and this was related to the fact that this material had the highest thermal conductivity of the carbon foams used in this study. According to differential scanning calorimetry (DSC) analyses, the materials in which erythritol was added (CF, CF-Ag, CF-Al, CF-CNT, and CF-G) showed lower latent heat values than pure erythritol, as a result of their supplementation with carbon foam. However, the latent heat loss of these supplemented materials was less than that of pure erythritol during thermal cycling tests because of capillary and surface tension forces.

Keywords: carbon foam; nanomaterials; phase change material; thermal conductivity; latent heat storage

---

1. Introduction

Thermal energy generated from fossil fuels is a type of high-quality energy that can be easily converted into mechanical energy and electric energy [1,2]. However, thermal energy of a low temperature, which remains after mechanical or electric use, is a low-level energy that is difficult to convert to other types of energy. For storage of high-quality thermal energy produced from low-level energy, the waste heat of low-level energy should be of a higher density to minimize energy loss [3–5]. Phase change materials (PCMs) capable of thermal energy storage (TES), especially latent heat storage, are increasingly receiving enormous amounts of attention as alternatives to physical thermal energy storage materials [6–8]. PCMs store energy in the form of latent heat during solid-liquid phase changes, and PCMs are widely used as thermal storage materials because they have a high thermal storage density, narrow range of temperatures for TES, and repeated utility [9–12]. Despite these advantages, PCMs typically have a low thermal conductivity, which can lead to disagreeable supercooling problems and phase segregation. To overcome this, high thermal conductive fillers have been widely used to enhance the thermal conductivity in PCMs, but high concentrations of fillers can reduce the TES capacity of the PCMs. Among fillers, porous thermal conducting materials are promising for resolving the above troubles related to the low thermal conductivity and reduction of TES; this can be accomplished through surface tension forces and capillary in the porous structure [13–17]. Especially, porous carbon materials can offer many advantages such as excellent physical, chemical, and thermal stability, non-toxicity, high specific surface areas, and fire resistance properties in addition to providing good heat storage capacities for PCMs [18,19]. Shin et al. [20] selected expanded graphites (EGs), which are one type of porous carbon material, as fillers and then researched the effects of the EGs on the thermal conductivity following impregnation of sodium acetate trihydrate (SAT) as the PCM. As shown by the results, the EGs increased the thermal conductivity of SAT composites and maximized the SAT impregnated between the interlayers of EGs. The resulting products exhibited excellent thermal cycling stabilities. Karthik et al. [21] prepared erythritol-graphite foam as a stable composite PCM and used the wetness impregnation method for the utilization and recovery of solar heat or industrial waste heat. Li et al. [22] researched the thermal performance enhancement of erythritol/carbon foam composites. By improving the wetting ability of the carbon foam surface by hydrogen peroxide, erythritol content was increased on the carbon foam surface, resulting in improvement of the thermal performances of the composites.

In this study, we prepared carbon foams as porous carbon materials to enhance the thermal conductivity and to maximize the latent heat of PCMs. Carbon foam has three-dimensional (3D) netlike construction and has been widely used in applications, such as electrode materials [23], heavy oil recovery [24], and sound energy absorption [25,26]. Here, carbon foam was prepared from carboxymethyl cellulose (CMC) derived from cellulose [27,28]. While CMC is an attractive material from eco-friendly and economic perspectives [29–31] to reduce the negative environmental effects [32], the thermal conductivity of CMC carbon foam is not very satisfactory. Therefore, to enhance the thermal conductivity of the CMC carbon foams used in the present study, we prepared various carbon foams containing various kinds of nano thermal conducting particles: Ag, Al, carbon nanotube (CNT), and graphene oxide (GO). We then investigated the enhancement effects on thermal conductivity. The reason for selecting these four nano thermal conducting particles was that carbon foams could not be prepared from CMC composites containing other nano thermal conducting particles due to decomposition by non-crosslinking between CMC and citric acid (CA) during electron beam irradiation (EBI). The erythritol as the PCM was added to the carbon foams containing the various nano thermal conducting particles via the vacuum method, and the different latent heat values and thermal properties were then evaluated. The carbon foams containing various nano thermal conducting particles and impregnated with erythritol were characterized by the use of differential scanning calorimetry (DSC) and scanning electron microscopy (SEM).

2. Experimental

2.1. Materials

The erythritol as the PCM was purchased from Cargill Co. (Minneapolis, MN, USA). The carboxymethyl cellulose used as the precursor of carbon foam was obtained from Sigma Aldrich Co. (St. Louis, MO, USA). The spherical Ag powder was received from HKK Solution Co. (Seoul, Korea); its degree of purity was 99.99%, with a particle size of 20 nm. The spherical Al powder was supplied from Ditto Technology Co. (Gunpo-si, Gyeonggi-do, Korea); its degree of purity was 99.99%, with a diameter in the range of 60–80 nm. The CNT powder was obtained from Nano Solution Co.; its degree of purity was above 95 wt %, with length in the range of 5–20 μm, diameter in the range of 8–15 nm, and tap density of 0.05 g cc^{-1}. The GO powder with 65.33% carbon and 31.31% oxygen content was supplied from Graphenall Co. (Siheung-si, Guionggi-do, Korea); its D/G ratio was 1.04, with pH 2.87.

The citric acid (CA) used as the cross-linking agent for preparing the carbon foam was of analytical grade and used as received.

2.2. Preparation of Simple CMC Carbon Foam and CMC Carbon Foams Containing Nano Thermal Conducting Materials

The simple CMC carbon foam and CMC carbon foams containing Ag, Al, CNT, and GO were respectively prepared through carbonization of CMC composites and CMC composites containing Ag, Al, CNT, and GO obtained via EBI treatments. For the simple CMC composite, 10 wt % CMC and 4 wt % CA were dissolved in distilled water at room temperature, which resulted in the production of a CMC paste. For CMC composites containing Ag, Al, CNT, and GO: 10 wt % CMC and 4 wt % CA were dissolved in distilled water at room temperature and then 2 wt % Ag, 2 wt % Al, 2 wt % CNT, and 2 wt % GO powder was added to the above pastes. The resulting pastes were stirred until the additives were uniformly dispersed. The obtained pastes were irradiated at a dose of 80 kGy of EBI, which was carried out by using a conveyor-type scanned beam with an accelerating voltage of 1.14 MeV and a beam current of 7.6 mA. The EBI treatment in this study was used for cross-linking of CMC and CA to prevent CMC composite decomposition at high temperature (over 1000 °C) during carbonization [27,28]. The EBI-treated samples were lyophilized. These lyophilized samples were carbonized at a high temperature of 1000 °C for 1 h under a nitrogen atmosphere in a tubular furnace to produce the carbon foams. These carbon foams were labeled as CF (CMC carbon foam), CF-Ag (CMC carbon foam containing 2 wt % Ag), CF-Al (CMC carbon foam containing 2 wt % Al), CNT (CMC carbon foam containing 2 wt % CNT), and CF-G (CMC carbon foam containing 2 wt % GO). In the case of the latter sample, GO powder in the CMC composite was reduced by thermal annealing during carbonization of CMC composite at 1000 °C, which changed it into reduced graphene oxide (rGO). Thermal reduction of GO is able to remove oxygen functionalities via a complicated mechanism that restores the π-conjugation arrangement characteristic of graphene. Thermal annealing at high temperature near 1000 °C has been proposed as the classic temperature for GO reduction resulting in rGO [33,34].

2.3. Impregnation of Erythritol into Carbon Foam and Carbon Foams Containing Various Nano Thermal Conducting Materials and Its Characterization

Figure 1 shows a schematic diagram of the process whereby erythritol was impregnated into the carbon foam, and carbon foams containing various nano thermal conducting materials, by the vacuum method to maximize thermal energy storage and minimize the loss of latent heat during repeated thermal cycling tests. The carbon foam and solid erythritol (1:100) were placed into an airtight glass container and this container was then heated in an oil bath. Molten erythritol at around 120 °C was impregnated into the carbon foam by removing the air, using a vacuum pump. The heating temperature was monitored by a thermocouple, which was located in a glass test tube filled with erythritol or erythritol mixed with carbon foams containing the various nano thermal conducting materials. The temperature variations of the erythritol in the glass test tubes during heating were passed through a temperature readout box, where the data were automatically recorded and saved onto a computer (Figure 1). The thermal conductivity measurements were conducted with a TPS 2500S instrument (Hot Disk, Göteborg, Sweden) and the resulting data obtained from the sensor, which was sandwiched between 2 same sample pieces. The samples were tested 10 times each, and the average value and the standard deviation of the obtained data were calculated. The latent heat values for pure erythritol and the various carbon foams impregnated with erythritol were determined by DSC (DSC25, TA instruments Co., New Castle, DE, USA) in the range of 30–140 °C. The samples were tested 10 times each, and the average value and the standard deviation of the obtained data were calculated. The microstructure for the morphology of various carbon foams and corresponding erythritol-impregnated carbon foams were evaluated with SEM (CX-200TA, COXEM, Daejeon, Korea).

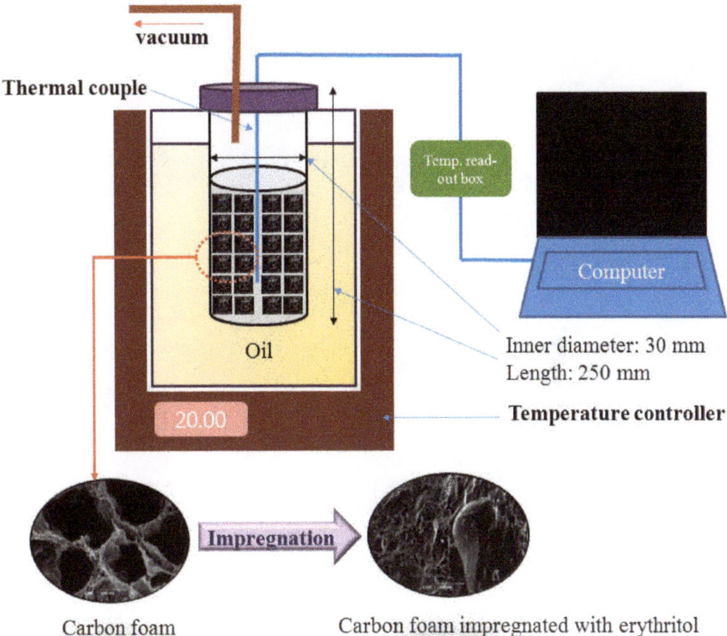

Figure 1. Schematic diagram of the temperature variation test unit for erythritol as the phase change material (PCM) and the vacuum impregnation of erythritol into carbon foam.

3. Results and discussion

3.1. Thermal Conductivity of the Various Carbon Foams

Figure 2 shows the thermal conductivity of CF, CF-Ag, CF-Al, CF-CNT, and CF-G. As shown in Figure 2, the thermal conductivity values of carbon foam containing various nano thermal conducting materials were higher than that of the simple CF without any nano thermal conducting materials. Generally, the thermal conductivity of non-graphitized carbon foam will have a low value below 1 W mK^{-1} [35], and here, the thermal conductivity of CF without nano thermal conducting materials was 0.12 W mK^{-1}. In contrast, the thermal conductivity values of CF-Ag, CF-Al, CF-CNT, and CF-G were approximately 0.18, 0.14, 0.22, and 0.27 W mK^{-1}, respectively. Compared with CF, these values represented enhancements of approximately 150%, 116.67%, 183.33%, and 225%, respectively. Individually, Ag (429 W mK^{-1}), Al (237 W mK^{-1}), CNT (~3500 W mK^{-1}), and graphene (4800~5300 W mK^{-1}) are excellent thermal conducting materials. Therefore, the addition of Ag, Al, CNT, and graphene as nano thermal conducting materials into the carbon foam supported the development of substantially higher thermal conductivities.

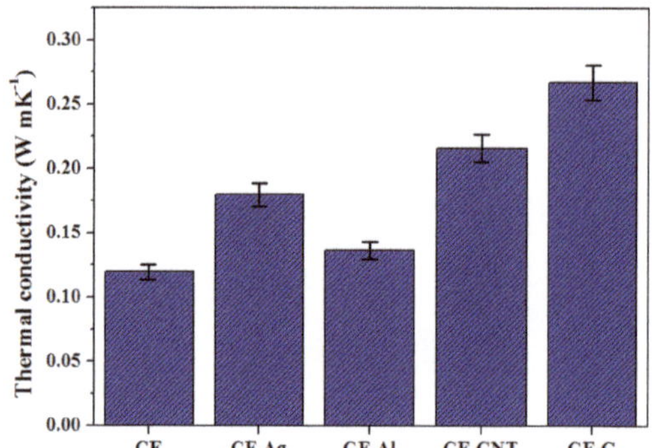

Figure 2. Thermal conductivity of simple carboxymethyl cellulose carbon foam (CF) and CF containing various nano thermal conducting materials.

3.2. Effects of Carbon Foams Containing Nano Thermal Conducting Materials on the Melting of Erythritol

Figure 3 shows the temperature changes with time during latent heat storage experiments with pure erythritol and erythritol added to simple CF, CF-Ag, CF-AL, CF-CNT, and CF-G. Here, we can see that erythritol initiated melting at a phase change temperature of around 120 °C and the time required to reach the phase change temperature of pure erythritol took around 1965 s. However, when erythritol was added to the simple CF and CFs with various nano thermal conducting materials, faster melting occurred than that with pure erythritol. In particular, as shown in Figure 3, when erythritol was added to the simple CF, the phase change occurred at approximately 1813 s. Even faster times of 1620 s, 1505 s, 1288 s, and 1046 s were achieved with CF-Ag, CF-Al, CF-CNT, and CF-G, respectively. These results indicated that the enhanced thermal conductivities of carbon foams led to reductions in the times required to reach the phase change temperature during heating with erythritol as the PCM, and among the carbon foams used in this study, CF-G showed the highest thermal conductivity and had the fastest phase change time, which was approximately 919 s less than that for pure erythritol.

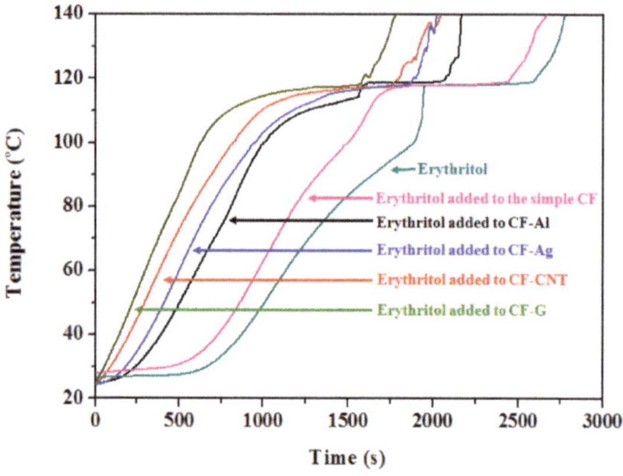

Figure 3. Temperature changes with time of erythritol and erythritol added to the simple CF and CF containing various nano thermal conducting materials.

3.3. Thermal Effects of Simple Carbon Foam and Carbon Foams Containing Various Nano Thermal Conducting Materials on the Latent Heat Storage of Erythritol

The latent heat for erythritol as the PCM was generally calculated by using the region of the endothermic peak near 120 °C during the DSC analysis. Figure 4a shows the DSC analysis results, and

Figure 4b presents the latent heat values calculated from the DSC analysis. In Figure 4a, it can be seen that the intensity of the endothermic peaks when erythritol was added to simple CF, CF-Ag, CF-Al, CF-CNT, and CF-G were smaller than that of pure erythritol. This was because the amount of erythritol was reduced following the supplementation of the carbon foam. By comparing the latent heat values in Figure 4b, it can be seen that the latent heat value for pure erythritol was 358.04 ± 8.95 J g^{-1}, but the latent heat values when erythritol was added to simple CF, CF-Ag, CF-Al, CF-CNT, and CF-G were generally similar, at around 290 and 300 J g^{-1}. This was because of the similar impregnation amounts of erythritol by the vacuum method, resulting from the similar porous structure of carbon foams, regardless of the types of nano thermal conducting materials.

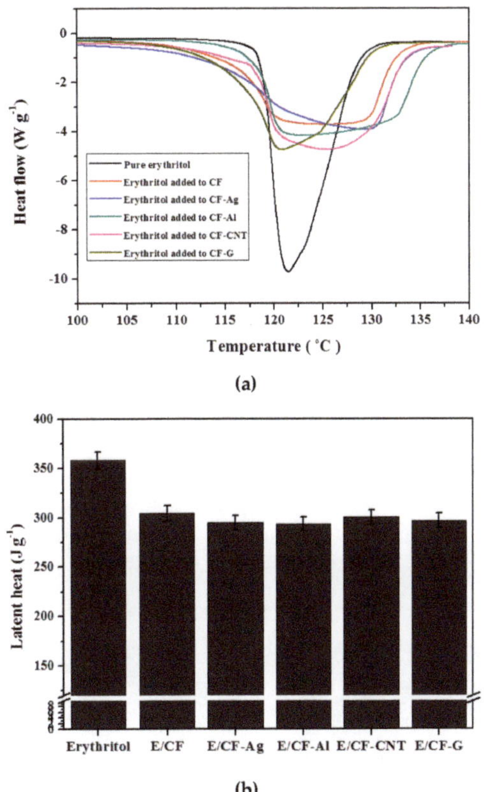

(a)

(b)

Figure 4. (a) Differential scanning calorimetry (DSC) analysis results and (b) comparison of latent heat values for pure erythritol and erythritol added to the simple CF and CF containing: Ag (CF-Ag); Al (CF-Al); CNT (CF-CNT); and graphene (CF-G).

Figure 5 shows the results of the thermal cycling test (10 cycles) for pure erythritol and erythritol added to simple CF, CF-Ag, CF-Al, CF-CNT, and CF-G. These tests results indicated the thermal stability for the changes of latent heat during repeated melting and solidification cycles of the PCM. In the case of pure erythritol, latent heat values decreased continuously during 10 cycles and up to ~5% of the initial latent heat value was lost. However, when erythritol was added to simple CF, CF-Ag, CF-Al, CF-CNT, and CF-G, the materials exhibited much less latent heat loss than the pure erythritol. These results likely occurred because the porosity of carbon foam can prohibit the leakage of erythritol and thus minimize the latent heat loss during thermal cycling tests through capillary and surface tension forces. Therefore, it can be concluded that CF, CF-Ag, CF-Al, CF-CNT, and CF-G demonstrated great thermal reliability as fillers for PCMs.

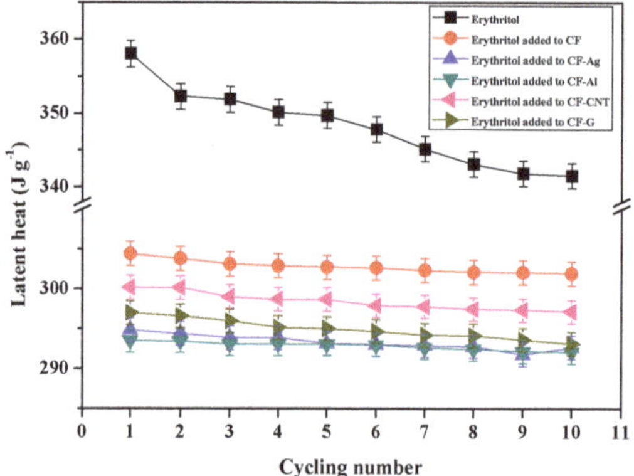

Figure 5. Thermal cycling test for pure erythritol and erythritol added to simple CF, CF-Ag, CF-Al, CF-CNT, and CF-G materials.

3.4. Scanning Electron Microscopy (SEM) Images for the Impregnation of Erythritol into Carbon Foams

The impregnation of the PCM into a filler, such as carbon foam, has a very import role in maximizing latent heat storage. If the PCM is not impregnated into fillers properly, latent heat storage is reduced. Figure 6 displays SEM images of simple CF, CF-Ag, CF-Al, CF-CNT, and CF-G and the corresponding images for the carbon foams impregnated with erythritol. As shown in Figure 6, the morphologies of simple CF, CF-Ag, CF-Al, CF-CNT, and CF-G consisted of the microstructures of the respective carbon foams with various sizes and non-uniform porosities. Here, all of the images confirmed that respective carbon foams, after impregnation with molten erythritol by the vacuum method at around 120 °C, were successfully impregnated with erythritol without any voids. Therefore, it was clear that erythritol could be effectively impregnated into the pores of the respective carbon foams through the vacuum method.

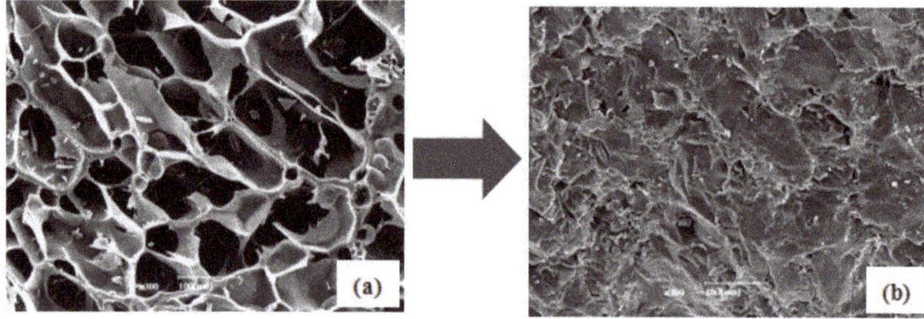

Figure 6. *Cont.*

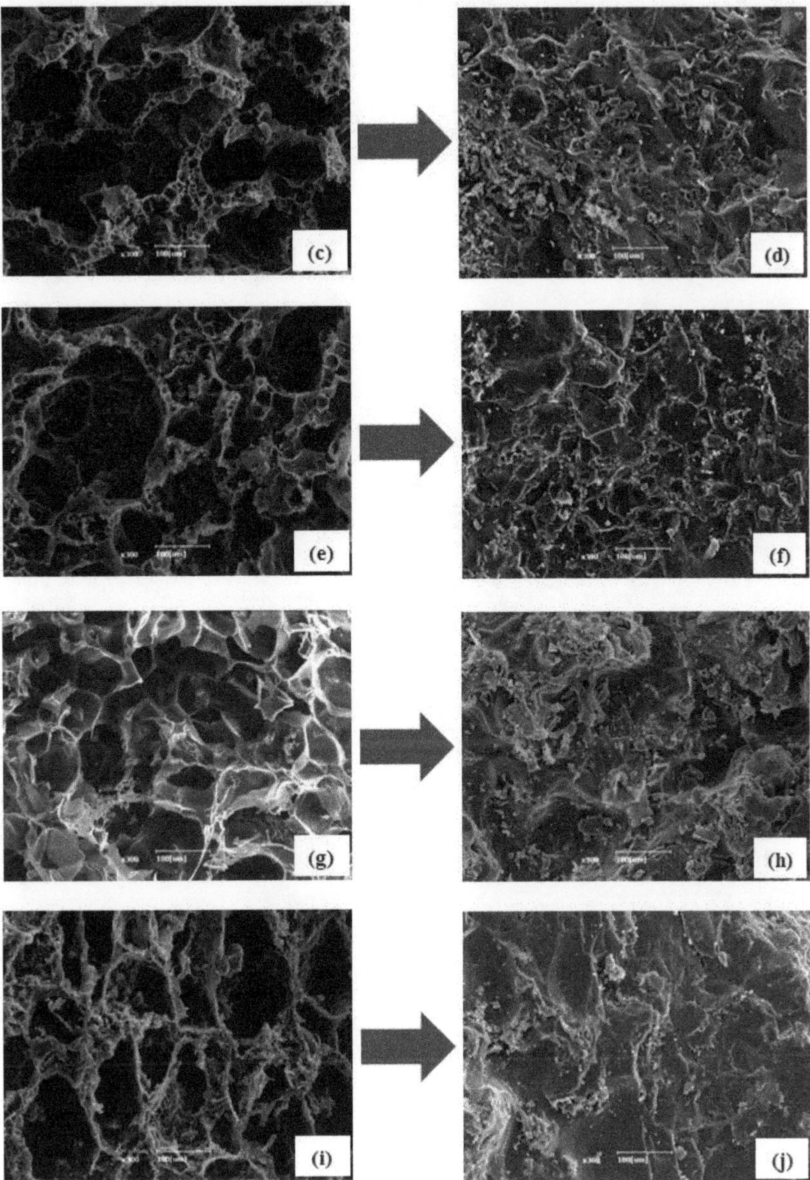

Figure 6. SEM images of (**a**) CF; (**c**) CF-Ag; (**e**) CF-Al; (**g**) CF-CNT; and (**i**) CF-G, as well as the SEM images for the corresponding carbon foams impregnated with erythritol (**b,d,f,h,j**, respectively).

4. Conclusions

In this study, we prepared both simple carbon foam without any nano thermal conducting materials and carbon foams containing Ag, Al, CNT, and graphene from CMC to enhance the thermal conductivity. We also used the vacuum method to impregnate erythritol as the PCM into the carbon foams to maximize the latent heat and to minimize the latent heat loss during thermal cycling. Firstly, we observed that CF-Ag (0.18 W mK^{-1}), CF-Al (0.14 W mK^{-1}), CF-CNT(0.22 W mK^{-1}), and CF-G (0.27

W mK^{-1}) carbon foams containing Ag, Al, CNT, and graphene, respectively, could support higher thermal conductivities than the simple CF (0.12 W mK^{-1}) alone. Next, erythritol was added to the simple CF, CF-Ag, CF-Al, CF-CNT, and CF-G and this resulted in a decrease in the time required to reach the phase change temperature. Especially, when erythritol was added with CF-G, this material reached phase change temperature fastest and the time was around 919 s lower than that of pure erythritol. However, the addition of erythritol to the simple CF, CF-Ag, CF-Al, CF-CNT, and CF-G resulted in latent heat values between about 290 and 300 J g^{-1} lower than pure erythritol (358.04 $\pm$ 8.95 J g^{-1}), which occurred as a result of the carbon foam supplementation. Regardless, the carbon foams displayed less latent heat loss than pure erythritol during thermal cycling tests. Lastly, with the SEM images, we could see that erythritol was effectively impregnated into the pores of the respective carbon foams through the vacuum method, and this helped to maximize the latent heat and minimize the latent heat loss during thermal cycling. Therefore, these carbon foams impregnated with erythritol represent promising materials for thermal energy storage applications. However, drawbacks in this study included: The difficulty in controlling the pore sizes uniformly of carbon foams; and the difficulty in preparing more varieties of carbon foams containing nano thermal conducting materials (because the CMC pastes containing some nano thermal conducting particles were not cross-linked between CMC and CA after treatment). Nevertheless, these carbon foams impregnated with erythritol represent promising materials for thermal energy storage applications.

Author Contributions: Conceptualization H.K.S.; Experimental work, H.K.S., Y.-S.K., S.O.L., and U.S.L.; formal analysis, H.K.S., Y.-S.K., and L.K.K.; resources, H.J.S. and H.G.K.; data curation, Y.-S.K., S.O.L., and U.S.L.; writing—original draft preparation, H.K.S.; writing—review and editing, H.K.S.; supervision, H.K.S. and H.G.K.; project administration, H.K.S. and H.G.K.

Funding: This work was supported by the National Research Foundation of Korea (NRF) grant funded by the Korea government (MSIP) (No. 2017R1A2B1008753). This research was also supported by Basic Science Research Program through the National Research Foundation of Korea (NRF) funded by the Ministry of Education (No. 2016R1A6A1A03012069).

Conflicts of Interest: The authors declare no conflict of interest.

References

1. Haillot, D.; Bauer, T.; Tamme, R. Thermal analysis of phase change materials in the temperature range 120–150 °C. *Themochim. Acta* **2011**, *513*, 49–59. [CrossRef]
2. Finck, C.; Li, R.L.; Kramer, R.; Zeirler, W. Quantifying demand flexibility of power-to-heat and thermal energy storage in the control of building heating systems. *Appl. Energy* **2018**, *209*, 409–425. [CrossRef]
3. Kjaizawa, A.; Kamano, H.; Kawai, A.; Jozuka, T.; Senda, T.; Maruoka, N.; Okinaka, N.; Akiyama, T. Tchnical feasibility study of waste heat transportation system using phase change material from industry to city. *ISIJ Int.* **2008**, *48*, 540–548. [CrossRef]
4. Nomura, T.; Okinaka, N.; Akiyama, T. Waste heat transportation system, using phase change material (PCM) from steelworks to chemical plant. *Resour. Conserv. Recycl.* **2010**, *54*, 1000–1006. [CrossRef]
5. Tyagi, V.V.; Buddhi, D. PCM thermal storage in buildings: A state of art. *Renew. Sustain. Energy Rev.* **2011**, *15*, 112–130. [CrossRef]
6. Sharma, A.; Tyagi, V.V.; Chen, C.R. Review on thermal energy storage with phase change materials and applications. *Renew. Sustain. Energy Rev.* **2009**, *13*, 318–345. [CrossRef]
7. Jensen, S.Ø.; Marszal-Pomianowska, A.; Lollini, R.; Pasut, W.; Knotzer, A.; Engelmann, P.; Stafford, A.; Reyndersm, G. IEA EBC annex 67 energy flexible buildings. *Energy Build.* **2017**, *155*, 25–34. [CrossRef]
8. Zalba, B.; Marin, J.M.; Cabeza, L.F.; Mehling, H. Review on thermal energy storage with phase change: Materials, heat transfer analysis and applications. *Appl. Therm. Eng.* **2003**, *23*, 251–283. [CrossRef]
9. Shin, H.K.; Rhee, K.Y.; Park, S.J. Effects of exfoliated graphite on the thermal properties of erythritol-based composites used as phase-change materials. *Compos. Part B-Eng.* **2016**, *96*, 350–353. [CrossRef]
10. Lee, S.Y.; Shin, H.K.; Park, M.; Rhee, K.Y.; Park, S.J. Thermal characterization of erythritol/expanded graphite composites for high thermal storage capacity. *Carbon* **2014**, *68*, 67–72. [CrossRef]

11. Wang, X.; Guo, Y.; Su, J.; Zhang, X.; Han, N.; Wang, X. Microstructure and thermal reliability of microcapsules containing phase change material with self-assembled graphene/organic nano-hybrid shells. *Nanomaterials* **2018**, *8*, 364. [CrossRef] [PubMed]

12. Marcos, M.A.; Cabaleiro, D.; Guimarey, M.J.G.; Comuñas, M.J.P.; Fedele, L.; Fernández, J.; Lugo, L. PEG 400-based phase change materials nano-enhanced with functionalized graphene nanoplatelets. *Nanomaterials* **2017**, *8*, 16. [CrossRef] [PubMed]

13. Deng, Z.; Liu, X.; Zhang, C.; Huang, Y.; Chen, Y. Melting behaviors of PCM in porous metal foam charterized by fractal geometry. *Int. J. Heat Mass Transf.* **2017**, *113*, 1031–1042. [CrossRef]

14. Al-Jethelah, M.; Ebadi, S.; Venkateshwar, K.; Tansnim, S.H.; Mahmud, S.; Dutta, A. Charging nanopartivle enhanced bio based PCM in open cell metallic foams: An experimental investigation. *Appl. Therm. Eng.* **2019**, *148*, 1029–1042. [CrossRef]

15. Andreozzi, A.; Buonomo, B.; Ercole, D.; Manca, O. Solar energy latent thermal storage by phase change materials (PCMs) in a honeycomb system. *Therm. Sci. Eng. Prog.* **2018**, *6*, 410–420. [CrossRef]

16. Li, Y.; Li, J.; Deng, Y.; Guan, W.; Wang, X.; Qian, T. Preparation of paraffin/porous TiO$_2$ foams with enhanced thermal conductivity as PCM, by covering the TiO$_2$ surface with a carbon layer. *Appl. Energy* **2016**, *171*, 37–45. [CrossRef]

17. Wu, Z.G.; Sheng, W.C.; Tao, W.Q.; Li, Z. A novel experimental-numerical method for studying the thermal behaviors of phase change material in a porous cavity. *Sol. Energy* **2018**, *169*, 325–334. [CrossRef]

18. Xiao, H.; Tao, C.; Li, Y.; Chen, X.; Huang, J.; Wang, J. Dopamine assisted one-step pyrolysis of glucose for the preparation of porous carbon with a high surface area. *Nanomaterials* **2018**, *8*, 854. [CrossRef]

19. Sagisaka, K.; Kondo, A.; Iiyama, T.; Kimura, M.; Fujimoto, H.; Touhara, H.; Hattori, Y. Hollow structured porous carbon fibers with the inherent texture of the cotton fibers. *Chem. Phys. Lett.* **2018**, *710*, 118–122. [CrossRef]

20. Shin, H.K.; Park, M.; Kim, H.Y.; Park, S.J. Thermal property and latent heat energy storage behavior of sodium acetate trihydrate composites containing expanded graphite and carboxymethyl cellulose for phase change materials. *Appl. Therm. Eng.* **2015**, *75*, 978–983. [CrossRef]

21. Karthik, M.; Faik, A.; Blanco-Rodriguez, P.; Rodriguez-Aseguinolaza, J.; D'Auanno, B. Preparation of erythritol-graphite foam phase change composite with enhanced thermal conductivity for thermal energy storage applications. *Carbon* **2015**, *94*, 266–276. [CrossRef]

22. Li, J.; Lu, W.; Luo, Z.; Zeng, Y. Thermal performance enhancement of erythritol/carbon foam composites via surface modification of carbon foam. *IOP Mater. Sci. Eng.* **2017**, *182*, 012009. [CrossRef]

23. Inagaki, M.; Konno, H.; Tanaike, O. Carbon materials for electrochemical capacitors. *J. Power Sources* **2010**, *195*, 7880–7903. [CrossRef]

24. Qu, J.; Han, Q.; Gao, F.; Qiu, J. Carbon foams produced from lignin-phenol-formaldehyde resin for oil/water separation. *New Carbon Mater.* **2017**, *32*, 86–91. [CrossRef]

25. Gao, N.; Cheng, B.; Hou, H.; Zhang, R. Mesophase pitch based carbon foams as sound absorbers. *Mater. Lett.* **2017**, *212*, 243–246. [CrossRef]

26. Wu, Y.; Sun, X.; Wu, W.; Liu, X.; Lin, X.; Shen, X.; Wang, Z.; Li, R.K.Y.; Yang, Z.; Lau, K.T.; et al. Graphene foam/carbon nanotube/poly(dimethyl siloxane) composites as excellent sound absorber. *Compos. Part A Appl. Sci. Manuf.* **2017**, *102*, 391–399. [CrossRef]

27. Kim, H.G.; Kim, Y.S.; Kwac, L.K.; Chae, S.H.; Shin, H.K. Synthesis of carbon foam from waste artificial marble powder and carboxymethyl cellulose via electron beam irradiation and its characterization. *Materials* **2018**, *11*, 469. [CrossRef] [PubMed]

28. Kim, H.G.; Kwac, L.K.; Kim, Y.S.; Shin, H.K.; Rhee, K.Y. Synthesis and characterization of eco-friendly carboxymethyl cellulose based carbon foam using electron beam irradiation. *Compos. Part B-Eng.* **2018**, *151*, 154–160. [CrossRef]

29. Shin, H.K.; Park, M.; Kang, P.H.; Rhee, K.Y.; Park, S.J. Role of electron beam irradiation on superabsorbent behaviors of carboxymethyl cellulose. *Res. Chem. Intermed.* **2015**, *41*, 6815–6823. [CrossRef]

30. Benslimane, A.; Bahlouli, I.M.; Bekkour, K.; Hammiche, D. Thermal gelation properties of carboxymethyl cellulose and bentonite-carboxymethyl cellulose dispersions: Rheo-logical considerations. *Appl. Clay Sci.* **2016**, *132–133*, 702–710. [CrossRef]

31. Lakshmi, D.S.; Trivedi, N.; Reddy, C.R.K. Synthesis and characterization of seaweed cellulose derived carboxymethyl cellulose. *Carbohydr. Polym.* **2017**, *157*, 1604–1610. [CrossRef] [PubMed]

32. Didaskalou, C.; Buyuktiryaki, S.; Kecili, R.; Fonte, C.P.; Szekely, G. Valorisation of agricultural waste with an adsorption/nanofiltration hybrid process: From materials to sustainable process design. *Green Chem.* **2017**, *19*, 3116–3125. [CrossRef]

33. Angrawal, P.R.; Kumar, R.; Teotia, S.; Kumari, S.; Mondal, D.P.; Dhakate, S.R. Lightweight, high electrical and thermal conducting carbon-rGO composites foam for superior electromagnetic interference shielding. *Compos. Part B-Eng.* **2019**, *160*, 131–139. [CrossRef]

34. Mittal, G.; Dhand, V.; Rhee, K.Y.; Park, S.J.; Lee, W.R. A review on carbon nanotubes and graphene as fillers in reinforced polymer nanocomposites. *J. Ind. Eng. Chem.* **2015**, *21*, 11–25. [CrossRef]

35. Jana, P.; Fierro, V.; Pizzi, A.; Celzard, A. Thermal conductivity improvement of composite carbon foams based on tannin-based disordered carbon matrix and graphite fillers. *Mater. Des.* **2015**, *83*, 635–643. [CrossRef]

 *nanomaterials*

Article

An Ultra-High-Energy Density Supercapacitor; Fabrication Based on Thiol-functionalized Graphene Oxide Scrolls

Janardhanan. R. Rani [1,†], Ranjith Thangavel [2,†], Se-I Oh [1], Yun Sung Lee [2] and Jae-Hyung Jang [1,3,*]

1   School of Electrical Engineering and Computer Science, Gwangju Institute of Science and Technology, Gwangju 61005, Korea; ranijnair@gmail.com (J.R.R.); ohseia@naver.com (S.-I.O.)
2   Faculty of Applied Chemical Engineering, Chonnam National University, Gwangju 61186, Korea; ranjith.cecri@gmail.com (R.T.); leeys@jnu.ac.kr (Y.S.L.)
3   Research Institute for Solar and Sustainable Energies, Gwangju Institute of Science and Technology, Gwangju 61005, Korea
*   Correspondence: jjang@gist.ac.kr; Tel.: +82-62-715-2209; Fax: +82-62-715-2204
†   These authors contributed equally to this work.

Received: 17 December 2018; Accepted: 18 January 2019; Published: 24 January 2019

Abstract: Present state-of-the-art graphene-based electrodes for supercapacitors remain far from commercial requirements in terms of high energy density. The realization of high energy supercapacitor electrodes remains challenging, because graphene-based electrode materials are synthesized by the chemical modification of graphene. The modified graphene electrodes have lower electrical conductivity than ideal graphene, and limited electrochemically active surface areas due to restacking, which hinders the access of electrolyte ions, resulting in a low energy density. In order to solve the issue of restacking and low electrical conductivity, we introduce thiol-functionalized, nitrogen-doped, reduced graphene oxide scrolls as the electrode materials for an electric double-layer supercapacitor. The fabricated supercapacitor exhibits a very high energy/power density of 206 Wh/kg (59.74 Wh/L)/496 W/kg at a current density of 0.25 A/g, and a high power/energy density of 32 kW/kg (9.8 kW/L)/9.58 Wh/kg at a current density of 50 A/g; it also operates in a voltage range of 0~4 V with excellent cyclic stability of more than 20,000 cycles. By suitably combining the scroll-based electrode and electrolyte material, this study presents a strategy for electrode design for next-generation energy storage devices with high energy density without compromising the power density.

Keywords: EDLC; rGO scrolls; thiol functionalization; supercapacitor; energy and power density

---

1. Introduction

The growing demand for energy storage systems in electric vehicles, load-leveling, and portable electronic devices has stimulated research into high-density electrochemical energy storage technologies that can deliver high power for long periods of time [1–3]. Electrochemical double-layer supercapacitors (EDLCs) have attracted considerable interest for such applications, due to their high specific power density and long cycle life [4,5]. The major drawback of EDLCs is their low energy density, which lies in the range of 3~5 Wh/kg [6,7], two orders of magnitude lower than that of the commercial lithium-ion batteries. Supercapacitors that can deliver high-energy density without sacrificing power density are critically needed for practical applications, such as hybrid electric vehicles [8], industrial forks [4], wind turbine energy storage [9], mobile electronics [1], fuel cells [10], regenerative braking [11], and power supply devices [12], etc. Several materials, including polymers [13–15] or reduced graphene oxide (rGO)-based materials have been introduced to overcome the EDLCs' low energy density issue. Among them, rGO-based EDLCs have been reported with high power density, high charge/discharge rates, and long cycle life performance. However, even though tremendous progress has been achieved in developing rGO-based EDLCs with high energy density using hydrazine reduction (85.6 Wh/kg) [16], microwave exfoliation (70 Wh/kg) [17], and other processes, their energy density values are still significantly lower than the level needed for many practical applications.

Two of the major drawbacks of the rGO-based EDLCs that lead to the lower energy density values are the restacking of rGO layers and the low electrical conductivity of rGO. The agglomeration and restacking effects lead to a reduced electrochemically active surface area and interlayer spacing, with the result that ions in the electrolyte become less accessible, which in turn leads to the low energy density of the rGO-based EDLCs. In order to effectively implement EDLCs, their specific energy storage capabilities must be further improved. Energy density can be significantly enhanced by (1) preventing the restacking of rGO sheets, (2) increasing the electrical conductivity of the rGO sheets, and (3) creating a narrow pore size distribution in the synthesized rGO electrode.

In the present study, in order to prevent restacking and to increase conductivity, we synthesized nitrogen-incorporated, thiol-functionalized rGO scrolls (hereafter referred to as NTGS). The NTGS have unique three-dimensional, interconnected networks with a continuous porous structure, narrow pore size distribution, and a large surface area, making them potentially excellent electrode materials for high-performance EDLCs. The NTGS represent a promising new design strategy for optimizing the electrochemical performance of EDLC materials. Suppressing the sheet restacking yields a highly open and porous structure that creates continuous ion transport channels and allows electrolyte solutions to easily access the surfaces of individual graphene sheets.

Thiol-functionalized groups are composed of sulfur and hydrogen atoms (–SH). By functionalizing rGOs with a thiol (–SH) group, it is possible to tailor the physical and chemical properties of the rGOs. The synthesized NTGS show high electrical conductivity, which is attributed to (i) the orbital overlap between sulfur 3s and 3p, with π- orbitals in the rGO sheets, and (ii) the presence of lone pair electrons in nitrogen. The chemical bonding between sulfur and carbon is stronger than the van der Waals force between adjacent π−π stacked rGO layers. Moreover, additional states are formed in the conduction band of the rGO because of the thiolation process. Furthermore, the nitrogen atoms provide additional free electrons to the conduction band. The two physical and chemical properties of the NTGS result in significant enhancement of the electrical conductivity of the electrode.

In the present study, we achieve high energy density by synthesizing the unique scroll structure, which provides a short and rapid transport pathway for electrons, leading to the outstanding performance of the cell. The fabricated NTGS-based EDLC bridges the energy density gap between conventional batteries and supercapacitors. The fabricated supercapacitor exhibits a very high energy/power density of 206 Wh/kg (59.74 Wh/L)/496 W/kg, at a current density of 0.25 A/g, and a high power/energy density of 32 kW/kg (9.8 kW/L)/9.58 Wh/kg at a current density of 50 A/g, and excellent stability (>20,000 cycles).

2. Experiment Section

2.1. Preparation of Reduced Graphene Oxide and Thiol-Functionalized Reduced Graphene Oxide Scrolls

We synthesized GO using the modified Hummer's method [18] by oxidizing graphite powder. Two-week-long dialysis was performed to completely remove the metal ions. The resulting GO solution was finally dried at 90 °C overnight to obtain GO powder. Then the powder was mixed well in an agate mortar for 30 min, followed by an annealing at 800 °C in ambient nitrogen for one hour. This rGO powder was used to fabricate the rGO cell.

For the preparation of thiol-functionalized rGO scrolls, the thiol-functionalized rGO powder (UniNanoTech Co., giheung-gu, Korea) was dispersed in 30 mL of water, and this mixture was suspended in water by sonication for 8 h, using an ultrasonic reactor operating at a frequency of 33 kHz. The mixture was centrifuged at 12,000 rpm for 15 min, which resulted in a homogenous suspension. Repeated sonication and centrifugation were carried out multiple times. The resulting solution was finally dried at 90 °C overnight to obtain the powder. The powder was mixed well in an agate mortar for 30 min, followed by an annealing at 800 °C in ambient nitrogen for one hour. The resulting powder consisted of nitrogen-incorporated, thiol–functionalized rGO nanoscrolls (NTGS).

The morphological, structural, and compositional properties of the NTGS powder were investigated using high-resolution transmission electron microscopy (HRTEM)/selective area electron diffraction (SAED) (JEM.ARM.200F), Raman spectroscopy (inVia Raman microscope), ultraviolet photoelectron spectroscopy (UPS), and X-ray photoelectron spectroscopy (XPS) measurements. The measurement of the nitrogen adsorption isothermal was carried out at 77 K by a low-temperature, nitrogen adsorption surface area analyzer (ASAP 2020, Micromeritics Ins., Norcross, GA, USA).

2.2. Cell Fabrication

Composite anodes were formulated with 80% active materials (NTGS or rGO), 10% Ketzen black (KB), and 10% teflonized acetylene black (TAB) using ethanol. The slurry was then pressed on a nickel mesh current collector with a 200-mm^2-area, and dried at 160 °C for 5 h in a vacuum oven.

To evaluate the electrochemical performance of the NTGS electrodes, supercapacitors were assembled with symmetrical cell geometry. The test cells were constructed in an argon-filled glove box by compressing the NTGS and rGO electrodes, and the devices were fabricated with an ionic liquid EMIMBF$_4$ electrolyte. The mass loading of electrodes is 2.5 mg cm^{-2}, and the thickness of the working electrode is 57 μm. In the present study, stainless steel is used as the current collector, because in a large operating voltage window, the current collector needs to be more stable and non-corrosive in nature to maintain a stable cycling performance.

The electrochemical performance of the supercapacitor cells in the form of CR2032 coin cells was measured in the range of 0~4 V using a battery tester (WBCS 3000, Won-A-Tech, Seoul, Korea). The cyclic voltammetry (CV) was carried out at scan rates ranging from 10 mV/s to 200 mV/s, and the Galvanostatic charge/discharge (GCD) cycling of the cells was performed at current densities ranging from 0.26 A/g to 50 A/g.

3. Results and Discussion

Figure 1a illustrates the NTGS formation procedure. The NTGS samples were formed by rolling thiolated rGO layers in one or more directions. Scrolling occurs because of the graphene oxide's (GO's) planar structure, which is extremely unfavorable to maintain. A detailed discussion regarding the formation of the GO scrolls has been reported in our previous work [19].

The driving force for the scrolling of the thiolated GO is the energy difference between the total surface energy of the system and the elastic energy associated with thiolated GO bending [20,21]. When the temperature increases during annealing, the GO layer bends to minimize the total surface energy of the system by reducing the exposed surfaces of the thiolated GO layers, and the subsequent rolling of the graphene oxide is driven by the reduction of the total area of the exposed GO surface. Consequently, the total surface energy of the system is minimized by tightly wrapping adjacent graphene layers of the scroll together, where the layers are interconnected to each other.

The morphology of the NTGS powder samples was analyzed using HRTEM images, as shown in Figure 1c,e. The HRTEM images of the NTGS clearly reveal the interconnected scroll structure. Additional TEM/HRTEM and selective area electron diffraction (SAED) images are shown in the Supplementary Figure S1. The TEM images show that the lateral diameter of the scrolls is in the range of 20~60 nm. The TEM images in Figure 1e indicate a high-quality interconnected carbon nanostructure, with randomly oriented rGO scrolls of various diameters and lengths that are connected and assembled together. Furthermore, the electron diffraction pattern in the high-magnification TEM image shows that the NTGS retain the highly crystalline graphitic structure (Figure S1i).

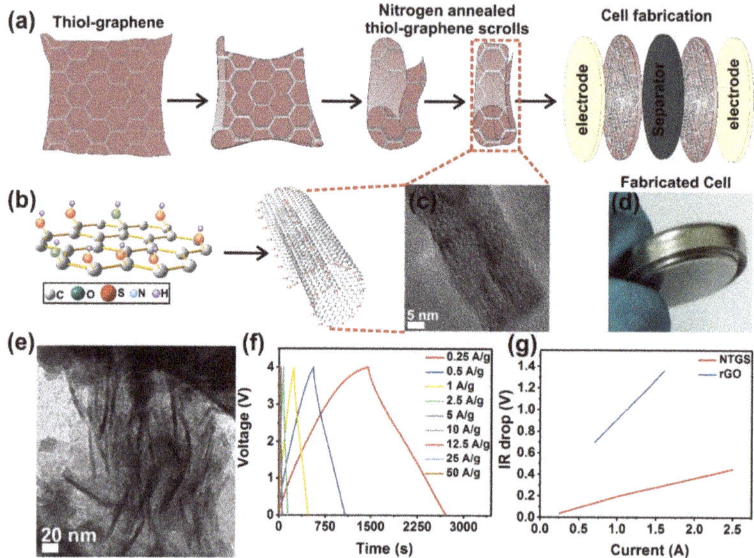

Figure 1. (**a**) Schematic showing the formation of the thiolated reduced graphene oxide (rGO) scrolls and cell fabrication, (**b**) structure of the planar N-doped thiol rGO and N-doped thiol rGO scroll, (**c**) high-resolution transmission electron microscopy (HRTEM) image of a single scroll, (**d**) digital image of a fabricated cell, (**e**) TEM image of an interconnected NTGS powder sample, (**f**) galvanostatic charge–discharge curves of the NTGS cell, and (**g**) ohmic voltage (IR) drop for the nitrogen-incorporated, thiol-functionalized rGO scrolls (NTGS) and rGO cells. The IR drop curve shows electrical conductivity was significantly enhanced in the NTGS compared to pure rGO.

The Raman spectrum was also recorded and analyzed to identify the formation of scroll structures. The Raman spectra of the NTGS shown in Supplementary Figure S2a clearly show that the scrolls are significantly different from those of the planar graphene. The inset of Figure S2a shows the Raman spectra in the lower vibrational frequency range. The Raman spectrum of the scrolled GO differs significantly from that of flat GO. The high-degree curvature in the scrolled GO causes the appearance of low-frequency radial breathing-like (RBLM) modes. The most important feature in the Raman spectrum of Carbon nanotubes (CNTs) is the radial breathing mode (RBM), which is usually located between 75 and 300 cm^{-1}. Figure S2a shows RBM modes around 100 and 172 cm^{-1}, which again confirms the tubular structure of the GO scrolls in our samples. Also, the bands indicating the formation of scrolls are at 920, 1800, and 2000 cm^{-1}, which are not observed in planar graphene/rGO [22,23]. The band observed at 630 cm^{-1} can be assigned to the C–S bond, confirming that there is strong interaction between carbon and sulfur [24]. The band observed at ~1280 cm^{-1} is consistent with sidewall functionalization, due to thiol-functional groups [25]. More explanations and the band assignment are given in the Supplementary Information. The functionalization of the thiol groups was also confirmed by X-ray photoelectron spectroscopy (XPS) analysis. The C1s, S2p, and N1s XPS spectra were fitted and deconvoluted, as shown in Supplementary Figure S2b–d. The XPS spectra clearly confirmed the strong chemical interaction between C–N and C–S in NTGS. The peak assignments and explanations are given in the Supplementary Information.

The electrochemical Galvanostatic charge/discharge (GCD) performance of the NTGS cell (in the voltage range between 0 and 4 V) was recorded at current densities ranging from 0.25 to 50 A/g (Figure 1f). The GCD curves clearly indicate excellent performance (The fabricated supercapacitor exhibits a very high energy /power density of 206 Wh/kg/496 W/kg, at a current density of 0.25 A/g and a high power/energy density of 32 kW/kg/9.58 Wh/kg at a current density of 50 A/g, specific capacitance of 360 F/g (104.4 F/cm^3) as well as the double layer capacitive behavior of the fabricated NTGS cell). It is worth pointing that the cell can withstand a current density of up to 50 A/g.

The devices were fabricated with ionic liquid EMIMBF$_4$ electrolyte. Compared with aqueous and organic electrolytes, ionic liquids (IL) are ideal electrolytes for supercapacitors because of their unique properties, such as high thermal stability and low flammability, which can improve the working temperature range and safety of supercapacitors. In particular, the wide electrochemical windows greater than 4 V can contribute greatly to improving the energy density of supercapacitors. A major drawback of the aqueous and organic electrolytes are safety issues, such as volatility and flammability. In addition, the working voltage window is very low, such as 1.2 V and 2.7 V for the aqueous and organic electrolytes, respectively. On the other hand, ILs are attractive electrolytes that can be an alternative to the conventional organic electrolytes because of their negligible vapor pressures, low flammability, and high electrochemical stability [26–28].

The GCD performance of the rGO cell is shown in the Supplementary Information (Figure S3). The rGO cell delivered a maximum energy density of 70 Wh/kg. The gravimetric specific capacitance (C_{gra}), the gravimetric energy, and power density were calculated using the following Equations [29,30]:

$$C_{gra} = 4\frac{i\Delta t}{m\Delta V} \tag{1}$$

$$E_{gra} = \frac{1}{8}C_{gra}\Delta V^2 \tag{2}$$

$$P_{gra} = \frac{E_{gra}}{t} \tag{3}$$

where i, m, Δt, and ΔV are the applied current (A); mass (g) of the active material, including both the anode and cathode in the cell; discharge time; and potential window, respectively.

The volumetric specific capacitance (C_{vol}) $_{can}$ be calculated using the following equation:

$$C_{vol} = C_{gra} \times \rho \tag{4}$$

The volumetric power density (P_{vol}) and volumetric energy density (E_{vol}) of the supercapacitor can be calculated as follows.

$$E_{vol} = E_{gra} \times \rho \tag{5}$$

$$P_{vol} = P_{gra} \times \rho \tag{6}$$

The packing density of total electrode material (ρ) was calculated according to the previous report, and the value is 0.29 g/cm^3 [31].

The NTGS cell delivered a maximum capacitance of 360 F/g (104.4 F/cm^3), an energy density of 206 Wh/kg (59.74 Wh/L), and a corresponding power density of 496 W/kg at a current density of 0.25 A/g. The cell delivered an energy density of 9.58 Wh/kg at a maximum power density of 32 kW/kg (9.8 kW/L) and at a current density of 50 A/g. Considering that the rGO cell delivered a maximum energy density of only 70 Wh/kg, the NTGS cell delivered three times higher energy density compared with the rGO cell.

A significant increase in the energy density of the NTGS electrode was observed, compared with those of other very recently reported graphene- and polymer-based supercapacitors. The gravimetric energy density of the NTGS electrode was higher than those of other very recently reported values, such as the conjugated indole-based macromolecule (30 Wh/kg) [32], reduced graphene oxide/mixed-valence manganese oxide composite (50 Wh/kg) [33], hollow particle-based nitrogen-doped carbon nanofiber (11 Wh/kg) [34], porous Ni$_3$S$_2$/CoNi$_2$S$_4$ three-dimensional (3D)-network structure (62.2 Wh/kg) [35], interlaced Ni(OH)$_2$ nanoflakes wrapped in zinc cobalt sulfide nanotube arrays (75.5 Wh/kg) [36], honeycomb-carbon frameworks (55 Wh/kg) [37].

The enhancement in energy density of the NTGS cell over that of the GO cell can be attributed to (i) the higher conductivity of the NTGS electrode, and (ii) the scroll structure of NTGS, which eliminates restacking of the rGO and increases the electrolyte/electrode contact surface area. The cell resistance of the NTGS and rGO cells was confirmed by the IR drop against the discharge current, as shown in Figure 1g. The IR drop is calculated from the GCD curve, as shown in the previously published paper [38]. The IR drop in the GCD curve at a current density of 0.25 A/g is shown in Supplementary Information (Figure S4). It is worth noting that our NTGS cell exhibited resistance values that were only 1/6 that of the pure rGO cell. In supercapacitors, charge storage occurs via ion transportation and the intercalation of ions into the active material. Therefore, higher electrical conductivity or lower resistance of the active material results in higher capacitance values. The NTGS cell exhibited six times higher conductivity than the pure rGO cell, which in turn resulted in the higher capacitance and the energy density values. Moreover, the higher contact area and narrow pore size distribution provided by the peculiar scroll structure resulted in an enhancement of the electron transfer process between the electrode and electrolyte.

The enhanced electrical conductivity, higher specific area, and the narrow pore size distribution of NTGS samples were further investigated using ultraviolet photoelectron spectroscopy (UPS) and Brunauer–Emmett–Teller (BET) surface area measurements.

Detailed study of the electronic structures of the powder samples was carried out using valence band UPS spectroscopy. UPS was performed in an ultrahigh vacuum (UHV) chamber with an He–I resonance line ($h\nu$ = 21.22 eV) as an excitation source. As UPS sources can excite only valence band electrons, UPS spectroscopy can provide information about the valence band electrons responsible for bonding. The Fermi level position is referred at binding energy of 0 eV. UPS spectra for the NTGS and rGO samples were measured in the binding energy, ranging from 0 to 20 eV, as shown in Figure S5. This UPS spectrum reveals valence band energy levels of NTGS and rGO powders with respect to the source emission line (He1α; 21.22 eV). Figure 2a corresponds to the fitted He–I UPS spectra of the samples after subtracting background noise in the binding energy ranging from 4 to 15 eV. This is because the peaks corresponding to the graphene oxide appear in this range.

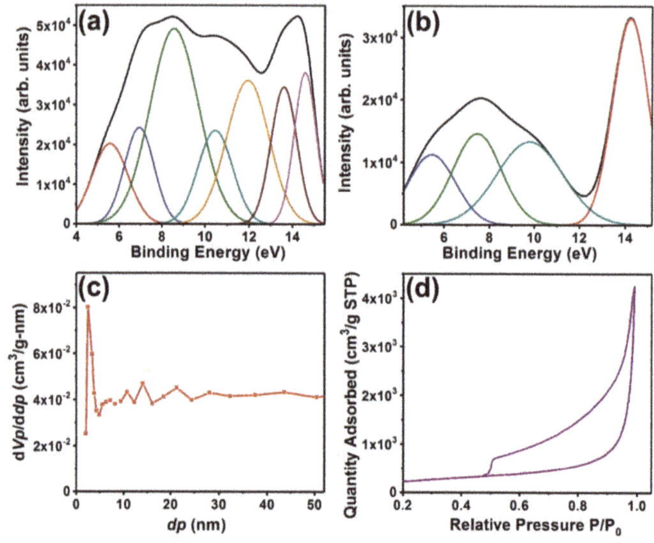

Figure 2. Fitted He-I ultraviolet photoelectron spectroscopy (UPS) spectra of (**a**) NTGS and (**b**) rGO powders after background subtraction. (**c**) Pore-size distribution and (**d**) nitrogen adsorption/ desorption plot at 77.4 K of the NTGS sample.

The fitted peaks and their corresponding assignments are shown in Supplementary Table S1. For comparison, the UPS spectra of pristine rGO samples were also recorded and deconvoluted, as shown in Figure 2b.

The UPS spectra of the rGO exhibited peaks at 5.4, 7.3, 9.9, and 14.3 eV. The UPS spectra of NTGS also showed similar peaks at 5.4, 7.0, and 14.6 eV. A noticeable difference was the additional 8.6, 10.5. 11.8, and 13.7 eV peaks in the NTGS UPS spectra. Interestingly, the peak observed at 9.9 eV in rGO, which was attributed to the orbital interaction between C2s and C2p, shifted towards a lower binding energy—8.6 eV in the NTGS [39]. We believe that this shift occurred because of the presence of lone pair electrons supplied by nitrogen. The additional peaks observed at 10.5, 11.8, and 13.7 eV correspond to C2p–S2p, C2p–S3s, and C2p–N2p interactions, respectively [39,40]. The UPS analyses confirmed that there were mixtures of the main orbital types of carbon and sulfur. The thiol functional group significantly improved the electrical conductivity because of the orbital overlap between sulfur 3s and 3p, with π- orbitals of carbon in the graphene sheets. The hybridization between carbon and sulfur increased the local charge density and the hybridization between S3p, S3s, and C2p states, resulting in the formation of impurity energy levels near the Fermi level, leading to higher conductivity. The hybridization between S3p states and C3s states also contributed to the formation of impurity energy levels.

The BET surface area measurement results are shown in Figure 2c,d. The interconnected NTGS exhibited a narrow pore size distribution, with an average pore diameter of 2.5 nm, as shown in Figure 2c. The scrolls had uniformly distributed mesopores with a narrow pore size distribution, resulting in easy accessibility for the electrolyte ions. Nitrogen adsorption/desorption plots of the samples at 77.4 K showed a type IV adsorption isothermal curve with a hysteresis loop (Figure 2d).

The interconnected NTGS had a large specific surface area of 803 m^2/g. The higher specific surface area provided by the nanoscrolls resulted in an increase in the number of sites actually accessible to electrolyte ions, which enhanced the electron transfer process between the electrode and electrolyte. This enhancement of the electron transfer process gives rise to higher energy density. The high-surface-area NTGS with well-defined pores provides fast electronic and ionic conducting channels, making the material ideal for electrodes in supercapacitor applications.

It is clear from Figure 3a that the cyclic voltammetry (CV) curves of the NTGS cells (scan rate in the range of 10–200 mV/s) are excellent and exhibit a typical rectangular shape, which confirms the formation of double layer capacitance. The variations in the C_{gra} values with respect to the current density of the NTGS cell is depicted in Figure 3b. The NTGS cell exhibited stable cycling performance and maintained 88% of its initial capacitance, even after 20,000 cycles at a current density of 5 Ag^{-1} (Figure 3c). Figure 3d shows the Ragone plots of the NTGS cell. It is apparent that the energy density values of the NTGS cells are superior to those of other recently reported supercapacitors.

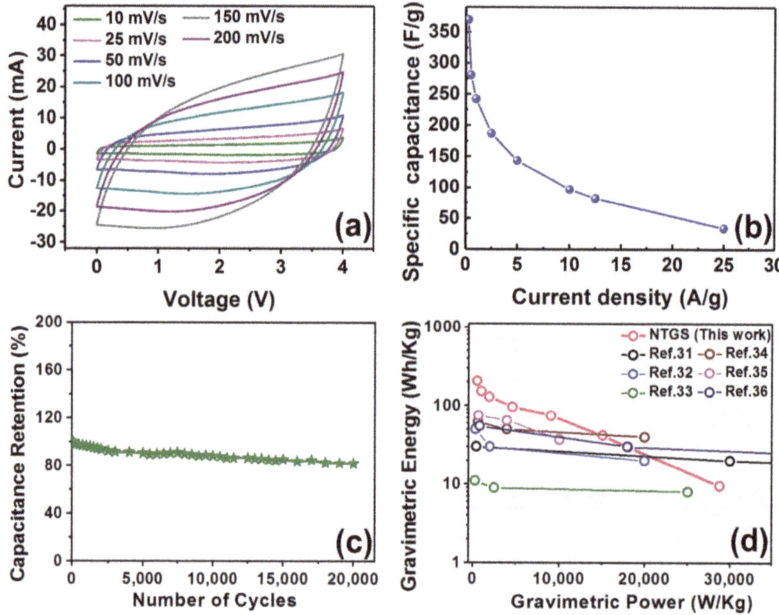

Figure 3. (**a**) Cyclic voltammetry (CV) curves of the NTGS from 10~200 mV/s; (**b**) specific capacitances of the NTGS cell as a function of the current density; (**c**) cycling performance of the NTGS cell; (**d**) the Ragone plots of the NTGS cells, compared with those of other recently reported supercapacitors [32–37].

The electrochemical impedance spectra (EIS) is a powerful tool for analyzing the internal resistance of the electrode material, and the EIS studies were conducted for the NTGS and pure GO cells separately, at a frequency range between 100 kHz and 100 MHz, with an amplitude of 10 mV in the open circuit condition. The corresponding EIS spectra are presented in Figure S6. The electrode resistance, bulk electrolyte resistance, diffuse layer resistance, and equilibrium differential capacitance can be retrieved directly from Nyquist plots. The Nyquist spectra of both cells consisted of semicircles in the high-to-medium frequency region, and a sloping line in the low-frequency region. The slope of the line represents the diffusion control process in the low-frequency region. The intercept of the semicircle with the real impedance axis (Z′) represents the equivalent series resistance (ESR), which is a combination of electrolyte and contact resistance. The intercept from the vertical line to the *x*-axis gives the total internal resistance of the cell. The ESR values for NTGS and GO cells, calculated [41] from a Nyquist plot, are 9.9 Ω and 19 Ω, respectively (Figure S6). Thus, Figure S6 clarifies that the ESR of the NTGS cell is lower than that of the pure GO cell. The total internal resistance for NTGS and GO cells are 13.5 Ω and 20.8 Ω, respectively (Figure S6). The diameter of the semicircle provides the charge transfer resistance, R_{ct}, resulting from the diffusion of electrons towards the electrode materials. The lower R_{ct} of the NTGS cell compared with that of the GO cell indicates the enhancement in the electron transfer process. The scroll structure provides a connected network, and it results in a lower resistance and a shorter electron diffusion pathway, which in turn leads to the enhanced electron transfer process. Furthermore, the open and connected architecture of the conductive scrolls provides a good interface for the electrolyte ions, which increases the electrochemical accessibility of the electrolyte ions into the NTGS electrode. Figure S4 clearly shows that the diffuse layer resistance is lower for NTGS compared to the GO cell. The formation of a narrow Helmholtz region in the NTGS cell is shown as a schematic in Figure 4. The unique structure of the NTGS cells, with their outstanding electrical conductivity, provides a short and rapid transport pathway for electrons, permitting high-capacity and high-energy density at high current densities. The excellent performance of the NTGS cell is suitable for applications that demand high gravimetric energy densities.

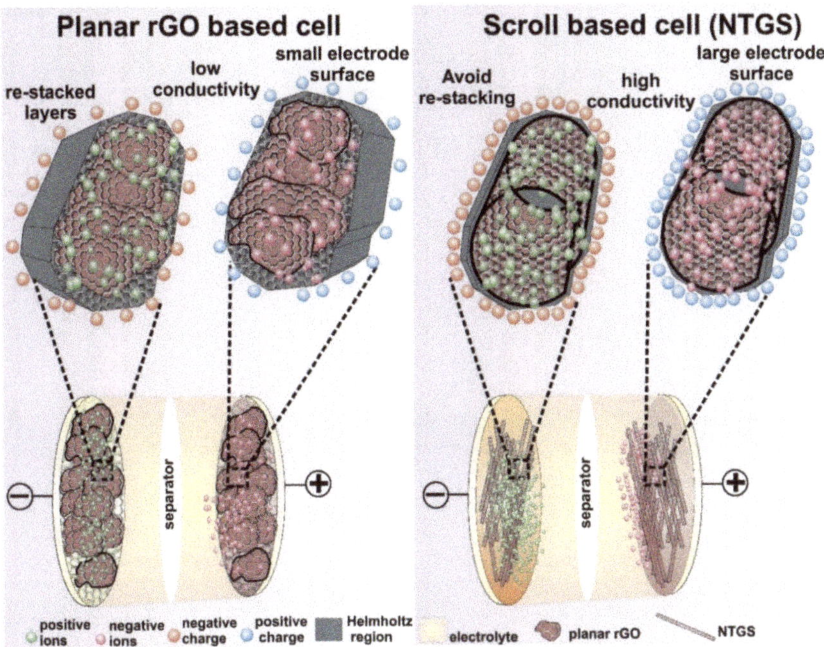

Figure 4. A schematic showing the electrochemical reactions in the rGO cell (**left**) and the NTGS cell (**right**).

4. Conclusions

In summary, to enhance the energy density of EDLCs, nitrogen-incorporated, thiol-functionalized, reduced graphene oxide scrolls (NTGS) were synthesized. In the voltage range from 0 to 4.0 V, a supercapacitor fabricated with NTGS powder exhibited high energy densities (206 Wh/kg, 59.74 Wh/L) compared to previously reported values (85–100 Wh/kg), and comparable to those of Li-ion batteries. The superior energy density values of the fabricated NTGS capacitor are attributed to the nitrogen doping, thiol functionalization, and scroll morphology. The NTGS cell also exhibited excellent cycling stability up to 20,000 cycles, and a power density of 32 kW/h (9.8 kW/L). Our strategy in developing NTGS using a combination of nitrogen incorporation, thiol-functionalization, and scroll formation was to offer a significant potential electrode material for energy storage systems, which can also be used to enhance the performance of the other electrode materials, including those for lithium-ion batteries and hybrid supercapacitors.

Supplementary Materials: The following are available online at http://www.mdpi.com/2079-4991/9/2/148/s1, Figure S1: (**a–g**) HRTEM images of an interconnected NTGS powder sample, (**h**) HRTEM image of a single scroll, (**i**) SAED pattern of NTGS sample from the area shown in (**g**). Figure S2: (**a**) Raman spectrum and fitted XPS spectra of (**b**) C1*s*, (**c**) N1*s*, and (**d**) S2*p* GFNS powder sample. The inset of (**a**) shows the Raman spectra of NTGS sample in a different range of vibrational frequencies. Figure S3: The Galvanostatic charge/discharge (GCD) performance of the rGO cell at current densities 1.6 A/g and 1.4 A/g. Figure S4: Ohmic voltage (IR) drop in Galvanostatic charge/discharge curve at a current density of 0.25 A/g. Figure S5: He–1 UPS spectra for the NTGS and rGO samples. Figure S6: Electrochemical impedance spectra of GO and NTGS powder samples. Table S1: The fitted UPS spectra peaks of rGO and NTGS powder samples and their corresponding assignments.

Author Contributions: Material preparation, J.R.R. and S.I.O.; fabrication and characterization of cell, R.T. and Y.S.L.; writing- original draft Preparation, J.R.R.; writing- review, editing and supervision, Y.S.L. and J.H.H.

Funding: This research was funded by by a GIST Research Institute (GRI) project through a grant provided by GIST, the National Research Foundation of Korea (NRF) grant (No.2017R1A2B3004049) funded by the Korea Government, Ministry of Science, ICT and Future Planning, Korea Institute of Energy Technology Evaluation and Planning(KETEP) and the Ministry of Trade, Industry & Energy(MOTIE) of the Republic of Korea (No. 20183010014310), as well as the Technology Innovation Program (No.10053010) funded by the Ministry of Trade, Industry and Energy (MI, Korea).

Acknowledgments: This work was supported by a GIST Research Institute (GRI) project through a grant provided by GIST, the National Research Foundation of Korea (NRF) grant funded by the Korea government (Ministry of Science and ICT) (No. 2017R1A2B3004049), Korea Institute of Energy Technology Evaluation and Planning(KETEP) and the Ministry of Trade, Industry & Energy(MOTIE) of the Republic of Korea (No. 20183010014310), as well as the Technology Innovation Program (No.10053010) funded by the Ministry of Trade, Industry and Energy (MI, Korea).

Conflicts of Interest: The authors declare no conflict of interest.

References

1. Miller, J.R.; Simon, P. Electrochemical Capacitors for Energy Management. *Science* **2008**, *321*, 651. [CrossRef] [PubMed]
2. Liu, C.; Li, F.; Ma, L.P.; Cheng, H.M. Advanced Materials for Energy Storage. *Adv. Mater.* **2010**, *22*, E28–E62. [CrossRef] [PubMed]
3. Rani, J.R.; Thangavel, R.; Oh, S.-I.; Woo, J.M.; Chandra Das, N.; Kim, S.-Y.; Lee, Y.-S.; Jang, J.-H. High Volumetric Energy Density Hybrid Supercapacitors Based on Reduced Graphene Oxide Scrolls. *ACS Appl. Mater. Inter.* **2017**, *9*, 22398–22407. [CrossRef] [PubMed]
4. Simon, P.; Gogotsi, Y. Materials for Electrochemical Capacitors. *Nat. Mater.* **2008**, *7*, 845. [CrossRef] [PubMed]
5. Chidembo, A.; Aboutalebi, S.H.; Konstantinov, K.; Salari, M.; Winton, B.; Yamini, S.A.; Nevirkovets, I.P.; Liu, H.K. Globular Reduced Graphene Oxide-Metal Oxide Structures for Energy Storage Applications. *Energy Environ. Sci.* **2012**, *5*, 5236–5240. [CrossRef]
6. Chen, Z.; Wen, J.; Yan, C.; Rice, L.; Sohn, H.; Shen, M.; Cai, M.; Dunn, B.; Lu, Y. High-Performance Supercapacitors Based on Hierarchically Porous Graphite Particles. *Adv. Energy Mater.* **2011**, *1*, 551–556. [CrossRef]
7. Kaempgen, M.; Chan, C.K.; Ma, J.; Cui, Y.; Gruner, G. Printable Thin Film Supercapacitors Using Single-Walled Carbon Nanotubes. *Nano Lett.* **2009**, *9*, 1872–1876. [CrossRef]
8. Paladini, V.; Donateo, T.; de Risi, A.; Laforgia, D. Super-Capacitors Fuel-Cell Hybrid Electric Vehicle Optimization and Control Strategy Development. *Energy Convers. Manage.* **2007**, *48*, 3001–3008. [CrossRef]
9. Abbey, C.; Joos, G. Supercapacitor Energy Storage for Wind Energy Applications. *IEEE Trans. Ind. Appl.* **2007**, *43*, 769–776. [CrossRef]
10. Winter, M.; Brodd, R.J. What Are Batteries, Fuel Cells, and Supercapacitors? *Chem. Rev.* **2004**, *104*, 4245–4270. [CrossRef]
11. Naseri, F.; Farjah, E.; Ghanbari, T. An Efficient Regenerative Braking System Based on Battery/Supercapacitor for Electric, Hybrid, and Plug-In Hybrid Electric Vehicles With BLDC Motor. *IEEE Trans. Veh. Technol.* **2017**, *66*, 3724–3738. [CrossRef]
12. Wu, Z.S.; Wang, D.W.; Ren, W.; Zhao, J.; Zhou, G.; Li, F.; Cheng, H.M. Anchoring Hydrous RuO_2 on Graphene Sheets for High-Performance Electrochemical Capacitors. *Adv. Funct. Mater.* **2010**, *20*, 3595–3602. [CrossRef]
13. Ramkumar, R.; Minakshi Sundaram, M. A biopolymer gel-decorated cobalt molybdate nanowafer: effective graft polymer cross-linked with an organic acid for better energy storage. *New J. Chem.* **2016**, *40*, 2863–2877. [CrossRef]
14. Ramkumar, R.; Sundaram, M.M. Electrochemical synthesis of polyaniline cross-linked NiMoO4 nanofibre dendrites for energy storage devices. *New J. Chem.* **2016**, *40*, 7456–7464. [CrossRef]
15. Ramkumar, R.; Minakshi, M. Fabrication of ultrathin CoMoO4 nanosheets modified with chitosan and their improved performance in energy storage device. *Dalton T.* **2015**, *44*, 6158–6168. [CrossRef] [PubMed]
16. Liu, C.; Yu, Z.; Neff, D.; Zhamu, A.; Jang, B.Z. Graphene-Based Supercapacitor with an Ultrahigh Energy Density. *Nano Lett.* **2010**, *10*, 4863–4868. [CrossRef] [PubMed]
17. Zhu, Y.; Murali, S.; Stoller, M.D.; Ganesh, K.J.; Cai, W.; Ferreira, P.J.; Pirkle, A.; Wallace, R.M.; Cychosz, K.A.; Thommes, M.; et al. Carbon-Based Supercapacitors Produced by Activation of Graphene. *Science* **2011**, *332*, 1537. [CrossRef]

18. Marcano, D.C.; Kosynkin, D.V.; Berlin, J.M.; Sinitskii, A.; Sun, Z.; Slesarev, A.; Alemany, L.B.; Lu, W.; Tour, J.M. Improved Synthesis of Graphene Oxide. *ACS Nano* **2010**, *4*, 4806–4814. [CrossRef]

19. Rani, J.R.; Oh, S.-I.; Woo, J.M.; Tarwal, N.L.; Kim, H.-W.; Mun, B.S.; Lee, S.; Kim, K.-J.; Jang, J.-H. Graphene Oxide–Phosphor Hybrid Nanoscrolls with High Luminescent Quantum Yield: Synthesis, Structural, and X-ray Absorption Studies. *ACS Appl. Mater. Inter.* **2015**, *7*, 5693–5700. [CrossRef]

20. Mirsaidov, U.; Mokkapati, V.R.S.S.; Bhattacharya, D.; Andersen, H.; Bosman, M.; Ozyilmaz, B.; Matsudaira, P. Scrolling Graphene into Nanofluidic Channels. *Lab Chip* **2013**, *13*, 2874–2878. [CrossRef]

21. Anoop, G.; Rani, J.R.; Lim, J.; Jang, M.S.; Suh, D.W.; Kang, S.; Jun, S.C.; Yoo, J.S. Reduced Graphene Oxide Enwrapped Phosphors for Long-Term Thermally Stable Phosphor Converted White Light Emitting Diodes. *Sci. Rep.* **2016**, *6*, 33993. [CrossRef] [PubMed]

22. Ferrari, A.C.; Robertson, J. Interpretation of Raman Spectra of Disordered and Amorphous Carbon. *Phys. Rev. B* **2000**, *61*, 14095–14107. [CrossRef]

23. Rani, R.J.; Oh, S.-I.; Jang, J.-H. Raman Spectra of Luminescent Graphene Oxide (GO)-Phosphor Hybrid Nanoscrolls. *Materials* **2015**, *8*, 8460–8466. [CrossRef] [PubMed]

24. Song, M.-K.; Zhang, Y.; Cairns, E.J. A Long-Life, High-Rate Lithium/Sulfur Cell: A Multifaceted Approach to Enhancing Cell Performance. *Nano Lett.* **2013**, *13*, 5891–5899. [CrossRef] [PubMed]

25. Alemany, L.B.; Zhang, L.; Zeng, L.; Edwards, C.L.; Barron, A.R. Solid-State NMR Analysis of Fluorinated Single-Walled Carbon Nanotubes: Assessing the Extent of Fluorination. *Chem. Mater.* **2007**, *19*, 735–744. [CrossRef]

26. Kado, Y.; Imoto, K.; Soneda, Y.; Yoshizawa, N. Highly enhanced capacitance of MgO-templated mesoporous carbons in low temperature ionic liquids. *J. Power Sources* **2014**, *271*, 377–381. [CrossRef]

27. Peng, C.; Wen, Z.; Qin, Y.; Schmidt-Mende, L.; Li, C.; Yang, S.; Shi, D.; Yang, J. Three-Dimensional Graphitized Carbon Nanovesicles for High-Performance Supercapacitors Based on Ionic Liquids. *ChemSusChem* **2014**, *7*, 777–784. [CrossRef] [PubMed]

28. Vu, A.; Li, X.; Phillips, J.; Han, A.; Smyrl, W.H.; Bühlmann, P.; Stein, A. Three-Dimensionally Ordered Mesoporous (3DOm) Carbon Materials as Electrodes for Electrochemical Double-Layer Capacitors with Ionic Liquid Electrolytes. *Chem. Mater.* **2013**, *25*, 4137–4148. [CrossRef]

29. Fan, L.-Q.; Liu, G.-J.; Wu, J.-H.; Liu, L.; Lin, J.-M.; Wei, Y.-L. Asymmetric Supercapacitor Based on Graphene Oxide/Polypyrrole Composite and Activated Carbon Electrodes. *Electrochim. Acta* **2014**, *137*, 26–33. [CrossRef]

30. Shao, Y.; El-Kady Maher, F.; Lin, C.W.; Zhu, G.; Marsh Kristofer, L.; Hwang Jee, Y.; Zhang, Q.; Li, Y.; Wang, H.; Kaner Richard, B. 3D Freeze-Casting of Cellular Graphene Films for Ultrahigh-Power-Density Supercapacitors. *Adv. Mater.* **2016**, *28*, 6719–6726. [CrossRef]

31. Sheng, L.; Jiang, L.; Wei, T.; Fan, Z. High Volumetric Energy Density Asymmetric Supercapacitors Based on Well-Balanced Graphene and Graphene-MnO$_2$ Electrodes with Densely Stacked Architectures. *Small* **2016**, *12*, 5217–5227. [CrossRef] [PubMed]

32. Xiong, T.; Lee, W.S.V.; Chen, L.; Tan, T.L.; Huang, X.; Xue, J. Indole-Based Conjugated Macromolecules as a Redox-Mediated Electrolyte for An Ultrahigh Power Supercapacitor. *Energy Environ. Sci.* **2017**, *10*, 2441–2449. [CrossRef]

33. Wang, Y.; Lai, W.; Wang, N.; Jiang, Z.; Wang, X.; Zou, P.; Lin, Z.; Fan, H.J.; Kang, F.; Wong, C.-P.; et al. A Reduced Graphene Oxide/Mixed-Valence Manganese Oxide Composite Electrode for Tailorable and Surface Mountable Supercapacitors with High Capacitance and Super-Long Life. *Energy Environ. Sci.* **2017**, *10*, 941–949. [CrossRef]

34. Chen, L.-F.; Lu, Y.; Yu, L.; Lou, X.W. Designed Formation of Hollow Particle-Based Nitrogen-Doped Carbon Nanofibers for High-Performance Supercapacitors. *Energy Environ. Sci.* **2017**, *10*, 1777–1783. [CrossRef]

35. He, W.; Wang, C.; Li, H.; Deng, X.; Xu, X.; Zhai, T. Ultrathin and Porous Ni$_3$S$_2$/CoNi$_2$S$_4$ 3D-Network Structure for Superhigh Energy Density Asymmetric Supercapacitors. *Adv. Energy Mater.* **2017**, *7*, 1700983. [CrossRef]

36. Syed Junaid, A.; Ma, J.; Zhu, B.; Tang, S.; Meng, X. Hierarchical Multicomponent Electrode with Interlaced Ni(OH)$_2$ Nanoflakes Wrapped Zinc Cobalt Sulfide Nanotube Arrays for Sustainable High-Performance Supercapacitors. *Adv. Energy Mater.* **2017**, *7*, 1701228. [CrossRef]

37. Sheng, L.; Jiang, L.; Wei, T.; Liu, Z.; Fan, Z. Spatial Charge Storage within Honeycomb-Carbon Frameworks for Ultrafast Supercapacitors with High Energy and Power Densities. *Adv. Energy Mater.* **2017**, *7*, 1700668. [CrossRef]
38. Wang, K.; Wu, H.; Meng, Y.; Zhang, Y.; Wei, Z. Integrated energy storage and electrochromic function in one flexible device: An energy storage smart window. *Energy Environ. Sci.* **2012**, *5*, 8384–8389. [CrossRef]
39. Gamot, T.D.; Bhattacharyya, A.R.; Sridhar, T.; Beach, F.; Tabor, R.F.; Majumder, M. Synthesis and Stability of Water-in-Oil Emulsion Using Partially Reduced Graphene Oxide as a Tailored Surfactant. *Langmuir* **2017**, *33*, 10311–10321. [CrossRef]
40. Umebayashi, T.; Yamaki, T.; Yamamoto, S.; Miyashita, A.; Tanaka, S.; Sumita, T.; Asai, K. Sulfur-Doping of Rutile-Titanium Dioxide by Ion Implantation: Photocurrent Spectroscopy and First-Principles Band Calculation Studies. *J. Appl. Phys.* **2003**, *93*, 5156–5160. [CrossRef]
41. Mei, B.-A.; Munteshari, O.; Lau, J.; Dunn, B.; Pilon, L. Physical Interpretations of Nyquist Plots for EDLC Electrodes and Devices. *J. Phys. Chem. C* **2018**, *122*, 194–206. [CrossRef]

 *nanomaterials*

Article

Facile and Green Synthesis of Graphene-Based Conductive Adhesives via Liquid Exfoliation Process

Jhao-Yi Wu [1], Yi-Chin Lai [2], Chien-Liang Chang [3], Wu-Ching Hung [3], Hsiao-Min Wu [3], Ying-Chih Liao [2,*], Chia-Hung Huang [4] and Wei-Ren Liu [1,*]

[1] Department of Chemical Engineering, Chung Yuan Christian University, R&D Center for Membrane Technology, 32023, No. 200, Chun Pei Rd., Chung Li District, Taoyuan 32023, Taiwan; james19931201@gmail.com

[2] Department of Chemical Engineering, National Taiwan University, No. 1, Sec. 4, Roosevelt Rd., Taipei 10617, Taiwan; yijin10221339@gmail.com

[3] National Chung Shan Institute of Science & Technology, Neighborhood, Sec. Jia'an, Zhongzheng Rd., Longtan Dist., Taoyuan 32546, Taiwan; rodin2005@yahoo.com.tw (C.-L.C.); 5000blackblue@gmail.com (W.-C.H.); prettysandy119@gmail.com (H.-M.W.)

[4] Metal Industries Research and Development Centre, Kaohsiung 81160, Taiwan; chiahung@mail.mirdc.org.tw

[*] Correspondence: liaoy@ntu.edu.tw (Y.-C.L.); WRLiu1203@gmail.com (W.-R.L.); Tel.: +886-3-265-4140 (W.-R.L.); Fax: +886-3-265-4199 (W.-R.L.)

Received: 11 December 2018; Accepted: 23 December 2018; Published: 28 December 2018

Abstract: In this study, we report a facile and green process to synthesize high-quality and few-layer graphene (FLG) derived from graphite via a liquid exfoliation process. The corresponding characterizations of FLG, such as scanning electron microscopy (SEM), transmission electron microscope (TEM), atomic force microscopy (AFM) and Raman spectroscopy, were carried out. The results of SEM show that the lateral size of as-synthesized FLG is 1–5 µm. The results of TEM and AFM indicate more than 80% of graphene layers is <10 layers. The most surprising thing is that D/G ratio of graphite and FLG are 0.15 and 0.19, respectively. The result of the similar D/G ratio demonstrates that little structural defects were created via the liquid exfoliation process. Electronic conductivity tests and resistance of composite film, in terms of different contents of graphite/polyvinylidene difluoride (PVDF) and FLG/PVDF, were carried out. Dramatically, the FLG/PVDF composite demonstrates superior performance compared to the graphite/PVDF composite at the same ratio. In addition, the post-sintering process plays an important role in improving electronic conductivity by 85%. The composition-optimized FLG/PVDF thin film exhibits 81.9 S·cm^{-1}. These results indicate that the developed FLG/PVDF composite adhesives could be a potential candidate for conductive adhesive applications.

Keywords: graphene; liquid exfoliation; polyvinylidene fluoride; conductive adhesives; flexiable

1. Introduction

Electronics or microelectronics industries have transitioned into making electronics lighter, smaller, thinner and more highly efficient [1,2]. Aside from this, they must employ techniques and materials that are environmentally friendly. As a result, they opt to replace lead-based solders for microelectronics packaging with electrically conductive adhesives (ECAs). A great deal of research has been carried out on the fabrication of polymer composites reinforced by metallic fillers ranging from metallic particles to carbon materials. These polymer composites have been used to develop new types of ECAs to reduce the cost to provide finer pitch capability [3]. Gold, silver [4–7], copper [8], palladium and nickel have been widely used as fillers for commercial ECAs due to their stability and excellent electrical conductivity [7,9]. However, a high concentration of filler is required to form a good electrical conductivity [5], which is the main reason for the high cost of ECAs. How to reduce the loading of metal fillers into the ECAs has become an important issue.

Many researches use carbon nanotubes (CNTs) [10–12] as fillers to improve the electrical properties of ECAs due to their high conductivity, flexibility, and large aspect ratios, and report the establishment of a percolated network at low filler content. However, the usage of CNTs has been limited by the challenges in processing, high cost and dispersion [13]. As a two-dimensional material, graphene has attracted wide attention due to its stable structure and excellent performance. These materials have drawn an intensive attention towards a variety of research fields to utilize their exceptional thermal, mechanical, optical and electrical properties [14–16]. Theoretical and experimental studies of graphene show they may possess high thermal conductivity ($5000 \ W \cdot m^{-1} \cdot K^{-1}$), high Young's modulus (~1 TPa), large surface area (~2600 $m^2 \cdot g^{-1}$) [17,18] and great electrical conductivity ($6000 \ S \cdot cm^{-1}$) [19]. Possessing the extremely high aspect ratio among all the nanostructured materials, graphene is a promising nanomaterial able to establish a percolated network at very low concentrations.

Xie et al. predicted that in comparison with nanotubes, graphene could provide higher conductivity when used as fillers for conductive adhesives. This is due to its large specific area [20–22]. Electrical conductivities of auxiliary forms of graphene have also been reported and studied as conductive fillers or reinforcements for polymers. The polypropylene/graphene oxide (PP/GO) nanocomposite prepared by Huang and colleagues through Zieglar-Natta polymerization provides a high conductivity of $0.3 \ S \cdot m^{-1}$ at a 4.9 wt.% loading of GO [23,24]. The chemically reduced graphene oxide/polystyrene (CRGO/PS) composite prepared by Stankovich [25] delivers $0.1 \ S \cdot m^{-1}$. Qi et al. [26] compared the electrical conductivity of MWCNT/PS and G/PS nanocomposites and found that at 0.69 wt.%, G/PS has a conductivity of $3.49 \ S \cdot m^{-1}$ while MWCNT/PS at the same content only has a conductivity of $3 \times 10^{-5} \ S \cdot m^{-1}$. Ansari et al. studied and compared the properties of functionalized graphene sheet/PVDF (FGS/PVDF) and exfoliated graphite/PVDF (EG/PVDF), which were prepared via a solution process with subsequent compression to fabricate conductive nanocomposites. According to their results, FGS remained well dispersed in PVDF and showed wrinkled topography with relatively thin graphene sheets that were bonded well in the matrix [27].

In order to obtain graphene, various methods, such as exfoliation and cleavage, liquid-phase exfoliation [28,29], growth on SiC [30,31], chemical vapor deposition [32,33], molecular beam epitaxy, chemical synthesis [34–37], and chemical routes, have been widely studied. Mass production of single-layer graphene or few-layer graphene is being hindered by the expensive cost and environmental threat of its conventional synthesis. From the environmentally friendly viewpoint, there are many researches that have studied the synthesis of graphene by a simple ultrasonication or jet cavitation method to obtain few-layer graphene [38]. In this paper, a low-temperature, high-pressure continuous flow cell disrupter was used as a delamination device to synthesize few-layer graphene. With this method, simple, scalable, low-cost graphene was swiftly produced without the incorporation of any toxic chemicals. The graphene obtained by this method was combined with polymer to make ECAs.

2. Experimental

2.1. Synthesis of FLG

Natural flake graphite (Øave = 500 μm) was used as a feed material for the delamination process while N-Methyl-2-pyrrolidone (NMP) and polyvinylidene fluoride (PVDF) purchased from UBIQ Technology Co. Ltd., Taiwan were used as a dispersing agent and a binder, respectively. The synthesis of FLG was described as follow: a 1 kg solution consisting of 10 wt.% natural graphite dispersed in de-ionized water (DI H_2O) was fed into a low-temperature, ultra-high-pressure continuous flow cell disrupter (LTHPD, JNBIO, JN 10C, Guangzhou, China). Prior to feeding the solution into the tank of the LTHPD, the solution was vigorously stirred to ensure uniform dispersion. The graphite solution was pumped into a nozzle with a high pressure and a high flow rate. Under the high-pressure, strong-impact and repeated cyclic stress, layers of graphite were exfoliated, obtaining few-layer graphene (FLG). LTHPD did not only delaminate the graphite flakes into FLG, but also ensures homogeneous dispersion in the solution. The operating condition was maintained at 1800 bar and 14–16 °C. The obtained product was dried in a vacuum oven at 30 °C overnight.

2.2. Conductive Adhesive Slurry Preparations

Different slurry compositions were prepared with various FLG:PVDF ratios, such as 1:99, 5:95, 8:92, 10:90, 30:70, 90:90, and 50:50 by dispersing PVDF with a designated amount into NMP. After ensuring the homogeneity of PVDF in NMP, FLG was added into the solution and stirred for 2 h. The slurry was coated onto glass slides and dried at 80 °C for 1 h. The graphite:PVDF slurry of the same compositions was also prepared for comparison.

2.3. Characterizations

The height profile of as-synthesized few-layer graphene was measured by using atomic force microscopy (AFM, Bruker Dimension Icon). The samples for AFM were prepared by dropping the dispersion directly onto a freshly cleaved mica wafer. Raman spectra were carried out by a micro Raman spectroscopy system with a laser frequency of 532 nm as the excitation source. The morphologies of the samples were analyzed using scanning electron microscopy (SEM) by Hitachi S-4100 (Hitachi, Tokyo, Japan) and high-resolution transmission electron microscopy (HRTEM) JEOL-JEM2000FXII (JEOL, Tokyo, Japan). The electrical conductivities of the graphene adhesives were measured using a resistivity meter (KeithLink TG2, KeithLink, Taipei, Taiwan) with a four-point probe.

3. Results and Discussion

3.1. Characterizations of FLG

Figure 1a,b show images of graphite and FLG with lateral sizes of 5–10 μm and 1–5 μm, respectively. After treatment by LTHPD, the graphite size was decreased and delaminated to be FLG. This shows that the thick sheets of graphite can be effectively exfoliated into thinner sheets. Figure 1c displays the HRTEM image of the FLG. This image of FLG is transparent and folded, which coincides well with the typical feature of the reported FLG. In order to observe the effect of the LTHPD, AFM characterization was carried out for its convenience in measuring flake thickness. The FLG was deposited on the silica wafer and dried in room temperature. More than 30 flakes were measured to determine the distribution. Figure 1d,e show the AFM image and the thickness distribution of FLG. They are believed to be monolayers according to the fact that FLG are often measured to be 0.4–1 nm by AFM due to some external factors such as the AFM equipment and substrates [39,40]. This shows that the range of FLG thickness is 1–4.5 nm, which is less than 10 layers. The delamination process used to obtain FLG was expected to cause defects, and thus, Raman analysis of graphite and FLG was determined and is depicted in Figure 1f. The result of similar D/G ratio demonstrates that little structural defects were created by the liquid exfoliation process. Additionally, the intensity ratio of D/G for the FLG is 0.19, which is much lower than that of the GO and chemically reduced graphene [41,42].

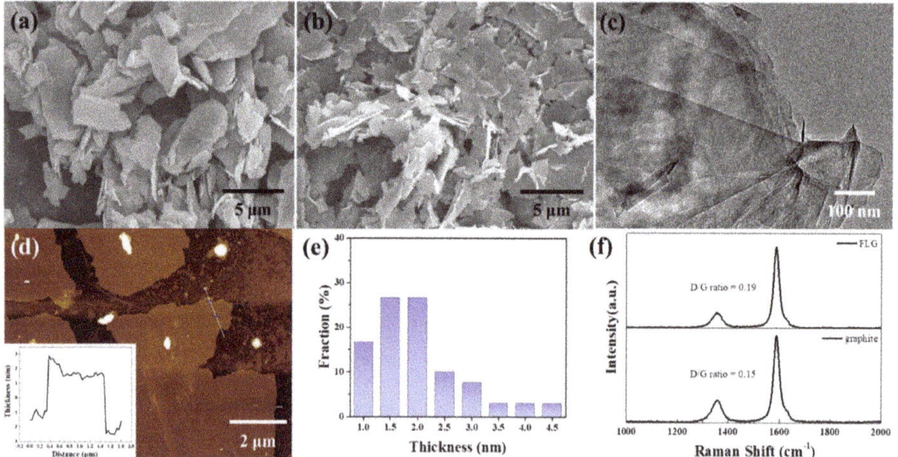

Figure 1. SEM images of (**a**) graphene and (**b**) FLG; (**c**) TEM image of FLG; (**d**) AFM image of FLG, insert: height profiles of FLG; (**e**) thickness distribution of FLG measured by AFM on 35 samples; (**f**) Raman spectra of graphite and FLG.

3.2. Characterizations of Conductive Adhesives

To formulate conductive adhesives, the conductive fillers, graphite and FLG were mixed with PVDF. After mixing, the composites inks were cast into a 2×2 mm well to create solid thin films (Figure 2). The dried thin films at various filler/PVDF ratios exhibit color variations. The FLG/PVDF composite also exhibits great conductivity compared to its graphite counterpart. As shown in Figure 3, at the same filler/PVDF ratio, the conductivity of the FLG/PVDF composite is always higher than that of the graphite/PVDF composite, which is possibly due to the higher aspect ratio of FLG. The lateral size of FLG is ~8 μm and the average thickness of FLG is ~4 nm. Thus, the aspect ratio of as-synthesized FLG is as high as ~2000. The higher aspect ratio provided a better percolation network for electron transfer, and thus resulted in the conductivity enhancement.

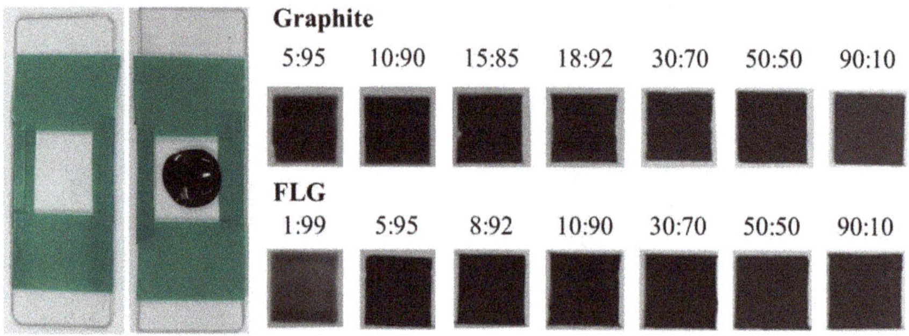

Figure 2. Photo images of graphite/PVDF and FLG/PVDF composites.

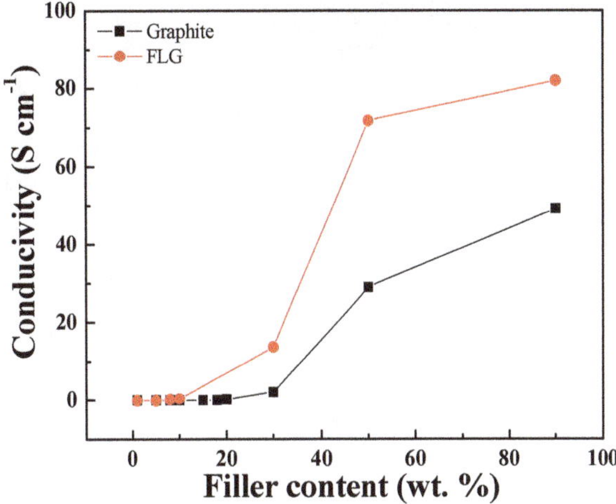

Figure 3. Electronic conductivity of graphite/PVDF and FLG/PVDF composite films measured by a four-point probe.

The conductivity of the FLG/PVDF thin film can be greatly enhanced after sintering. As shown in Figure 4, when the temperature increases, the resistance decreases. This reduction of resistivity comes from relaxation of the polymer chains at higher temperatures and thus leads to a decrease in the mean distance between graphene [43–45]. The shortened distance between graphene layers helps the electron transfer and thus increases the conductivity [46]. However, as the temperature reaches the glass transition temperature of PVDF from ~400 to 425 °C, the relaxation process facilitates the re-arrangement of the dispersed graphene [45] and thus the resistance plateau occurs at ~400 °C.

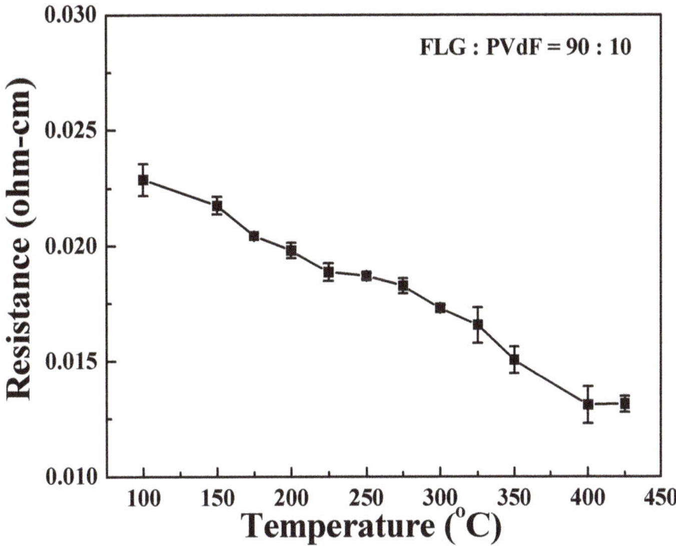

Figure 4. Variation in sheet resistance of FLG/PVDF composite films with temperature. The FLG/PVDF thin film is coated on glass with a thickness of 0.02 mm and is placed on a hot plate for 30 min.

The FLG/PVDF composite also exhibits great flexibility and mechanical strength after being coated on flexible plastic sheets. The electrical resistance of the FLG/PVDF thin film on the PET film under the bending test (inset picture in Figure 5a) shows a fairly small variation after thousands of bending cycles. The small increase of ~5% in resistance is possibly attributed to the crack of film under bending [47]. This great mechanical strength makes the FLG/PVDF a great conductive adhesive for flexible electronic applications. For demonstration, two pieces of printed FLG/PVDF thin films on the PET film were connected to an LED, and the printed conductive tracks were twisted or bent with different degrees of curvature (Figure 5b,d). As shown in Figure 5c, the PET film with slight curvature could remain bright. The bulb shone constantly after bending the film into a spiral shape, as shown in Figure 5d, indicating the great bending property of the graphene/PVDF composites on the PET film.

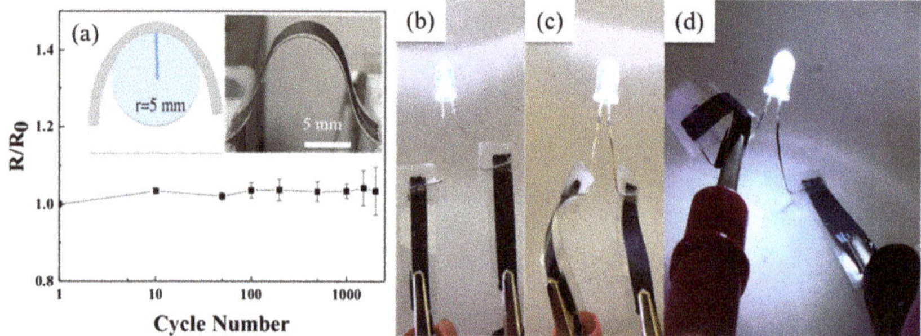

Figure 5. (**a**) The resistance increase ratio (R/R$_0$) of graphene/PVDF thin film on the PET film under the bending performance test with a radius of curvature of 5 mm. After connecting the graphene/polymer composite to an LED, the light remained bright at various degrees of bending: (**b**) flat, (**c**) bending, and (**d**) spiral shape.

4. Discussion

In this study, a facile liquid exfoliation process was developed to synthesize FLG. Electron microscopic examination shows the as-synthesized FLG has a lateral size of 1–5 µm. AFM indicates more than 80% of FLG has less than 10 layers. Raman spectroscopy also shows a great D/G ratio of 0.19, which is close to that of graphite (0.15), indicating that little structural defects were created via the liquid exfoliation process. After being mixed with PVDF, the composite FLG/PVDF thin film exhibits superior electronic conductivity compared to the graphite/PVDF composite with the same ratio. Moreover, the post-sintering process can further improve electronic conductivity by 85%. The composition-optimized FLG/PVDF composite exhibits 81.9 S·cm^{-1}. The printed FLG/PVDF thin film patterns also show great flexibility and mechanical strength under bending conditions. The electrical conductivity remains nearly the same after thousands of bending cycles. These results indicate that the formulated FLG/PVDF composite adhesives have a great potential in conductive adhesive applications.

Author Contributions: J.-Y.W. synthesize few layer graphene. Y.-C.L. assemble different ratio of graphene/PVDF as composite film. J.-Y.W. and Y.-C.L. write the manuscript. C.-L.C., W.-C.H., C.-H.H. and H.-M.W. help to characterize the properties of graphene/PVDF film. Y.-C.L. and W.-R.L. are supervisors to guide experiment and manuscript refine.

Funding: This research is supported by the National Science Council of Taiwan under contract numbers: 102-2221-E-033-050- MY2, 102-3011-P-033-003, and MOST 106-2218-E-002-038.

Conflicts of Interest: The authors declare no conflicts of interest.

References

1. Périchaud, M.-G.; Delétage, J.-Y.; Frémont, H.; Danto, Y.; Faure, C. Reliability evaluation of adhesive bonded SMT components in industrial applications. *Microelectron. Reliab.* **2000**, *40*, 1227–1234. [CrossRef]

2.   Daniel Lu, D.; Grace Li, Y.; Wong, C. Recent advances in nano-conductive adhesives. *J. Adhes. Sci. Technol.* **2008**, *22*, 815–834. [CrossRef]

3.   Li, Y.; Moon, K.-S.; Wong, C. Electronics without lead. *Science* **2005**, *308*, 1419–1420. [CrossRef] [PubMed]

4.   Ye, L.; Lai, Z.; Liu, J.; Tholen, A. Effect of Ag particle size on electrical conductivity of isotropically conductive adhesives. *IEEE Trans. Electron. Packag. Manuf.* **1999**, *22*, 299–302.

5.   Kotthaus, S.; Gunther, B.H.; Hang, R.; Schafer, H. Study of isotropically conductive bondings filled with aggregates of nano-sited Ag-particles. *IEEE Trans. Electron. Packag. Manuf.* **1997**, *20*, 15–20. [CrossRef]

6.   Lee, H.-H.; Chou, K.-S.; Shih, Z.-W. Effect of nano-sized silver particles on the resistivity of polymeric conductive adhesives. *Int. J. Adhes. Adhes.* **2005**, *25*, 437–441. [CrossRef]

7.   Wu, H.; Liu, J.; Wu, X.; Ge, M.; Wang, Y.; Zhang, G.; Jiang, J. High conductivity of isotropic conductive adhesives filled with silver nanowires. *Int. J. Adhes. Adhes.* **2006**, *26*, 617–621. [CrossRef]

8.   Zhao, H.; Liang, T.; Liu, B. Synthesis and properties of copper conductive adhesives modified by SiO_2 nanoparticles. *Int. J. Adhes. Adhes.* **2007**, *27*, 429–433. [CrossRef]

9.   Jia, W.; Tchoudakov, R.; Joseph, R.; Narkis, M.; Siegmann, A. The role of a third component on the conductivity behavior of ternary epoxy/Ag conductive composites. *Polym. Compos.* **2002**, *23*, 510–519. [CrossRef]

10.   Chun, K.-Y.; Oh, Y.; Rho, J.; Ahn, J.-H.; Kim, Y.-J.; Choi, H.R.; Baik, S. Highly conductive, printable and stretchable composite films of carbon nanotubes and silver. *Nat. Nanotechnol.* **2010**, *5*, 853–857. [CrossRef]

11.   Ji, S.Y.; Ajmal, C.M.; Kim, T.; Chang, W.S.; Baik, S. Laser patterning of highly conductive flexible circuits. *Nanotechnology* **2017**, *28*, 165301. [CrossRef]

12.   Gorrasi, G.; Sarno, M.; Di Bartolomeo, A.; Sannino, D.; Ciambelli, P.; Vittoria, V. Incorporation of carbon nanotubes into polyethylene by high energy ball milling: Morphology and physical properties. *J. Polym. Sci. Part B Polym. Phys.* **2007**, *45*, 597–606. [CrossRef]

13.   Kim, J.; Yim, B.-S.; Kim, J.-M.; Kim, J. The effects of functionalized graphene nanosheets on the thermal and mechanical properties of epoxy composites for anisotropic conductive adhesives (ACAs). *Microelectron. Reliab.* **2012**, *52*, 595–602. [CrossRef]

14.   Giubileo, F.; Di Bartolomeo, A.; Martucciello, N.; Romeo, F.; Iemmo, L.; Romano, P.; Passacantando, M. Contact resistance and channel conductance of graphene field-effect transistors under low-energy electron irradiation. *Nanomaterials* **2016**, *6*, 206. [CrossRef]

15.   Alvarado Chavarin, C.; Strobel, C.; Kitzmann, J.; Di Bartolomeo, A.; Lukosius, M.; Albert, M.; Bartha, J.W.; Wenger, C. Current Modulation of a Heterojunction Structure by an Ultra-Thin Graphene Base Electrode. *Materials* **2018**, *11*, 345. [CrossRef]

16.   Luongo, G.; Di Bartolomeo, A.; Giubileo, F.; Alvarado Chavarin, C.; Wenger, C. Electronic properties of graphene/p-silicon Schottky junction. *J. Phys. D Appl. Phys.* **2018**, *51*. [CrossRef]

17.   Balandin, A.A.; Ghosh, S.; Bao, W.; Calizo, I.; Teweldebrhan, D.; Miao, F.; Lau, C.N. Superior thermal conductivity of single-layer graphene. *Nano Lett.* **2008**, *8*, 902–907. [CrossRef] [PubMed]

18.   Lee, C.; Wei, X.; Kysar, J.W.; Hone, J. Measurement of the elastic properties and intrinsic strength of monolayer graphene. *Science* **2008**, *321*, 385–388. [CrossRef]

19.   Du, X.; Skachko, I.; Barker, A.; Andrei, E.Y. Approaching ballistic transport in suspended graphene. *Nat. Nanotechnol.* **2008**, *3*, 491–495. [CrossRef]

20.   Du, J.; Cheng, H.M. The fabrication, properties, and uses of graphene/polymer composites. *Macromol. Chem. Phys.* **2012**, *213*, 1060–1077. [CrossRef]

21.   Khanam, P.N.; Ponnamma, D.; Al-Madeed, M. Electrical properties of graphene polymer nanocomposites. In *Graphene-Based Polymer Nanocomposites in Electronics*; Springer: Berlin, Germany, 2015; pp. 25–47.

22.   Xie, S.; Liu, Y.; Li, J. Comparison of the effective conductivity between composites reinforced by graphene nanosheets and carbon nanotubes. *Appl. Phys. Lett.* **2008**, *92*, 243121. [CrossRef]

23.   Huang, Y.; Qin, Y.; Zhou, Y.; Niu, H.; Yu, Z.-Z.; Dong, J.-Y. Polypropylene/Graphene Oxide Nanocomposites Prepared by In Situ Ziegler−Natta Polymerization. *Chem. Mater.* **2010**, *22*, 4096–4102. [CrossRef]

24.   Gorrasi, G.; Bugatti, V.; Milone, C.; Mastronardo, E.; Piperopoulos, E.; Iemmo, L.; Di Bartolomeo, A. Effect of temperature and morphology on the electrical properties of PET/conductive nano fillers composites. *Compos. Part B Eng.* **2018**, *135*, 149–154. [CrossRef]

25.   Stankovich, S.; Dikin, D.A.; Dommett, G.H.; Kohlhaas, K.M.; Zimney, E.J.; Stach, E.A.; Piner, R.D.; Nguyen, S.T.; Ruoff, R.S. Graphene-based composite materials. *Nature* **2006**, *442*, 282–286. [CrossRef] [PubMed]

26. Qi, X.-Y.; Yan, D.; Jiang, Z.; Cao, Y.-K.; Yu, Z.-Z.; Yavari, F.; Koratkar, N. Enhanced electrical conductivity in polystyrene nanocomposites at ultra-low graphene content. *ACS Appl. Mater. Interfaces* **2011**, *3*, 3130–3133. [CrossRef] [PubMed]

27. Ansari, S.; Giannelis, E.P. Functionalized graphene sheet—Poly (vinylidene fluoride) conductive nanocomposites. *J. Polym. Sci. Part B Polym. Phys.* **2009**, *47*, 888–897. [CrossRef]

28. Hernandez, Y.; Nicolosi, V.; Lotya, M.; Blighe, F.M.; Sun, Z.; De, S.; McGovern, I.; Holland, B.; Byrne, M.; Gun'Ko, Y.K. High-yield production of graphene by liquid-phase exfoliation of graphite. *Nat. Nanotechnol.* **2008**, *3*, 563–568. [CrossRef]

29. Lotya, M.; Hernandez, Y.; King, P.J.; Smith, R.J.; Nicolosi, V.; Karlsson, L.S.; Blighe, F.M.; De, S.; Wang, Z.; McGovern, I. Liquid phase production of graphene by exfoliation of graphite in surfactant/water solutions. *J. Am. Chem. Soc.* **2009**, *131*, 3611–3620. [CrossRef]

30. Virojanadara, C.; Syväjarvi, M.; Yakimova, R.; Johansson, L.; Zakharov, A.; Balasubramanian, T. Homogeneous large-area graphene layer growth on 6 H-SiC (0001). *Phys. Rev. B* **2008**, *78*, 245403. [CrossRef]

31. Sprinkle, M.; Ruan, M.; Hu, Y.; Hankinson, J.; Rubio-Roy, M.; Zhang, B.; Wu, X.; Berger, C.; De Heer, W.A. Scalable templated growth of graphene nanoribbons on SiC. *Nat. Nanotechnol.* **2010**, *5*, 727–731. [CrossRef]

32. Reina, A.; Jia, X.; Ho, J.; Nezich, D.; Son, H.; Bulovic, V.; Dresselhaus, M.S.; Kong, J. Large area, few-layer graphene films on arbitrary substrates by chemical vapor deposition. *Nano Lett.* **2008**, *9*, 30–35. [CrossRef] [PubMed]

33. Wei, D.; Liu, Y.; Wang, Y.; Zhang, H.; Huang, L.; Yu, G. Synthesis of N-doped graphene by chemical vapor deposition and its electrical properties. *Nano Lett.* **2009**, *9*, 1752–1758. [CrossRef] [PubMed]

34. Chen, J.; Yao, B.; Li, C.; Shi, G. An improved Hummers method for eco-friendly synthesis of graphene oxide. *Carbon* **2013**, *64*, 225–229. [CrossRef]

35. Eigler, S.; Enzelberger-Heim, M.; Grimm, S.; Hofmann, P.; Kroener, W.; Geworski, A.; Dotzer, C.; Röckert, M.; Xiao, J.; Papp, C. Wet chemical synthesis of graphene. *Adv. Mater.* **2013**, *25*, 3583–3587. [CrossRef] [PubMed]

36. Zhang, Y.; Fugane, K.; Mori, T.; Niu, L.; Ye, J. Wet chemical synthesis of nitrogen-doped graphene towards oxygen reduction electrocatalysts without high-temperature pyrolysis. *J. Mater. Chem.* **2012**, *22*, 6575–6580. [CrossRef]

37. Chen, L.; Hernandez, Y.; Feng, X.; Müllen, K. From nanographene and graphene nanoribbons to graphene sheets: Chemical synthesis. *Angew. Chem. Int. Ed.* **2012**, *51*, 7640–7654. [CrossRef] [PubMed]

38. Liang, S.; Shen, Z.; Yi, M.; Liu, L.; Zhang, X.; Cai, C.; Ma, S. Effects of Processing Parameters on Massive Production of Graphene by Jet Cavitation. *J. Nanosci. Nanotechnol.* **2015**, *15*, 2686–2694. [CrossRef]

39. Nacken, T.; Damm, C.; Walter, J.; Rüger, A.; Peukert, W. Delamination of graphite in a high pressure homogenizer. *RSC Adv.* **2015**, *5*, 57328–57338. [CrossRef]

40. Novoselov, K.S.; Geim, A.K.; Morozov, S.V.; Jiang, D.; Zhang, Y.; Dubonos, S.V.; Grigorieva, I.V.; Firsov, A.A. Electric field effect in atomically thin carbon films. *Science* **2004**, *306*, 666–669. [CrossRef]

41. Stankovich, S.; Piner, R.D.; Chen, X.; Wu, N.; Nguyen, S.T.; Ruoff, R.S. Stable aqueous dispersions of graphitic nanoplatelets via the reduction of exfoliated graphite oxide in the presence of poly (sodium 4-styrenesulfonate). *J. Mater. Chem.* **2006**, *16*, 155–158. [CrossRef]

42. Stankovich, S.; Dikin, D.A.; Piner, R.D.; Kohlhaas, K.A.; Kleinhammes, A.; Jia, Y.; Wu, Y.; Nguyen, S.T.; Ruoff, R.S. Synthesis of graphene-based nanosheets via chemical reduction of exfoliated graphite oxide. *Carbon* **2007**, *45*, 1558–1565. [CrossRef]

43. Palza, H.; Garzón, C.; Arias, O. Modifying the electrical behaviour of polypropylene/carbon nanotube composites by adding a second nanoparticle and by annealing processes. *Express Polym. Lett.* **2012**, *6*, 639–646. [CrossRef]

44. Palza, H.; Kappes, M.; Hennrich, F.; Wilhelm, M. Morphological changes of carbon nanotubes in polyethylene matrices under oscillatory tests as determined by dielectrical measurements. *Compos. Sci. Technol.* **2011**, *71*, 1361–1366. [CrossRef]

45. Cipriano, B.H.; Kota, A.K.; Gershon, A.L.; Laskowski, C.J.; Kashiwagi, T.; Bruck, H.A.; Raghavan, S.R. Conductivity enhancement of carbon nanotube and nanofiber-based polymer nanocomposites by melt annealing. *Polymer* **2008**, *49*, 4846–4851. [CrossRef]

46. Garzón, C.; Palza, H. Electrical behavior of polypropylene composites melt mixed with carbon-based particles: Effect of the kind of particle and annealing process. *Compos. Sci. Technol.* **2014**, *99*, 117–123. [CrossRef]
47. Lewis, J. Material challenge for flexible organic devices. *Mater. Today* **2006**, *9*, 38–45. [CrossRef]

 *nanomaterials*

Communication

Electrochemical Enantiomer Recognition Based on sp^3-to-sp^2 Converted Regenerative Graphene/Diamond Electrode

Jingyao Gao [1,†], Haoyang Zhang [2,†], Chen Ye [1], Qilong Yuan [1,3], Kuan W. A. Chee [1,3], Weitao Su [2], Aimin Yu [4], Jinhong Yu [1], Cheng-Te Lin [1], Dan Dai [1,*] and Li Fu [2,*]

[1] Key Laboratory of Marine Materials and Related Technologies, Zhejiang Key Laboratory of Marine Materials and Protective Technologies, Ningbo Institute of Materials Technology and Engineering (NIMTE), Chinese Academy of Sciences, Ningbo 315201, China gaojingyao@nimte.ac.cn (J.G.); yechen@nimte.ac.cn (C.Y.); Qilong.Yuan@nottingham.edu.cn (Q.Y.); Kuan.Chee@nottingham.edu.cn (K.W.A.C.); yujinhong@nimte.ac.cn (J.Y.); linzhengde@nimte.ac.cn (C.-T.L.)
[2] College of Materials and Environmental Engineering, Hangzhou Dianzi University, Hangzhou 310018, China; zhanghaoyang@hdu.edu.cn (H.Z.); suweitao@hdu.edu.cn (W.S.)
[3] Department of Electrical and Electronic Engineering, Faculty of Science and Engineering, University of Nottingham, Ningbo 315100, China
[4] Department of Chemistry and Biotechnology, Faculty of Science, Engineering and Technology, Swinburne University of Technology, Hawthorn, VIC 3122, Australia; aiminyu@swin.edu.au
* Correspondence: daidan@nimte.ac.cn (D.D.); fuli@hdu.edu.cn (L.F.)
† These authors contributed equally to this work.

Received: 22 October 2018; Accepted: 3 December 2018; Published: 14 December 2018

Abstract: It is of great significance to distinguish enantiomers due to their different, even completely opposite biological, physiological and pharmacological activities compared to those with different stereochemistry. A sp^3-to-sp^2 converted highly stable and regenerative graphene/diamond electrode (G/D) was proposed as an enantiomer recognition platform after a simple β-cyclodextrin (β-CD) drop casting process. The proposed enantiomer recognition sensor has been successfully used for D and L-phenylalanine recognition. In addition, the G/D electrode can be simply regenerated by half-minute sonication due to the strong interfacial bonding between graphene and diamond. Therefore, the proposed G/D electrode showed significant potential as a reusable sensing platform for enantiomer recognition.

Keywords: graphene; chemical vapor deposition; electronic materials; enantiomer recognition; phenylalanine

---

1. Introduction

In stereochemistry, the characteristic of a substance that cannot coincide with its image is called chirality. Chirality is ubiquitous in nature, which is the most basic characteristics of life processes. Most of the organic compounds that constitute the life body are chiral molecules, such as proteins, polysaccharides and nucleic acids. A number of metabolic and regulatory processes in biological systems are closely related to stereochemistry, so chiral molecules participate in a series of enzymatic reactions, host and guest effects as well as other biological phenomena. In addition, chiral drugs often have only one enantiomer that shows a therapeutic effect, and the other may be ineffective or even have toxic side effects.

Nowadays, considerable attention has been paid to the research of chiral recognition. Several methods of chiral recognition have been developed including spectrum based [1], chromatography based [2] and sensor based technologies [3]. Although spectrum based and chromatography based methods could combine enantiomer separation and determination together, the operation process is still sophisticated, and detection time is also quite long. Therefore, the development of sensor based chiral recognition methods, specifically electrochemical sensors, is of great practical significance due to their simplicity, fast response, low cost, on-line and real-time detection. Electrochemical chiral sensors are based on the differences in the effects of chiral selectors and enantiomers, and this difference can be identified by conversion of redox probe signals into electrical signals. However, surface modification of the electrode by a chiral selector or its composite is still a complex process. For example, glassy carbon electrodes (GCE) are reusable, but the pre-polishing process has great influence on the performance of the bare electrode. At the same time, the chiral selectors or molecular imprinted polymers often affects the conductivity and further amplifies the instability of the GCE [4]. Screen printed electrode (SPE) are low cost and disposable electrodes, but it is difficult to use for molecular recognition after the chiral selector modification because of the poor electrical conductivity of the printed ink. In the past decade, the growing interest in graphene chemistry provided many graphene based fascinating composites for electrochemical sensor fabrication due to their excellent electrical and morphological properties [5–8]. The preparation of reusable graphene electrodes is an important topic in electrochemical sensing.

L-phenylalanine is an essential amino acid that human beings are unable to produce, and must be supplemented from food sources. L-phenylalanine deficiency can affect tyrosine synthesis, resulting in the decrease of thyroxine level and its metabolic activity [9]. Its enantiomer, D-phenylalanine, is commonly recommended to promote muscle stiffness, help with walking and speaking, and is therefore a useful health care product for treatment of Parkinson's disease [10].

In this work, we proposed a new strategy for electrochemical chiral sensor fabrication, which transforms a diamond surface into a graphene film via sp^3-sp^2 conversion. The fabricated graphene/diamond (G/D) was successfully applied to D- and L-phenylalanine recognition after a simple β-CD drop casting process. Due to the extremely strong interfacial forces between graphene and diamond, the proposed G/D electrode can be simply regenerated via a half-minute of ultrasonic cleaning treatment.

2. Materials and Experimental

All reagents were analytical grade and used without any further purification process. Phosphate buffer solutions (PBS) were made by mixing 0.1 M KH_2PO_4 with Na_2HPO_4. High pressure high temperature (HPHT) synthetic diamond substrate with size of $3.5 \times 3.5 \times 1$ mm^3 was purchased from Shenzhen Tiantian Xiangshang Diamond Co. Ltd. (Shenzhen, China).

For graphene-diamond (G/D) electrode preparation, a HPHT diamond substrate was placed on a Cu plate surface and annealed at 1100 °C with 8 sccm H_2 flow ($\approx$0.14 Torr). Residue Cu was etched by immersing the G/D into a mixed solution ($CuSO_4$:HCl:H_2O = 1 g:50 mL:50 mL). The G/D was then partially encapsulated by silicone resin and with the graphene surface left open. The electrode was connected to Cu foil via silver paint (Figure 1A). For β-CD surface medication, 10 μL of β-CD solution (1 mM) was drop casted on the G/D electrode surface and dried naturally (β-CD/G/D).

All electrochemical tests were performed on a CHI760E Electrochemical Analyzer. A β-CD/G/D, Pt wire and 3 M Ag/AgCl electrode were used as a working electrode, counter electrode and reference electrode, respectively.

3. Results and Discussion

Figure 1 shows the detail of graphene film growth via sp^3-to-sp^2 conversion process. Under high temperature, carbon atoms on HPHT defects centers will continue to diffuse into melted copper. Due to the low solubility of carbon in copper [11], the saturated carbon atoms will gradually precipitate and rearrange within the interface between copper and HPHT to form a graphene film. When graphene film completely blocked the contact between HPHT and copper, the entire catalytic reaction ended. As a result, a HPHT substrate with graphene surface was obtained, which can be used as a highly stable and regenerative electrode after encapsulation (Figure 1B,C,D) [12].

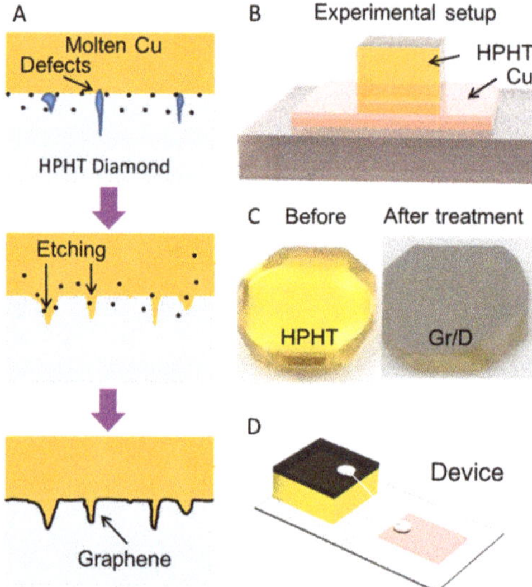

Figure 1. (**A**) Schematic diagram of graphene formation. (**B**) Scheme of preparation of graphene/diamond (G/D) electrode using chemical vapor deposition (CVD) method. (**C**) Photograph of sample before and after thermal treatment. (**D**) Scheme of the device after encapsulation

The surface morphology change of HPHT can be observed via SEM characterization. As shown in Figure 2A,B, due to the continuous diffusion of carbon atoms into copper, the surface of HPHT was etched into many island structures. XPS was used to characterize the chemical state of carbon on the surface of HPHT and G/D. As shown in Figure 2C, the surface of HPHT was basically composed of sp^3 bonds, except for a small number of defective oxygen functional groups. In contrast, the sp^3 bond (C—C) on the surface of G/D had nearly decreased by half, while the sp^2 bond (C=C) had formed clearly, indicating the sp^3 bond had successfully converted to sp^2. The quality of formed graphene was evaluated using Raman mapping. As shown in Figure 2D, the ratio of I_{2D}/I_G of graphene in most areas of G/D was between 0.9–1.2, indicating that it contained only a few layers. A full Raman spectrum is shown in Figure 2E. Since the graphene formed on diamond was only a few layers, the diamond peak was still visible. The sheet resistance of formed G/D was around 400 Ω/Y.

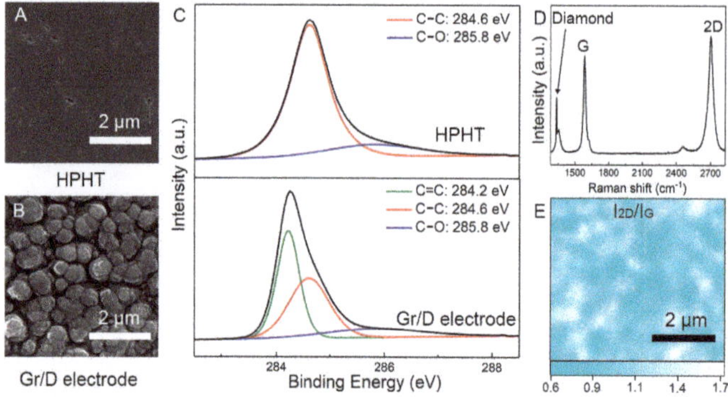

Figure 2. (**A,B**) SEM images and (**C**) XPS spectra of the high pressure high temperature (HPHT) and G/D electrode. (**D**) Raman mapping profile of I_{2D}/I_G in a 5×5 μm area. (**E**) A typical Raman spectrum of G/D.

In this work, L and D-phenylalanine were chosen for evaluating the feasibility of a G/D electrode for electrochemical enantiomer recognition. β-CD was chosen as a chiral selector due to its excellent solubility and chiral recognition property [13]. Linear sweep voltammetry (LSV) was used for electrochemical enantiomers recognition. Figure 3 shows the LSV profiles of G/D and β-CD/G/D electrodes for L and D-phenylalanine recognition. It can be seen that 0.1 mM L, D-phenylalanine oxidized at the same potential on the G/D electrode without β-CD modification (Figure 3A). Although the oxidation current of L-phenylalanine was slightly higher than that of D-phenylalanine, it could not be used to identify the type of phenylalanine with an unknown concentration. In contrast, L-phenylalanine and D-phenylalanine were oxidized at different potentials on the β-CD/G/D electrode (Figure 3B), which confirms the G/D electrode can be successfully used for L, D-phenylalanine recognition after a simple β-CD drop casting process. More specifically, the oxidation of L-phenylalanine was observed at 0.576 V with an enhanced current response. This enhancement could be ascribed to the strong interaction between β-CD molecules and L-phenylalanine, which concentrated more L-phenylalanine on electrode surface. This phenomenon is in agreement with other works [14,15]. In contrast, a suppressed oxidation of D-phenylalanine was observed at 0.550 V.

The reproducibility of the G/D electrode was tested by five individually fabricated electrodes. After the β-CD modification, these five electrodes all exhibited different oxidation potentials for L,D-phenylalanine. More specifically, the oxidation potential of the D-phenylalanine was always lower than the potential of L-phenylalanine oxidation. Because of the strong interaction between graphene and HPHT, G/D electrodes can be reused after a half-minute ultrasonic cleaning process in water with performance remaining very stable. The G/D electrode can be reused more than twenty times without sensing degradation. Currently, fabrication of graphene-based electrodes is a very hot area for biosensor development due to the outstanding properties of graphene. In most cases, the graphene or graphene-based composites are formed in a dispersion form and used as a modifier for commercial electrode modification, such as glassy carbon electrode or screen printed electrodes. These biosensors cannot be reused because the cleaning process could remove the modifier as well. In our reports, the G/D electrode can be simply regenerated by a half-minute sonication due to the strong interfacial bonding between the graphene and diamond. Therefore, the G/D enantiomer recognition sensor proposed in this work has more practical potential compared to modifier based electrochemical enantiomer recognition sensors.

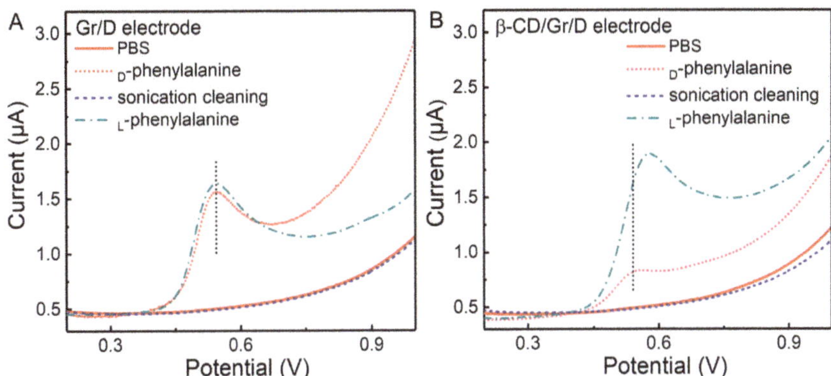

Figure 3. Linear sweep voltammetry (LSV) responses of (**A**) G/D and (**B**) β-CD/G/D in the absence and presence of 0.1 mM L- and D-phenylalanine in 0.1 M PBS (pH 7.0). Amplitude: 50 mV; Pulse width: 0.05 s; Pulse period: 0.5 s.

4. Conclusions

In conclusion, a highly stable and regenerative G/D electrode was formed via a carbon sp^3-to-sp^2 conversion process. After a simple β-CD drop casting, the G/D electrode can be used as an electrochemical platform for L and D-phenylalanine recognition. Due to the strong interaction between graphene and HPHT, the proposed G/D electrode can be regenerated using a simple sonication cleaning process. Therefore, the G/D electrode has more practical potential compared to modifier based electrochemical sensors.

Author Contributions: Conceptualization, C.-T.L. D.D. and L.F.; methodology, L.F., J.Y. and A.Y.; investigation, J.G. and H.Z.; resources, C.Y., W.S. and K.W.A.C.; data curation, X.X.; writing—original draft preparation, L.F.; writing—review and editing, J.G., C.Y. and Q.Y.; project administration, J.Y.

Funding: This work was supported by the National Natural Science Foundation of China (51573201, 51501209, 6165011017 and 201675165), NSFC-Zhejiang Joint Fund for the Integration of Industrialization and Informatization (U1709205), Scientific Instrument Developing Project of the Chinese Academy of Sciences (YZ201640), the Project of the Chinese Academy of Sciences (KFZD-SW-409), Science and Technology Major Project of Ningbo (2016S1002 and 2016B10038), Natural Science Foundation of Ningbo (2014A610154 and 2017A610095) and International S&T Cooperation Program of Ningbo (2017D10016) for financial support. We also thank the Chinese Academy of Sciences for Hundred Talents Program, Chinese Central Government for Thousand Young Talents Program, 3315 Program of Ningbo, and the Key Technology of Nuclear Energy (CAS Interdisciplinary Innovation Team, 2014).

Conflicts of Interest: The authors declare no conflict of interest.

References

1. Liu, T.; Abrahams, I.; Dennis, T.J.S. Structural Identification of 19 Purified Isomers of the OPV Acceptor Material bisPCBM by 13C NMR and UV–Vis Absorption Spectroscopy and High-Performance Liquid Chromatography. *J. Phys. Chem. A* **2018**, *122*, 4138–4152. [CrossRef] [PubMed]

2. Malik, P.; Bhushan, R. Development of liquid chromatographic methods for enantioseparation and sensitive detection of β-adrenolytics/β₂-agonists in human plasma using a single enantiomer reagent. *J. Chromatogr. B* **2017**, *1061*, 117–122. [CrossRef] [PubMed]

3. Ou, S.-H.; Pan, L.-S.; Jow, J.-J.; Chen, H.-R.; Ling, T.-R. Molecularly imprinted electrochemical sensor, formed on Ag screen-printed electrodes, for the enantioselective recognition of d and l phenylalanine. *Biosens. Bioelectron.* **2018**, *105*, 143–150. [CrossRef] [PubMed]

4. Canfarotta, F.; Rapini, R.; Piletsky, S. Recent advances in electrochemical sensors based on chiral and nano-sized imprinted polymers. *Curr. Opin. Electrochem.* **2018**, *7*, 146–152. [CrossRef]

5. Batalla, P.; Martín, A.; López, M.A.N.; González, M.A.C.; Escarpa, A. Enzyme-based microfluidic chip coupled to graphene electrodes for the detection of D-amino acid enantiomer-biomarkers. *Anal. Chem.* **2015**, *87*, 5074–5078. [CrossRef] [PubMed]

6. Upadhyay, S.S.; Kalambate, P.K.; Srivastava, A.K. Enantioselective analysis of Moxifloxacin hydrochloride enantiomers with graphene-β-Cyclodextrin-nanocomposite modified carbon paste electrode using adsorptive stripping differential pulse Voltammetry. *Electrochim. Acta* **2017**, *248*, 258–269. [CrossRef]

7. Dong, S.; Bi, Q.; Qiao, C.; Sun, Y.; Zhang, X.; Lu, X.; Zhao, L. Electrochemical sensor for discrimination tyrosine enantiomers using graphene quantum dots and β-cyclodextrins composites. *Talanta* **2017**, *173*, 94–100. [CrossRef] [PubMed]

8. Muñoz, J.; González-Campo, A.; Riba-Moliner, M.; Baeza, M.; Mas-Torrent, M. Chiral magnetic-nanobiofluids for rapid electrochemical screening of enantiomers at a magneto nanocomposite graphene-paste electrode. *Biosens. Bioelectron.* **2018**, *105*, 95–102. [CrossRef] [PubMed]

9. Doepel, L.; Hewage, I.; Lapierre, H. Milk protein yield and mammary metabolism are affected by phenylalanine deficiency but not by threonine or tryptophan deficiency. *J. Dairy Sci.* **2016**, *99*, 3144–3156. [CrossRef] [PubMed]

10. Singh, V.; Rai, R.K.; Arora, A.; Sinha, N.; Thakur, A.K. Therapeutic implication of L-phenylalanine aggregation mechanism and its modulation by D-phenylalanine in phenylketonuria. *Sci. Rep.* **2014**, *4*, 3875. [CrossRef] [PubMed]

11. Okamoto, H.; Massalski, T. The Au-C (gold-carbon) system. *J. Phase Equilib.* **1984**, *5*, 378–379. [CrossRef]

12. Yuan, Q.; Liu, Y.; Ye, C.; Sun, H.; Dai, D.; Wei, Q.; Lai, G.; Wu, T.; Yu, A.; Fu, L.; et al. Highly stable and regenerative graphene–diamond hybrid electrochemical biosensor for fouling target dopamine detection. *Biosens. Bioelectron.* **2018**, *111*, 117–123. [CrossRef] [PubMed]
13. Wei, Y.; Li, H.; Hao, H.; Chen, Y.; Dong, C.; Wang, G. β-Cyclodextrin functionalized Mn-doped ZnS quantum dots for the chiral sensing of tryptophan enantiomers. *Polym. Chem.* **2015**, *6*, 591–598. [CrossRef]
14. Freeman, R.; Finder, T.; Bahshi, L.; Willner, I. β-cyclodextrin-modified CdSe/ZnS quantum dots for sensing and chiroselective analysis. *Nano Lett.* **2009**, *9*, 2073–2076. [CrossRef] [PubMed]
15. Zaidi, S.A. Facile and efficient electrochemical enantiomer recognition of phenylalanine using β-Cyclodextrin immobilized on reduced graphene oxide. *Biosens. Bioelectron.* **2017**, *94*, 714–718. [CrossRef] [PubMed]

 *nanomaterials*

Article

Effect of Defects on the Mechanical and Thermal Properties of Graphene

Maoyuan Li [1], Tianzhengxiong Deng [1], Bing Zheng [1], Yun Zhang [1,*], Yonggui Liao [2] and Huamin Zhou [1]

[1]	State Key Laboratory of Material Processing and Die & Mold Technology, Huazhong University of Science and Technology, Wuhan 430074, China; limaoyuan@hust.edu.cn (M.L.); uezuuezu@hust.edu.cn (T.D.); zhengbing@hust.edu.cn (B.Z.); hmzhou@hust.edu.cn (H.Z.)

[2]	Key Laboratory of Material Chemistry for Energy Conversion and Storage, School of Chemistry and Chemical Engineering, Huazhong University of Science and Technology, Ministry of Education, Wuhan 430074, China; ygliao@mail.hust.edu.cn

*	Correspondence: marblezy@hust.edu.cn; Tel.: +86-27-87543492

Received: 26 January 2019; Accepted: 21 February 2019; Published: 3 March 2019

Abstract: In this study, the mechanical and thermal properties of graphene were systematically investigated using molecular dynamic simulations. The effects of temperature, strain rate and defect on the mechanical properties, including Young's modulus, fracture strength and fracture strain, were studied. The results indicate that the Young's modulus, fracture strength and fracture strain of graphene decreased with the increase of temperature, while the fracture strength of graphene along the zigzag direction was more sensitive to the strain rate than that along armchair direction by calculating the strain rate sensitive index. The mechanical properties were significantly reduced with the existence of defect, which was due to more cracks and local stress concentration points. Besides, the thermal conductivity of graphene followed a power law of $\lambda \sim L^{0.28}$, and decreased monotonously with the increase of defect concentration. Compared with the pristine graphene, the thermal conductivity of defective graphene showed a low temperature-dependent behavior since the phonon scattering caused by defect dominated the thermal properties. In addition, the corresponding underlying mechanisms were analyzed by the stress distribution, fracture structure during the deformation and phonon vibration power spectrum.

Keywords: mechanical properties; thermal properties; graphene; defect; molecular dynamic

---

1. Introduction

Due to its excellent mechanical, thermal and other physical properties, Graphene (Gr) has attracted great attention from researchers since it was first prepared by Geim and Novoselov [1] in 2004. Experimental studies have proven that Gr presents superior thermal conductivity (TC) of ~4840–5300 W/mK [2], which is far more than other common thermal management materials, such as copper (~400 W/mK), silver (~429 W/mK), etc. [3]. By nanoindentation in an atomic force microscopy, the Young's modulus and fracture strength of Gr are reported as 1.0 ± 0.1 TPa and 123.5 GPa [4], respectively. The outstanding physics properties are mainly due to the two-dimensional structure consisting of sp^2 bonded carbon atoms.

However, Gr fabricated by various experimental methods is not perfect and different types of defects are unavoidable. The common types of defects include single vacancy (SV) [5,6], double vacancy (DV) [5,6], Stone–Wales (SW) [7], grain boundaries [8], carbon adatoms [9], etc. The pristine structure would be destroyed by the existence of defects, which have a significant impact on the mechanical and thermal properties of Gr. For instance, by using Raman spectral, Zandiatashbar et al. [10] found that the Young's modulus of Gr was maintained and fracture strength decreased by only ~14% even at a high concentration of sp^3-type defects; however, it decreased significantly with the existence of vacancy defect. Using the molecular dynamic (MD) simulations, Mortazavi et al. [11] indicated that the TC of Gr decreased ~50% at a defect concentration of ~0.25%, and the Young's modulus, fracture strength and fracture strain decreased with the increase of defect concentration. Zhao et al. [12] found that the oxygen plasma treatment could reduce the TC of Gr significantly at a low defect concentration (~83% reduction for ~0.1% defect concentration) through MD simulations and non-contact optothermal Raman measurement. Jing et al. [13] indicated that the vacancy defect could decrease the Young's modulus while the reconstruction of vacancy could stabilize the modulus. Although there are some pioneering reports on the mechanical and thermal properties of Gr, as described above, there is still a lack of comprehensive study about some important factors, e.g., how the type and concentration of defects are related to the mechanical and thermal properties, especially at different temperatures. The reduction or enhancement mechanisms for the TC of defective Gr at different temperature were also not well understood. Meanwhile, the Gr is commonly used as nanofiller to enhance the mechanical and thermal properties of polymer materials, e.g., Gr/epoxy nanocomposites [14,15], etc., thus it is of great significance to fully understand the mechanical and thermal properties of Gr and effects of various defects.

In this study, we conducted a series of MD simulations to investigate the mechanical and thermal properties of Gr. The effects of temperature and strain rate on the Young's modulus, fracture strength and fracture strain were first studied. Three typical defects, SV, DV and SW, were considered in detail. Besides, the influences of temperature, system size and defects on the TC of Gr were investigated by non-equilibrium molecular dynamic (NEMD) simulations. By calculating the phonon vibration power spectrum and phonon scattering, the related mechanisms of TC of Gr with/without defect at different temperature were clarified.

2. Computational Methods

2.1. Molecular Model of Gr

MD simulations were conducted to evaluate the mechanical and thermal properties of Gr, and the molecular models were first constructed. The molecular model of Gr for the uniaxial tensile test consisted of 960 carbon atoms with the dimension around 50 Å × 50 Å, which has been proven to successfully calculate the mechanical properties of Gr [16], as shown in Figure 1a. Figure 1b shows the molecular model for calculating the TC of Gr with a size of 59 Å × 200 Å containing 4512 carbon atoms.

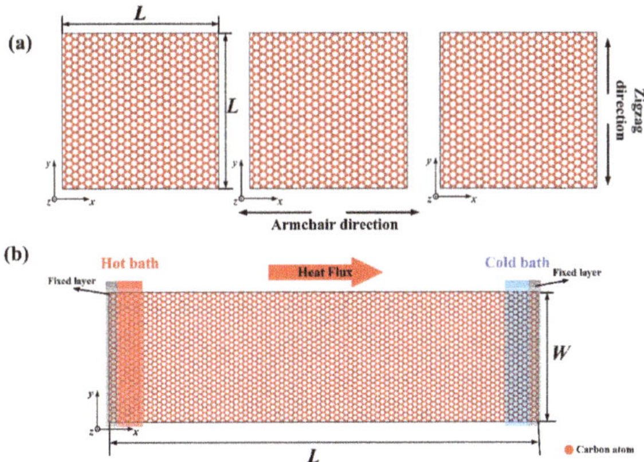

Figure 1. Molecular model of: (**a**) the armchair and zigzag Gr for uniaxial tensile test; and (**b**) Gr for the calculating TC.

To investigate the effect of different defects, Gr with three common types of defects, namely SV, DV and SW, was constructed (as shown in Figure 2). The SV and DV were created by removing one carbon atom or two adjacent carbon atoms from the pristine Gr, respectively. The SW was created by rotating one of the C–C bonds by 90°. Due to the different sizes of the models, the defect concentration varied from 0% to 3.125% for calculating mechanical properties, and from 0% to 0.24% in the study of TC. The defect concentrations of SV and DV were defined as the number density of atoms removed from the pristine Gr. The concentration of SW was defined by considering two defective atoms for each defect. The three kinds of defects were randomly distributed on the Gr sheets, respectively.

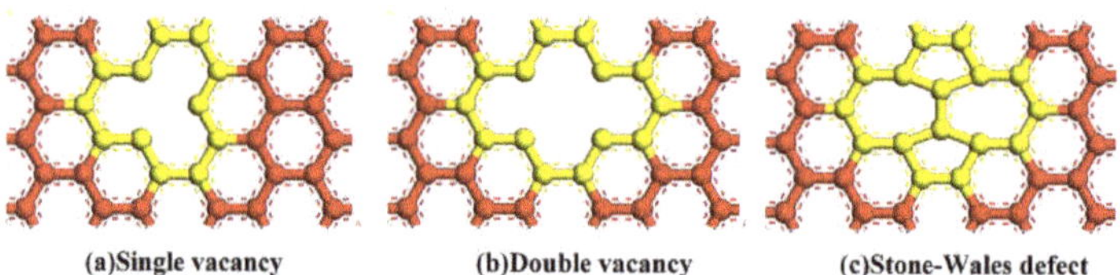

(a)Single vacancy   **(b)Double vacancy**   **(c)Stone-Wales defect**

Figure 2. Types of defect studied in this work: (**a**) Single vacancy (SV); (**b**) Double vacancy (DV); and (**c**) Stones–Wales (SW).

All MD simulations were conducted by The Large-scale Atomic/Molecular Massively Parallel Simulator (LAMMPS) [17], and the velocity-Verlet method was used to integrate the equations of motion. The adaptive intermolecular reactive bond order (AIREBO) [18] potential was used to simulate the C–C interaction, as it has successfully investigated the thermal and mechanical properties of carbon-based system, e.g., Gr [16], graphene [15], etc. A time step of 1 fs was used in the whole stages.

2.2. Calculation of Mechanical and Thermal Properties

After the initial models were established, the mechanical properties were calculated by uniaxial tensile test. As shown in Figure 1a, the chirality of Gr was considered, i.e., the armchair (*x*) and zigzag (*y*) directions. To eliminate the boundary effect, the periodic boundary conditions were applied along the x and y directions. The system was first relaxed to reach an equilibrium state; the relaxation involved two steps. At the beginning, an energy minimization was performed using the conjugate gradient algorithm. The system was then relaxed in a canonical NVT ensemble (i.e., constant number of atoms, volume and temperature) at temperature T = 300 K for 10^6 timesteps followed by a microcanonical NPT ensemble along x/y directions (i.e., constant number of atoms, pressure and energy) for another 10^6 timesteps. Followed by the equilibration, a constant uniaxial strain was applied along the x- or y-direction with a strain rate of 5×10^{-3} ps^{-1}. The atomic stress of the Gr during the uniaxial tension were calculated by the viral theorem using Equation (1) [19]:

$$\sigma_i^{\alpha\beta} = \frac{1}{\Omega_i}\left\{ -m_i v_i^{\alpha} v_i^{\beta} + \frac{1}{2}\sum_{j\neq i} F_{ij}^{\alpha} r_{ij}^{\beta} \right\} \tag{1}$$

where Ω_i, m_i and v_i represent the volume, mass and velocity of atom *i*, respectively; F_{ij} and r_{ij} are the force and distance between atom *i* and *j*, respectively; and indices α and β denote the Cartesian coordinate components. The thickness of Gr was determined by van der Waals interaction between the single layers, i.e., 3.35 Å. Then, the mechanical properties including Young's modulus, fracture strength and fracture strain could be obtained from the stress–strain curves.

In this study, the direct non-equilibrium molecular dynamic (NEMD) [20] was applied to calculate the TC. In the NEMD method, the atoms near the left/right end were treated as the heat/cold baths, the temperature of which was set to $T_H = T_0 (1 + \Delta)$ and $T_C = T_0 (1 - \Delta)$ by Langevin thermostat, respectively (as shown in Figure 1b). T_0 is the average temperature and Δ is the normalized temperature difference. To evaluate the effect of temperature, T_0 varied from 300 K to 900 K, while Δ was fixed at 0.03. During the NEMD simulations, the energies removed from the cold bath and added to the hot bath as a function of time were calculated. The sum of added/removed energy is equal to zero, thus the total energy is conserved. The heat flux along the x-direction J_x can be expressed by:

$$J_x = \frac{dE/dt}{A} \tag{2}$$

where E is the accumulated energy, t is the simulation time in NVE ensemble and A is the cross-section area obtained by the width multiplied by thickness. Once the steady-state temperature profile along the heat flux was reached, the TC could be calculated by Fourier law:

$$\lambda = \frac{J_x}{\frac{dT}{dx}} \tag{3}$$

where λ and dT/dx are the TC and temperature gradient along the x direction, respectively. After the temperature profile was stable, another 10 ns of NEMD simulations were applied. The final TC was the average value of the last ten different time blocks (every 10^6 timesteps), the error bars were determined by the standard deviation.

3. Results and Discussion

3.1. Validation of Models

To evaluate the molecular model and the AIREBO potential, the mechanical and thermal properties of pristine Gr were calculated. Figure 3 shows the stress–strain curves and the total energy variations along the armchair and zigzag directions. According to the stress–strain curves, the Young's modulus could be calculated by linear fitting the curves when the strain <2%; the value of fracture strength was defined as the maximum stress while the corresponding strain was the fracture strain. The calculated results and the relevant values obtained by previous simulations or experiments are displayed in Table 1. It can be seen in Table 1 that the calculated results, including Young's modulus (i.e., 961 GPa and 911 GPa for Gr along the armchair and zigzag direction, respectively), fracture strength (i.e., 93 GPa and 106 GPa for Gr along the armchair and zigzag directions, respectively) and fracture strain (0.14 and 0.20 for Gr along the armchair and zigzag directions, respectively) were in agreement with the experimental values [4] (i.e., ~1000 GPa, 130 ± 10 GPa, and 0.25). The simulation results were also consistent with previous studies [16,21,22], specific differences being determined by the selection of simulation conditions such as potentials, system size, etc.

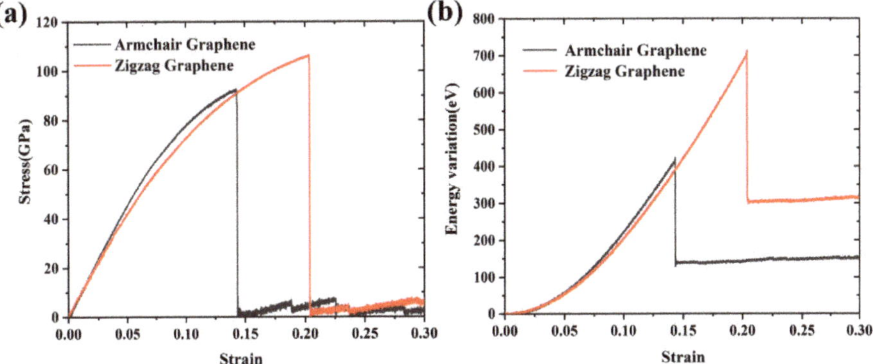

Figure 3. (**a**) Stress–strain curves for Gr along the armchair and zigzag directions; and (**b**) the total energy variations during the loading process.

Table 1. The calculated results and relevant values obtained by previous simulations or experiment.

References	Method	Direction	Young's Modulus (GPa)	Fracture Strength (GPa)	Fracture Strain
Lee (2008) [4]	Experiment	/	1000	130 ± 10	0.25
Liu (2007) [23]	DFT	Armchair	1050	110	0.19
		Zigzag	1050	121	0.26
Q.X. Pei (2010) [16]	MD (AIREBO)	Armchair	890	105	0.17
		Zigzag	830	137	0.27
Ansari (2012) [21]	MD (Tersoff)	Armchair	790	123	0.23
		Zigzag	807	127	0.22
This paper	MD (AIREBO)	Armchair	961	93	0.14
		Zigzag	911	106	0.20

The Young's modulus along the armchair direction was less than that along the zigzag direction, while the fracture strength and strain were greater than those along the zigzag direction, which indicated a typical anisotropic behavior. The fracture process and the distribution of von Mises along the armchair and zigzag directions are shown in Figure 4. The von Mises was calculated as:

$$\sigma_m = \sqrt{\frac{1}{2}\left[(\sigma_{11} - \sigma_{22})^2 + (\sigma_{11} - \sigma_{33})^2 + (\sigma_{22} - \sigma_{33})^2 + 6(\sigma^2{}_{11} + \sigma^2{}_{22} + \sigma^2{}_{33})\right]} \tag{4}$$

where σ represents stress, and subscripts 1–3 represent the coordinate directions (i.e., x, y and z directions). The Gr was subjected to simple elongation deformation in the elastic deformation stage, and the carbon rings remained hexagon. As the load increased, the carbon ring at the boundary began to be irregular. When the load exceeded a certain value, Gr began to crack. For the armchair direction, the break of the C–C bonding occurred at the both sides of boundary and propagated inward rapidly. For the zigzag direction, the crack propagated rapidly and showed a remarkable zigzag shape. Comparing the fracture process, it was found that the fracture along the armchair direction belonged to Mode I (i.e., the tensile stress was perpendicular to the crack), while the fracture along the zigzag direction belonged to a mixture of Modes I and II (i.e., the angle between tensile stress and crack was ~60°). Based on the force analysis, the force parallel to the loading direction F_A was greater than that with an angle of 60° to the loading direction F_Z, thus Mode I ruptured first, leading a greater fracture strength (106 GPa) and strain (0.20) along the zigzag direction than those (93 GPa and 0.14, respectively) along the armchair direction.

The TC of pristine Gr was also calculated. Figure 5 shows the steady-state temperature profile along the heat flux direction and the calculated energies added to the hot bath and removed from the cold bath according to the time. Based on Equations (2) and (3), the TC value of pristine Gr was 181.97 ± 0.007 $\mathrm{Wm^{-1}\,K^{-1}}$ at the temperature of 300 K. The measured value of TC with respect to time during the steady-state simulations is provided in the Supplementary Materials. The TC and relevant values obtained by simulations or experiments are listed in Table 2. The results show that the TC of Gr was related with the calculation method, system size, potentials, etc. The present results were consistent with the previous simulation results [24,25] (with same potential and size). Therefore, the above results confirmed the validity of the AIREBO potential, molecular models and calculation method.

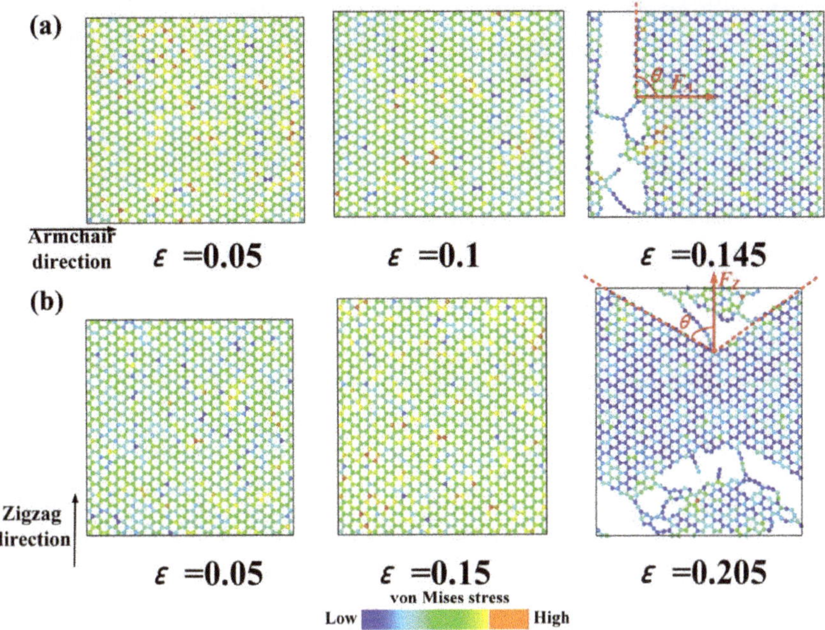

Figure 4. The fracture process and distributions of von Mises along: (**a**) the armchair direction and; (**b**) the zigzag direction.

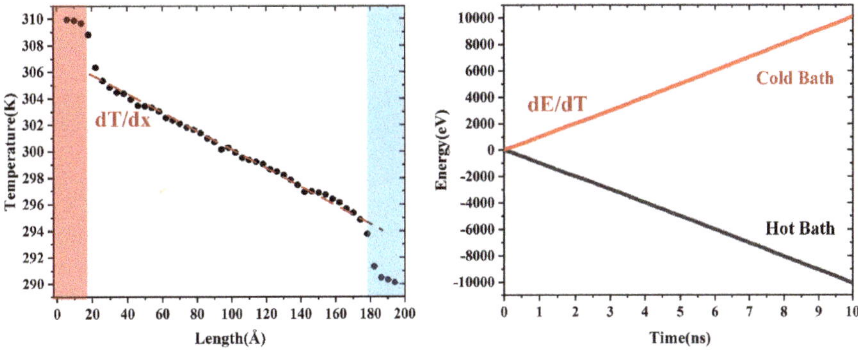

Figure 5. (**a**) Steady-state temperature profile of Gr without defect obtained using the NEMD simulations at 300 K. The color bars highlight the hot/cold bath in simulation. (**b**) Energies added to the hot bath and removed from the cold bath with respect to the time.

Table 2. The TC and relevant values obtained by simulations or experiment.

References	Potentials	Method	Size	TC at 300 K (Wm^{-1} K^{-1})
Balandin (2008) [2]	/	Experiment	~0.5–1 um	~4840–5300
Wei (2011) [25]	AIREBO	RNEMD	102×102Å^2	77.3
Yang (2012) [26]	AIREBO	EMD	$(90{\sim}270) \times (40{\sim}180)$ Å^2	~3200–5200
Xu (2014) [27]	Tersoff	NEMD	$50 \times (2 \times 150)$ Å^2	~400–1800
Zhang (2012) [24]	AIREBO	RNEMD	61×200 Å^2	~170
This paper	AIREBO	NEMD	60×200 Å^2	182

3.2. Effects of Temperature and Strain Rate on the Mechanical Properties

The temperature/strain rate-dependent effects are of great importance for mechanical properties of low-dimension materials, thus the effects of temperature and strain rate were investigated. The temperature varied from 300 K to 1000 K, and the stress–strain curves were along different directions, as shown in Figure 6. The curves indicated that the Gr showed a similar deformation behavior, i.e., a brittle fracture behavior, indicating the stable molecular structure of Gr.

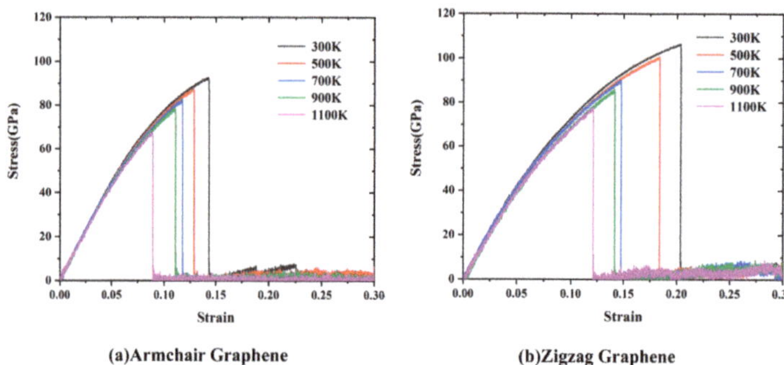

Figure 6. The stress–strain curves of Gr at different temperature along: (**a**) armchair; and (**b**) zigzag.

As shown in Figure 7a, the Young's modulus, fracture strength and strain of Gr along different directions decreased with the increase of temperature. When the temperature increased from 300 K to 1100 K, the Young's modulus along the armchair direction decreased ~6.8% (from 961.61 GPa to 896.56 GPa), fracture strength decreased ~27% (from 92.67 GPa to 67.68 GPa) and fracture strain decreased ~37.6% (from 0.14 to 0.089), while it decreased ~8.5% (from 911.08 GPa to 834 GPa), ~26.7% (from 106 GPa to 77.73 GPa) and ~40.49% (from 0.204 to 0.121) along the zigzag direction, respectively. The temperature-dependent mechanical properties along the different directions was similar. Compared with the fracture strength and strain, the Young's modulus was less sensitive to temperature. The total kinetic energy increased with the increase of temperature; the thermal vibration of carbon atoms was more vigorous, leading a larger vibration amplitude around the equilibrium position. Under the action of external forces, the atoms were more likely to be away from their original equilibrium position, resulting in a softer and less rigid structure. Meanwhile, at the high temperature, in addition to the formation of cracks at the boundary, the cracks also occurred from inside, resulting in more defect cracks. Therefore, the fracture strength and strain were significantly reduced. Tang et al. [28] explained the mechanism from the perspective of energy: the deformation process of Gr was determined by the strain energy and thermal energy; the thermal energy increased with the temperature, thus the strain energy required for the fracture reduced.

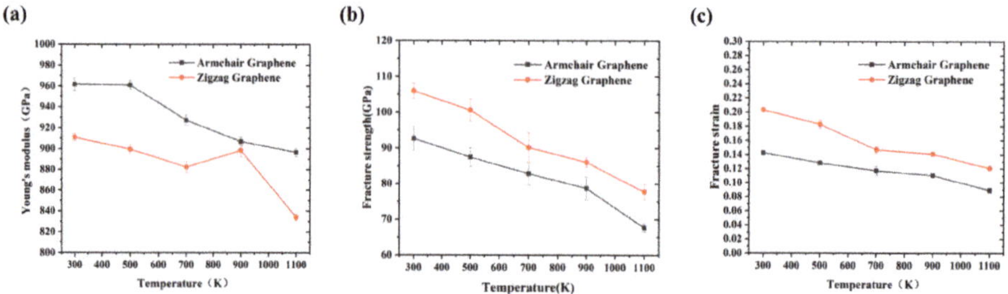

Figure 7. (**a**) The Young's modulus; (**b**) the fracture strength; and (**c**) the fracture strain of Gr along the armchair and zigzag directions at different temperatures.

As shown in Figure 8, at different strain rate (0.5×10^{-5}–3×10^{-3} ps^{-1}), the Young's modulus was insensitive to the strain rate, while the fracture strength and strain slightly increased at higher strain rate. The lower strain rate means a longer response time for Gr, which would increase the number of atoms that could overcome the energy barrier required for fracture, leading a lower fracture strength and strain. However, the effect of strain rate was less significant than that of temperature. The relation between fracture strength and strain rate can be described by Arrhenius equation [29]:

$$\dot{\varepsilon} = A\sigma^{\frac{1}{m}} \exp(-\frac{Q}{RT}) \tag{5}$$

where $\dot{\varepsilon}$, σ, Q, R, T and m represent the strain rate, fracture strength, the activation energy, universal gas consent, temperature and the strain-rate sensitivity, respectively, and A is a constant. By taking the natural logarithm of both sides of Equation (6):

$$\ln(\dot{\varepsilon}) = \ln(A) + \frac{1}{m}\ln(\sigma) - \frac{Q}{RT} \tag{6}$$

At the constant temperature, the partial differentiation of Equation (7):

$$m = \frac{\ln(\sigma)}{\ln(\dot{\varepsilon})} \tag{7}$$

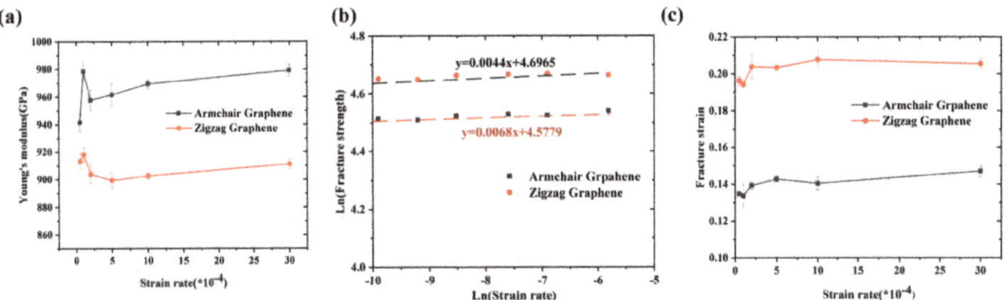

Figure 8. (a) The Young's modulus; (b) the fracture strength; and (c) the fracture strain of Gr along the armchair and zigzag directions at different strain rates.

The strain-rate sensitivity m can be obtained from the slope of the ln (σ) versus ln ($\dot{\varepsilon}$). The results show that the m along the armchair and zigzag directions were 0.0044 and 0.0068, respectively. Therefore, the fracture strength of Gr along the zigzag direction was more sensitive than to strain rate that along the armchair direction.

3.3. Effect of Defects on the Mechanical Properties

In this section, the effect of defects was investigated. As described above, three types of defects, i.e., SV, DV and SW, were considered. As shown in Figure 9, the mechanical properties of Gr decreased remarkably with the increase of defect concentrations, and it showed a similar behavior along different direction. For simplicity, only the mechanical properties of Gr along the zigzag are discussed. At the same defect concentration, Gr was more sensitive to the SV than DV. For example, when the concentration was 1.67%, the Young's modulus of Gr containing SV decreased ~47.6% (from 106 GPa to 55.50 GPa), while it decreased ~39.1% (from 106 GPa to 64.53 GPa) for the Gr containing DV. The main reason was the fact that the SV produced more dangling bonding than DV, resulting in more cracks. The literature [13] confirms that, at the same missing carbon atoms, the mechanical properties of Gr are mainly determined by the number of dangling bonding. Zandiatashbar et al. [10]

also quantitatively investigated the effect of defects on the mechanical properties using the Raman spectroscopy; their results show that the Young's modulus and strength of Gr is insensitive to sp^3-type defect, while decreases significantly with the increase of vacancy defect. Meanwhile, further increase of defect concentration has little effect on the fracture strength and strain.

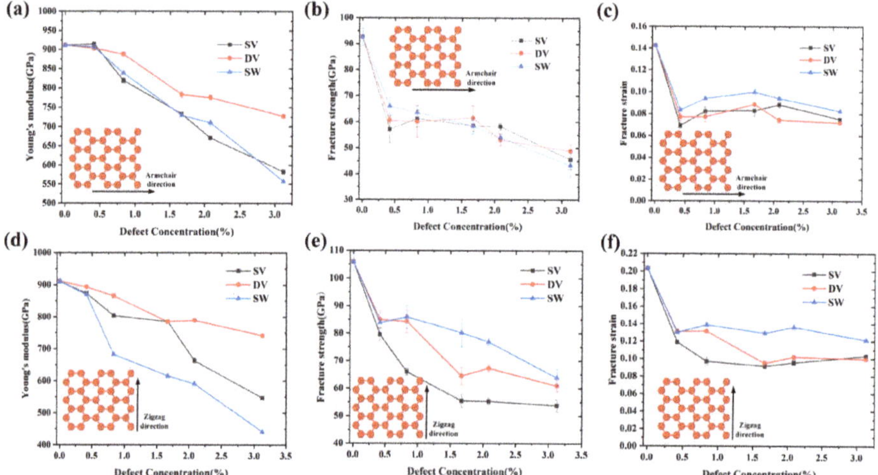

Figure 9. Under different defect types and concentrations, the variations of: (**a**) Young' modulus; (**b**) fracture strength; and (**c**) fracture strain of Gr along the armchair direction. The variations of: (**d**) Young' modulus; (**e**) fracture strength; and (**f**) fracture strain of Gr along the zigzag direction with respect to defect concentration at different defect types.

Figure 10 shows the fracture process of Gr containing different defects. Compared with the fracture process of pristine Gr, the overall trend was similar for the defective Gr. However, the initial crack did not occur near the boundary, but rather formed inside and gradually spread until the fracture. Before a crack occurs, the existence of defects will cause the local stress concentration point, leading to a lower fracture strength. Due to the limited size (50 Å × 50 Å) and high temperature (>300 K) in present study, the defect concentration was limited since the structure with high defect concentration was unstable at high temperature. When the defect concentration was larger, the interaction between defects, including the merging of defect and crack blocking, could result in more complicated fracture behavior. For example, Xu et al. found that when the concentration exceeded ~7%, sp-sp^2 and sp^2-sp^3 network structures were formed, and fracture behavior for Gr changed from brittle to ductile.

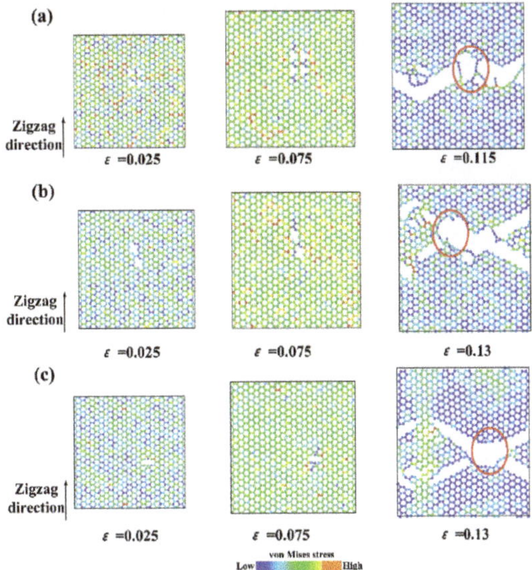

Figure 10. The fracture process of Gr containing: (**a**) SV; (**b**) DV; and (**c**) SW. The red ring indicates the location of defect.

3.4. Effects of Temperature and System Size on the Thermal Properties

The thermal properties of Gr and other low-dimension materials showed significant relationship with temperature and system size. As shown in Figure 11, the TCs of Gr with lengths of 20, 40, 80, 120 and 160 nm were calculated. The results indicate that the TC of Gr increased monotonically with increase of length. By fitting the data, the TC was found to follow the power law of $\lambda \sim L^{0.28}$, which was similar to the previous results of $\lambda \sim L^{0.35}$ obtained by Guo et al. [30]. Such phenomenon was also consistent with the previous studies about the carbon nanotube [31], $\lambda \sim L^{\beta}$ and $\beta \sim 0.3$–0.4. The size-dependent behavior was due to the long mean free path of phonons (MFP) of Gr, ~775 nm [32], which was much longer than that used in MD. Therefore, apart from the phonon–phonon scattering, the phonon scattering existed at the boundary of Gr. With the increase of length, more phonons will be excited and contribute to the increase of TC. We also compared the TCs calculated at different boundary conditions along the width direction (y), as shown in Figure 11a. The fixed boundary condition means that the particle could not interact across the boundary and move from one side of the box to the other, i.e., the length of Gr was finite in the width direction. This clearly indicated that the TCs obtained at the fixed boundary condition were smaller than those obtained at the periodic boundary condition. For instance, when the length was 80 nm, the TCs at the periodic and fixed boundary condition were 273.1 and 168.6 W/mK, respectively. At the fixed boundary condition, the power law between the TCs and length, $\lambda \sim L^{0.18}$, the index of power law was smaller than that at the periodic boundary condition. This was because, at the periodic boundary condition, the phonons can across the regions perpendicular to the direction of the heat flow without boundary scattering, thus the width of Gr had negligible effect on the calculation results. In addition, the vibration density of states (VDOS) of Gr at the two different boundary conditions were calculated. The VDOS can be obtained by calculating the Fourier transformation of atomic velocities autocorrelation function at the equilibrium state:

$$D(\omega) = \int_0^{\tau} \Gamma(t) \cos(\omega t)\, dt \tag{8}$$

where ω is frequency, $D(\omega)$ is vibration density of state at frequency ω, τ is the total time, and $\Gamma(t)$ is the velocity autocorrelation function of atoms, which is given by:

$$\Gamma(t) = \langle v(t) \cdot v(0) \rangle \tag{9}$$

where $v(t)$ is the atom velocity at time t, and $<\cdots>$ denotes time and atom number-averaged velocity autocorrelation function. In this study, the velocity was correlated every 5 fs with a total integration time $\tau = 25$ ps. Compared with the periodic boundary condition, the VDOS of Gr was suppressed at the fixed boundary condition, especially at the high frequency (~50 THZ), leading to a lower TC.

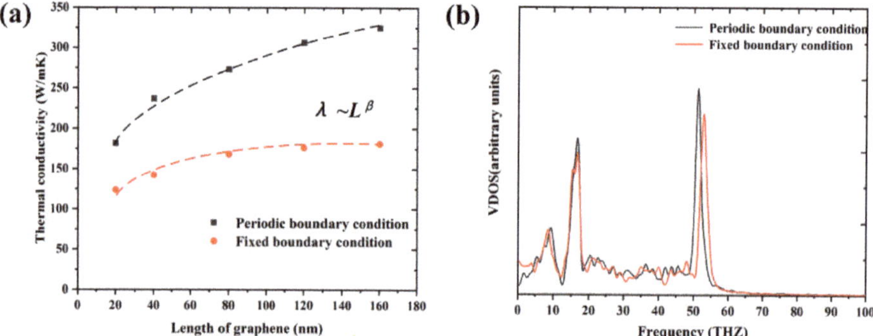

Figure 11. (a) The variation of TCs of pristine Gr with respect to length; and (b) the VDOS of Gr under periodic/fixed boundary conditions.

The TCs of Gr at different temperature are shown in Figure 12a. The TCs of Gr decreased with increase of temperature, consistent with previous results obtained by Hu et al. [33]. Notably, Seol et al. [34] found that, at the low temperature (<300 K), the lattice vibration increased with the increase of temperature, leading to a longer MFP and higher TC. As the temperature continued to rise, the TC decreased. We also calculated the energies added to the hot bath and removed from the cold bath at different temperatures and the accumulated energy with respect to simulation times, i.e., the corresponding heat flux decreased with the increase of temperature. At high temperature, the phonon motion became more vigorous and the interaction and collision between phonons increased, resulting in stronger inelastic phonon scattering and lower TC.

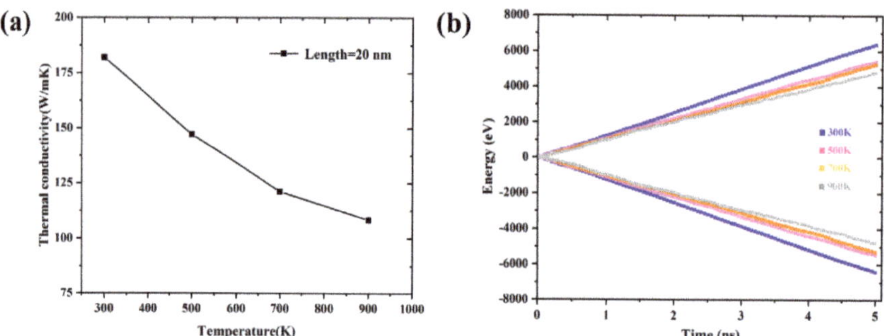

Figure 12. (a) The TCs of Gr; and (b) the energies added to the hot bath and removed from the cold bath during the NEMD at different temperature.

3.5. Effect of Defects on the Thermal Properties

The effect of defects, including SV, DV and SW, on the TC of Gr were investigated. As shown in Figure 13, the TCs of Gr decreased with the increase of defect concentration. When the defect concentration varied from 0% to 0.24%, the TCs of Gr containing SV, DV and SW varied from 181.97 to 77.21 W/mK (~57.6%), 102.97 W/mK (~43.4%) and 123.86 W/mK (~31.9%), respectively. This agreed with the previous simulation results of TCs of Gr obtained by Mortazavi et al. [11] and that of carbon nanotube obtained by Sevik [35]. Zhao et al. [12] also found that the defect introduced by oxygen-plasma treatment could decrease significantly the TC of Gr (~83% reduction at a defect concentration of ~0.1%) using the non-contact optothermal Raman technique. The results indicate that the TC of Gr decreased even at a low defect concentration (~0.24%), which was mainly due to the scattering caused by defect. Based on the theory of classical TC, the TC can be obtained by:

$$\lambda = \frac{1}{3}CVl \tag{10}$$

where C, V and l denoted the specific heat capacity, group velocity of sound wave in solid and the MFP, respectively. The MFP of pristine Gr was determined by the phonon–phonon scattering. While the defects were introduced, the effective MFP was:

$$\frac{1}{l} = \frac{1}{l_{phonon-phonon}} + \frac{1}{l_{defect-phonon}} \tag{11}$$

where $l_{phonon\text{-}phonon}$ and $l_{defect\text{-}phonon}$ are the length of phonon–phonon scattering and scattering caused by defects, respectively. According to Equations (10) and (11), the TC of defective Gr can be obtained:

$$\frac{1}{\lambda} \propto \frac{1}{l} \propto \frac{1}{l_{phonon-phonon}} + \frac{1}{l_{defect-phonon}} \tag{12}$$

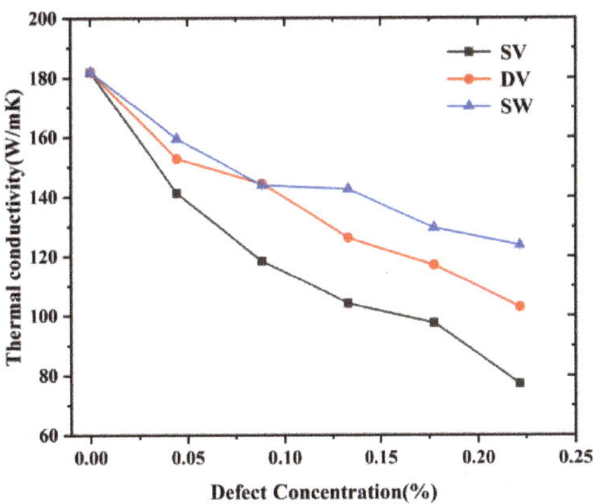

Figure 13. The TCs of Gr with different concentration at different types of defects.

According to Equation (12), the existence of defects could decrease the MFP of pristine Gr, resulting in a lower TC. Comparing different defect types, the TC of Gr containing SV was the smallest (a reduction of ~57.6%) at the same defect concentration. Such phenomenon can be explained as follows. At the same concentration, the Gr containing SV had more defect scatters than the Gr containing DV, resulting in more phonon defect scattering [24]. Meanwhile, a SV was formed by removing one atom leaving three carbon atoms two-coordinated, effectively breaking the sp^2 structure of the local lattice, while a DV was formed by removing two adjusted atoms as the local structure can rearrange to restore the three-coordinated sp^2 bonding by creating an octagon and two pentagon structures [35]. The two-coordinated atoms were less stable, leading to higher level of defect scattering. Previously, Haskin et al. [35] found that the Gr containing SV decreased ~80% at a concentration of 0.1%, while the Gr containing DV and SW decreased ~70%. Zhang et al. [24] also found that different types of defects had different effects on TC at a low concentration ($\leq$0.2%). When the concentration increased (>0.2%), they had similar effect on TC, which was mainly because the heat transport mechanism changed from propagating to diffusive, and the TC was insensitive to the defect type in diffusive form.

Next, we investigated the TCs of Gr with/without defect at different temperatures. The TCs of Gr containing different defects at different temperatures are shown in Figure 14a, at the concentration of 0.13%. It indicated that the TCs of Gr with/without decreased with the increase of temperature. For instance, when the temperature increased from 300 K to 900 K, the TCs of Gr containing SV, DV and SW varied from 104.12 to 72.82 W/mK (~30%), 126.28 W/mK to 83.26 W/mK (~34%) and 142.65 W/mK to 106.53 W/mK (~25%), respectively. The TCs were fitted by power law, $\lambda \sim T^{-\alpha}$; the corresponding index is shown in Figure 14b. Compared with the pristine Gr, the power law index of defective Gr decreased significantly, i.e., showing a weak temperature-dependent behavior. The main reason was that, compared with the phonon scattering caused by temperature, the scattering caused by defects was the dominant factor for the TC of defective Gr, which was also consistent with previous results obtained by Zhang [24] and Hu et al. [33].

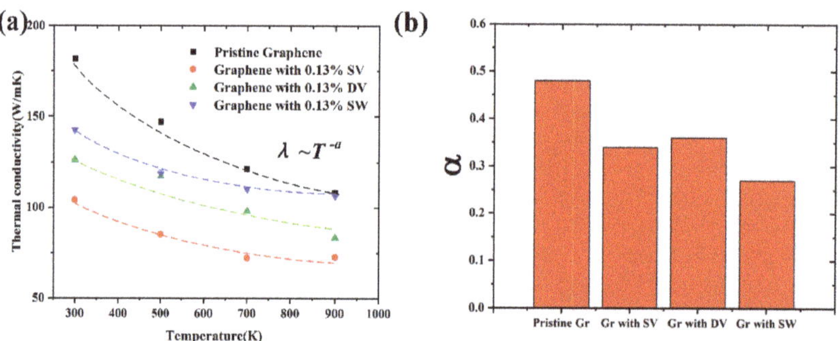

Figure 14. (**a**) The TCs of Gr with/without defect at different temperature; and (**b**) the corresponding power law index α.

4. Conclusions

In summary, we investigated the mechanical and thermal properties of Gr by conducting MD simulations. For the mechanical properties, the effects of temperature, strain rate and defects were studied. The results show that the mechanical properties, including Young's modulus, fracture strength and fracture strain, decreased with the increase of temperature. By calculating the strain rate sensitivity index, it was found that the mechanical properties of Gr along the zigzag direction were more sensitive to strain rate than those along the armchair direction. Then, the existence of defects, including SV, DV and SW, was found to significantly reduce the mechanical properties, which were more sensitive to SV than DV at the same concentration. Meanwhile, the local stress point caused by defects reduced the fracture strength.

For the thermal properties, the effects of temperature, system and defect were investigated. It was found that the TC followed a power law of $\lambda \sim L^{0.28}$: more phonons were excited with the increase of length, contributing to the TC. The results also indicate that the TC of Gr containing SV, DV and SW decreased ~57.6%, ~43.4% and ~31.9% even at the low concentration of 0.23%, respectively, which was mainly due to the reduced MFP caused by defect scattering. Besides, the TCs of Gr with/without defect at different temperature were calculated, showing that, compared with the phonon scattering caused by temperature, the phonon scattering caused by defect dominated the thermal properties.

The TCs of defective Gr showed a low temperature-dependent behavior. The above findings can provide a comprehensive understanding in the mechanical and thermal properties of Gr.

Supplementary Materials: The following are available online at http://www.mdpi.com/2079-4991/9/3/347/s1.

Author Contributions: Conceptualization, M.L.; methodology, M.L.; software, M.L. and T.D.; validation, M.L., T.D. and B.Z.; formal analysis, M.L.; investigation, M.L.; resources, Y.Z, Y.L. and H.Z.; data curation, Y.Z. and Z.H.; writing—original draft preparation, M.L. and T.D.; writing—review and editing, M.L.; visualization, M.L.; supervision, M.L., T.D. and B.Z.; project administration, M.L.; and funding acquisition, Y.Z., Y.L and H.Z.

Funding: The authors would like to acknowledge financial support from the National Key Research and Development Program of China (Grant No. 2018YFB1106700), National Natural Science Foundation of China (Grant Nos. 51675199, 51635006, and 51575207), and Fundamental Research Funds for the Central Universities (Grant Nos. 2016YXZD059 and 2015ZDTD028).

Acknowledgments: The authors thank the National Supercomputing Center of China in Shenzhen for providing computing resources.

Conflicts of Interest: The authors declare no conflict of interest.

References

1. Novoselov, K.S.; Geim, A.K.; Morozov, S.V.; Jiang, D.; Zhang, Y.; Dubonos, S.V.; Grigorieva, I.V.; Firsov, A.A. Electric Field Effect in Atomically Thin Carbon Films. *Science* **2004**, *306*, 666–669. [CrossRef] [PubMed]
2. Balandin, A.A.; Ghosh, S.; Bao, W.; Calizo, I.; Teweldebrhan, D.; Miao, F.; Lau, C.N. Superior Thermal Conductivity of Single-Layer Graphene. *Nano Lett.* **2008**, *8*, 902–907. [CrossRef] [PubMed]
3. Huang, X.; Jiang, P.; Tanaka, T. A review of dielectric polymer composites with high thermal conductivity. *IEEE Electr. Insul. Mag.* **2011**, *27*, 8–16. [CrossRef]
4. Lee, C.; Wei, X.; Kysar, J.W.; Hone, J. Measurement of the elastic properties and intrinsic strength of monolayer graphene. *Science* **2008**, *321*, 385–388. [CrossRef] [PubMed]
5. Meyer, J.C.; Kisielowski, C.; Erni, R.; Rossell, M.D.; Crommie, M.; Zettl, A. Direct imaging of lattice atoms and topological defects in graphene membranes. *Nano Lett.* **2008**, *8*, 3582–3586. [CrossRef] [PubMed]
6. Gass, M.H.; Bangert, U.; Bleloch, A.L.; Wang, P.; Nair, R.R.; Geim, A. Free-standing graphene at atomic resolution. *Nat. Nanotechnol.* **2008**, *3*, 676. [CrossRef] [PubMed]
7. Wei, Y.; Wu, J.; Yin, H.; Shi, X.; Yang, R.; Dresselhaus, M. The nature of strength enhancement and weakening by pentagon–heptagon defects in graphene. *Nat. Mater.* **2012**, *11*, 759–763. [CrossRef] [PubMed]
8. Yazyev, O.V.; Louie, S.G. Topological defects in graphene: Dislocations and grain boundaries. *Phys. Rev. B* **2010**, *81*, 195420. [CrossRef]
9. Cretu, O.; Krasheninnikov, A.V.; Rodríguez-Manzo, J.A.; Sun, L.; Nieminen, R.M.; Banhart, F. Migration and localization of metal atoms on strained graphene. *Phys. Rev. Lett.* **2010**, *105*, 196102. [CrossRef] [PubMed]
10. Zandiatashbar, A.; Lee, G.-H.; An, S.J.; Lee, S.; Mathew, N.; Terrones, M.; Hayashi, T.; Picu, C.R.; Hone, J.; Koratkar, N. Effect of defects on the intrinsic strength and stiffness of graphene. *Nat. Commun.* **2014**, *5*, 3186. [CrossRef] [PubMed]
11. Mortazavi, B.; Ahzi, S. Thermal conductivity and tensile response of defective graphene: A molecular dynamics study. *Carbon* **2013**, *63*, 460–470. [CrossRef]
12. Zhao, W.; Wang, Y.; Wu, Z.; Wang, W.; Bi, K.; Liang, Z.; Yang, J.; Chen, Y.; Xu, Z.; Ni, Z. Defect-Engineered Heat Transport in Graphene: A Route to High Efficient Thermal Rectification. *Sci. Rep.* **2015**, *5*, 11962. [CrossRef] [PubMed]
13. Jing, N.; Xue, Q.; Ling, C.; Shan, M.; Zhang, T.; Zhou, X.; Jiao, Z. Effect of defects on Young's modulus of graphene sheets: A molecular dynamics simulation. *RSC Adv.* **2012**, *2*, 9124–9129. [CrossRef]
14. Li, M.; Zhou, H.; Zhang, Y.; Liao, Y.; Zhou, H. Effect of defects on thermal conductivity of graphene/epoxy nanocomposites. *Carbon* **2018**, *130*, 295–303. [CrossRef]
15. Li, M.; Zhou, H.; Zhang, Y.; Liao, Y.; Zhou, H. The effect of defects on the interfacial mechanical properties of graphene/epoxy composites. *RSC Adv.* **2017**, *7*, 46101–46108. [CrossRef]
16. Pei, Q.X.; Zhang, Y.W.; Shenoy, V.B. A molecular dynamics study of the mechanical properties of hydrogen functionalized graphene. *Carbon* **2010**, *48*, 898–904. [CrossRef]
17. Plimpton, S. Fast Parallel Algorithms for Short-Range Molecular Dynamics. *J. Comput. Phys.* **1995**, *117*, 1–19. [CrossRef]

18. Donald, W.B.; Olga, A.S.; Judith, A.H.; Steven, J.S.; Boris, N.; Susan, B.S. A second-generation reactive empirical bond order (REBO) potential energy expression for hydrocarbons. *J. Phys. Condens. Matter* **2002**, *14*, 783.

19. Diao, J.; Gall, K.; Dunn, M.L. Atomistic simulation of the structure and elastic properties of gold nanowires. *J. Mech. Phys. Solids* **2004**, *52*, 1935–1962. [CrossRef]

20. Jund, P.; Jullien, R. Molecular-dynamics calculation of the thermal conductivity of vitreous silica. *Phys. Rev. B* **1999**, *59*, 13707. [CrossRef]

21. Ansari, R.; Ajori, S.; Motevalli, B. Mechanical properties of defective single-layered graphene sheets via molecular dynamics simulation. *Superlattices Microstruct.* **2012**, *51*, 274–289. [CrossRef]

22. Zhang, Y.Y.; Gu, Y.T. Mechanical properties of graphene: Effects of layer number, temperature and isotope. *Comput. Mater. Sci.* **2013**, *71*, 197–200. [CrossRef]

23. Liu, F.; Ming, P.; Li, J. Ab initio calculation of ideal strength and phonon instability of graphene under tension. *Phys. Rev. B* **2007**, *76*, 064120. [CrossRef]

24. Zhang, Y.Y.; Cheng, Y.; Pei, Q.X.; Wang, C.M.; Xiang, Y. Thermal conductivity of defective graphene. *Phys. Lett. A* **2012**, *376*, 3668–3672. [CrossRef]

25. Wei, N.; Xu, L.; Wang, H.-Q.; Zheng, J.-C. Strain engineering of thermal conductivity in graphene sheets and nanoribbons: A demonstration of magic flexibility. *Nanotechnology* **2011**, *22*, 105705. [CrossRef] [PubMed]

26. Yang, D.; Ma, F.; Sun, Y.; Hu, T.; Xu, K. Influence of typical defects on thermal conductivity of graphene nanoribbons: An equilibrium molecular dynamics simulation. *Appl. Surf. Sci.* **2012**, *258*, 9926–9931. [CrossRef]

27. Xu, X.; Pereira, L.F.; Wang, Y.; Wu, J.; Zhang, K.; Zhao, X.; Bae, S.; Bui, C.T.; Xie, R.; Thong, J.T. Length-dependent thermal conductivity in suspended single-layer graphene. *Nat. Commun.* **2014**, *5*, 3689. [CrossRef] [PubMed]

28. Tang, C.; Guo, W.; Chen, C. Molecular dynamics simulation of tensile elongation of carbon nanotubes: Temperature and size effects. *Phys. Rev. B* **2009**, *79*, 155436. [CrossRef]

29. Sellars, C.M.; McTegart, W. On the mechanism of hot deformation. *Acta Metall.* **1966**, *14*, 1136–1138. [CrossRef]

30. Guo, Z.; Zhang, D.; Gong, X.-G. Thermal conductivity of graphene nanoribbons. *Appl. Phys. Lett.* **2009**, *95*, 163103. [CrossRef]

31. Zhang, G.; Li, B. Thermal conductivity of nanotubes revisited: Effects of chirality, isotope impurity, tube length, and temperature. *J. Chem. Phys.* **2005**, *123*, 114714. [CrossRef] [PubMed]

32. Balandin, A.A. Thermal properties of graphene and nanostructured carbon materials. *Nat. Mater.* **2011**, *10*, 569–581. [CrossRef] [PubMed]

33. Hu, S.; Chen, J.; Yang, N.; Li, B. Thermal transport in graphene with defect and doping: Phonon modes analysis. *Carbon* **2017**, *116*, 139–144. [CrossRef]

34. Seol, J.H.; Jo, I.; Moore, A.L.; Lindsay, L.; Aitken, Z.H.; Pettes, M.T.; Li, X.; Yao, Z.; Huang, R.; Broido, D. Two-dimensional phonon transport in supported graphene. *Science* **2010**, *328*, 213–216. [CrossRef] [PubMed]

35. Haskins, J.; Kınacı, A.; Sevik, C.; Sevinçli, H.; Cuniberti, G.; Çağın, T. Control of Thermal and Electronic Transport in Defect-Engineered Graphene Nanoribbons. *ACS Nano* **2011**, *5*, 3779–3787. [CrossRef] [PubMed]

 *nanomaterials*

Article

Atomistic Study of Mechanical Behaviors of Carbon Honeycombs

Huaipeng Wang [1], Qiang Cao [1,*], Qing Peng [1,2,*] and Sheng Liu [1,3]

1   The Institute of Technological Sciences, Wuhan University, Wuhan 430072, China;
    wanghuaipeng@whu.edu.cn
2   Nuclear Engineering and Radiological Sciences, University of Michigan, Ann Arbor, MI 48109, USA
3   School of Power and Mechanical Engineering, Wuhan University, Wuhan 430072, China;
    victor_liu63@126.com
*   Correspondence: caoqiang@whu.edu.cn (Q.C.); Qing.Peng@whu.edu.cn (Q.P.);
    Tel.: +86-137-0129-2834 (Q.C.); 734-763-3866 (Q.P.)

Received: 11 December 2018; Accepted: 14 January 2019; Published: 18 January 2019

Abstract: With an ultralarge surface-to-volume ratio, a recently synthesized three-dimensional graphene structure, namely, carbon honeycomb, promises important engineering applications. Herein, we have investigated, via molecular dynamics simulations, its mechanical properties, which are inevitable for its integrity and desirable for any feasible implementations. The uniaxial tension and nanoindentation behaviors are numerically examined. Stress–strain curves manifest a transformation of covalent bonds of hinge atoms when they are stretched in the channel direction. The load–displacement curve in nanoindentation simulation implies the hardness and Young's modulus to be 50.9 GPa and 461±9 GPa, respectively. Our results might be useful for material and device design for carbon honeycomb-based systems.

Keywords: carbon honeycomb; molecular dynamics; LAMMPS; uniaxial tension; nanoindentation

---

1. Introduction

Graphene is known widely, due to its excellent mechanical nature, as a so-called "miracle material", with many of its characteristics measured experimentally or theoretically exceeding those obtained in other materials: A Young's modulus of 1 TPa and intrinsic strength of 130 GPa [1,2], high stiffness [3] and fracture strain [4,5], a normal-auxeticity mechanical phase transition [5], etc. Due to these predominant properties, graphene has promising potential to be employed in various applications: Paints and coatings of nanocomposites [4,6], flexible electronics [7], and bioapplications [8–10]. Nevertheless, considering the difficulty in engineering synthesis of large-area and high-quality graphene, so far, there is little practical application of industrially produced graphene. Consequently, it is worth focusing on allotropes of graphene [11–13], which are more stable and feasible to produce. Recently, a stable carbon allotrope, namely, carbon honeycomb (CHC), a three-dimensional graphene, was synthesized. There are a few reports on of its physical absorption [14,15], electronic band structure [16], thermoelectric performance [17], thermal transport properties [18,19], and phonon properties [15].

According to the cell patterns in the plane perpendicular to the cell axis, carbon honeycombs can be categorized into two sets: Armchair CHC (ac_n CHC) and zigzag CHC (zz_m CHC) [20]. When the number of armchair or zigzag lines (n in ac_n or m in zz_m) varies, the cell sizes of CHCs and, thus, their mechanical properties change correspondently. Krainyukova et al. [14] synthesized this carbon allotrope by deposition of vacuum-sublimated graphite and proposed periodic and random structures of CHCs. Regarding the mechanical properties of CHC, some analytical studies have been carried out [16,19–24]. Karfunkel et al. first proposed a family of hypothetical zz_m–sp^2–sp^3 CHC structures and revealed that the carbon modifications are as stable as diamond by solid-state semi-empirical SCF methods using the modified neglect of diatomic overlap (MNDO) Hamiltonian [23]. Park et al. studied the mechanical properties of zz_m–sp^2–sp^3 CHCs with different m (where m can be an integer or a half integer) using ab initio pseudopotential as well as the environment-dependent tight-binding method, making clear that the carbon allotrope is elastically stable and has a fairly high shear modulus [24]. Pang et al. made a systematical analysis of failure strength and strong anisotropic Poisson's effect of ac_n–sp^2–sp^3 CHCs with different cell sizes via molecular dynamics simulation using optimized reactive empirical bond-order potential [19]. Gu et al. used molecular dynamics (MD) simulation with modified reactive empirical bond-order potential to give the stress–strain curves of symmetrical and asymmetrical ac_2–sp^2–sp^3 CHCs, ac_3–sp^2–sp^3 CHCs, and zz_2–sp^2–sp^3 and $zz_{2.5}$–sp^2–sp^3 CHCs [21]. Zhang et al. [16] built the in-plane compression and out-of-plane nanoindentation tests of ac_n–sp^2 and zz_m–sp^2–sp^3 CHCs via MD analysis with the Adaptive Intermolecular Reactive Empirical Bond Order potential (AIREBO) [25] potential. Meng et al. investigated the out-of-plane compression behaviors of both ac_n–sp^2–sp^3 and zz_m–sp^2–sp^3 CHCs using MD simulation with AIREBO potential [20]. Despite these studies, a full and clear understanding of the inner mechanism of the deformation of CHCs is still lacking but desirable due to its various promising applications as an engineering material. In this study, we used MD simulation with AIREBO potential to investigate the transformation from an ac_2–sp^2–sp^3 CHC to an ac_2–sp^2 CHC, which causes the change of tensile deformation behavior of ac_2 CHCs. In the meanwhile, we gave the stress–strain curves, Young's modulus, and Poisson's ratio of ac_2 CHCs, which show a great agreement with previous studies. We also built an out-of-plane nanoindentation test to investigate the plastic deformation behaviors of ac_2–sp^2–sp^3 CHCs, giving the hardness and Young's modulus.

2. Crystal Structure of Carbon Honeycomb

Carbon honeycomb, just as its name implies, is a kind of tubular structure with a honeycomb-like pattern through a top–down perspective. The tube wall can be regarded as a graphene-like monolayer, thereby deemed to be a 3D version of graphene. One thing about its structure that remains to be discussed is whether three adjacent tube walls are bound up with each other via sp^3 or sp^2 bonding (Figure 1b,c). Undoubtedly, atoms that build up the tube shells are combined with neighboring carbon atoms by an sp^2 bond, as is the case in graphene, whereas those who pull contiguous graphene-like walls together, which are called hinge atoms, give rise to more options, one kind of typical sp^3 bonds with non-hinge atoms (tube atoms) or one variant of diamond-like sp^3 bonds with both adjacent hinge and non-hinge atoms. In the case of sp^2 bonds, hinge atoms do not interact with each other, allowing for electron exchange with only adjacent tube atoms. In the case of sp^3 bonds, it is not only nearby tube atoms but also nearest-neighbor hinge atoms that get bonded with the center hinge atom, resulting in a diamond-like atom-hinge that ties up three around the atomic walls. According to the carbon patterns in the plane perpendicular to the axis of cell channel, carbon honeycombs can be categorized into two sets: Armchair CHC (ac_n CHC) and zigzag CHC (zz_m CHC) [20], as shown in Figure 1. Furthermore, carbon honeycombs can be subdivided into sp^2 and sp^2–sp^3 CHCs. In zz_m CHCs, carbon atoms of graphene-like walls are bonded by sp^2 bonds, while hinge atoms can be bonded with adjacent atoms only by sp^3 bonds. Therefore, zz_m CHCs always refer to zz_m–sp^2–sp^3 CHCs. However, the other case is different. In ac_n CHCs, hinge atoms can be bonded with atoms around them by sp^2 or sp^3 bonds. Therefore, ac_n CHCs can be subdivided into ac_n–sp^2 CHCs and ac_n–sp^2–sp^3 CHCs. In the case of ac_n–sp^2–sp^3 CHCs, hinge atoms can be distributed symmetrically or asymmetrically in a crystal cell, according to which ac_n–sp^2–sp^3 CHCs can also be subdivided into a symmetrical one and asymmetrical one, as shown as Figure 1b,c,e,f.

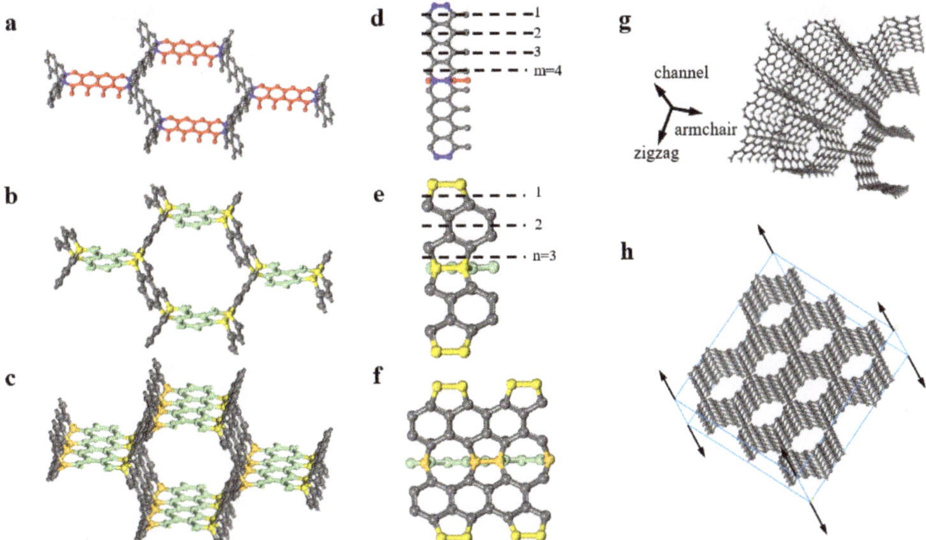

Figure 1. Honeycomb structures in (**a**) zigzag sp^2–sp^3-bonded carbon honeycomb (zz_4–sp^2–sp^3 CHC), (**b**) symmetrical armchair sp^2–sp^3-bonded carbon honeycomb (sym-ac_3–sp^2–sp^3 CHC), and (**c**) asymmetrical armchair sp^2–sp^3-bonded carbon honeycomb (asym-ac_3–sp^2–sp^3 CHC). (**d**), (**e**), and (**f**) are the zoom-in views exhibiting detailed carbon patterns and bonding ways of hinge atoms shown in (**a**), (**b**), and (**c**), respectively. For convenience, a coordinate system attached to sym-ac_3–sp^2–sp^3 CHC is displayed in (**g**). Uniaxial tension is applied on sym-ac_3–sp^2 and sym-ac_3–sp^2–sp^3 CHCs along the axis of cell channel. All the simulation supercells are cubic and periodical in three directions.

Herein, we implemented a density functional theory (DFT) calculation for both primitive cells of sym-ac_3–sp^2 and sym-ac_3–sp^2–sp^3 CHCs, as shown in Figure 2b,c, using PWmat [26,27], to examine their stabilities. For both types, cell relaxation and atom relaxation were exerted in subsequence to determine their respective energy-minimized configurations. Subsequently, self-consistent calculation was carried out to optimize the electron interaction in both primitive cells, as visualized in Figure 1d,e. As for the sym-ac_3–sp^2–sp^3 CHC, stronger sp3 bonds between the hinge atom and nearest-neighbor tube atoms can be observed than those between two adjacent hinge atoms, the bond angle of which is measured as 103.35°. Meanwhile, exactly strong sp^2 bonds among graphene-like atoms can be distinguished from sp^3 bonds, with denser electron density around tube atoms observed (red-yellow hexagons in Figure 1d). As for the sym-ac_3–sp^2 CHC, hinge atoms are not attached to each other, with weaker sp2 bonds between hinge and non-hinge atoms observed than those between only tube atoms. The DFT calculation results, i.e., that the cohesive energy is –154.19 eV/atom of the sym-ac_3–sp^2 CHC, –154.84 eV/atom of sym-ac_3–sp^2–sp^3 CHC, respectively, indicate that two types of configurations are both likely to exist stably. Additionally, the slightly stronger cohesive interaction implies the sym-ac_3–sp^2–sp^3 CHC structure is relatively more stable.

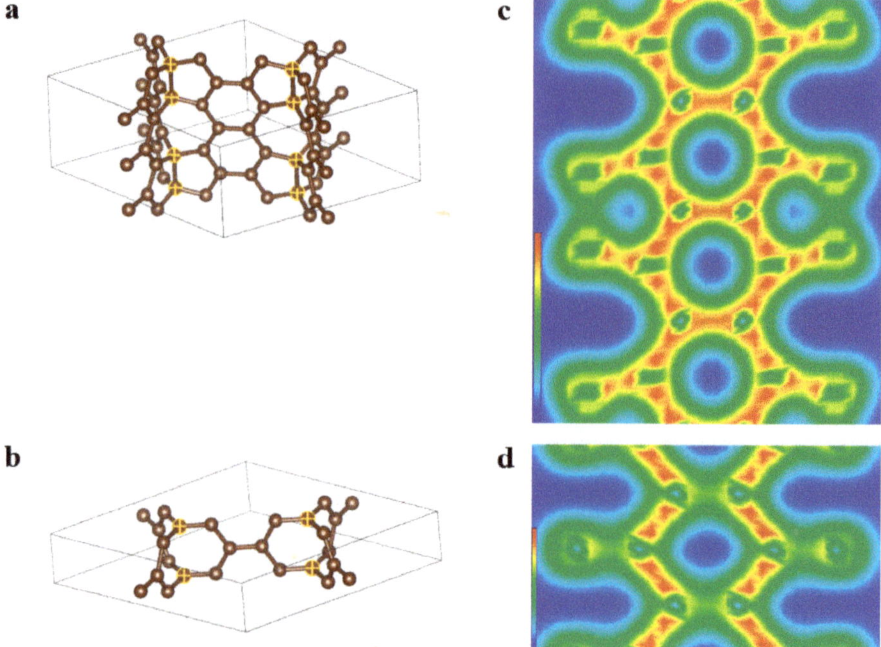

Figure 2. The primitive cells of sym-ac$_3$–sp^2 and sym-ac$_3$–sp^2–sp^3 CHCs are shown in (**a**) and (**b**), respectively. (**c**) and (**d**) display the slice contours of electron density from density functional theory (DFT) calculation of structures shown in (**a**) and (**b**), respectively.

3. Molecular Dynamics Simulations

We used molecular dynamics simulations to do the investigations of mechanical behaviors of the sym-ac$_3$–sp^2 CHC and sym-ac$_3$–sp^2–sp^3 CHC. The reason we focused on them is that in the work of Krainyukova et al. [14], a periodic carbon structure identical to the sym-ac$_3$–sp^2 CHC according to their synthetic CHC samples, proving that ac$_3$ CHCs are more likely to exist in the nature. The method of MD simulations is well established to investigate the elastic and plastic behaviors of sym-ac$_3$–sp^2 and sym-ac$_3$–sp^2–sp^3 CHCs. Initially, we carried out uniaxial tension tests for both types of CHCs, simultaneously determining the Young's moduli of the sym-ac$_3$–sp^2–sp^3 CHC to be 551 $\pm$ 4 GPa and of sym-ac$_3$–sp^2 CHC to be 542$\pm$4 GPa in the channel direction (Figure 1g,h), following which we carried out tension tests in the channel direction of two kinds of CHCs, giving stress–strain curves, respectively. Interestingly, a transformation of hinge atoms from sp^3 to sp^2 was observed on the atomic scale, which can account for the yield stage that can be observed in the sym-ac$_3$–sp^2–sp^3 CHC, while cannot be observed in the sym-ac$_3$–sp^2 CHC. Furthermore, a nanoindentation simulation was devised and the load–displacement curve was plotted to determine the hardness and Young's modulus of the CHC, which are 50.9 GPa of hardness and 461 $\pm$ 9 GPa of Young's modulus, respectively. Meanwhile, we discussed the plastic behavior of CHC in the process of nanoindentation.

We used a molecular dynamics (MD) software, LAMMPS [28], which is an open-source code, to simulate two typical mechanical tests, uniaxial tension tests of the sym-ac$_3$–sp^2 CHC and sym-ac$_3$–sp^2–sp^3 CHC and nanoindentation of the ac$_3$–sp^2–sp^3 CHC. As for the tension simulation, given that we have set the boundary conditions as periodic in three directions, and thus, the mechanical response of the system is independent from the simulation scale, in order to accelerate the calculation, the simulation box was chosen as 3.408 nm × 2.951 nm × 4.181 nm, containing 2856 atoms. Both end-faces perpendicular to the direction of tension were freed from all kinds of forces. The whole system was minimized for energy and relaxed at specified temperatures (10 K and 300 K) and zero pressure in the NPT ensemble (where the number of particles, pressure, and temperature are constant) using AIREBO [25] potential (cutoff radius = 2.5 Å), to eliminate inner stress. The strain rate is set as 0.01 ps^{-1}. Considering that temperature effect will cause random distribution of atoms, which will prevent us from observing the bonding transformation of hinge atoms, in addition to at room temperature, uniaxial tension tests were also carried out at 10K, at which little thermal motion is allowed so that a perfect configuration might remain.

We also simulated the entire process of nanoindentation on an ac_3–sp^2–sp^3 CHC along the channel direction. A thick CHC plate measured as 17.82 nm × 15.43 nm × 11.60 nm was set up in a simulation box with a vacuum layer 9.00 nm thick in addition (Figure 2). The indenter is a cubic diamond sphere, 8.00 nm in diameter, with [001] direction aligned with the channel direction. The velocity of this indenter is set at 0.25 Å/ps. There is a 0.50-nm-thick layer on the bottom of the slab of CHC, all forces on which are zeroed. Nonperiodic conditions were selected in three directions in order to simulate an isolated system. Before the mechanical test, an energy minimization had been finished and the total ensemble had been relaxed efficiently without applied stress at 300 K, using an NVT ensemble (where the number of particles, volume and temperature are constant) and NPT ensemble with zero pressure successively, with AIREBO-morse [29] potential (cutoff radius = 2.5 Å).

4. Results and Discussion

4.1. Uniaxial Tension

Strain–Stress Curve

From uniaxial tension, we obtained the engineering strain–stress curves of both the sym-ac_3–sp^2 CHC and sym-ac_3–sp^2–sp^3 CHC (Figure 3), which can depict the elastic pattern directly and characterize the plastic properties indirectly, gaining an insight into the similarities and differences between the mechanical behaviors of two kinds of CHCs. Our results are in great agreement with the stress–strain curves from a previous study [21], in which the fracture strength and strain equal to our results are exhibited. Although the stress–strain curves of ac_n–sp^2 CHCs with different sizes have been studied systematically, there have been a limited number of studies of the yield process, i.e., the transformation from sp^3 to sp^2, which is what we focus on. We also calculated the Young's moduli of both CHCs, 542 ± 4 GPa for the sym-ac_3–sp^2 CHC and 551 ± 4 GPa for the sym-ac_3–sp^2–sp^3 CHC (300 K). From a previous study [16], the Young's modulus of the CHC, which is assumed to be a nanoscale cell solid, can be calculated by the cell wall width via an analytic method. In our cases (wall width is determined as 7 Å), the analytical result is 560 GPa, in a great agreement with our simulation results.

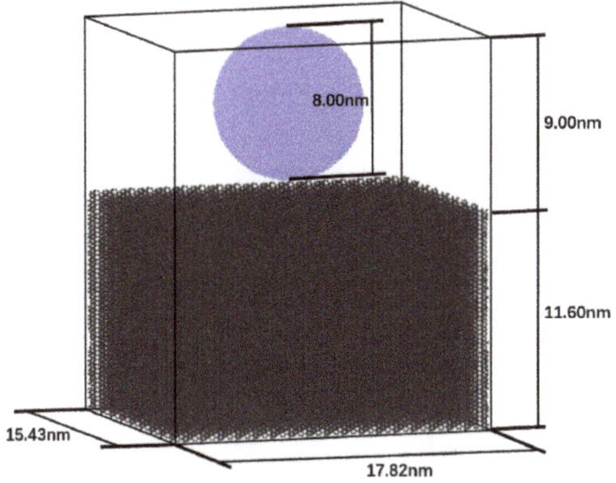

Figure 3. Schematic of simulation model of nanoindentation along the channel direction.

When the tension starts, both CHCs are stretched quasilinearly under homogeneous and elastic deformation; herein, the initial curve of the sym-ac$_3$–sp^2–sp^3 CHC deviates a little from that of the sym-ac$_3$–sp^2 CHC. Then, in a certain strain range (from 0.072 to 0.132 for case at 10 K), a yield stage occurs in the deformation of the sym-ac$_3$–sp^2–sp^3 CHC, which indicates a plastic deformation has initiated. Later in this article, the mechanism therein will be discussed in detail. After the yield step, the sym-ac$_3$–sp^2–sp^3 curve behaves just like a sym-ac$_3$–sp^2 curve does, even breaking off at the same fracture strength, which is around 61 GPa (56 GPa for case at 300 K), with a retardation measured as around 0.06 in strain for the case at 10 K, equal to the length of the preceding yield stage. Thus, the mechanical behavior of the sym-ac$_3$–sp^2–sp^3 CHC can be distinguished from that of the sym-ac$_3$–sp^2 CHC by the fracture retardation caused by yield terrace mentioned above.

In order to illustrate the mechanism of this yield stage, a slice through some sheets of graphene-like walls has been exhibited in Figure 4. Both cases at 10 K and 300 K exhibit the same yield stage, so in order to avoid random distribution of atoms caused by thermal motion, which will prevent us from observing the detailed transfomation of hinge atoms, we explain the yield process based on the tension at 10 K. The color of atoms characterizes the channel-direction component of the stress on each atom, of which the colormap is linearly related to the value of stress. Through the variation of the channel-direction stress, we can justify whether the hinge atoms interact with other hinge atoms, so that we can determine the bond type of hinge atoms. For hinge atoms with an sp^3 bond, the neighboring hinge atoms must be bonded with each other, thus exhibiting a high channel-direction stress; while for the case of sp^2, hinge atoms do not interact with each other, so there will be a low and even zero channel-direction stress on them. In addition, sp^2 bonding is in-plane (two-dimentional), while sp^3 is three-dimentional; thus, we can also tell them apart by their geometry and nearest neighbors. Three snapshots at different states of strain, corresponding to three characteristic points in Figure 4, respectively, are displayed. At the beginning of tension (Figure 4a), hinge atoms share most of the distortion energy through the extension of the sp^3 bond between two hinge atoms, while the sp^3 configuration remains. However, when the engineering strain reaches around 0.072, some sp^3 pairs of hinge atoms break up, the distortion energy of graphene-like atoms begins to increase gradually, and simultaneously, the yield stage starts. When the strain attains 0.132, almost all sp^3 pairs are disconnected, most of the deformation is shared by the atomic cellular walls, and the yield stage terminates at the same time. By contrast, the sym-ac$_3$–sp^2 CHC, without sp^3 interaction, does not experience such a transformation, and atomic tubes instead of hinge atoms bear most proportions of tensile force. Compared with graphene, although CHC has a lower fracture strength and modulus, it can obtain a higher fracture strain than graphene through this transformation, which may be applied in flexible coating.

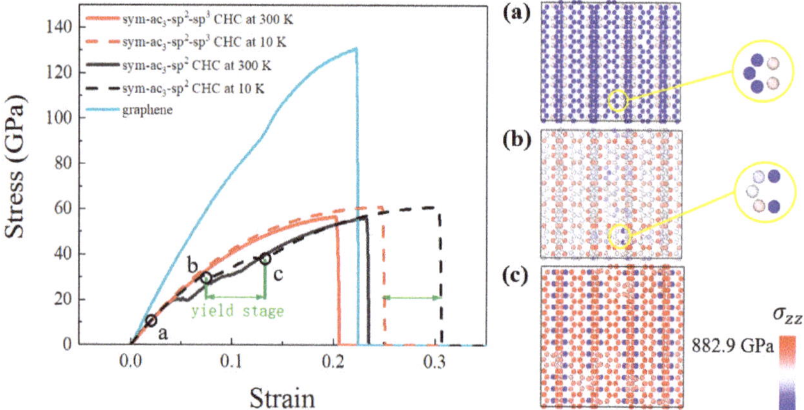

Figure 4. Stress–strain curves of the sym-ac$_3$–sp^2–sp^3 CHC and sym-ac$_3$–sp^2 CHC. There are three characteristic points representing three different stress states, and corresponding snapshots display the

transformation of hinge atoms from sp^3 to sp^2, (**a**) for minuscular strain state, (**b**) for the start point of yield terrace, (**c**) for the termination of the yield terrace.

4.2. Nanoindentation

At present, nanoindentation is commonly used for the research of mechanical properties of materials on the nanoscale [30–32]. There are two main reasons this methodology has been put into widespread usage [33]. As for nanoindentation, the stress applied by indenter is nonhomogeneous and can thus create an elastically physically isolated volume. Secondly, not only the location of elastic instability, but also the slip behaviors can be predicted via nanoindentation.

In a previous study [16], nanoindentation behaviors of different sizes of ac$_n$–sp^2 and zz$_m$–sp^2–sp^3 CHCs were reported. However, the indent depth was small (4 Å) and the unloading process was not displayed. Herein, we simulated the process of nanoindentation in the channel direction of the sym-ac$_3$–sp^2–sp^3 CHC, an indentation depth of about 40 Å, containing two steps, acting load and unloading, via LAMMPS [28], under the confinement mentioned in the section "Molecular Dynamics Simulations". The result is displayed in Figure 5, which we used to determine the hardness and Young's modulus of the CHC in the channel direction.

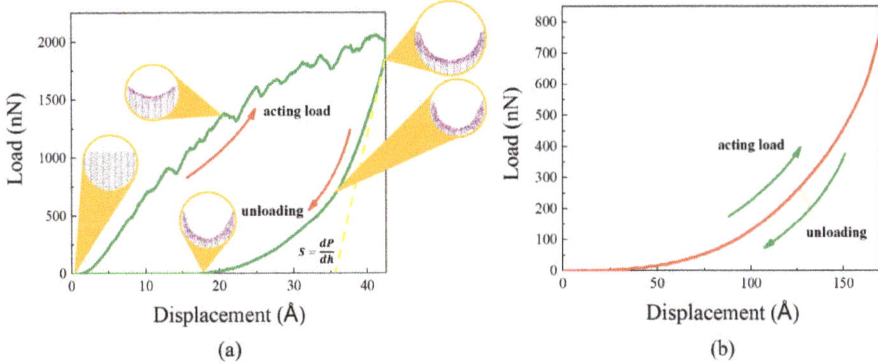

Figure 5. (**a**) The acting–load curve and unloading curve. The slope of tangent line at the initial stage of unloading refers to the contact stiffness $S = \frac{dP}{dh}$. In addition, several snapshots of the morphology of in situ indentation are exhibited here, in which the color of atoms distinguishes different crystal lattices, with grey referring to normal CHC lattice, while purple refers to simple cube lattice, which is a more compact type than the normal one. For comparison between 2D and 3D graphene, the displacement–load curve of graphene under the same simulation condition is displayed in (**b**).

Nanoindentation hardness is the average pressure the material can support under an external load, defined as the ratio of indentation load P and projected contact area A_p [30]. From the load–displacement curve, the hardness H can be attained using the maximum load as:

$$H = \frac{P_{max}}{A_p}. \tag{1}$$

In addition, Young's modulus can also be calculated indirectly via reduced modulus E_r, which can be expressed as:

$$E_r = \frac{\sqrt{\pi}}{2\beta} \frac{S}{\sqrt{A_p}}, \tag{2}$$

where S is termed as contact stiffness, which can be obtained as the slope at the tip of unloading curve, β is a constant that depends on the geometry of indenter, for a spherical indenter, $\beta = 1$. Furthermore, we can determine the substrate's modulus using following relationship that reflects impacts of both indenter's and substrate's moduli:

$$\frac{1}{E_r} = \frac{1 - v_i^2}{E_i} + \frac{1 - v_s^2}{E_s},$$
(3)

where the subscript i refers to *indenter*, while s refers to *substrate*.

The nanoindentation hardness is determined as 50.9 GPa in Tables 1 and 2, which is higher than quartz (9.25 GPa) and even some ion-beam-irradiation hardened metal (over 20 GPa) [34]. In order to analyze the plastic behavior of the ac_3–sp^2–sp^3 CHC in the process of nanoindentation, five snapshots are displayed in Figure 5. In the case of loading, the amounts of atoms which are identified as simple cube lattice are increasing with the dent produced by spherical diamond indenter deepening. This phenomenon indicates a transformation of covalent bonds from a previous normal type to a more compact one. Subsequently, although the indenter retreated and the dent recovering partially in an elastic way, the number of simple cubic atoms is maintained almost the same number as there is when the indenter penetrates the deepest part, which illustrates that a plastic deformation has occurred and cannot rebound.

Table 1. Hardness and Young's modulus calculated from nanoindentation. Calculated hardness on the nanoscale.

Projected Contact Area ($Å^2$)	Maximum Load (nN/ $Å$)	Hardness (GPa)
3689	1878.1	50.9

Table 2. Hardness and Young's modulus calculated from nanoindentation. Comparison between Young's moduli from tension test and nanoindentation.

Contact Stiffness (nN/$Å$)	Reduced Modulus (GPa)	Young's Moudulus (From Tension Simulation) (GPa)	Young's Modulus (From Nanoindentation) (GPa)
228 ± 3	285 ± 4	551 ± 4	461 ± 9

5. Conclusions

We discussed the elastic and plastic behaviors of sym-ac_3–sp^2 CHCs and especially the sym-ac_3–sp^2–sp^3 CHC which is slightly more stable according to our DFT calculation and previous experimental study, using LAMMPS to carry out a typical uniaxial tension and nanoindentation simulation. In the case of uniaxial tension, we discussed the yield stage in the stress–strain curve of the sym-ac_3–sp^2–sp^3 CHC caused by a intriguing transformation of covalent bonds of hinge atoms from strong sp^3 to comparatively weak sp^2, and we also determined the Young's moduli at room temperature, 542 ± 4 GPa for the sym-ac_3–sp^2 CHC and 551 ± 4 GPa for the sym-ac_3–sp^2–sp^3 CHC, which are in great agreement with a previous analytical study. Then, in the discussion of nanoindentation, we determined the hardness (50.9 GPa) and Young's modulus (461 ± 9 GPa) via the load–displacement curve and gave an insight into the plastic deformation caused by the indenter on the nanoscale, which can be explained by the permanent conversion of covalent bonds from sp^3 type to sp^2 type. These findings provide an insight into the relationship between the covalent bond type of hinge atoms in carbon honeycomb and the plastic behavior of carbon honeycomb.

Author Contributions: Q.C. and Q.P. conceived the idea of the paper; H.W. conceived, designed, and performed the simulations; Q.C. and H.W. wrote the paper; S.L. discussed the results and modified the manuscript. All the authors had a full discussion and commented on the paper.

Funding: This research was funded by the National Natural Science Foundation of China (No. 51727901).

Acknowledgments: The numerical calculations in this paper have been done on the supercomputing system in the Supercomputing Center of Wuhan University.

Conflicts of Interest: There is no conflict in this research.

References

1. Lee, C.; Wei, X.; Kysar, J.W.; Hone, J. Measurement of the elastic properties and intrinsic strength of monolayer graphene. *Science* **2008**, *321*, 385–388. [CrossRef] [PubMed]
2. Liu, F.; Ming, P.; Li, J. Ab initio calculation of ideal strength and phonon instability of graphene under tension. *Phys. Rev. B* **2007**, *76*, 064120. [CrossRef]
3. Lee, G.-H.; Cooper, R.C.; An, S.J.; Lee, S.; van der Zande, A.; Petrone, N.; Hammerberg, A.G.; Lee, C.; Crawford, B.; Oliver, W.; et al. High-strength chemical-vapor–deposited graphene and grain boundaries. *Science* **2013**, *340*, 1073–1076. [CrossRef] [PubMed]
4. Novoselov, K.S.; Fal'ko, V.I.; Colombo, L.; Gellert, P.R.; Schwab, M.G.; Kim, K. A roadmap for graphene. *Nature* **2012**, *490*, 192–200. [CrossRef] [PubMed]
5. Binghui, D.; Jie, H.; Hanxing, Z.; Sheng, L.; Emily, L.; Yunfeng, S.; Qing, P. The normal-auxeticity mechanical phase transition in graphene. *2D Mater.* **2017**, *4*, 021020. [CrossRef]
6. Young, R.J.; Kinloch, I.A.; Gong, L.; Novoselov, K.S. The mechanics of graphene nanocomposites: A review. *Compos. Sci. Technol.* **2012**, *72*, 1459–1476. [CrossRef]
7. Bae, S.; Kim, H.; Lee, Y.; Xu, X.; Park, J.-S.; Zheng, Y.; Balakrishnan, J.; Lei, T.; Ri Kim, H.; Song, Y.I.; et al. Roll-to-roll production of 30-inch graphene films for transparent electrodes. *Nat. Nanotechnol.* **2010**, *5*, 574–578. [CrossRef]
8. Nayak, T.R.; Andersen, H.; Makam, V.S.; Khaw, C.; Bae, S.; Xu, X.; Ee, P.-L.R.; Ahn, J.-H.; Hong, B.H.; Pastorin, G.; et al. Graphene for controlled and accelerated osteogenic differentiation of human mesenchymal stem cells. *ACS Nano* **2011**, *5*, 4670–4678. [CrossRef]
9. Nair, R.R.; Blake, P.; Blake, J.R.; Zan, R.; Anissimova, S.; Bangert, U.; Golovanov, A.P.; Morozov, S.V.; Geim, A.K.; Novoselov, K.S.; et al. Graphene as a transparent conductive support for studying biological molecules by transmission electron microscopy. *Appl. Phys. Lett.* **2010**, *97*, 153102. [CrossRef]
10. Kuila, T.; Bose, S.; Khanra, P.; Mishra, A.K.; Kim, N.H.; Lee, J.H. Recent advances in graphene-based biosensors. *Biosens. Bioelectron.* **2011**, *26*, 4637–4648. [CrossRef]
11. Zhang, Y.; Huang, H. Stability of single-wall silicon carbide nanotubes – molecular dynamics simulations. *Comput. Mater. Sci.* **2008**, *43*, 664–669. [CrossRef]
12. Hong, S.; Lundstrom, T.; Ghosh, R.; Abdi, H.; Hao, J.; Jeoung, S.K.; Su, P.; Suhr, J.; Vaziri, A.; Jalili, N.; et al. Highly anisotropic adhesive film made from upside-down, flat, and uniform vertically aligned cnts. *ACS Appl. Mater. Interfaces* **2016**, *8*, 34061–34067. [CrossRef] [PubMed]
13. Peng, Q.; Ji, W.; De, S. Mechanical properties of graphyne monolayers: A first-principles study. *Phys. Chem. Chem. Phys.* **2012**, *14*, 13385–13391. [CrossRef] [PubMed]
14. Krainyukova, N.V.; Zubarev, E.N. Carbon honeycomb high capacity storage for gaseous and liquid species. *Phys. Rev. Lett.* **2016**, *116*, 055501. [CrossRef] [PubMed]
15. Gao, Y.; Chen, Y.; Zhong, C.; Zhang, Z.; Xie, Y.; Zhang, S. Electron and phonon properties and gas storage in carbon honeycombs. *Nanoscale* **2016**, *8*, 12863–12868. [CrossRef] [PubMed]
16. Zhang, Z.; Kutana, A.; Yang, Y.; Krainyukova, N.V.; Penev, E.S.; Yakobson, B.I. Nanomechanics of carbon honeycomb cellular structures. *Carbon* **2017**, *113*, 26–32. [CrossRef]
17. Yang, Z.; Lan, G.; Ouyang, B.; Xu, L.-C.; Liu, R.; Liu, X.; Song, J. The thermoelectric performance of bulk three-dimensional graphene. *Mater. Chem. Phys.* **2016**, *183*, 6–10. [CrossRef]
18. Wei, Z.; Yang, F.; Bi, K.; Yang, J.; Chen, Y. Thermal transport properties of all-sp2 three-dimensional graphene: Anisotropy, size and pressure effects. *Carbon* **2017**, *113*, 212–218. [CrossRef]
19. Pang, Z.; Gu, X.; Wei, Y.; Yang, R.; Dresselhaus, M.S. Bottom-up design of three-dimensional carbon-honeycomb with superb specific strength and high thermal conductivity. *Nano Lett.* **2016**, *17*, 179–185. [CrossRef]
20. Meng, F.; Chen, C.; Hu, D.; Song, J. Deformation behaviors of three-dimensional graphene honeycombs under out-of-plane compression: Atomistic simulations and predictive modeling. *J. Mech. Phys. Solids* **2017**, *109*, 241–251. [CrossRef]
21. Gu, X.; Pang, Z.; Wei, Y.; Yang, R. On the influence of junction structures on the mechanical and thermal properties of carbon honeycombs. *Carbon* **2017**, *119*, 278–286. [CrossRef]
22. Liu, Y.; Liu, J.; Yue, S.; Zhao, J.; Ouyang, B.; Jing, Y. Atomistic simulations on the tensile deformation behaviors of three-dimensional graphene. *Phys. Status Solidi* **2018**, *255*, 1700680. [CrossRef]

23. Karfunkel, H.R.; Dressler, T. New hypothetical carbon allotropes of remarkable stability estimated by mndo solid-state scf computations. *J. Am. Chem. Soc.* **1992**, *114*, 2285–2288. [CrossRef]

24. Park, N.; Ihm, J. Electronic structure and mechanical stability of the graphitic honeycomb lattice. *Phys. Rev. B* **2000**, *62*, 7614–7618. [CrossRef]

25. Stuart, S.J.; Tutein, A.B.; Harrison, J.A. A reactive potential for hydrocarbons with intermolecular interactions. *J. Chem. Phys.* **2000**, *112*, 6472–6486. [CrossRef]

26. Jia, W.; Cao, Z.; Wang, L.; Fu, J.; Chi, X.; Gao, W.; Wang, L.-W. The analysis of a plane wave pseudopotential density functional theory code on a gpu machine. *Comput. Phys. Commun.* **2013**, *184*, 9–18. [CrossRef]

27. Jia, W.; Fu, J.; Cao, Z.; Wang, L.; Chi, X.; Gao, W.; Wang, L.-W. Fast plane wave density functional theory molecular dynamics calculations on multi-gpu machines. *J. Comput. Phys.* **2013**, *251*, 102–115. [CrossRef]

28. Plimpton, S. Fast parallel algorithms for short-range molecular dynamics. *J. Comput. Phys.* **1995**, *117*, 1–19. [CrossRef]

29. O'Connor, T.C.; Andzelm, J.; Robbins, M.O. Airebo-m: A reactive model for hydrocarbons at extreme pressures. *J. Chem. Phys.* **2015**, *142*, 024903. [CrossRef]

30. Li, X.; Bhushan, B. A review of nanoindentation continuous stiffness measurement technique and its applications. *Mater. Charact.* **2002**, *48*, 11–36. [CrossRef]

31. Peng, Q.; Zhang, X.; Huang, C.; Carter, E.A.; Lu, G. Quantum mechanical study of solid solution effects on dislocation nucleation during nanoindentation. *Model. Simul. Mater. Sci. Eng.* **2010**, *18*, 075003. [CrossRef]

32. Peng, Q.; Zhang, X.; Lu, G. Quantum mechanical simulations of nanoindentation of al thin film. *Comput. Mater. Sci.* **2010**, *47*, 769–774. [CrossRef]

33. Li, J.; Van Vliet, K.J.; Zhu, T.; Yip, S.; Suresh, S. Atomistic mechanisms governing elastic limit and incipient plasticity in crystals. *Nature* **2002**, *418*, 307–310. [CrossRef] [PubMed]

34. Takayama, Y.; Kasada, R.; Sakamoto, Y.; Yabuuchi, K.; Kimura, A.; Ando, M.; Hamaguchi, D.; Tanigawa, H. Nanoindentation hardness and its extrapolation to bulk-equivalent hardness of F82H steels after single- and dual-ion beam irradiation. *J. Nucl. Mater.* **2013**, *442*, S23–S27. [CrossRef]

 *nanomaterials*

Article

Facile Synthesis of SnO$_2$ Aerogel/Reduced Graphene Oxide Nanocomposites via in Situ Annealing for the Photocatalytic Degradation of Methyl Orange

Taehee Kim [1], Vinayak G. Parale [1], Hae-Noo-Ree Jung [1], Younghun Kim [1], Zied Driss [2], Dorra Driss [3], Abdallah Bouabidi [2], Souhir Euchy [2] and Hyung-Ho Park [1,*]

[1] Department of Materials Science and Engineering, Yonsei University, Seoul 03722, Korea; taehee-kim@yonsei.ac.kr (T.K.); vinayakparale3@gmail.com (V.G.P.); nuri_j@yonsei.ac.kr (H.-N.-R.J.); younghun_kim@yonsei.ac.kr (Y.K.)

[2] Laboratory of Electromechanical Systems (LASEM), National School of Engineers of Sfax (ENIS), University of Sfax, PO Box 1173, Route Soukra km 3.5, 3038 Sfax, Tunisia; zied.driss@enis.tn (Z.D.); bouabidi_abdallah@yahoo.fr (A.B.); echi.souhir@yahoo.com (S.E.)

[3] Laboratory of Molecular and Cellular Screening Processes, Centre of Biotechnology of Sfax (CBS), University of Sfax, PO Box 1177, Road Sidi Mansour km 6, 3018 Sfax, Tunisia; dorra_driss@yahoo.fr

* Correspondence: hhpark@yonsei.ac.kr; Tel.: +82-2-2123-2853

Received: 31 January 2019; Accepted: 27 February 2019; Published: 4 March 2019

Abstract: SnO$_2$ aerogel/reduced graphene oxide (rGO) nanocomposites were synthesized using the sol–gel method. A homogeneous dispersion of graphene oxide (GO) flakes in a tin precursor solution was captured in a three-dimensional network SnO$_2$ aerogel matrix and successively underwent supercritical alcohol drying followed by the in situ thermal reduction of GO, resulting in SnO$_2$ aerogel/rGO nanocomposites. The chemical interaction between aerogel matrix and GO functional groups was confirmed by a peak shift in the Fourier transform infrared spectra and a change in the optical bandgap of the diffuse reflectance spectra. The role of rGO in 3D aerogel structure was studied in terms of photocatalytic activity with detailed mechanism of the enhancement such as electron transfer between the GO and SnO$_2$. In addition, the photocatalytic activity of these nanocomposites in the methyl orange degradation varied depending on the amount of rGO loading in the SnO$_2$ aerogel matrix; an appropriate amount of rGO was required for the highest enhancement in the photocatalytic activity of the SnO$_2$ aerogel. The proposed nanocomposites could be a useful solution against water pollutants.

Keywords: SnO$_2$ aerogel; sol–gel method; graphene oxide; nanocomposite; photocatalysis

---

1. Introduction

The demand for solutions against several environmental issues has recently increased; in particular, the presence of air pollutants and organic dyes in water are global concerns. Since the latters are often highly toxic and have mutagenic properties, their removal is a major problem in the industry [1–3]. Until now, many efforts have been made to decompose the organic pollutants by using various processes involving, e.g., (homogeneous and/or heterogeneous) catalysts, adsorbents, and ozone [4–8] and photocatalysis is one of the most effective and economical paths for their removal [9–13]. Semiconducting metal oxides such as TiO$_2$, ZnO, and SnO$_2$ have been widely used as photocatalysts due to their ability to generate electron–hole pairs when photon energy is provided [12–19]. Among them, SnO$_2$ has gained much attention because of its high natural abundance, optical transparency, and physicochemical stability, relatively high electrical conductivity, and lack of toxicity [20–22]. SnO$_2$ is an n-type semiconductor with a wide bandgap (3.6 eV) and a rutile-type crystal structure, but it exhibits a low photocatalytic activity due to such wide bandgap and its high photogenerated electron–hole pair recombination rate [23]. Also, the economic aspects of the preparation of SnO_2 photocatalyst for the mass production is still challenging. The cost and compatibility should be considered in photocatalytic degradation [24–28].

Many previous studies aimed to enhance the photocatalytic activity of SnO_2-based semiconductors by introducing nanostructures and composites [29–33]. Different nanostructures for SnO_2-based photocatalysts, such as nanoparticles [29], flower-like structures [30], and simonkolleite nanopetals [31], have been reported so far. In addition, the use of SnO_2 composites with carbon materials (in particular, carbon core-shell particles [34], graphene oxide (GO) [35], activated carbon [36], and fullerene [37]) instead of pure SnO_2 to decrease the electron–hole recombination rate, which would enhance the photocatalytic activity, has been investigated. Moreover, some studies have reported chemical interactions between these carbon materials and the SnO_2 matrix, which decrease the bandgap and improve the photocatalytic activity [38,39].

In the present work, we synthesized porous SnO_2 aerogel/reduced GO (rGO) nanocomposites via an epoxide-assisted sol–gel process. GO flakes were uniformly dispersed in a tin solution and captured in a colloidal SnO_2 three-dimensional (3D) network. Then, the in situ thermal reduction of GO was performed continuously in autoclave during the supercritical drying of the resulting nanocomposites. To the best of our knowledge, this is the first report on the in situ annealing of SnO_2 aerogel/GO nanocomposites in an autoclave, providing SnO_2 aerogel/rGO nanocomposites as the final product. Furthermore, the dispersion of 2D GO sheets would hinder the network formation. In this study, the reaction condition was carefully controlled to achieve the uniform distribution as well as the reaction between oxygen containing functional groups on GO sheet and Sn metal center and confirmed by various characterizations. In addition, the synthesized nanocomposites were analyzed in detail and their performance in the photocatalytic degradation of methyl orange (MO) was quantified. As per our knowledge, SnO_2 aerogel and rGO composite has not been studied in application of photocatalysis.

2. Materials and Methods

2.1. Sample Preparation

Tin tetrachloride pentahydrate ($SnCl_4 \cdot 5H_2O$) was added to a mixture of water and ethanol (3:1, v/v) and stirred for 1 h; after complete dissolution, 0.05, 0.1, and 0.3 wt.% of GO flakes were added to each solution separately and their uniform distribution was ensured via ultrasonic processing for 1 h. When the GO flakes were suspended in the tin salt solution, this was chilled in an ice bath. Excess propylene oxide (C_3H_6O, 143 mmol; Sigma-Aldrich, St. Louis, MO, USA) was added to the solution dropwise by using a syringe. The chemistry of the epoxide-initiated gelation method is well described in [40]; the gelation took place in 10 min and the alcogel was left for 20 min to complete the reaction. The resulting alcogel was added with ethanol and aged at room temperature for 3 days by exchanging with fresh ethanol every 24 h. Then, the solvent exchange was performed with methanol for 24 h. The so-obtained SnO_2 alcogel underwent supercritical methanol drying at 265 °C and 105 bar in an autoclave equipped with a 2 L vessel (Parr Instruments, Moline, IL, USA). After complete supercritical fluid extraction, in situ drying was performed in the same autoclave by heating the vessel to 300 °C under N_2 flow and the dried aerogel was annealed for 1 h to induce the reduction of GO to rGO.

2.2. Characterization

A Fourier transform infrared spectroscopy (FTIR) system (Perkin Elmer, Waltham, MA, USA) was used to monitor the reaction and characterize the impurities. The specific surface area of nanoporous aerogel composites was measured using a Brunauer–Emmett–Teller (BET) analyzer (Quantachrome, Boynton Beach, FL, USA) and the Barrett–Joyner–Halenda (BJH) method. Their crystallinity and structure were investigated with an X-ray diffraction (XRD) system (Rigaku Ultima, Tokyo, Japan) using Cu Kα radiation (1.5418 Å) in the 20–80° 2θ range. The surface morphology of the SnO_2 aerogel/rGO nanocomposites was analyzed using a field emission scanning electron microscopy

(FESEM) system (JEOL JSM 7001F, Tokyo, Japan). The photoluminescence (PL) analysis was performed on a LabRam Aramis system (Horriba Jobin Yvon, Madrid, Spain) at room temperature and with a laser excitation wavelength of ~325 nm. Ultraviolet diffuse reflectance spectra (UV-DRS) were recorded on a spectrometer (JASCO 780, Tokyo, Japan) was performed using powder DRS kit at room temperature.

2.3. Photocatalyst Properties

The photocatalytic activity of both pristine SnO_2 aerogel and SnO_2 aerogel/rGO nanocomposites was determined based on the degradation degree of an aqueous MO dye solution. The photocatalytic decolorization of the MO solution was derived from its absorption (absorption peak at 464 nm by using a UV–visible (Vis) spectrophotometer (JASCO 570, Tokyo, Japan) in the 300–800 nm range. The initial concentration of the MO dye in the solution was 1×10^{-5} M and the photocatalyst (1 mg mL^{-1}) was dispersed in it. The photocatalytic degradation was performed using a UV lamp (40 W; Philips TL-K) with peak intensity at 370 nm irradiating directly the solution after achieving the adsorption/desorption equilibrium for 30 min. The solution was stirred continuously with a magnetic stirrer. After centrifugation with a microcentrifuge (DAIHAN CF−10, Seoul, Korea), the absorbance spectra of the solution with the photocatalyst suspension (approximately 1.2 mL aliquots) were recorded over time at ambient conditions.

3. Results and Discussion

Metal alkoxides are generally used as non-silica-based precursors in the synthesis of aerogels [41–43], but they are costly and their reactivity is hard to control [44]. In this study, cost effective tin chloride was selected as the precursor and propylene oxide was used to initialize the sol–gel process for the synthesis of the SnO_2 aerogel/rGO nanocomposites. Propylene oxide acts as proton scavenger and ring-strained epoxide, increasing the hydrolysis and condensation rate [45]; its addition usually speeds up the gelation (less than a minute), leading to the formation of an opaque white alcogel [40,46]. However, in this study, the tin precursor solution and propylene oxide were chilled in an ice bath to obtain uniform pores in the SnO_2 aerogel, giving a clear transparent alcogel. In addition, due to the relatively fast gelation (approximately 10 min), the GO flakes homogeneously dispersed via ultrasonication were trapped within the SnO_2 matrix without large agglomerations. The porous SnO_2 aerogel/rGO nanocomposites were obtained by removing the solvent in the 3D wet gel via drying at supercritical pressure and temperature. The thermal annealing method was used to reduce GO to rGO by decomposing the oxygen-containing functional groups. The BET and XRD results about the textural and crystalline properties of the as-synthesized pristine SnO_2 aerogel (PTO) and SnO_2 aerogel/rGO nanocomposites (named as TGO05, TGO1, and TGO3 according to the rGO loading of 0.05, 0.1, and 0.3 wt.%, respectively) are shown in Figure 1 and Table 1.

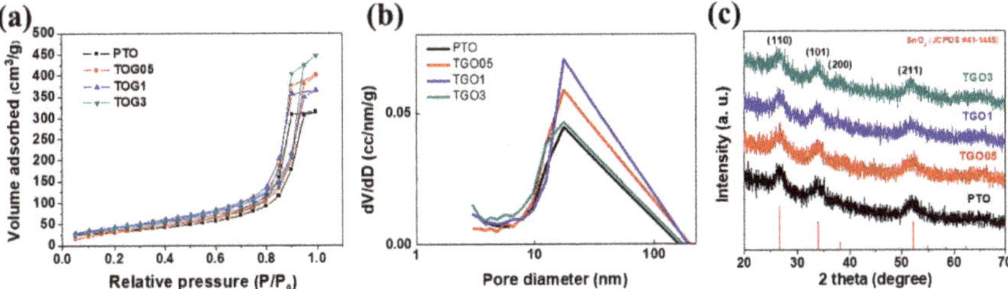

Figure 1. (**a**) Brunauer–Emmett–Teller isotherms, (**b**) Barrett–Joyner–Halenda graph and (**c**) X-ray diffraction spectra of pristine SnO_2 aerogel (PTO) and SnO_2 aerogel/reduced graphene oxide (rGO) nanocomposites with different rGO loadings (0.05 wt.%: TGO05; 0.1 wt.%: TGO1; 0.3 wt.%: TGO3).

The nanoporous structure of the various samples was determined via the BET analysis (Figure 1a). All isotherms showed type V according to the IUPAC classification, revealing the mesoporous structure of the SnO_2 aerogel. PTO exhibited a slightly smaller surface area compared to the rGO-added nanocomposites, while TGO3 showed the largest one (157 $m^2\ g^{-1}$); the specific surface area of the SnO_2 aerogel increased with the rGO loading. The rGO presence could clearly enhance this parameter without hindering the colloidal aerogel formation and this was due to the high initial surface area of the GO flakes. Moreover, the well-distributed rGO sheets with a high mechanical strength could reduce the thermal stress induced during the thermal annealing step. Moreover, the average pore volumes and pore sizes, calculated via the BJH method (Figure 1b), slightly increased in the SnO_2 aerogel/rGO nanocomposites compared to the pristine sample.

Figure 1c shows the XRD patterns of the SnO_2 aerogel/rGO nanocomposite after the heat treatment. The SnO_2 aerogel exhibited a rutile-type tetragonal crystal structure and the XRD peaks at 26.61°, 33.89°, 37.95°, and 51.78° corresponded to the (110), (101), (200), and (211) crystal planes of the SnO_2 aerogel/rGO nanocomposites, respectively, with the following cell parameters: a, b = 4.738 Å and c = 3.187 Å (JCPDS #41–1445) [47]. However, the graphene diffraction pattern could not be indexed because of the low amount of GO flakes. The average crystallite size of the aerogel nanocomposites, calculated using the Scherrer equation, was approximately 3.4 nm for all the samples and no distinctive influence of the rGO addition was observed on the crystallinity of the SnO_2 aerogel.

Table 1. Textural properties and crystallite size of pristine SnO_2 aerogel and SnO_2 aerogel/reduced graphene oxide (rGO) nanocomposites with different rGO loadings.

Graphene Content (wt.%)	Surface Area ($m^2\ g^{-1}$)	Pore Volume ($cm^3\ g^{-1}$)	Pore diameter (nm)	Average Crystallite Size (nm)
0	117	0.49	17	3.48
0.05	131	0.62	19	3.42
0.1	147	0.69	18	3.46
0.3	157	0.57	15	3.35

Exfoliated GO can be reduced to rGO via simple annealing under oxidizing or inert atmosphere [48], but annealing SnO_2 aerogel/GO composites under air atmosphere would reduce the number of oxygen vacancies in the aerogel, lowering its photocatalytic activity. Therefore, after drying, the reaction vessel was heated to 300 °C to induce the reduction of the GO flakes [48]; this is a facile method for in situ thermal annealing with desirable atmosphere consecutively just after the supercritical alcohol drying. For comparison, the BET specific surface areas of PTO and the various SnO_2 aerogel/rGO nanocomposites were measured before, during, and after (i.e., after the supercritical drying) the in situ annealing (Table 2). The results suggested that the in situ annealing method can minimize the thermal stress. Therefore, the GO reduction in the SnO_2 aerogel matrix via in situ annealing could introduce the enhancement of the specific surface area.

Table 2. Comparison of the effect of in situ annealing on the specific surface area of SnO_2 aerogel, with and without the addition of graphene oxide.

Graphene Content (wt.%)	Surface Area ($m^2\ g^{-1}$) Before Annealing	Surface Area ($m^2\ g^{-1}$) Facile, in-situ Annealing	Surface Area ($m^2\ g^{-1}$) Post Annealing
0	199	117	84
0.05	277	131	75
0.1	314	147	73
0.3	359	157	92

Figure 2 shows the FTIR spectra of PTO and the various SnO_2 aerogel/rGO nanocomposites. All samples show weak and broad absorption peaks of hydroxyl (–OH) group around 3400 cm^{-1} owing to the surface hydroxyl groups on the SnO_2 aerogel primary particles. This represents the hydrophilic surface which would favor to absorb aqueous solution for photocatalytic organic dye degradation [49]. In addition, Figure 2a shows a broad absorption peak at around 495 cm^{-1}, for all samples, that indicates the formation of Sn–O–Sn bonds via the epoxide-initiated sol–gel process. However, this peak shifted toward higher wavenumbers with increasing the GO content, as shown in the magnified spectra in Figure 2b. This broad peak was probably a combination of Sn–O–Sn vibration and Sn–O–C stretching vibration (520 cm^{-1}) [50,51]. The presence of the latter confirmed the chemical bond formation between SnO$_2$ aerogel and residual oxygen-containing functional groups in GO during the synthesis.

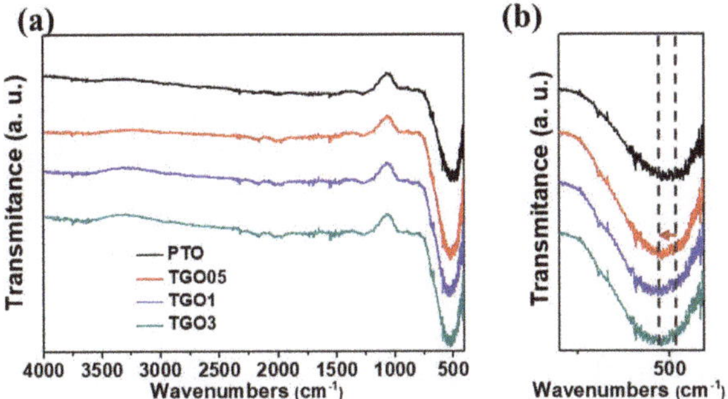

Figure 2. (**a**) Fourier transform infrared spectra and (**b**) their magnification in the low-wavenumber region for the pristine SnO$_2$ aerogel (PTO) and the SnO$_2$ aerogel/reduced graphene oxide (rGO) nanocomposites with different rGO loadings (0.05 wt.%: TGO05; 0.1 wt.%: TGO1; 0.3 wt.%: TGO3).

PL emission is an important tool to determine the charge transportation and separation of electron–hole pairs. Figure 3 shows the photoluminescence spectra of PTO and the various SnO$_2$ aerogel/rGO nanocomposites, measured at 325 nm. PTO exhibited a strong emission at around 554 nm due to the crystal defect of the SnO$_2$ matrix [52], which significantly decreased in TGO05, TGO1, and TGO3; however, these samples showed similar intensities regardless the rGO amount added. This phenomenon indicated that the rGO addition could inhibit the electron–hole pair recombination, which implied that the rGO-based nanocomposites acted as electron trapping sites, benefiting the charge transfer in the enhanced photocatalytic activity. In addition, the PL quenching effect was also caused by the excellent electrical conductivity of rGO which will favor the photocatalytic activity [23].

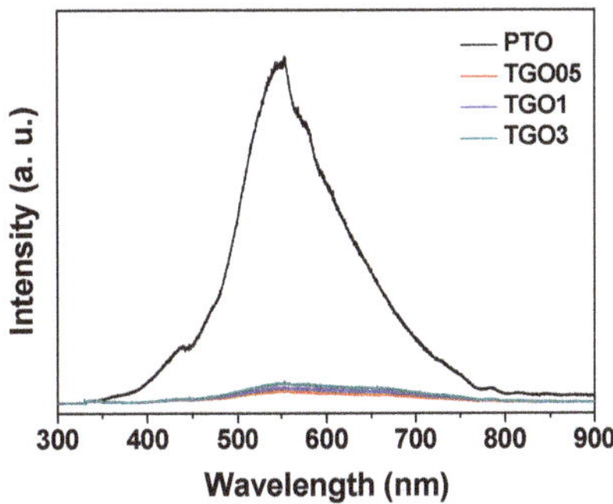

Figure 3. Photoluminescence spectra of pristine SnO$_2$ aerogel (PTO) and SnO$_2$ aerogel/reduced graphene oxide (rGO) nanocomposites with different rGO loadings (0.05 wt.%: TGO05; 0.1 wt.%: TGO1; 0.3 wt.%: TGO3).

The FESEM images of PTO and TGO1 are shown in Figure 4. In regards to the various rGO-added samples, only the results for TGO1 are presented because they all exhibited similar surface morphologies but more rGO graphene incorporation sites are found. The increase in the rGO loading did not influence the surface morphology of the nanocomposites. The colloidal porous nature of PTO was confirmed by a mesoporous aerogel network (Figure 4a). Figure 4d confirms the incorporation of rGO in the SnO_2 matrix, which clearly depicts the growth of primary particles of aerogel on the surface of rGO. The visible well deposition of the aerogel particles allowed us to assume that there was a chemical interaction between the aerogel matrix and the functional groups on the rGO surface.

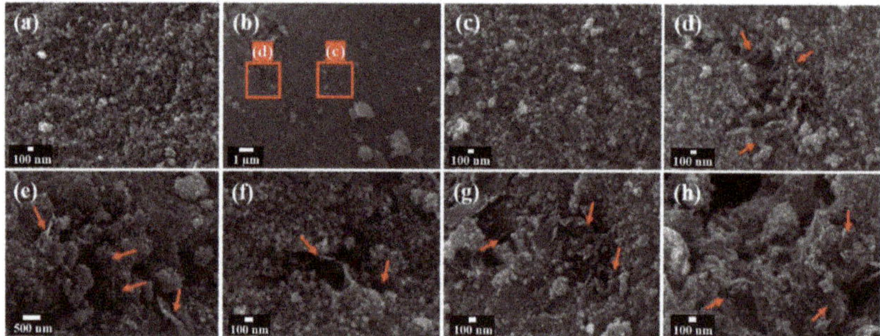

Figure 4. Field-emission scanning electron microscopy images of: (**a**) pristine SnO_2 aerogel (PTO); (**b**) SnO_2 aerogel/reduced graphene oxide (rGO) nanocomposites with an rGO loading of 0.1 wt.% (TGO1); (**c,d**) magnification of red square areas in (**b**), showing the rGO incorporation in the SnO_2 aerogel matrix; (**e–h**). The red arrows represent rGO sheets.

The chemical bond formation between SnO_2 matrix and the functional groups on GO was further confirmed by the UV-DRS analysis. Figure 5a shows the DRS spectra of PTO and TGO1 samples, revealing a considerable difference after the GO addition. Figure 5b shows their Kubelka–Munk plots; the calculated optical bandgaps of the semiconductor powders were approximately 3.56 and 3.29 eV for PTO and TGO1, respectively. The bandgap of PTO well agreed with the value for SnO_2 nanoparticles reported in literature [53]. The rGO incorporation corresponded to a narrower bandgap, which could be attributed to the Sn–O–C bond formation between SnO_2 and the rGO functional groups [50]. Since new impurity energy levels above the valence-band edge would be generated by the rGO introduction, a smaller input energy would be required for the charge carrier excitation. Therefore, the SnO_2 aerogel/rGO nanocomposite exhibited a decrease in the optical bandgap with the Sn–O–C bond formation. Although this phenomenon has been previously observed in the case of TiO_2-graphene composites, no such mechanism has been reported for SnO_2-reduced graphene composites so far [54].

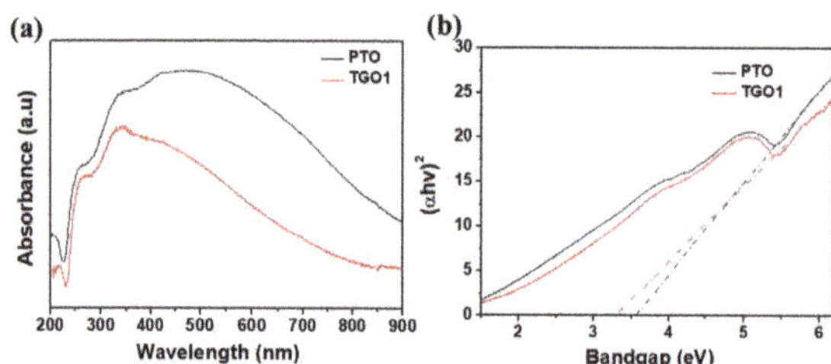

Figure 5. (**a**) Diffuse reflectance spectra and (**b**) Kubelka–Munk plots for the calculated bandgaps of pristine SnO_2 aerogel (PTO) and SnO_2 aerogel/reduced graphene oxide (rGO) nanocomposites with an rGO loading of 0.1 wt.% (TGO1).

The photocatalytic activity of both pristine SnO_2 aerogel and its rGO nanocomposites was investigated by the degradation of an MO solution. The MO concentration was determined by the absorbance of the solution at 464 nm using UV–Vis spectrophotometer after the centrifugal separation of the photocatalyst powder. All the MO solutions (with pristine and nanocomposite catalysts) were left in dark for 30 min to achieve the adsorption/desorption equilibrium state on the photocatalyst surface. Then, the photocatalytic activity of the samples was initiated via UV light irradiation under continuous magnetic stirring. The absorbance of the MO solution was measured starting from time t = 0, every 10 min of irradiation. Figure 6 shows the changes in the concentration ratio C/C_0, where C_0 and C are the MO concentrations at the initial time t_0 and the irradiation time t, respectively, for the various samples. PTO exhibited a moderately reasonable photocatalytic activity, with a 56% MO degradation after 60 min of UV irradiation. Moreover, the rGO addition resulted in enhanced photocatalytic activity, reaching with TGO1 an 84% MO degradation in the same time. The activity of TGO05 was also relatively higher compared to PTO, but the difference was negligible. The enhancement in the photocatalytic activity of TGO1 was discussed by a comparison with PTO, confirming that the rGO introduction can considerably enhance the photocatalytic degradation of MO.

Figure 6b shows the photocatalytic reaction rate constant k values, derived from the slope of the $\ln(C_0/C)$ versus time plot, for PTO, TGO05, TGO1, and TGO3. All samples exhibited a first-order rate law with a linear behavior. The highest (2.9×10^{-2} min^{-1}) and lowest (1.2×10^{-2} min^{-1}) k values were observed with TGO1 and PTO, respectively, meaning that TGO1 attained a 2.4-fold higher photocatalytic degradation rate compared to PTO.

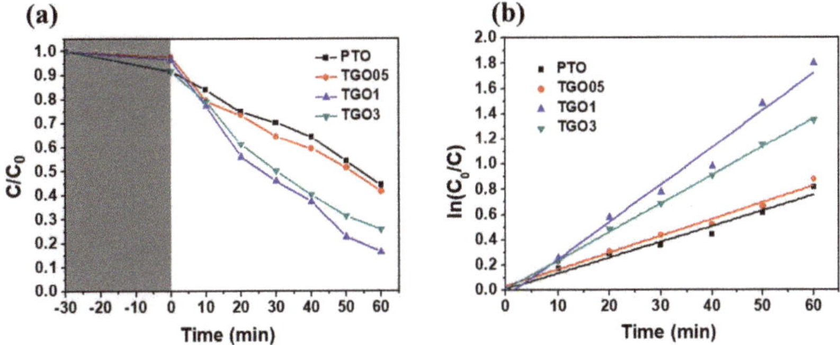

Figure 6. Photocatalytic activities of pristine SnO_2 aerogel (PTO) and SnO_2 aerogel/reduced graphene oxide (rGO) nanocomposites with different rGO loadings (0.05 wt.%: TGO05; 0.1 wt.%: TGO1; 0.3 wt.%: TGO3), represented as (**a**) the degradation of methyl orange (MO) dye (in terms of concentration ratio C/C_0, where C_0 and C are the MO concentrations at initial time t_0 and a given ultraviolet irradiation time t) and (**b**) its reaction rate during time.

In general, the three main factors influencing the photocatalytic activity are the light absorption intensity, the specific surface area, and the separation/recombination rate of photoexcited electron–hole pairs [23]. In this study, the rGO addition to SnO_2 sol increased the specific surface area of the final aerogel nanocomposites by introducing more reactive surface sites and, hence, generating more electron–hole pairs [55]. In addition, the photoexcited electrons could move to the conduction band of rGO to interact with absorbed O_2 and generate reactive radicals for the MO degradation; this phenomenon was hindered by the electron–hole pair recombination, enhancing the photocatalytic activity of the SnO_2 aerogel/rGO nanocomposites. The detailed mechanisms for the formation of reactive radical intermediates during the photoactivation of the SnO_2/rGO nanocomposites are described by the Equations (1)–(6) and the illustration in Figure 7. The photogenerated electron could move to rGO and produce reactive superoxide anions, eventually leading to the MO degradation.

$$SnO_2 + hv \rightarrow SnO_2(e^- + h^+) \tag{1}$$

$$SnO_2(e^- + h^+) + rGO \rightarrow SnO_2(h^+) + rGO(e^-) \tag{2}$$

$$SnO_2(h^+) + H_2O \rightarrow SnO_2 + OH^{\cdot} + H^+ \tag{3}$$

$$rGO(e^-) + O_2 \rightarrow rGO + O_2^{\cdot -} \tag{4}$$

$$O_2^{\cdot -} + H_2O \rightarrow H_2O^{\cdot} + OH^- \tag{5}$$

$$MO + OH^{\cdot}/O_2^{\cdot -}/H_2O^{\cdot} \rightarrow CO_2 + H_2O \tag{6}$$

On the other hand, the addition of excess graphene to the SnO_2 aerogel matrix (i.e., the TGO3 case) reduced the enhancement of the photocatalytic activity, resulting in a 74% MO degradation in 60 min of UV irradiation. The photocatalytic activity was in the following order: TGO1 > TGO3 > TGO05 > PTO. Hence, a suitable amount of rGO loading would benefit the photocatalytic activity while its excess would decrease the photon absorption efficiency by SnO_2 [56,57]; furthermore, such enhancement could be possible only at a certain extent. At the beginning state of the study, the experiment was designed to have 0.5–2.0 wt.% of rGO in SnO_2 aerogel. However, authors have found that large amount of rGO in SnO_2 aerogel retarded the photocatalytic activity as well as the surface area. This would happen because an excess rGO loading could increase the probability of collisions between photogenerated electrons and holes, favoring the electron–hole pair recombination and, thus affecting the photocatalytic activity [58]. Moreover, SnO_2 aerogel/rGO nanocomposites are compared with previous reported SnO_2 based photocatalysts in Table 3.

Table 3. Comparison of photocatalytic performance of SnO_2 aerogel/rGO nanocomposite photocatalyst in this work with previously reported composite photocatalysts.

Photocatalyst	Light Source/Pollutant	Experimental Condition	Photodegradation Efficiency	Ref.
SnO_2 nanoparticles coated on MWCNT	4×6 W fluorescent halogen lamps (254 nm), methyl orange	C: 1000 mg L^{-1}; P: 20 mg L^{-1}	45 min/94%	[17]
Simonkolleite nanopetals with SnO_2	3×8 W UV lamps (254 nm), rhodamine 6G	C: 500 mg L^{-1}; P: 10 mg L^{-1}	32 min /100%	[25]
SnO_2–graphene nanocomposite (solid state)	300 W mercury lamp, methyl orang and rhodamine B	C: 500 mg L^{-1}; P: 20 mg L^{-1}	Methyl orange 40 min/95% Rhodamine B 60 min/97%	[49]
SnO_2–CNT nanocomposites	9 W eight UV–vis lamps (365 nm), methylene blue and methyl orange	C: 200 mg L^{-1}; P: 20 ppm (methylene blue), 10 ppm (methyl orange)	Methylene blue 180 min/93% Methyl orange 180 min/79%	[1]
SnO_2 aerogel/rGO nanocomposite	40 W UV lamp (370 nm), methyl orange	C: 1×10^{-5} M; P: 100 mg L^{-1}	Methyl orange 60 min/84%	This work

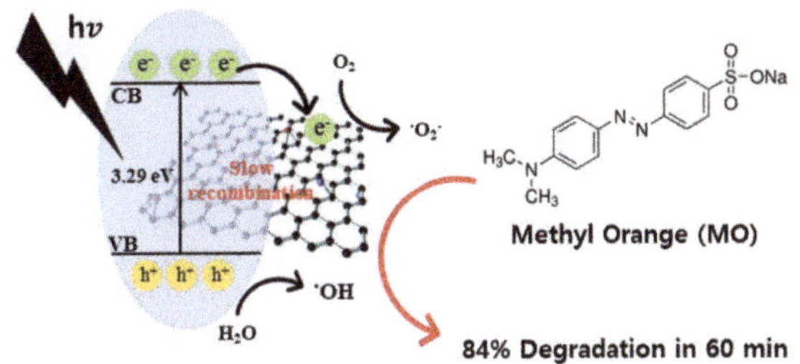

Figure 7. Schematic of MO dye photodegradation by using SnO$_2$ aerogel/reduced graphene oxide nanocomposites under ultraviolet light.

4. Conclusions

We introduced GO into an SnO$_2$ aerogel matrix followed by its reduction to rGO via a facile in situ annealing process in a supercritical autoclave. During the sol–gel synthesis, the dispersed rGO flakes were captured in the SnO$_2$ matrix and the colloidal SnO$_2$ reacted with their functional groups, as confirmed by the FTIR and UV-DRS spectra. The so-obtained nanocomposites exhibited enhanced photocatalytic activity in the degradation of an MO dye solution by reducing the electron–hole pair recombination rate; the photogenerated electrons were trapped in rGO, leading to the formation of reactive superoxide anions, and the photocatalytic activity varied depending on the rGO loading in the SnO$_2$ aerogel matrix. This enhancement in photocatalytic activity with rGO addition in aerogel structure was due to the movement of photogenerated electrons to conduction band in rGO which reduces the recombination rate. In 60 min under UV irradiation, the pristine SnO$_2$ aerogel and the SnO$_2$ aerogel/rGO (0.1 wt.%) nanocomposites achieved an MO degradation of 56% and 84% degradation, respectively. Therefore, SnO$_2$ aerogel/rGO nanocomposites could be good candidates for the photodegradation of organic pollutants in industrial wastewater.

Author Contributions: Methodology, T.K. and H.-H.P.; investigation, T.K.; formal analysis, T.K., Y.K., H.-N.-R.J., Z.D., D.D., A.B., S.E., and V.G.P.; writing—original draft preparation, T.K.; writing—review and editing, V.G.P. and H.-H.P.

Funding: This work was supported by 'Korea-Africa Joint Research Programme' grant funded by the Korea government (Ministry of Science, Technology & ICT) in 2017K1A3A1A09085891. This work was supported (researched) by the third Stage of Brain Korea 21 Plus Project in 2019.

Conflicts of Interest: The authors declare no conflict of interest.

References

1. Kim, S.P.; Choi, M.Y.; Choi, H.C. Characterization and photocatalytic performance of SnO$_2$—CNT nanocomposites. *Appl. Surf. Sci.* **2015**, *357*, 302–308. [CrossRef]
2. Zhu, X.; An, S.; Liu, Y.; Hu, J.; Liu, H.; Tian, C.; Dai, S.; Yang, X.; Wang, H.; Abney, C.W.; et al. Efficient removal of organic dye pollutants using covalent organic frameworks. *Aiche J.* **2017**, *63*, 3470–3478. [CrossRef]
3. Sreekanth, T.V.M.; Shim, J.-J.; Lee, Y.R. Degradation of organic pollutants by bio-inspired rectangular and hexagonal titanium dioxide nanostructures. *J. Photochem. Photobiol. B Biol.* **2017**, *169*, 90–95. [CrossRef] [PubMed]
4. Wang, W.; Tadé, M.O.; Shao, Z. Research progress of perovskite materials in photocatalysis- and photovoltaics-related energy conversion and environmental treatment. *Chem. Soc. Rev.* **2015**, *44*, 5371–5408. [CrossRef] [PubMed]
5. Upadhyay, R.K.; Soin, N.; Roy, S.S. Role of graphene/metal oxide composites as photocatalysts, adsorbents and disinfectants in water treatment: A review. *RSC Adv.* **2014**, *4*, 3823–3851. [CrossRef]
6. Qin, C.; Li, Z.; Chen, G.; Zhao, Y.; Lin, T. Fabrication and visible-light photocatalytic behavior of perovskite praseodymium ferrite porous nanotubes. *J. Power Sour.* **2015**, *285*, 178–184. [CrossRef]

7. Yang, Z.; Chen, S.; Fu, K.; Liu, X.; Li, F.; Du, Y.; Zhou, P.; Cheng, Z. Highly efficient adsorbent for organic dyes based on a temperature- and pH-responsive multiblock polymer. *J. Appl. Polym. Sci.* **2018**, *135*, 46626. [CrossRef]

8. Zucker, I.; Lester, Y.; Avisar, D.; Hübner, U.; Jekel, M.; Weinberger, Y.; Mamane, H. Influence of Wastewater Particles on Ozone Degradation of Trace Organic Contaminants. *Environ. Sci. Technol.* **2015**, *49*, 301–308. [CrossRef] [PubMed]

9. Lee, D.-S.; Lee, S.-Y.; Rhee, K.Y.; Park, S.-J. Effect of hydrothermal temperature on photocatalytic properties of TiO$_2$ nanotubes. *Curr. Appl. Phys.* **2014**, *14*, 415–420. [CrossRef]

10. Pant, A.; Tanwar, R.; Kaur, B.; Mandal, U.K. A magnetically recyclable photocatalyst with commendable dye degradation activity at ambient conditions. *Sci. Rep.* **2018**, *8*, 14700. [CrossRef] [PubMed]

11. Wang, M.; You, M.; Zhang, K.; Zhang, Y.; Han, J.; Fu, R.; Liu, S.; Zhu, T. Bi$_{3.64}$Mo$_{0.36}$O$_{6.55}$/Bi$_2$MoO$_6$ heterostructure composite with enhanced photocatalytic activity for organic pollutants degradation. *J. Alloy. Compd.* **2018**, *766*, 1037–1045. [CrossRef]

12. Zhang, Y.; Park, M.; Kim, H.-Y.; El-Newehy, M.; Rhee, K.Y.; Park, S.-J. Effect of TiO$_2$ on photocatalytic activity of polyvinylpyrrolidone fabricated via electrospinning. *Compos. Part B Eng.* **2015**, *80*, 355–360. [CrossRef]

13. Wang, H.; Qiu, X.; Liu, W.; Yang, D. Facile preparation of well-combined lignin-based carbon/ZnO hybrid composite with excellent photocatalytic activity. *Appl. Surf. Sci.* **2017**, *426*, 206–216. [CrossRef]

14. Anjum, M.; Miandad, R.; Waqas, M.; Gehany, F.; Barakat, M.A. Remediation of wastewater using various nano-materials. *Arab. J. Chem.* **2016**. [CrossRef]

15. Karri, R.R.; Tanzifi, M.; Tavakkoli Yaraki, M.; Sahu, J.N. Optimization and modeling of methyl orange adsorption onto polyaniline nano-adsorbent through response surface methodology and differential evolution embedded neural network. *J. Environ. Manag.* **2018**, *223*, 517–529. [CrossRef] [PubMed]

16. Pastrana-Martínez, L.M.; Morales-Torres, S.; Likodimos, V.; Figueiredo, J.L.; Faria, J.L.; Falaras, P.; Silva, A.M.T. Advanced nanostructured photocatalysts based on reduced graphene oxide–TiO$_2$ composites for degradation of diphenhydramine pharmaceutical and methyl orange dye. *Appl. Catal. B Environ.* **2012**, *123–124*, 241–256. [CrossRef]

17. Khalid, N.R.; Ahmed, E.; Hong, Z.; Zhang, Y.; Ullah, M.; Ahmed, M. Graphene modified Nd/TiO$_2$ photocatalyst for methyl orange degradation under visible light irradiation. *Ceram. Int.* **2013**, *39*, 3569–3575. [CrossRef]

18. Filice, S.; D'Angelo, D.; Libertino, S.; Nicotera, I.; Kosma, V.; Privitera, V.; Scalese, S. Graphene oxide and titania hybrid Nafion membranes for efficient removal of methyl orange dye from water. *Carbon* **2015**, *82*, 489–499. [CrossRef]

19. Bhattacharjee, A.; Ahmaruzzaman, M.; Sinha, T. A novel approach for the synthesis of SnO$_2$ nanoparticles and its application as a catalyst in the reduction and photodegradation of organic compounds. *Spectrochim. Acta Part A Mol. Biomol. Spectrosc.* **2015**, *136*, 751–760. [CrossRef] [PubMed]

20. Kumar, M.; Mehta, A.; Mishra, A.; Singh, J.; Rawat, M.; Basu, S. Biosynthesis of tin oxide nanoparticles using Psidium Guajava leave extract for photocatalytic dye degradation under sunlight. *Mater. Lett.* **2018**, *215*, 121–124. [CrossRef]

21. Zhu, L.-P.; Bing, N.-C.; Yang, D.-D.; Yang, Y.; Liao, G.-H.; Wang, L.-J. Synthesis and photocatalytic properties of core–shell structured α-Fe$_2$O$_3$@SnO$_2$ shuttle-like nanocomposites. *Cryst. Eng. Comm.* **2011**, *13*, 4486–4490. [CrossRef]

22. Oh, T. Effect of double junctions in nano structure oxide materials and gas sensitivity. *Trans. Electr. Electron. Mater.* **2018**, *19*, 382–386. [CrossRef]

23. Wang, N.; Xu, J.; Guan, L. Synthesis and enhanced photocatalytic activity of tin oxide nanoparticles coated on multi-walled carbon nanotube. *Mater. Res. Bull.* **2011**, *46*, 1372–1376. [CrossRef]

24. Wang, C.; Wang, X.; Xu, B.-Q.; Zhao, J.; Mai, B.; Peng, P.; Sheng, G.; Fu, J. Enhanced photocatalytic performance of nanosized coupled ZnO/SnO$_2$ photocatalysts for methyl orange degradation. *J. Photochem. Photobiol. A Chem.* **2014**, *168*, 47–52. [CrossRef]

25. Li, Y.; Li, X.; Li, J.; Yin, J. Photocatalytic degradation of methyl orange by TiO$_2$-coated activated carbon and kinetic study. *Water Res.* **2006**, *40*, 1119–1126. [CrossRef] [PubMed]

26. Tian, C.; Zhang, Q.; Wu, A.; Jiang, M.; Liang, Z.; Jiang, B.; Fu, H. Cost-effective large-scale synthesis of ZnO photocatalyst with excellent performance for dye photodegradation. *Chem. Commun.* **2012**, *48*, 2858–2860. [CrossRef] [PubMed]

27. Jiwan, S.; Prasanna, L.L.; Yoon-Young, C.; Jae-Kyu, Y.; Reddy, K.J. Degradation and mechanism of methyl orange by nanometallic particles under a fenton-like process. *Environ. Eng. Sci.* **2017**, *34*, 350–356. [CrossRef]

28. Singh, J.; Chang, Y.-Y.; Koduru, J.R.; Yang, J.-K.; Singh, D.P. Rapid Fenton-like degradation of methyl orange by ultrasonically dispersed nano-metallic particles. *Environ. Eng. Res.* **2017**, *22*, 245–254. [CrossRef]

29. Tammina, S.K.; Mandal, B.K.; Kadiyala, N.K. Photocatalytic degradation of methylene blue dye by nonconventional synthesized SnO_2 nanoparticles. *Environ. Nanotechnol. Monit. Manag.* **2018**, *10*, 339–350. [CrossRef]

30. Dai, S.; Yao, Z. Synthesis of flower-like SnO_2 single crystals and its enhanced photocatalytic activity. *Appl. Surf. Sci.* **2012**, *258*, 5703–5706. [CrossRef]

31. Pal, M.; Bera, S.; Jana, S. Sol–gel based simonkolleite nanopetals with SnO_2 nanoparticles in graphite-like amorphous carbon as an efficient and reusable photocatalyst. *RSC Adv.* **2015**, *5*, 75062–75074. [CrossRef]

32. Hejazi Juybari, S.A.; Milani Moghaddam, H. Facile fabrication of porous hierarchical SnO_2 via a self-degraded template and their remarkable photocatalytic performance. *Appl. Surf. Sci.* **2018**, *457*, 179–186. [CrossRef]

33. Dhanalakshmi, M.; Saravanakumar, K.; Lakshmi Prabavathi, S.; Abinaya, M.; Muthuraj, V. Fabrication of novel surface plasmon resonance induced visible light driven iridium decorated SnO_2 nanorods for degradation of organic contaminants. *J. Alloys. Compd.* **2018**. [CrossRef]

34. Zhang, P.; Wang, L.; Zhang, X.; Shao, C.; Hu, J.; Shao, G. SnO_2-core carbon-shell composite nanotubes with enhanced photocurrent and photocatalytic performance. *Appl. Catal. B Environ.* **2015**, *166–167*, 193–201. [CrossRef]

35. Liang, Y.; Wu, W.; Wang, P.; Liou, S.-C.; Liu, D.; Ehrman, S.H. Scalable fabrication of SnO_2/eo-GO nanocomposites for the photoreduction of CO_2 to CH_4. *Nano Res.* **2018**, *11*, 4049–4061. [CrossRef]

36. Begum, S.; Ahmaruzzaman, M. Biogenic synthesis of SnO_2/activated carbon nanocomposite and its application as photocatalyst in the degradation of naproxen. *Appl. Surf. Sci.* **2018**, *449*, 780–789. [CrossRef]

37. Ding, S.-S.; Huang, W.-Q.; Zhou, B.-X.; Peng, P.; Hu, W.-Y.; Long, M.-Q.; Huang, G.-F. The mechanism of enhanced photocatalytic activity of SnO_2 through fullerene modification. *Curr. Appl. Phys.* **2017**, *17*, 1547–1556. [CrossRef]

38. Hung-Low, F.; Ramirez, D.A.; Peterson, G.R.; Hikal, W.M.; Hope-Weeks, L.J. Development of a carbon-supported Sn–SnO_2 photocatalyst by a new hybridized sol-gel/dextran approach. *RSC Adv.* **2016**, *6*, 21019–21025. [CrossRef]

39. Chen, Y.; Li, H.; Ma, Q.; Zhang, Z.; Wang, J.; Wang, G.; Che, Q.; Yang, P. A novel electrospun approach for highly-dispersed carambola-like SnO_2/C composite microparticles with superior photocatalytic performance. *Mater. Lett.* **2017**, *202*, 17–20. [CrossRef]

40. Mahadik, D.B.; Lee, Y.K.; Park, C.-S.; Chung, H.-Y.; Hong, M.-H.; Jung, H.-N.-R.; Han, W.; Park, H.-H. Effect of water ethanol solvents mixture on textural and gas sensing properties of tin oxide prepared using epoxide-assisted sol-gel process and dried at ambient pressure. *Solid State Sci.* **2015**, *50*, 1–8. [CrossRef]

41. Jung, H.-N.-R.; Parale, V.G.; Kim, T.; Cho, H.H.; Park, H.-H. Zirconia-based alumina compound aerogels with enhanced mesopore structure. *Ceram. Int.* **2018**, *44*, 10579–10584. [CrossRef]

42. Parale, V.G.; Jung, H.-N.-R.; Han, W.; Lee, K.-Y.; Mahadik, D.B.; Cho, H.H.; Park, H.-H. Improvement in the high temperature thermal insulation performance of Y_2O_3 opacified silica aerogels. *J. Alloy. Compd.* **2017**, *727*, 871–878. [CrossRef]

43. Huang, R.; Hou, L.; Zhou, B.; Zhao, Q.; Ren, S. Formation and characterization of tin oxide aerogel derived from sol-gel process based on Tetra(n-butoxy)tin(IV). *J. Non-Cryst. Solids* **2005**, *351*, 23–28. [CrossRef]

44. Zhao, Z.; Chen, D.; Jiao, X. Zirconia aerogels with high surface area derived from sols prepared by electrolyzing zirconium oxychloride solution: Comparison of aerogels prepared by freeze-drying and supercritical CO_2(I) Extraction. *J. Phys. Chem. C* **2007**, *777*, 18738–18743. [CrossRef]

45. Kido, Y.; Nakanishi, K.; Miyasaka, A.; Kanamori, K. Synthesis of monolithic hierarchically porous iron-based xerogels from Iron(III) salts via an epoxide-mediated sol-gel process. *Chem. Mater.* **2012**, *24*, 2071–2077. [CrossRef]

46. Baumann, T.F.; Kucheyev, S.O.; Gash, A.E.; Satcher, J.H. Facile synthesis of a crystalline, high-surface-area SnO_2 aerogel. *Adv. Mater.* **2005**, *17*, 1546–1548. [CrossRef]

47. Rakibuddin, M.; Ananthakrishnan, R. A novel Ag deposited nanocoordination polymer derived porous SnO_2/NiO heteronanostructure for the enhanced photocatalytic reduction of Cr(VI) under visible light. *New J. Chem.* **2016**, *40*, 3385–3394. [CrossRef]

48. Wang, Z.-l.; Xu, D.; Huang, Y.; Wu, Z.; Wang, L.-m.; Zhang, X.-B. Facile, mild and fast thermal-decomposition reduction of graphene oxide in air and its application in high-performance lithium batteries. *Chem. Commun.* **2012**, *48*, 976–978. [CrossRef] [PubMed]

49. Da Cunha, C.R.; Toffolo, G.H.; dos Santos, C.E.I.; Pezzi, R.P. Structural, optical and chemical characterizations of sol-gel grown tin oxide aerogels. *J. Non-Cryst. Solids* **2013**, *380*, 48–52. [CrossRef]

50. Zhang, H.; Lv, X.; Li, Y.; Wang, Y.; Li, J. P25-Graphene composite as a high performance photocatalyst. *ACS Nano* **2010**, *4*, 380–386. [CrossRef] [PubMed]

51. Wan, F.; Lü, H.-Y.; Wu, X.-L.; Yan, X.; Guo, J.-Z.; Zhang, J.-P.; Wang, G.; Han, D.-X.; Niu, L. Do the bridging oxygen bonds between active Sn nanodots and graphene improve the Li-storage properties? *Energy Storage Mater.* **2016**, *5*, 214–222. [CrossRef]

52. Chen, F.-J.; Cao, Y.-L.; Jia, D.-Z. A room-temperature solid-state route for the synthesis of graphene oxide–metal sulfide composites with excellent photocatalytic activity. *Cryst. Eng. Comm.* **2013**, *15*, 4747–4754. [CrossRef]

53. Wang, C.; Shao, C.; Zhang, X.; Liu, Y. SnO_2 nanostructures-TiO_2 nanofibers heterostructures: Controlled fabrication and high photocatalytic properties. *Inorg. Chem.* **2009**, *48*, 7261–7268. [CrossRef] [PubMed]

54. Wang, P.; Zhan, S.; Xia, Y.; Ma, S.; Zhou, Q.; Li, Y. The fundamental role and mechanism of reduced graphene oxide in rGO/Pt-TiO_2 nanocomposite for high-performance photocatalytic water splitting. *Appl. Catal. B Environ.* **2017**, *207*, 335–346. [CrossRef]

55. Cao, Y.; Li, Y.; Jia, D.; Xie, J. Solid-state synthesis of SnO_2—Graphene nanocomposite for photocatalysis and formaldehyde gas sensing. *RSC Adv.* **2014**, *4*, 46179–46186. [CrossRef]

56. Zhu, Y.; Wang, Y.; Yao, W.; Zong, R.; Zhu, Y. New insights into the relationship between photocatalytic activity and TiO_2–GR composites. *RSC Adv.* **2015**, *5*, 29201–29208. [CrossRef]

57. Yadav, H.M.; Kim, J.-S. Solvothermal synthesis of anatase TiO_2—Graphene oxide nanocomposites and their photocatalytic performance. *J. Alloy. Compd.* **2016**, *688*, 123–129. [CrossRef]

58. Jo, W.-K.; Won, Y.; Hwang, I.; Tayade, R.J. Enhanced photocatalytic degradation of aqueous nitrobenzene using graphitic carbon–TiO_2 composites. *Ind. Eng. Chem. Res.* **2014**, *53*, 3455–3461. [CrossRef]

Article

Origin of Room-Temperature Ferromagnetism in Hydrogenated Epitaxial Graphene on Silicon Carbide

Mohamed Ridene, Ameneh Najafi and Kees Flipse *

Molecular Materials and Nanosystems, Eindhoven University of Technology,
5600 MB Eindhoven, The Netherlands; ridane_m@outlook.com (M.R); ameneh.na@gmail.com (A.N.)
* Correspondence: c.f.j.flipse@tue.nl; Tel.: +32-402-474-118

Received: 4 January 2019; Accepted: 30 January 2019; Published: 8 February 2019

Abstract: The discovery of room-temperature ferromagnetism of hydrogenated epitaxial graphene on silicon carbide challenges for a fundamental understanding of this long-range phenomenon. Carbon allotropes with their dispersive electron states at the Fermi level and a small spin-orbit coupling are not an obvious candidate for ferromagnetism. Here we show that the origin of ferromagnetism in hydrogenated epitaxial graphene with a relatively high Curie temperature (>300 K) lies in the formation of curved specific carbon site regions in the graphene layer, induced by the underlying Si-dangling bonds and by the hydrogen bonding. Hydrogen adsorption is therefore more favourable at only one sublattice site, resulting in a localized state at the Fermi energy that can be attributed to a pseudo-Landau level splitting. This $n = 0$ level forms a spin-polarized narrow band at the Fermi energy leading to a high Curie temperature and larger magnetic moment can be achieved due to the presence of Si dangling bonds underneath the hydrogenated graphene layer.

Keywords: hydrogenated epitaxial graphene; electronic structure; ferromagnetism

---

1. Introduction

In the last decade, chemical functionalization of graphene attracted a good deal of attention among scientists for various reasons [1–23]. The common motivation is the modification of electronic properties of graphene. The electronic properties are much more affected under the influence of covalent functionalization, which encourages researchers to focus on studying the decoration of various defects in graphene [3,5,6,10,13,15,16]. In this way, it is also possible to make graphene spintronic devices with the aim of utilizing spintronics technology [3,4,6,8–10,12,15,22]. An important issue in exploring graphene-based spintronics is how to introduce magnetic order in graphene and in particular ferromagnetism. In this approach, ferromagnetism was predicted theoretically for hydrogenated graphene [4,18,24]. For example, using spin-polarized density functional theory (DFT), Zhou *et al.* [17] have shown that semihydrogenated graphene, a layer of graphene in which half of the carbon atoms (all belongs to one sublattice) are bonded to an hydrogen atom, is a ferromagnet with an estimated Curie temperature between 278 K and 417 K. The suggested mechanism is the coupling of unpaired localized electrons, present at non-hydrogenated carbon atoms due to the breaking of the π bonding network of graphene. Their calculations show that a ferromagnetic configuration is the ground state of such a system. In principle, defects are associated with lattice distortions, vacancies or chemical functionalization, introducing localized states at the Fermi level in graphene [2]. Onsite Coulomb electron–electron interactions lead to spin-polarization of these dispersionless flat states at the Fermi level and thus the existence of local magnetic moments. The magnetic properties of graphene are determined by the interaction of the mentioned localized states. In the literature, several different mechanisms have been proposed for ferromagnetism of graphene. In Ref. [4] the authors calculated the coupling of defect-induced extended states, assuming two different types of defects, the hydrogen chemisorption defect and the vacancy defect by first principles electronic structure calculations, showing either ferromagnetic or antiferromagnetic ordering depending on whether the defects are distributed on the same or different sublattices of the graphene lattice. In this last approach, room-temperature ferromagnetism is explored theoretically for hydrogenated graphene, but experimentally it was only obtained for epitaxial graphene on silicon carbide (SiC) [25]. This was the motivation to study the interplay between the structural properties and the electronic structure of hydrogenated epitaxial graphene.

We will show that in the case of epitaxial graphene specifically on SiC, the presence of the Si-dangling bonds creates preferential sites for hydrogen adsorption [26]. This leads to a preferred sublattice carbon occupation of H atoms, which is necessary for ferromagnetic behaviour. Importantly, our ab-initio calculations show that the created localized state at the Fermi energy can be attributed to a pseudo-Landau level splitting. This $n = 0$ level forms a spin-polarized narrow band at the Fermi energy leading to a high Curie temperature and larger magnetic moment can be achieved due to the presence of Si dangling bonds underneath the hydrogenated graphene layer. Hence, at first the structural properties of epitaxial graphene on SiC and the effect of hydrogen chemisorption on magnetic ordering will be described, followed by our ab-initio calculations of the electronic structure of the system will be discussed.

2. Materials and Methods

Ab-initio calculations were performed based on the density functional theory (DFT) within the generalized gradient approximation (GGA) as implemented in the SIESTA computational code [27] within the Perdew-Burke-Ernzerhof (PBE) form [28]. The core electrons were replaced by Troulier-Martins pseudopotentials [29]. A double-ζ basis set of localized atomic orbitals was used for the valence electrons. Sampling of the Brillouin zone has been performed using by a $(10 \times 10 \times 1)$ shifted Monhorst-Pack grid [30], while a mesh cut-off energy of 200 Ry has been imposed for real-space integration. All structures have been relaxed until forces were less than 0.05 eV/ Å. The influence of the SiC substrate was introduced by an extra Hubbard-U term to account for electron correlation as a result of the Coulomb interaction between the Si localized electrons [31]. Therefore, this correction which is an effective potential defined by $U_{eff} = U - J$, where U and J are the Coulomb and the exchange parameters [32] are only applied to the 3p orbitals of the Si-dangling bonds. In the calculations, U_{eff} is considered equal to 3 eV which is in a good agreement with the experimental results [31]. A vacuum interval of more than 10 Å was used to avoid the interaction between the periodic slabs.

3. Results and Discussion

3.1. Preferential Hydrogen Adsorption Sites in Epitaxial Graphene on SiC(0001)

Upon annealing SiC at high temperature, growing graphene on the Si face of 6H-SiC substrate, a first carbon layer strongly bonded to the substrate is formed. This layer does not present the electronic properties of graphene since it is interacting with the substrate by means of the covalent bonds between the C atoms of this layer and the Si atoms of the substrate surface. However not all the C atoms bond to the Si atoms, so there are some Si-dangling bonds existing underneath. This layer is well known as the buffer layer. The first layer of the C sheet on top of the buffer layer is considered as the first layer of graphene which shows the linear dispersion relation at the Dirac point. Characteristic for scanning tunneling microscopy (STM) images of epitaxial graphene on SiC(0001) is the observed Moiré pattern due to the presence of the buffer layer below the graphene layer [33]. Instead of a $(6\sqrt{3} \times 6\sqrt{3})R30°$ reconstruction, expected from a reconstruction point of view with respect to SiC lattice and observed in low energy electron diffraction (LEED) pattern of epitaxial graphene [34], STM images show only a superstructure with a quasi $- (6 \times 6)$ SiC periodicity (~ 1.8 nm) due to electron interference effects [see Figure 2(d)] [35]. This superstructure is related to the existence of intrinsic curved regions in epitaxial graphene [35]. Ab-initio calculations have shown that the Si-dangling bonds in the buffer layer induce these curved regions and, consequently, the graphene layer mimics this morphology to form a quasi—(6×6)SiC pattern of bright spots observed in the STM image (Figure 2d). To clarify that, the total charge density of the buffer layer states is shown in Figure 1a–c. As can be seen from Figure 1(a), an apparent quasi − (6×6)SiC modulation, denoted by the full line diamond cell, is recognized in the total charge density of buffer layer states.

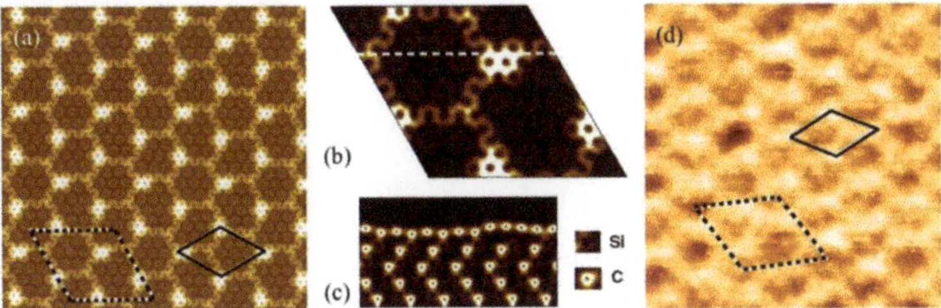

Figure 1. Buffer layer structure on top of the Si-terminated SiC surface (reprinted figure from Ref. [35], with the permission of APS Publishing). (**a**) 11 × 11 nm^2 image of the total ab-initio charge density; the common $(6\sqrt{3}\times6\sqrt{3})R30°$ SiC unit cell is shown by the dashed line diamond cell and the incommensurate (6×6) SiC modulation is denoted by full line diamond cell; (**b**) The irregular hexagonal patterns inside a $(6\sqrt{3}\times6\sqrt{3})R30°$ SiC unit cell with a cross section along the defined dashed line shown in (**c**); (**d**) 12 × 12 nm^2 STM image of the buffer layer (V = −2.0 V and I = 0.5 nA); the superstructure (6×6)SiC is clearly visible.

Patterns with dark regions are separated from each other by brighter borders. The dark regions represent the C atoms which are covalently bounded to the Si atoms of the substrate and the bright spots are associated with the C atoms with no interaction with the substrate, so leaving the Si atom with a dangling bond underneath. Figure 1b shows a few of these irregular hexagonal shapes and the cross-section of those is depicted in Figure 1c. Here one can see the curving of the surface due to the existence of Si-dangling bond. These convex regions are mimicked by the first layer of graphene, as shown in Figure 2a.

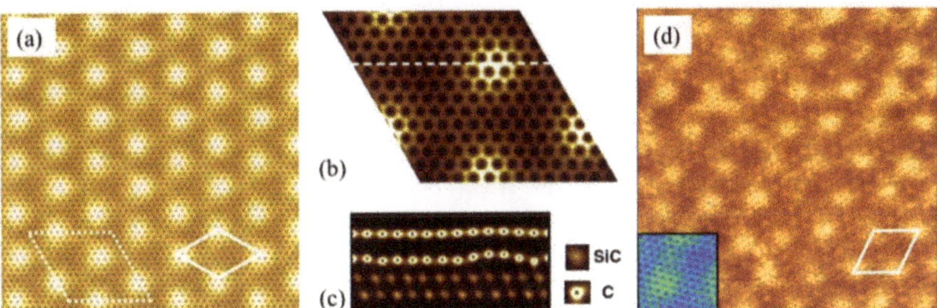

Figure 2. The first graphene layer on top of the buffer layer of Si-terminated SiC surface (reprinted figure from Ref. [35], with the permission of APS Publishing). (**a**) 11 × 11 nm^2 image of the total ab-initio charge density; the common $(6\sqrt{3}\times6\sqrt{3})R30°$ SiC unit cell and the incommensurate (6×6)SiC modulation are defined by a dashed line and a full line diamond cell respectively; (**b**) The total charge density inside the $(6\sqrt{3}\times6\sqrt{3})R30°$ SiC unit cell; (**c**) The cross section of the dashed line defined in (**b**); (**d**) 12 × 12 nm^2 STM image of the monolayer graphene (V= −0.2 Vand I= 0.1 nA); the periodicity of (6×6)SiC is indicated by a full line.

Therefore, the total charge modulation density of graphene layer states exhibits a honeycomb lattice plane containing corrugations [see cross section in Figure 2c] with the quasi − (6×6)SiC periodicity. The corrugations in graphene layer are smoother than those of the buffer layer. There is some evidence which indicates that these convex regions are favourable sites for hydrogen adsorption [26]. STM images of an epitaxial graphene on SiC before and after hydrogenation are displayed in Figure 3. According to Ref. [26], hydrogenation is performed with a very low coverage of $0.76 \pm 0.17\%$ of the surface. Comparing Figure 3a,b, they clearly show protrusions at the corners of (6×6)SiC superstructures after hydrogenation process. This observation indicates that hydrogen atoms preferentially attach to convexly curved regions of monolayer graphene grown on SiC(0001) but not to the concave areas.

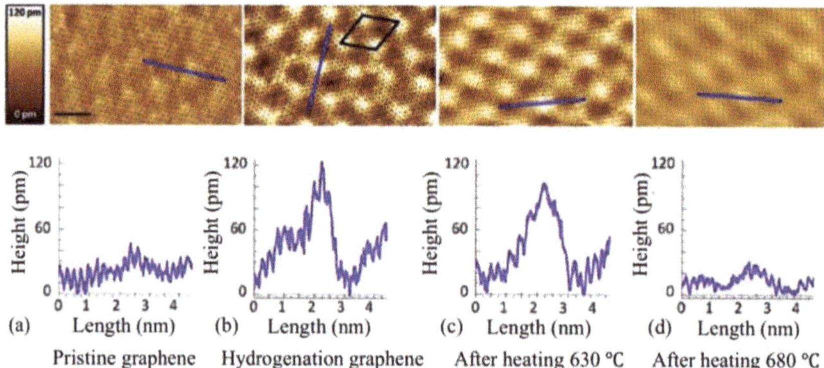

Figure 3. Room-temperature STM images of pristine monolayer graphene and hydrogenated graphene on Si terminated SiC (reprinted figure from Ref. [26], with the permission of ACS Publishing). (a) Pristine graphene (V = 115 mV and I = 0.3 nA); (b) hydrogenated graphene with a low coverage of 0.76±0.17% (V = 50 mV and I = 0.3 nA); (c) hydrogenated graphene after annealing for 5 min at 630 °C (V = 50 mV and I = 0.3 nA); (d) hydrogenated graphene after annealing for 5 min at 680 °C (V = 50 mV and I = 0.3 nA). The cross sections, shown below each STM image, indicate the increase of corrugation at the corners of the quasi $-$ (6×6) superstructure after hydrogenation and decreasing of them by annealing the hydrogenated graphene due to desorption of hydrogen atoms. Scale bar is 2 nm and the sizes of images as well as the cross section lines are the same.

Thereafter, the sample is heated and it is shown that above ∼630 °C, hydrogen atoms start to desorb from the surface, and as a result the height of the protrusions decreases and at the point where the sample is annealed up to 680 °C, the corrugations came back to the levels of pristine graphene. A (6×6)SiC superstructure with respect to SiC surface has a periodicity of ∼1.8 nm, corresponding to ∼7.5 times the lattice parameter of graphene (0.24 nm), and recent theoretical work has shown that hydrogen adsorption is more favourable at only one sublattice (A or B) with sublattice distances larger than 1 nm [36]. Since magnetic moments arise from this hydrogen adsorption on graphene (see below), an imbalance of occupied sublattice sites will lead to a ferromagnetic ground state obeying the Lieb theorem [37]. This defines the ground state total spin angular momentum of the system with a bipartite lattice and a half-filled band based on the difference in the number of sites in the A and B sublattices, i.e., $S = 1/2||A| - |B||$, yielding itinerant-electron ferromagnetism with $|A| \neq |B|$. In the case of hydrogenated graphene, the difference is between the occupation of A or B sublattice sites with hydrogen. In epitaxial graphene the Si-dangling bond induced curving of the graphene layer creates a superlattice for hydrogen adsorption, preventing a random occupation of both sublattices which would be the case in hydrogenated quasi-free standing epitaxial graphene that does not show a ferromagnetic response [25].

3.2. Room-Temperature Ferromagnetism in Hydrogenated Epitaxial Graphene on SiC(0001)

We performed electronic structure calculations on hydrogenated graphene on SiC based on density functional theory (DFT) calculations using the SIESTA code. In first instance order the influence of hydrogen chemisorption on curved graphene was studied. The curved graphene layer with H adsorbed system was relaxed, so the initial curvature is not necessarily kept. An effective electron doping of the graphene layer, as is the case for graphene on SiC, was simulated by a shift of the Dirac point of 0.5 eV below the Fermi Energy; after adding a hydrogen atom and keeping the same value of the doping, the state is shifted towards the Fermi energy by ~0.3 eV, as is shown in Figure 4b. This indicates a possible hydrogen induced p type doping. The role of SiC and the buffer layer will be considered explicitly in Section 3.3. We considered the graphene layer as a curved layer to retain the induced rippling with a (6×6)SiC superstructure of the underlying buffer layer. In order to simulate the quasi − (6×6)SiC pattern, a curved $(7\times7)_G$ cell containing 98 carbon atoms was applied (the subscript "G" is used to indicate that the lattice parameter is seven times the lattice parameter of graphene, i.e., 0.24 nm). This is a reasonable choice for a DFT calculation of the electronic structure since it has a periodicity close to the quasi − (6×6) SiC periodicity observed in STM images of epitaxial graphene on SiC (Figure 2d). According to the observations of Ref. [26], hydrogen atoms preferentially bond to sites where the lattice is maximally convexly curved, which enhances the transition from sp^2 to sp^3. Therefore, we added one hydrogen atom to the highest carbon atom in the lattice unit cell of calculations. The relaxed structure of the curved (7×7)G graphene cell with one H adatom is shown in Figure 4a.

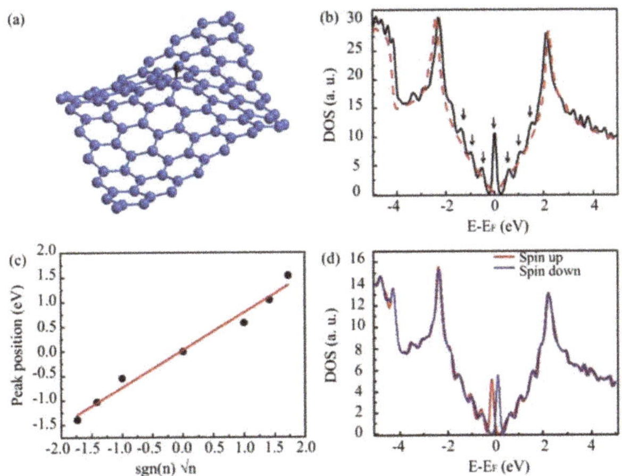

Figure 4. Pseudo-Landau levels in hydrogenated epitaxial graphene. (**a**) Relaxed curved (7×7) graphene cell upon hydrogen adsorption; (**b**) The total density of states of the structure in (a) (black line) which exhibit changes compared to the total DOS of curved graphene (red dashed line); (**c**) Fitting of the localized state appearing at the Dirac point and other states symmetric to it using the Landau levels dispersion in grapheme; (**d**) Spin-polarized DOS showing the pseudo-Landau levels exchanged split.

The adsorbed H atom breaks a π bond and slightly pulls out the carbon atom below due to a covalent bond interaction. This carbon atom changes its hybridization state from sp^2 to sp^3 and the overall structure changes, causing a structural modulation, as was discussed in Reference [4]. The corresponding DOS (black solid line in Figure 4b) exhibits relatively intense peaks compared to the DOS of the curved $(7\times7)_G$ (red dot line). The state at the Fermi energy (which is the position of the Dirac point [E_D]) has been obtained and reported by previous work [4] as arising from the carbon hydrogen bond, but the presence of states symmetrically located around it reflects another nature. Interestingly, similar as in Ref. [38], the contributing peaks can be fitted to a series of Landau levels in graphene (Figure 4c), satisfying the relation:

$$E_n = E_D \pm \sqrt{2ev_F\hbar B_s|n|} \tag{1}$$

where, E_D is Dirac energy, e is elementary charge, v_F is Fermi velocity, $\hbar$ is reduced Planck constant (h/2π) and B_s is a pseudo-magnetic field. This indicates that the observed peaks in the DOS can be assigned to pseudo-Landau levels (PLLs). It is worth to note that the H-induced PLLs are not only caused by "mechanical" strain (as is clear from red dashed line of Figure 4b which belongs to curved graphene cell) but the redistribution of the hopping integral values is also changed by the chemical change from sp^2 to sp^3 in a similar way as for strained graphene with a triangular symmetry [38,39]. As interpreted in Refs. [40,41], structural modulations cause a perturbation in the hopping parameter such as t→t+δt for a C atom experiencing strain. As a result, the energy eigenvalues of the system change with the variation of the wave vector $\vec{k} \rightarrow \vec{k} + \delta\vec{k}$. In analogy to a 2D epitaxial graphene submitted to a perpendicular magnetic field where the momenta of the electrons are shifted like $\vec{k} \rightarrow \vec{k} + (e/h)\vec{A}$ ($\vec{A}$ is the vector potential), the modulation in the hopping parameter t due to the C-H sp^3 hybridization induces an effective vector potential (pseudo-vector potential) and, hence, a pseudo-magnetic field. Using a Fermi velocity of 1.0×10^6 ms^{-1}, the resulting pseudo-magnetic field extracted from the fit is ~250 T and its value increases for higher hydrogen coverage and vice versa. This is in agreement with previous theoretical results [39].

From the value of the pseudo-magnetic field, it is noteworthy that the magnetic length ($l_B \cong 26/\sqrt{B[T]}$nm) is comparable to the distance between the hydrogen atoms, i.e., the unit cell used in the calculations. This is important since it indicates that the entire region between hydrogen atoms is affected by the magnetic field. Nevertheless, the dependence of the magnetic field magnitude on the cell leads to an extremely inhomogeneous magnetic field distribution on the surface that contributes to the localization of graphene quasiparticles [42]. By including exchange interactions between C atom and H atoms into the calculations, similar as what was shown by Yazyev and Helm [4], the half-filled localized state at the Fermi level (PLL $n = 0$) will split into two states; filled state (spin up) and empty state (spin down) denoted by red and blue lines in Figure 4d, respectively. In this case, the spin-splitting is ~200 meV corresponding to the coverage of ~1%, which is expected to increase for higher coverage [4]. So, we can concluded that the exchange interaction, which leads to ferromagnetic behavior, arises from the C–H chemisorption induced pseudo-Landau level states at the Fermi level in hydrogenated graphene. According to Lieb's theorem [37], the condition to have magnetic order is an unequal occupation of A and B sublattices by hydrogen. This statement is clearly illustrated in Figure 5.

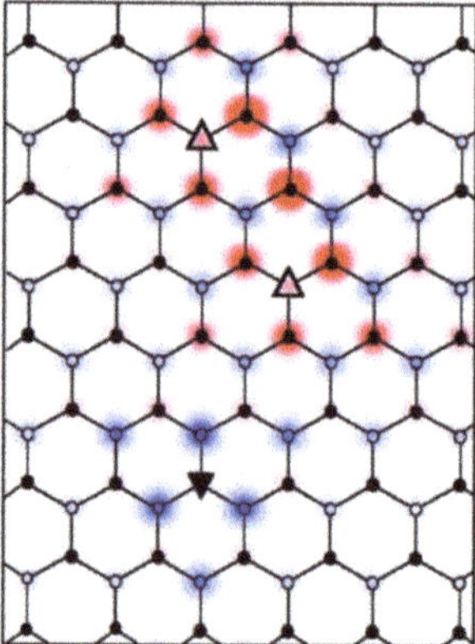

Figure 5. Spin density distribution of hydrogenated graphene with two hydrogen attachments on sublattice A and one hydrogen chemisorption on sublattice B of graphene (reprinted figure from Ref. [4] with the permission of APS Publishing). Red and blue represents opposite spin directions.

The exchange interaction between localized electron states leads to a texture of magnetic states on the H site and the odd nearest neighbours of the carbon atom below H, which exhibits a triangular symmetry and extends to similar regions around the H atoms. The preference for H adsorption on certain sites in epitaxial graphene on SiC is due to the existence of Si-dangling bonds underneath, i.e., the corners of the quasi − (6×6) SiC superstructure. Our calculation results for the isosurface of the spin density distribution of four hydrogen adsorbed atoms, occupying an A sublattice at the corners of the quasi − (6×6) SiC superstructure, are shown in Figure 6a.

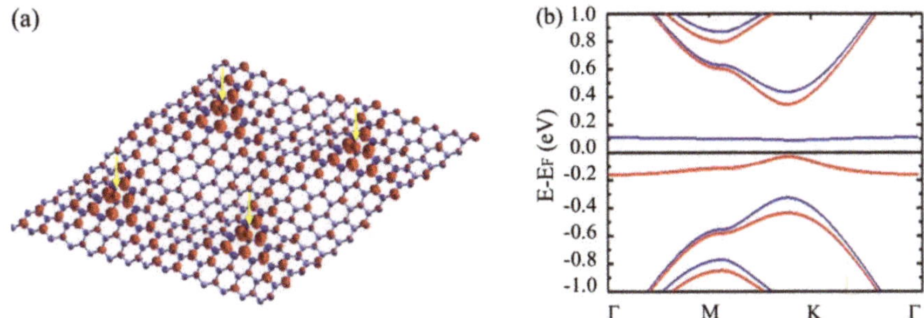

Figure 6. (a) Isosurface representation of the spin density distribution around four hydrogen adatoms on sublattice A sites (indicated by arrows); (b) Spin-polarized band structure of hydrogenated (7×7) graphene cell. Spin-up bands are in red and spin-down bands are in blue.

The attached hydrogen atoms (remarked by arrows in the figure) are separated from each other by ~1.7 nm, and, as clearly seen, the localized electron states extend over a few graphene unit cells around the H atoms. The long-range magnetic coupling is therefore attributed to these connected regions of spin-polarized quasi-localized states.

Very unexpectedly, experimental results of hydrogenated epitaxial graphene show room-temperature ferromagnetism [25,43]. However, s-p electron systems can generate ferromagnetism with relatively high Curie temperatures if a narrow band of localized states is formed in a gap, or a pseudo-gap, which leads to a high value of magnon stiffness so that the low-lying spin fluctuations do not reduce the critical ordering temperature in such systems [44]. The (spin-polarized) band structure of hydrogenated graphene is shown in Figure 6b. It is obvious that the hydrogenation process leads to the disappearence of the linear dispersion of pristine graphene at the K point and opens a gap. In addition, the PLL $n = 0$ gives rise to two exchange split states lying in a band gap which is defined by the energy difference between the PLLs $n = 1$ and $n = -1$. The bandwidth of the filled (spin up) band is ~0.1 eV and the empty (spin down) band is almost flat. This is in agreement with Ref. [44], therefore, the Curie temperature can be deduced from the exchange splitting of the PLL $n = 0$ which is ~200 meV. This value is well above k_BT (k_B is the Boltzmann constant) at room temperature.

3.3. Substrate Effect on the Electronic Properties of Hydrogenated Epitaxial Graphene

In the previous sections, the origin of the room-temperature ferromagnetic property of hydrogenated graphene on SiC was, for simplicity, discussed excluding the effect of the SiC substrate and buffer layer in the electronic structure calculations, just to indicate the connection between the role of local strain and the existence of pseudo-Landau levels. So far, a ferromagnetic behaviour is reported only for hydrogenated graphene, which is grown on a SiC substrate [25,43]. So it is important to include the effect of SiC as well as buffer layer, explicitly, in the calculations to find out their specific role. For the most realistic calculation a $(6\sqrt{3}\times6\sqrt{3})R30°$ unit cell with respect to SiC should be taken into account, which means a cell of (13×13) graphene units. This is a large cell and a time-consuming calculation, therefore it is very common in calculations to replace the $(6\sqrt{3}\times6\sqrt{3})R30°$ cell periodicity with a $(\sqrt{3}\times\sqrt{3})R30°$ SiC cell which contains (2×2) graphene units [45,46]. In order to get commensurability between a (2×2) graphene cell and a $(\sqrt{3}\times\sqrt{3})$ SiC cell rotated by 30 degrees, stretching of the graphene layer by 8% is needed. Adsorbing a hydrogen atom will overestimate the C-C bond, which will break for sp^2 to sp^3 hybridization. In order to avoid this, we compress the $(\sqrt{3}\times\sqrt{3})$R30∘ SiC by 7% instead of stretching it. This leads to a very high coverage of hydrogen ($\sim$12%) atoms in the unit cell and therefore it was replaced by a (4×4) graphene cell, commensurate with a $(2\sqrt{3}\times2\sqrt{3})$R30∘ SiC substrate, while keeping the 7% compression of the substrate to get the commensurability between the two lattices. This was also used for epitaxial two-layer graphene on the Si-face of SiC by Gao and Tosatti [47]. Adding one H atom to this unit cell results in a $\sim$3% hydrogen coverage. The relaxed structure in the presence of H adatom is shown in Figure 7. It is worth mentioning that the geometry relaxation took place in different steps. Firstly, the SiC layers were saturated by hydrogen atoms on the C-face (bottom face) and were relaxed until forces were less than 0.05 eV/Å. Secondly, the first graphene layer is added on top and the structure was free to relax except the bottom SiC layer; this led to the buffer layer formation. The second graphene layer is then added and relaxed and finally the hydrogen atom is added on top of the highest carbon atom of the graphene layer (which coincides with the position of the Si atom with a dangling bond in the buffer layer); as previously mentioned, all the atoms were free to move except the bottom SiC layer and the structure is relaxed until forces were less than 0.05 eV/Å.

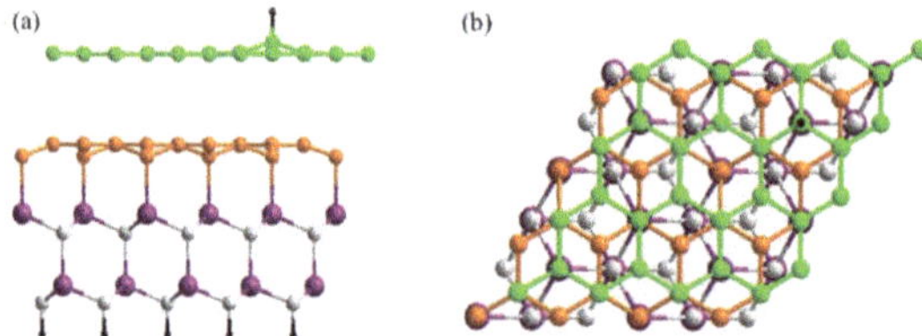

Figure 7. Ball and stick images of hydrogenated Si-face epitaxial graphene with 3% hydrogen coverage. (**a**) Side view and (**b**) top view of hydrogenated epitaxial graphene. The $((2\sqrt{3}\times2\sqrt{3})$R30° SiC substrate (violet for Si atoms and gray for carbon atoms) is saturated by hydrogen atoms (black) and compressed to host a (4×4) graphene cell to model the buffer layer (orange atoms). The epitaxial graphene layer (green atoms) is in a Bernal stacking with the buffer layer and has the same dimensions (4×4).

From the side view of the structure in Figure 7a, it is clear that two compressed bilayers of 6H-SiC were used with the dangling bonds on the bottom C-face saturated with the hydrogen atoms. On top of SiC layers, a (4×4) graphene honeycomb lattice is added to model the buffer layer, and then a second (4×4) graphene lattice is added in a Bernal stacking with the first one to form the epitaxial graphene layer. The curving of the graphene layer is not so clear due to the small cell that we used. Ab-initio calculations were performed based on the density functional theory (DFT) within the generalized gradient approximation (GGA) as implemented in the SIESTA computational code within the Perdew–Burke–Ernzerhof (PBE) form.

The spin unrestricted calculated band structure of epitaxial graphene and hydrogenated epitaxial graphene on SiC are shown in Figure 8a,b, respectively. In Figure 8a, the n-doped characteristic of epitaxial graphene on SiC is clearly evidenced by the position of Dirac point shifted to $\sim$0.5 eV below the Fermi level at the K point, this is consistent with the previous experimental and theoretical results [34,46]. However, unlike other results [45], our calculations result in a spin-polarized band structure for epitaxial graphene. This could be interpreted as a side effect of the defined unit cell in the calculations in which the SiC substrate is compressed.

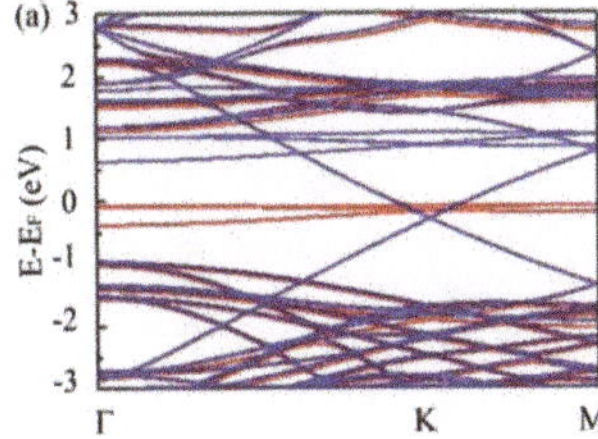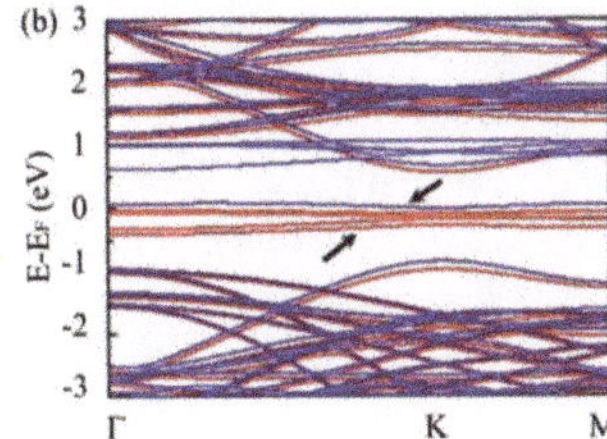

Figure 8. Influence of the SiC substrate and the buffer layer on the band structure of epitaxial graphene and hydrogenated epitaxial graphene. (**a**) Spin-polarized band structure of epitaxial graphene (red curve for up-spin, blue curve for down-spin). The Dirac point is situated 0.5 eV below Fermi energy; (**b**) Spin-polarized band structure of hydrogenated epitaxial graphene (red curve for up-spin, blue curve for down-spin). The arrows indicates the spin-splitted pseudo-Landau level $n = 0$.

However, it is very likely to be attributed to Coulomb interactions between the localized electrons of the Si-dangling bonds. Looking more closely, while no spin polarization was obtained for the graphene π-electrons, each localized state attributed to the Si-dangling bonds [45,46] splits into two spin-polarized states, one filled and one empty. This may explain the paramagnetic response of the buffer layer ascribed to the presence of the Si-dangling bond localized states in the buffer layer [25], resulting in a total magnetic moment of $\sim4\mu B$ per unit cell considered in the calculations. Comparing the spin-polarized band structure of hydrogenated epitaxial graphene (Figure 8b) with epitaxial graphene (Figure 8a) shows notable differences. It is shown that the linear dispersion relation disappears at the Dirac point. Additional narrow spin-polarized bands (indicated by arrows in Figure 8b) appear besides the remaining intact (or with a small position change) spin-polarized bands of the Si-dangling bonds. These two extra spin-polarized bands correspond to the split PLL $n = 0$. In the geometry we utilized, the hydrogen coverage is $\sim3\%$; this leads to the observation of a filled (majority spin) pseudo-Level $n = 0$ at about 0.3 eV below the Fermi energy at the K point. This value may decrease if the hydrogen coverage is less and vice versa [4]. As stated above, the energy "gap" between the PLLs n = 0 and $n = -1$ decreases when the hydrogen coverage is reduced since it is related to the pseudo-magnetic field.

In addition, the calculated total magnetic moment without the SiC nor the buffer layer contribution but with the simulated effective electron doping as discussed in 3.2, is 0.5 μB per H adatom. This value is smaller than the one obtained when the Si-dangling bonds of the buffer layer are present in 3.3, which is 0.7 μB per H adatom. This indicates that Coulomb interactions between localized Si-dangling bond states and the quasi-localized PLL $n = 0$ could play a role leading to an increase in the exchange splitting of this latter and hence gives rise to a larger magnetic moment.

4. Conclusion

In conclusion, our electronic structure calculations in combination with published experimental results provides a coherent picture to understand the observed room-temperature ferromagnetism in hydrogenated epitaxial graphene on SiC. We have theoretically shown that specific carbon sites, due to the presence of Si-dangling bonds in the underlying buffer layer, result in a quasi (6×6)SiC superstructure of epitaxial graphene, favouring one sublattice site hydrogen occupation. The chemical modification of the graphene layer by H adatoms and strain-like effects generate a pseudo-magnetic field and consequently localized pseudo-Landau levels in the electron density of states. Extended spin-polarized electron states from the pseudo-Landau level $n = 0$ give rise to long-range ferromagnetic coupling. The relative high Curie temperature is obtained due the narrow bandwidth of these states formed in a bulk energy gap leading to a high magnon stiffness. Furthermore, Coulomb interactions between the pseudo-Landau level and the Si-dangling bond states can lead to an increase of the total magnetic moment. The presence of localized electron states close to the Fermi level is evidenced by photoemission spectroscopy measurements, which also indicates the p-doping of epitaxial graphene due to hydrogenation.

Author Contributions: M.R. performed the calculations; all the authors analyzed the data and wrote the paper.

Funding: The authors are thankful to the Dutch Organization for Scientific Research (NWO) under Project No. NWO-nano11447, and by the European Regional Development Fund under Project No. CEITEC- CZ.1.05/1.1.00/02.0068.

Conflicts of Interest: The authors declare no conflicts of interest.

References

1. Deng, W.Q.; Xu, X.; Goddard, W.A. New alkali doped pillared carbon materials designed to achieve practical reversible hydrogen storage for transportation. *Phys. Rev. Lett.* **2004**, *92*, 166103. [CrossRef] [PubMed]
2. Pereira, V.M.; Guinea, F.; Lopes dos Santos, J.M.B.; Peres, N.M.R.; Castro Neto, A.H. Disorder induced localized states in graphene. *Phys. Rev. Lett.* **2006**, *96*, 036801. [CrossRef] [PubMed]
3. Gunlycke, D.; Li, J.; Mintmire, J.W.; White, C.T. Altering low-bias transport in zigzag-edge graphene nanostrips with edge chemistry. *Appl. Phys. Lett.* **2007**, *91*, 112108. [CrossRef]
4. Yazyev, O.V.; Helm, L. Defect-induced magnetism in graphene. *Phys. Rev. B* **2007**, *75*, 125408. [CrossRef]
5. Filho, R.N.C.; Farias, G.A.; Peeters, F.M. Graphene ribbons with a line of impurities: Opening of a gap. *Phys. Rev. B* **2007**, *76*, 193409. [CrossRef]
6. Kan, E.J.; Li, Z.; Yang, J.; Hou, J.G. Half-metallicity in edge-modified zigzag graphene nanoribbons. *J. Am. Chem. Soc.* **2008**, *130*, 4224–4225. [CrossRef] [PubMed]
7. Robinson, J.P.; Schomerus, H.; Oroszlány, L.; Fal'ko, V.I. Adsorbate-limited conductivity of graphene. *Phys. Rev. Lett.* **2008**, *101*, 196803. [CrossRef]
8. Zanella, I.; Guerini, S.; Fagan, S.B.; Filho, J.M.; Filho, A.G.S. Chemical doping-induced gap opening and spin polarization in graphene, *Phys. Rev. B* **2008**, *77*, 073404. [CrossRef]
9. Boukhvalov, D.W.; Katsnelson, M.I. Chemical functionalization of graphene with defects. *Nano Lett.* **2008**, *8*, 4373–4379. [CrossRef]
10. Cervantes-Sodi, F.; Csányi, G.; Piscanec, S.; Ferrari, A.C. Edge-functionalized and substitutionally doped graphene nanoribbons: Electronic and spin properties. *Phys. Rev. B* **2008**, *77*, 165427. [CrossRef]
11. Chan, K.T.; Neaton, J.B.; Cohen, M.L. First-principles study of metal adatom adsorption on graphene. *Phys. Rev. B* **2008**, *77*, 235430. [CrossRef]
12. Boukhvalov, D.W.; Katsnelson, M.I.; Lichtenstein, A.I. Hydrogen on graphene: Electronic structure, total energy, structural distortions and magnetism from first-principles calculations. *Phys. Rev. B* **2008**, *77*, 035427. [CrossRef]
13. Lin, Y.; Ding, F.; Yakobson, B.I. Hydrogen storage by spillover on graphene as a phase nucleation process. *Phys. Rev. B* **2008**, *78*, 041402. [CrossRef]
14. Miwa, R.H.; Martins, T.B.; Fazzio, A. Hydrogen adsorption on boron doped graphene: An ab initio study. *Nanotechnology* **2008**, *19*, 155708. [CrossRef] [PubMed]
15. Huang, B.; Liu, F.; Wu, J.; Gu, B.L.; Duan, W. Suppression of spin polarization in graphene nanoribbons by edge defects and impurities, *Phys. Rev. B* **2008**, *77*, 153411. [CrossRef]
16. Marconcini, P.; Cresti, A.; Triozon, F.; Fiori, G.; Biel, B.; Niquet, Y.-M.; Macucci, M.; Roche, S. Atomistic boron-doped graphene field-effect transistors: A route toward unipolar characteristics. *ACS Nano* **2012**, *6*, 7942–7947. [CrossRef] [PubMed]
17. Boukhvalov, D.W.; Katsnelson, M.I. Tuning the gap in bilayer graphene using chemical functionalization: Density functional calculations. *Phys. Rev. B* **2008**, *78*, 085413. [CrossRef]
18. Zhou, J.; Wang, Q.; Sun, Q.; Chen, X.S.; Kawazoe, Y.; Jena, P. Ferromagnetism in Semihydrogenated Graphene Sheet. *Nano Lett.* **2009**, *9*, 3867–3870. [CrossRef]
19. Elias, D.C.; Nair, R.R.; Mohiuddin, T.M.G.; Morozov, S.V.; Blake, P.; Halsall, M.P.; Ferrari, A.C.; Boukhvalov, D.W.; Katsnelson, M.I.; Geim, A.K.; et al. Control of graphene's properties by reversible hydrogenation: evidence for graphane. *Science* **2009**, *323*, 610–613. [CrossRef]
20. Bostwick, A.; McChesney, J.L.; Emtsev, K.V.; Seyller, T.; Horn, K.; Kevan, S.D.; Rotenberg, E. Quasiparticle transformation during a metal-insulator transition in graphene. *Phys. Rev. Lett.* **2009**, *103*, 056404. [CrossRef]

21. Bostwick, A.; Ohta, T.; McChesney, J.L.; Emtsev, K.V.; Speck, F.; Seyller, T.; Horn, K.; Kevan, S.D.; Rotenberg, E. The interaction of quasi-particles in graphene with chemical dopants. *New J. Phys.* **2010**, *12*, 125014. [CrossRef]

22. Liu, Y.; Tang, N.; Wan, X.; Feng, Q.; Li, M.; Xu, Q.; Liu, F.; Du, Y. Realization of ferromagnetic graphene oxide with high magnetization by doping graphene oxide with nitrogen. *Sci. Rep.* **2013**, *3*, 2566. [CrossRef] [PubMed]

23. Li, H.; Daukiya, L.; Haldar, S.; Lindblad, A.; Sanyal, B.; Eriksson, O.; Aubel, D.; Hajjar-Garreau, S.; Simon, L.; Leifer, K. Site-selective local fluorination of graphene induced by focused ion beam irradiation. *Sci. Rep.* **2016**, *6*, 19719. [CrossRef] [PubMed]

24. Yazyev, O.V. Magnetism in disordered graphene and irradiated graphite. *Phys. Rev. Lett.* **2008**, *101*, 037203. [CrossRef]

25. Giesbers, A.J.M.; Uhlírová, K.; Konečny, M.; Peters, E.C.; Burghard, M.; Aarts, J.; Flipse, C.F.J. Interface-induced room-temperature ferromagnetism in hydrogenated epitaxial graphene. *Phys. Rev. Lett.* **2013**, *111*, 166101. [CrossRef]

26. Goler, S.; Coletti, C.; Tozzini, V.; Piazza, V.; Mashoff, T.; Beltram, F.; Pellegrini, V.; Heun, S. Influence of graphene curvature on hydrogen adsorption: toward hydrogen storage devices. *J. Phys. Chem. C* **2013**, *117*, 11506–11513. [CrossRef]

27. Soler, J.M.; Artacho, E.; Gale, J.D.; García, A.; Junquera, J.; Ordejón, P.; Sánchez-Portal, D. The SIESTA method for ab initio order-N materials simulation. *J. Phys. Condens. Matter* **2002**, *14*, 2745–2779. [CrossRef]

28. Perdew, J.P.; Burke, K.; Ernzerhof, M. Generalized gradient approximation made simple. *Phys. Rev. Lett.* **1996**, *77*, 3865–3868. [CrossRef]

29. Troullier, N.; Martins, J.L. Efficient pseudopotentials for plane-wave calculations. *Phys. Rev. B* **1993**, *43*, 1993–2006. [CrossRef]

30. Monkhorst, H.J.; Pack, J.D. Special points for Brillouin-zone integrations. *Phys. Rev. B* **1976**, *13*, 5188–5192. [CrossRef]

31. Ridene, M.; Kha, C.S.; Flipse, C.F.J. Role of silicon dangling bonds in the electronic properties of epitaxial graphene on silicon carbide. *Nanotechnology* **2016**, *27*, 125705. [CrossRef]

32. Dudarev, S.L.; Botton, G.A.; Savrasov, S.Y.; Humphreys, C.J.; Sutton, A.P. Electron-energy-loss spectra and the structural stability of nickel oxide: An LSDA+U study. *Phys. Rev. B* **1998**, *57*, 1505–1509. [CrossRef]

33. First, P.N.; de Heer, W.A.; Seyller, T.; Berger, C.; Stroscio, J.A.; Moon, J.-S. Epitaxial graphene on silicon carbide. *MRS Bull.* **2010**, *35*, 296–305. [CrossRef]

34. Riedl, C.; Coletti, C.; Iwasaki, T.; Zakharov, A.A.; Starke, U. Quasi-free-standing epitaxial graphene on SiC obtained by hydrogen intercalation. *Phys. Rev. Lett.* **2009**, *103*, 246804. [CrossRef]

35. Varchon, F.; Mallet, P.; Veuillen, J.-Y.; Magaud, L. Ripples in epitaxial graphene on the Si-terminated SiC(0001) surface. *Phys. Rev. B* **2008**, *77*, 235412. [CrossRef]

36. Moaied, M.; Alvarez, J.V.; Palacios, J.J. Hydrogenation-induced ferromagnetism on graphite surfaces. *Phys. Rev. B* **2014**, *90*, 115441. [CrossRef]

37. Lieb, E.H. Two theorems on the Hubbard model. *Phys. Rev. Lett.* **1989**, *62*, 1201–1204. [CrossRef]

38. Levy, N.; Burke, S.A.; Meaker, K.L.; Panlasigui, M.; Zettl, A.; Guinea, F.; Neto, A.H.C.; Crommie, M.F. Strain-induced pseudo–magnetic fields greater than 300 tesla in graphene nanobubbles. *Science* **2010**, *329*, 544–547. [CrossRef]

39. Guinea, F.; Katsnelson, M.I.; Geim, A.K. Energy gaps and a zero-field quantum Hall effect in graphene by strain engineering. *Nat. Phys.* **2010**, *6*, 30–33. [CrossRef]

40. Guinea, F.; Katsnelson, M.I.; Vozmediano, M.A.H. Midgap states and charge inhomogeneities in corrugated graphene. *Phys. Rev. B* **2008**, *77*, 075422. [CrossRef]

41. Wehling, T.O.; Balatsky, A.V.; Tsvelik, A.M.; Katsnelson, M.I.; Lichtenstein, A.I. Midgap states in corrugated graphene: Ab initio calculations and effective field theory. *EPL* **2008**, *84*, 17003. [CrossRef]

42. Dell'Anna, L.; de Martino, A. Multiple magnetic barriers in graphene. *Phys. Rev. B* **2009**, 045420. [CrossRef]

43. Wang, Y.; Huang, Y.; Song, Y.; Zhang, X.; Ma, Y.; Liang, J.; Chen, Y. Room-temperature ferromagnetism of graphene. *Nano Lett.* **2009**, *9*, 220–224. [CrossRef]

44. Edwards, D.M.; Katsnelson, M.I. High-temperature ferromagnetism of sp electrons in narrow impurity bands: application to CaB_6. *J. Phys. Condens. Matter* **2006**, *18*, 7209–7725. [CrossRef]

45. Mattausch, A.; Pankratov, O. *Ab Initio* study of graphene on SiC. *Phys. Rev. Lett.* **2007**, *99*, 076802-1-4. [CrossRef]
46. Varchon, F.; Feng, R.; Hass, J.; Li, X.; Nguyen, B.N.; Naud, C.; Mallet, P.; Veuillen, J.-Y.; Berger, C.; Conrad, E.H.; Magaud, L. Electronic structure of epitaxial graphene layers on SiC: effect of the substrate. *Phys. Rev. Lett.* **2007**, *99*, 126805. [CrossRef]
47. Gao, Y.; Cao, T.; Cellini, F.; Berger, C.; de Heer, W.A.; Tosatti, E.; Riedo, E.; Bongiorno, A. Ultrahard carbon film from epitaxial graphene. *Nature Nanotechnol.* **2018**, *13*, 133–139. [CrossRef]

Review

Recent Progress on Graphene-Functionalized Metasurfaces for Tunable Phase and Polarization Control

Jierong Cheng [1], Fei Fan [1] and Shengjiang Chang [1,2,*]

[1] Institute of Modern Optics, Nankai University, Tianjin 300350, China; chengjr@nankai.edu.cn (J.C.); 014054@nankai.edu.cn (F.F.)

[2] Tianjin Key Laboratory of Optoelectronic Sensor and Sensing Network Technology, Tianjin 300350, China

* Correspondence: sjchang@nankai.edu.cn; Tel.: +86-22-23502275

Received: 22 January 2019; Accepted: 17 February 2019; Published: 8 March 2019

Abstract: The combination of graphene and a metasurface holds great promise for dynamic manipulation of the electromagnetic wave from low terahertz to mid-infrared. The optical response of graphene is significantly enhanced by the highly-localized fields in the meta-atoms, and the characteristics of meta-atoms can in turn be modulated in a large dynamic range through electrical doping of graphene. Graphene metasurfaces are initially focused on intensity modulation as modulators and tunable absorbers. In this paper, we review the recent progress of graphene metasurfaces for active control of the phase and the polarization. The related applications involve, but are not limited to lenses with tunable intensity or focal length, dynamic beam scanning, wave plates with tunable frequency, switchable polarizers, and real-time generation of an arbitrary polarization state, all by tuning the gate voltage of graphene. The review is concluded with a discussion of the existing challenges and the personal perspective of future directions.

Keywords: graphene; metasurface; phase shift; polarization; wavefront shaping; tunability

---

1. Introduction

The electromagnetic response of an optical device, from basic lenses and mirrors to advanced metasurface alternatives, is governed by two factors: material and structure. Beam engineering in conventional optics relies on separate contributions from the material refractive index n and the interface profiles $d(x, y)$, leading to bulky size and limited functionality. Driven by the continuous demand on minimization and the progress of nano-fabrication, metasurfaces offer a promising platform for beam transformation over a thin layer of meta-atoms [1–3]. Meta-atoms with deep subwavelength structures couple to the light with strong field localization such that the effective material property is modified locally as $n_{eff}(x, y)$, which is quite different from that of the natural material. With such spatially-designable $n_{eff}(x, y)$, beam molding is possible over an ultrathin layer of meta-atoms with fixed thickness d. Metasurfaces significantly reduce the size of optical devices with comparable or even unavailable novel functionalities, such as a metalens with close-to-unity numerical aperture and a planar profile [4–6], polarization-multiplexed holography [7–10], the synthesis of multiple functions [11–13], and coupling of spin-angular momentum [14,15], to name a few.

Actively-tunable or reconfigurable responses are highly desired in practical applications including dynamic wavefront shaping, adaptive optics, and single-pixel detection, which can be facilitated by including functional materials in the metasurfaces with dynamic modulation. Graphene, a single layer of carbon atoms arranged in a honey-comb lattice, is an outstanding candidate with extraordinary electrical and optical properties [16,17], as its Fermi level and the corresponding charge carrier density can be drastically modified by DC gate biasing. In graphene metasurfaces, the material and structure promote each other such that the resonator strengthens the interaction of light with the atomic thin layer, and the tunable material property of graphene in turn modulates the response of meta-atoms. Therefore, the structured graphene or integration of graphene with passive metasurfaces is an effective solution to enhance the dynamic modulation of beam engineering.

Compared to metasurfaces employing other active materials with different tuning mechanisms, such as electrically-tuned liquid crystals [18,19], thermally-modulated phase change materials [20], and mechanically-deformed elastic materials [21–23], graphene metasurfaces have several advantages. Graphene as the thinnest active material is naturally compatible with planar metasurfaces and silicon photonics. The electrical tuning speed is very high, from MHz [24,25] to GHz [26], with a large dynamic tuning range or tuning depth. More importantly, the material property of graphene can be electrically modulated in an extremely wide spectrum from sub-terahertz (sub-THz) to near-infrared (near-IR). Interestingly, the conductivity modulation in different sub-bands is quite different, with modulation focused on the real part at sub-THz and on the imaginary part at higher frequencies, leading to various adjustable properties in metadevices.

The development of graphene metasurfaces starts from intensity modulation. In 2011, the intensity modulation of the THz wave in graphene micro-ribbon array was reported [27], followed by several designs as modulators and switches [24,25,28–31], tunable metasurface absorbers [26,32]. and spatial light modulators [33]. Besides intensity modulation, increasing attention has focused in recent years on the phase modulation for dynamic wavefront shaping and tunable polarization control, with plenty of theoretical designs and several experimental breakthroughs. This paper emphasizes recent development of graphene metasurfaces for active control of phase and polarization, which has not been systematically summarized in related review articles [3,34,35].

The paper is organized as follows. Section 2 introduces the distinct material properties of graphene in different frequency bands with a tunable Fermi level, which lays the physical foundation for graphene-based active metasurfaces and offers different mechanisms of the tunability. Section 3 summarizes the dynamic graphene phase shifters with the target to achieve a large range of dynamic phase response for desired wavefront shaping working in transmission, reflection, in-plane, and out-of-plane manners. Section 4 is focused on the polarization manipulation to achieve tunable polarization selectivity and polarization conversion. Conclusions and personal remarks on the challenges and development directions of graphene metasurfaces are given in Section 5.

2. Material Properties of Graphene

The optical response of graphene is governed fundamentally by its surface conductivity. Since the intuitive optical characterization is usually done through the permittivity or the mode index, we summarize in this section the conductivity, the effective permittivity, and the mode index of the surface plasmon (SP) wave supported by graphene, so as to perceive the transformation of optical response with Fermi level and in different frequency bands.

The surface conductivity of graphene can be modeled by the well-known Kubo formula derived with the random phase approximation (RPA) theory [36–38].

$$\sigma = \frac{-ie^2}{\pi\hbar^2(\omega+i/\tau)}\int_0^\infty \epsilon\left[\frac{\partial f_d(\epsilon)}{\partial\epsilon} - \frac{\partial f_d(-\epsilon)}{\partial\epsilon}\right]d\epsilon + \frac{ie^2(\omega+i/\tau)}{\pi\hbar^2}\int_0^\infty \frac{f_d(\epsilon)-f_d(-\epsilon)}{(\omega+i/\tau)^2-4(\epsilon/\hbar)^2}d\epsilon \quad (1)$$

where e is the electron charge, $\hbar$ is the reduced Planck constant, and ω is the angular frequency. $f_d(\epsilon)$ is the Fermi–Dirac distribution as $f_d(\epsilon) = (e^{(\epsilon-E_F)/k_BT} + 1)^{-1}$. k_B is the Boltzmann constant. T is the temperature. E_F is the Fermi level. τ is the carrier relaxation time defined by $\tau = \mu E_F/(ev_f^2)$. $v_f = 10^6$ m/s is the Fermi velocity, and μ is the carrier mobility fluctuating in a large range determined by the fabrication process. μ is around 0.1 m^2/Vs in the chemical vapor deposition (CVD) graphene [39], 1 m^2/Vs in the mechanically-exfoliated graphene [40], and even 100 m^2/Vs in the suspend graphene at low temperature [41]. The carrier density n is related to the Fermi level by $n = E_F^2/(\pi\hbar^2 v_f^2)$. The dynamic modulation of n with E_F makes graphene either metallic, semiconductive, or dielectric.

The first term in Equation (1) is the intraband contribution, which can be explicitly expressed as [38]:

$$\sigma^{intra} = \frac{2k_B T e^2}{\pi\hbar^2}\ln\left(2\cosh\frac{E_F}{2k_B T}\right)\frac{i}{\omega + i/\tau} \tag{2}$$

When $E_F \gg k_B T$, Equation (2) is simplified as the Drude model $\sigma^{intra} = \frac{ie^2 E_F}{\pi\hbar^2(\omega+i/\tau)}$. The second term in Equation (1) originates from the interband contribution and can be approximated, if $E_F \gg k_B T$, as [38]:

$$\sigma^{inter} = \frac{ie^2}{4\pi\hbar}\ln\left(\frac{2E_F - (\omega + i/\tau)}{2E_F + (\omega + i/\tau)}\right) \tag{3}$$

If the thickness d is considered, the effective permittivity of graphene relative to the vacuum permittivity is defined as:

$$\epsilon_r = \epsilon_{av} + \frac{i\sigma}{\epsilon_0 \omega d} \tag{4}$$

where ϵ_0 is the vacuum permittivity and ϵ_{av} is the average relative permittivity of the surrounding environment. The mode index of the SP wave supported in the graphene monolayer is derived from the dispersion equation as [42,43]:

$$n_{sp} = \sqrt{\epsilon_{av} - (\frac{2\epsilon_{av}}{\sigma\eta})^2} \tag{5}$$

with η being the wave impedance.

In Figure 1, the three parameters above are plotted at different frequencies associated with different types of light–matter interaction. The parameters are specified in the caption. In the sub-THz region (Figure 1a), the real part of the conductivity σ', corresponding to the imaginary part of the permittivity ϵ_r'', is dominant, indicating graphene as a lossy film. The loss is weakly dependent on the frequency, but is very sensitive to the variation of E_F. Electric gating of the graphene-loaded metallic metasurface in this frequency band is mainly embodied in the amplitude modulation of the local resonances together with a limited range of phase variation due to the transition between underdamped and overdamped phases. Graphene at this frequency is seldom patterned as it does not support the SP wave with n_{sp} less than one.

In the THz to mid-IR band (Figure 1b), the most important feature is the SP wave with very large n_{sp}' and very small loss. The variation of E_F changes σ'', ϵ_r', and n_{sp}', but only has a small effect on the loss. The metasurface with patterned graphene shows tunable SP resonance frequency together with a large dynamic range of phase variation. Even in the hybrid metasurface where the resonance originates from the metallic antennas, variation of E_F shifts the resonance frequency because graphene modulates the dielectric constant of the environment.

When $\hbar\omega$ is comparable to $2E_F$ around the step in σ' and ϵ_r'' (Figure 1c), both intraband and interband transitions contribute to the optical response, such that the metasurface with graphene still has tunability by changing the Fermi level. In the near-IR and visible frequencies where $\hbar\omega \gg 2E_F$, the interband contribution dominates. ϵ_r' becomes positive with a constant ϵ_r''. Graphene behaves as a dielectric layer with universal absorption of 2.3% [44].

The variation of key material properties from sub-THz to near-IR is summarized in Figure 1d. It dictates different modulation mechanisms in the graphene-based metasurfaces, which endows the design with diverse tunability with E_F, such as shifting the operation frequency, modulating the efficiency of wavefront shaping, switching among multiple functionalities, and dynamic polarization conversion, as will be detailed in Sections 3 and 4.

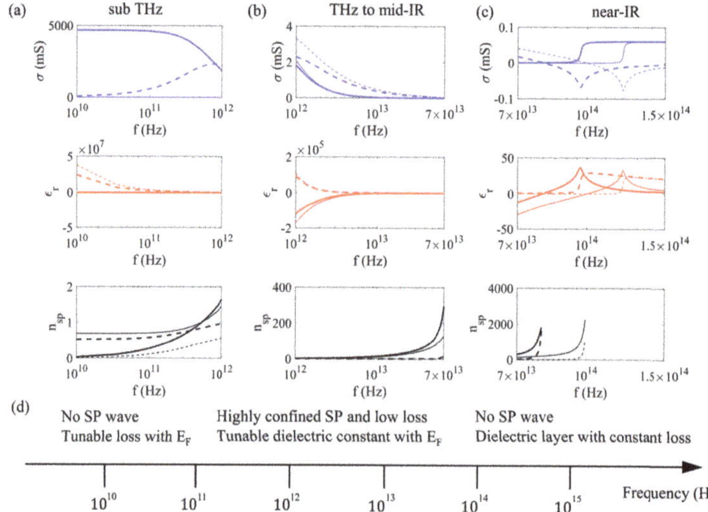

Figure 1. σ (from Equations (2) and (3)), ϵ_r (from Equation (4)), and n_{sp} (from Equation (5)) at sub-THz (**a**), THz to mid-IR (**b**), and near-IR frequencies (**c**). $\mu = 1 \text{ m}^2/\text{Vs}$, $d = 0.35$ nm, $\epsilon_{av} = 1$. The thick lines are for $E_F = 0.2$ eV, and the thin lines are for $E_F = 0.25$ eV. The solid lines are the real parts, and the dash lines are the imaginary parts. The key material properties of graphene are summarized in (**d**) at different frequencies.

3. Phase Modulation in Graphene Metasurfaces

Metasurfaces have been applied with great success for complicated wavefront shaping because of their compactness and flexibility when compared to conventional bulky devices. Dynamic wavefront shaping is a highly-desirable feature towards practical applications including, but not limited to adaptive optics and satellite communication. Ideally, the phase shifters should cover the complete 2π phase range with high efficiency or even independent control of the amplitude and phase responses to implement arbitrary kinds of wavefront shaping. In reality, it is not easily achievable by simply tuning the optical properties of an atomic layer. The combination of graphene and a metasurface presents an effective solution to enhancing the interaction with increased phase shifting. The constitutive phase shifters in the graphene metasurface array are either graphene micro-/nano-structures with separate biasing or graphene-hybridized metallic/dielectric resonators with a single-location biasing. The former offers the maximum degree of freedom for dynamic wavefront shaping, and the latter is more practical in implementation.

3.1. Wavefront Shaping in Transmission

In 2012, Min's group experimentally demonstrated terahertz switching [28] by attaching graphene to a layer of hexagonal metallic meta-atoms with top and bottom electrodes for static biasing (inset of Figure 2a). Working in the low THz range similar to the case of Figure 1a, the graphene layer acts as the surrounding medium with tunable loss in close proximity to the meta-atoms. The charge carriers accumulated at the edges of the hexagon increasingly leak into the neighboring element through the graphene layer with increased conductivity. Therefore, the resonance becomes weaker and slightly shifted as the biasing voltage is applied from the charge neutral point (CNP). The measured phase response in Figure 2a based on the THz time-domain spectroscopy (THz-TDS) system shows a dynamic phase shift of 32° at 0.65 THz through 850-V voltage variation from the CNP. However, the limited phase coverage is not enough for wavefront shaping. In 2018, more than 90° phase modulation together with the 50-dB amplitude modulation was experimentally demonstrated [45] in a bilayer-graphene-loaded split ring resonator array, as shown in Figure 2b,c.

When graphene itself is patterned as a phase shifter, such as a graphene ribbon or a graphene patch, it only provides a dynamic phase range of 180° around the plasmonic resonance by tuning the Fermi level or the dimension due to the intrinsic Lorentz-shape electric dipole resonance (Figure 2d). In fact, the effect of the material loss and the environment leads to an even smaller phase range [46,47]. Additionally the transmission intensity is very small away from the resonance. To improve the efficiency of such a phase shifter, two graphene nanoribbons are combined in parallel in a unit cell [46]. By individually tuning the Fermi level of each nanoribbon, phase delay varies from 0 to 180° with comparable transmission intensity. Numerical simulations validate an anomalous deflector at IR composed of such elements with tunable beam direction and a flat lens with tunable focal length in Figure 2e. In order to cover the full 360° phase range, a stack of three graphene ribbons with proper dielectric separation was proposed as the metasurface unit cell for efficient beam focusing [48]. The phase shifters achieve a high transmission amplitude above 0.7 with full control over the phase of the transmitted wave when the outer and inner graphene ribbons are independently biased. However, separate gate biasing among elements in three layers is too complicated to implement at the studied mid-IR frequencies.

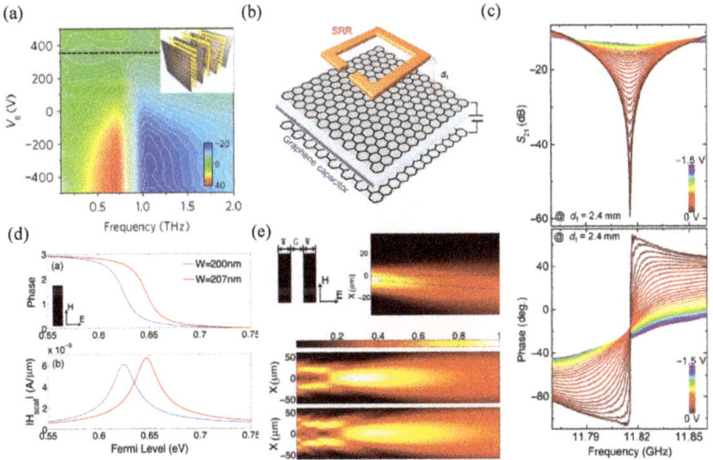

Figure 2. Transmissive graphene phase shifters for dynamic wavefront shaping. (**a**) Transmission phase variation with biasing voltage and frequency in the hexagonal metallic metasurface integrated with graphene. The inset is a schematic of the structure. The dashed line at 350 V corresponds to the charge neutral point. At a fixed frequency 0.65 THz, dynamic phase shift of 32° is observed by tuning the loss of graphene when V_g is from 350 V to −500 V. Reproduced with permission from [28], Copyright Springer Nature, 2012. (**b**) Unit cell of the hybrid metasurface consisting of a split-ring resonator and double-layer graphene capacitor. (**c**) Transmission magnitude and phase spectra at various biasing voltages. The dynamic phase shift is due to the tunable material loss of graphene. Reproduced with permission from [45], Copyright The authors, 2018. (**d**) Phase coverage and scattered magnetic field of uniform graphene ribbons by tuning the Fermi level. The phase coverage is limited to 180°. (**e**) Two parallel graphene ribbons are grouped in a unit cell to achieve nearly constant transmission intensity for different phase shifters. An array of such unit cells with individual biasing is simulated for anomalous refraction (top) and focusing with tunable focal length (bottom). Reproduced with permission from [46], Copyright WILEY-VCH Verlag GmbH & Co. KGaA, Weinheim, 2014.

Recently, the geometric phase has been utilized widely in metasurfaces for wavefront shaping by rotating the asymmetric inclusions [49,50]. In Figure 3a, through an array of graphene nanocross resonators with different orientations [51], the transmitted circularly-polarized beam with the opposite handedness of the excitation gains the geometric phase as twice the rotation angle while keeping the transmission amplitude of 0.4 at the optimum frequency. The phase profile for anomalous refraction is solely determined by the orientation of the graphene inclusions irrelevant to the conductivity. Therefore, highly efficient anomalous refraction is maintained over a wide bandwidth from 14.5 THz to 17 THz when the conductivity is dynamically adjusted, as shown in Figure 3b. The increased carrier density of graphene blue shifts the optimum frequency together with a dynamic scan of the beam direction. Although all the graphene elements experience the same voltage, the biasing is complicated due to the separate arrangement. The related study stops at the numerical validation. Until 2018, a similar idea was experimentally demonstrated by Min and Zhang [52] around 1.15 THz with a more feasible structure in Figure 3c. The geometric phase profile is provided by the metallic resonators with proper orientation, on top of which is a graphene monolayer biased at the periphery. For anomalous refraction and focusing, the phase profile does not change with gate voltage, while the intensity does (Figure 3d). Gate-induced increase in the carrier density of graphene results in a stronger absorption of THz waves through the intraband transitions. Therefore, the bending and focusing are most efficient at the CNP and experience an intensity decrease with further biasing (Figure 3e,f). Comparison of Figure 3a,c shows that graphene is a frequency-tunable resonator in the former and loss-tunable medium close to the metallic resonator in the latter, corresponding to the material properties of Figure 1a,b, respectively. Therefore, the optimum working frequency varies with the Fermi level in Figure 3b and stays the same in Figure 3d.

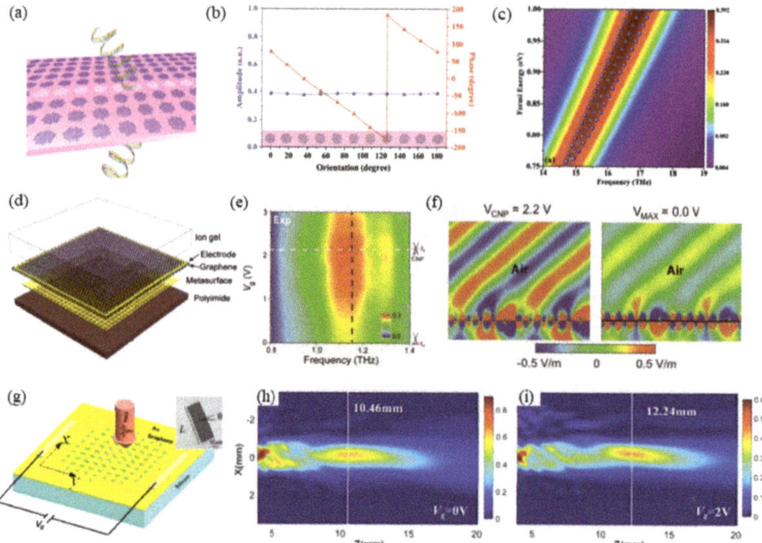

Figure 3. Transmissive active metasurfaces for dynamic wavefront shaping based on the geometric phase. (**a**) Graphene nanocross metasurface with space-variant orientation for deflection of the circularly-polarized beam. (**b**) The transmitted circularly-polarized beam with the opposite handedness of the excitation gains the geometric phase as twice the rotation angle while keeping the transmission amplitude of 0.4 at the optimum frequency. (**c**) Deflection efficiency spectrum when the graphene nanocrosses have different Fermi levels. The large deflection efficiency is well maintained over a wide bandwidth and a large range of the Fermi level. Reproduced with permission from [51], Copyright WILEY-VCH Verlag GmbH & Co. KGaA, Weinheim, 2015. (**d**) Schematic of the active metasurface composed of a single layer of graphene deposited on a U-shaped metallic aperture where the spatial phase function is defined by the orientation of the U-shaped aperture. (**e**) Variation of the efficiency of anomalous refraction with biasing voltage and frequency. (**f**) Electric field distribution at two different gate voltages when the metasurface is designed for anomalous refraction. The metasurface works best at the CNP and shows reduced intensity with additional biasing due to increased loss in graphene. Reproduced with permission from [52], Copyright WILEY-VCH Verlag GmbH & Co. KGaA, Weinheim,

2017. (**g**) Monolayer graphene attached to gold aperture antennas with different orientations for geometric phase response and different lengths for resonance phase response. The resonance phase is tunable by voltage to carefully compensate the variation of the phase function at different frequencies. (**h,i**) The phase profile is dynamically tuned to satisfy two parabolic functions at two Fermi levels, leading to different focal lengths with comparable focusing efficiency. Reproduced with permission from [53], Copyright Chinese Laser Press, 2018.

As the geometric phase does not change with the Fermi level, the active tunability of the device is limited to intensity modulation at a fixed frequency. In more scenarios, one wants to tune the focal point or the steering angle dynamically with the maximum possible intensity. For this purpose, Wang's group proposed a strategy in 2017 to achieve a transmissive metalens with a dynamically-tunable focal length [54]. The experimental demonstration was done very recently [53]. The structure consists of monolayer graphene on top of gold aperture antennas with different orientations and lengths (Figure 3g). The phase profile is determined by both the geometric phase and resonance phase, the latter of which is adjustable by the Fermi level of graphene. With a careful design, the phase profile can be dynamically tuned to satisfy two parabolic functions at two Fermi levels, leading to an electrically-tunable focal length in a large range of 4.45λ in experiments with comparable efficiency, as shown in Figure 3h,i.

3.2. Wavefront Shaping in Reflection

When a metallic mirror is introduced to the active phase shifter with a dielectric spacer, the phase modulation range can be significantly enlarged. The whole structure can be considered as an asymmetric Fabry–Perot resonator, where the large phase range is gained from the multi-path propagation between the top and bottom layers. In other words, the electric current circulating between the top metasurface and the bottom plate forms a magnetic resonance with $360°$ phase variation across the resonance frequency.

In order to access the $360°$ phase range dynamically at a fixed working frequency, a loss tuning mechanism is required in the top layer to tune the structure through the underdamped-critical coupling-overdamped transition. Tuning the resistive loss of the top layer dictates how far the energy goes through the dielectric spacer and how strong the coupling is with the bottom plate, and is therefore an effective way to tune the phase response. The resistive loss changes the reflection intensity and phase, but does not shift the resonance frequency. One can only access half of the $360°$ phase range at a fixed frequency with the structure in Figure 4a, where the metallic patches are covered by graphene. The $360°$ phase range can be covered by two types of metallic patches with shifted resonance frequencies. The working frequency is chosen in the shaded area in Figure 4b between the two resonances [55]. The THz-TDS measurement has demonstrated a maximum phase modulation range of $243°$ at 0.48 THz, which can be further increased if the carrier scattering loss is reduced in graphene.

The above design works at sub-THz, where graphene is a lossy surrounding medium, as analyzed in Figure 1a. The necessity of two types of unit cells to access $360°$ phase response originates from the inability of graphene to tune the resonance frequency. Another approach to accessing the full phase range is to work at higher frequencies where graphene directly modulates the resonance frequency with different biasing. In 2017, Atwater's group demonstrated more than $230°$ phase modulation using the structure in Figure 4c in mid-IR frequencies [56]. Here, graphene is attached to a uniform array of gold antennas with very small gaps. In mid-IR, graphene shows low loss and a tunable dielectric constant with the carrier density, as discussed in Figure 1b. The resonance frequency shifts with the environmental dielectric constant, resulting in a widely-tunable phase range of more than $200°$ from 8.5 μm to 8.7 μm, as shown in Figure 4d,e. Although the measured reflectivity is quite low due to absorption of the SiN_x substrate and the low mobility of graphene charge carriers, the proposed design is one of the most promising designs for dynamic beam steering. Despite the similar configurations in Figure 4a,c, graphene plays different roles in the dynamic adjustment depending on the working frequencies.

Alternatively, graphene resonators can replace the graphene–metallic resonator structure in the top layer for reflective phase modulation, as summarized in Figure 5. All of them are targeted at mid-IR beam shaping, where the graphene pattern is a plasmonic resonator with tunable resonance frequency. Among various shapes, the graphene ribbons are the most extensively-studied ones, due to their simplicity and ease of biasing at one end of the structure. Specifically, Figure 5a utilizes the spatially-variant graphene ribbons to achieve desired phase profile at the center frequency and to change the Fermi level for dynamic modulation [47,57]. Figure 5b shows the phase variation with the ribbon width, which serves as the guidance for designing the metasurface array for focusing and steering applications.

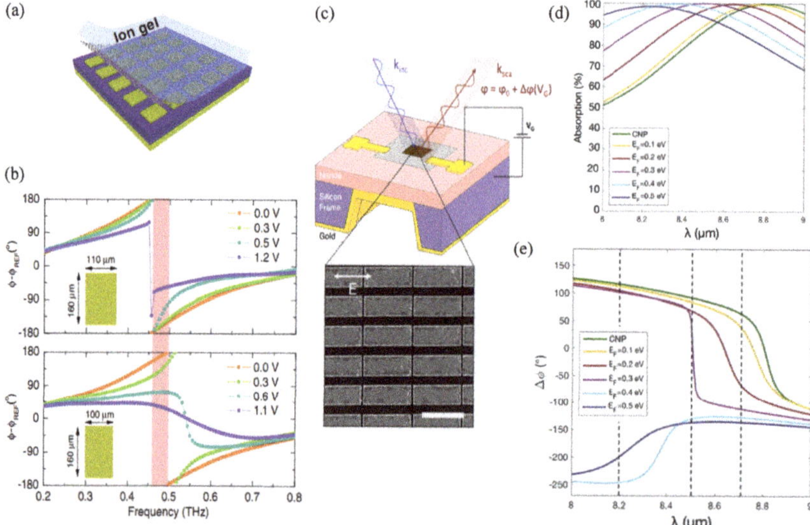

Figure 4. Reflective phase shifters with monolayer graphene for dynamic phase modulation. (**a**) Monolayer graphene is attached to the array of Al patches separated with the bottom Al film with an SU-8 spacer. (**b**) Gate-dependent reflection phase spectra of the combined structure when the Al patches have two sizes. Here, graphene as a lossy surrounding medium in sub-THz frequencies tunes the resonance intensity and phase. In each unit cell, the maximum phase change is 180° around the resonance frequency by tuning the biasing voltage. By utilizing two types of unit cells with different resonance frequencies and working in the shaded frequencies, a 360° phase change is accessible by tuning the biasing voltage. Reproduced with permission from [55], Copyright American Physical Society, 2015. (**c**) Graphene-tuned gold antenna array with back mirror. The zoom-in plot is the SEM image of the gold antennas with nanometer gaps. (**d,e**) Tunable absorption (**d**) and tunable reflection phase (**e**) for different graphene Fermi levels. The dielectric constant of graphene is sensitively modulated in the mid-IR, leading to a widely-tunable phase range of more than 200° from 8.5 μm to 8.7 μm. Reproduced with permission from [56], Copyright American Chemical Society, 2017.

When the graphene ribbons are separately biased, more degrees of freedom are gained for controlling the functionality temporally and spatially [58,59]. Figure 5c shows that the smooth 360° phase variation is covered by tuning the Fermi level, and it is well maintained from 4 THz to 6 THz. Beyond this frequency range, the phase variation is either too sharp or too weak to use (Figure 5d). It leads to switching of the functionality by frequency, for example, from an anomalous reflector at 5 THz to a normal reflector at 3.5 THz. With the Fermi level of the graphene ribbon tuned spatially and temporally, a metalens of a large numerical aperture is numerically demonstrated with either a fixed or variable focal point over a wide bandwidth [58]. Completely different functionalities, such as cloaking, illusion, and focusing, can be implemented in one metasurface [60]. In such kinds of active designs, the efficiency is mainly governed by the mobility of graphene carriers. The study shows that the carrier relaxation time τ seldom affects the reflection phase, but mainly controls the local reflectivity. τ is around 1 ps in the designs of Figure 5, leading to an overall efficiency of around 60 to 70%. The efficiency can be improved to 90% if the relaxation time is increased to 5 ps [59].

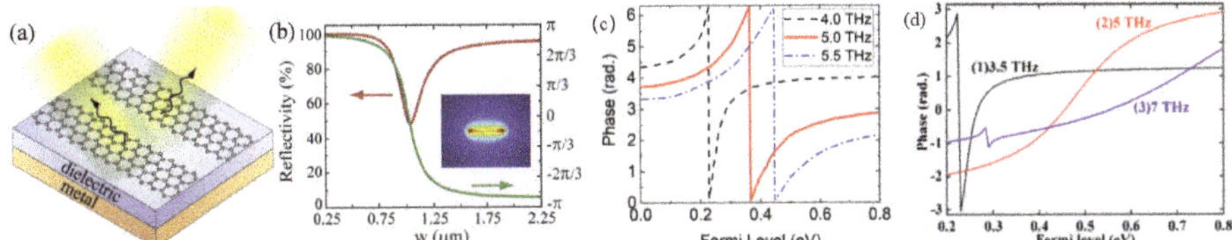

Figure 5. Graphene nanoribbons for reflective wavefront shaping. (**a**) Graphene ribbon element on top of the dielectric/metal substrate. (**b**) Variation of reflectivity and phase with the ribbon width. Reproduced with permission from [47], Copyright The authors, 2015. (**c**) Smooth 360° phase variation is covered by tuning the Fermi level of graphene , and it is shifted with a well-maintained shape from 4 THz to 6 THz. Adapted from [58], Copyright IOP Publishing Ltd., 2017. (**d**) Far away from the optimum frequency, the phase variation is either too sharp or too weak to use. Reproduced with permission from [59], Copyright IEEE, 2018. Wavefront shaping is possible by properly selecting the ribbon width at the fixed Fermi level from (**b**), or by individual gate control over an uniform graphene nanoribbon array following (**c,d**).

Similar to the transmission cases, uniform biasing of either monolayer graphene or graphene ribbons is promising for fabrication, but limited to efficiency modulation or on-off switching of a specific function. In contrast, individual biasing of each ribbon is extremely difficult to implement in device integration, although at-will wavefront shaping is theoretically possible in real time with satisfactory efficiency. A compromise solution utilizes both the geometric phase and resonance phase with a similar configuration as Figure 3g, but working in the reflection side with a back mirror [61]. The geometric phase governs the general phase profile, and the resonance phase is tuned by the Fermi level to compensate minor phase variations needed for different focal lengths. However, in both transmission and reflection designs, the lens aperture is limited by the small tuning range of the resonance phase with the Fermi level, leading to an inaccurate focal point as compared to the design.

In addition, the metasurface is endowed with increased power when more complicated graphene elements and the corresponding gate configuration are considered. For example, when two graphene patches are combined in a unit cell in-plane [62] or stacked on top of the substrate with individual gate control [63], the wavefront engineering has been studied with an increased bandwidth. Taking advantage of the highly-confined surface plasmon in graphene, a dual-band focusing lens is proposed by stacking two layers of graphene nanoribbons above the gold reflector with negligible crosstalk [64]. The wavefront can be independently engineered in two distinct frequencies with each layer being responsible for one frequency.

3.3. In-Plane Plasmonic Wavefront Shaping and Coupling with Out-of-Plane Propagation

Apart from wavefront shaping in free space, control of the SP waves propagating along the graphene surface holds significant promises towards minimization of the photonic integrated circuits. Compared to the noble metal counterparts, graphene plasmonics show low loss and strong confinement together with the tunability via electrical or chemical doping. As will be shown in the following literature, in-plane and out-of-plane beam shaping focus on applications in the THz to mid-IR frequencies taking advantage of the spatially- and temporally-tunable n_{sp} of the SP wave with the Fermi level following the discussion in Figure 1b.

Fundamentally, the plasmonic wavefront is manipulated via a nonuniform conductivity pattern along graphene. Several ways are proposed to achieve such a conductivity pattern [65]. Figure 6a uses the uneven ground plane to create different distances between the flat graphene layer and the highly-doped silicon ground plane [66]. When a fixed biasing voltage is applied, the electric fields experienced by the graphene inside and outside of the double-convex region are different, which in turn produces an inhomogeneous conductivity pattern for spatial Fourier transformation in Figure 6b. With a proper design of the uneven ground plane, the focal length is even variable with the voltage [67]. Chemical doping, though without dynamic control, is another popular way to change the optical properties of graphene. Figure 6c shows that the adsorption of proper molecules on graphene leads to chemical doping through charge transfer. By patterning two types of organic molecules on graphene, a plasmonic metasurface can be realized with any gradient effective refractive index profile to manipulate SP beams as desired [43]. A multiscale theoretical approach combining the first-principles electronic structure calculations and finite-difference time-domain simulations is developed to reconcile the band structure modification by the molecules and the mesoscopic effect on the SP wave propagation. The designed plasmonic Luneburg lens and SELFOC lens are in a notably subwavelength size of around one tenth of the free space wavelength. By creating vacancies with the designed shape in the graphene layer (Figure 6d), this enables excitation of the SP wave from free-space illumination and manipulation of the SP wavefront [68]. Figure 6e,f shows the plasmonic superfocusing and plasmonic vortex beam generation with a well-designed vacancy geometry. The size of the focusing hotspot is far below the diffraction limit considering the strong plasmonic localization in graphene.

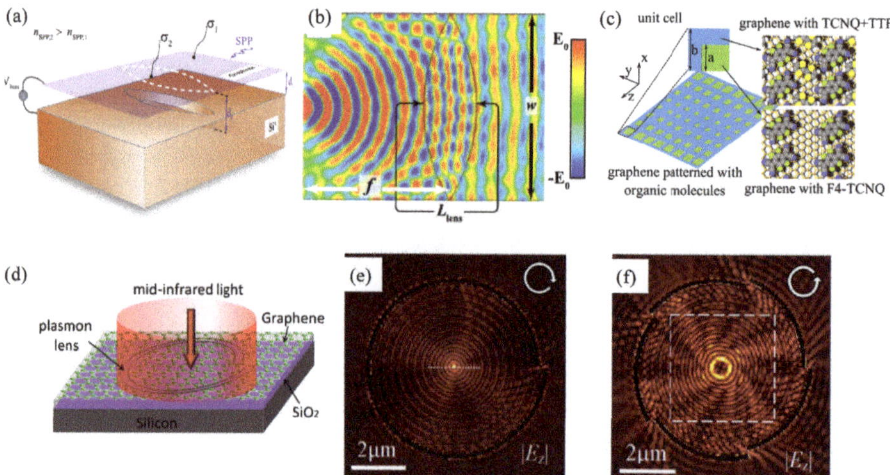

Figure 6. In-plane plasmonic wavefront engineering. (**a**) Monolayer graphene on top of the uneven ground plane made of highly-doped silicon. Different distances between the graphene layer and the ground plane lead to two conductivities and two refractive indices of the SP wave in and out of the lentil area. (**b**) Simulated electric field of the SP wave through the graphene Fourier lens with point source excitation. Reproduce with permission from [66], Copyright Aptara Inc., 2012. (**c**) Schematic representation of the graphene layer patterned with two types of organic molecules (TCNQ + TTF and F4 - TCNQ) to achieve an inhomogeneous conductivity profile through charge transfer. Reproduced with permission from [43], Copyright American Chemical Society, 2014. (**d**) Schematic of the spiral-shaped graphene vacancy for excitation and focusing of the SP wave. (**e**) Superfocusing using a spiral-shaped graphene vacancy together with a circularly-polarized incident beam. (**f**) Generation of plasmonic vortex beam with five segmented graphene vacancies. All the in-plane wavefront shaping is

achieved in a notable subwavelength area and beyond the diffraction limit considering the strong localization of the graphene plasmon. Reproduced with permission from [68], Copyright Optical Society of America, 2014.

On the other hand, the strong mismatch of the wave vectors in the plasmonic wave and free space wave causes a great obstacle in exciting and leaking the SP waves in graphene. Different mechanisms have been reported to excite the strongly-localized SP waves in graphene through diffraction gratings [69], sharp tips [70,71], and vacancies [68]. Similarly, the coupling from the SP wave to the propagation wave is the reverse process, which can be done in the same structure. To further increase the flexibility, coupling the SP wave to the free space beam with steerable directions is a highly-desired feature in many applications including communications, remote sensing, and image scanning, which is not easily available especially in the THz region. In the design of Figure 7a, by densely distributing gate pads underneath the monolayer graphene, the surface reactance can be sinusoidally modulated with adequate biasing voltage to each pad [72]. The surface wave interacts with the sinusoidal modulation to produce the leaky wave radiation. Since the radiation direction is determined by the modulation periodicity, which can be dynamically varied by grouping different numbers of pads in one period, the radiation direction can be steered in real time at a fixed frequency, as calculated in Figure 7b. Alternatively, Figure 7c utilizes the periodic metal–dielectric–graphene plasmonic grating to achieve electrically-controllable THz beam scanning [73]. The graphene gratings are biased with the same voltage, which modulates the effective refractive index of the SP wave supported in graphene, leading to varied radiation directions based on the grating equation. Figure 7d shows the radiated near-field from $13°$ to $-18°$ when the biasing voltage is 256.5 mV (left) and 53.2 mV (right).

The above leaky wave scanning is limited to one plane. Scanning in arbitrary directions in the full space is more difficult, as pixel-by-pixel biasing makes the configuration a huge challenge for device integration. Figure 7e is a simplified design to achieve such functionality. The dynamic bean scanning in both elevation and azimuth planes is achieved by applying two groups of one-dimensional biasing pads underneath the graphene sheet [74]. They are orthogonal and decoupled. One group offering monotonic impedance variation along the y direction mainly determines the radiation in the azimuthal plane, while the other provides sinusoidal impedance modulation along the x direction to decide the elevation angle of the radiation. Figure 7f shows examples of the two-dimensional radiation pattern towards different directions with the adequate choice of the voltages in the two groups of gating pads.

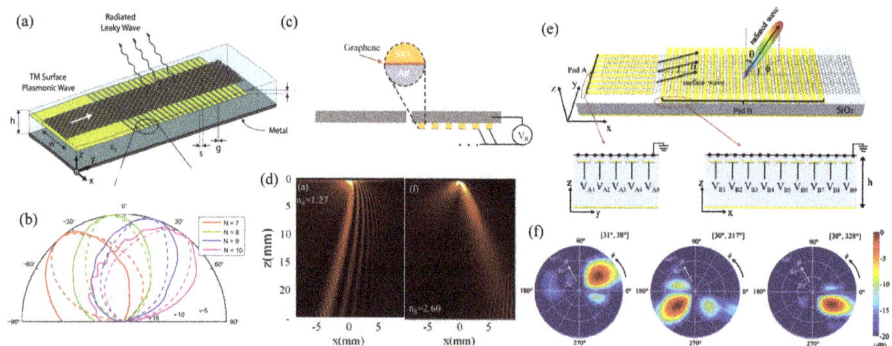

Figure 7. Leaky waves from graphene plasmonic structures with a dynamically-steerable direction. (**a**) A graphene sheet on the back-metalized substrate with the isolated poly-silicon gating pads for space-dependent DC biasing. The surface reactance is sinusoidally modulated with adequate biasing voltage to each pad. The periodic modulation offers effective momentum to transfer the surface wave into free-space radiation. (**b**) The radiation direction of the leaky wave is dynamically shifted when different numbers of pads are contained in a modulation period. Reproduced with permission from [72], Copyright IEEE, 2014. (**c**) The silica-graphene grating with the silver substrate and a slit for THz beam scanning. All the graphene ribbons are biased with the same voltage. The leaky beam direction is determined and tuned by the effective refractive index of the SP wave in graphene. (**d**) The near-field plot of the radiation with different biasing voltages applied. Biasing voltage is 256.5 mV (left) and 53.2 mV (right). The refractive index of the SP wave is 1.27 and 2.60, resulting in the radiation towards 13° and −18°, respectively. Reproduced with permission from [73], Copyright Elsevier B.V., 2015. (**e**) Graphene leaky wave antenna for two-dimensional beam scanning with the simplified two groups of gating pads. One group on the left offering monotonic impedance variation along the y direction mainly determines the radiation in the azimuthal plane, and the other provides sinusoidal impedance modulation along the x direction to decide the elevation angle of the radiation. (**f**) Radiation pattern in different directions via simulation by simply changing the two groups of biasing voltages. Reproduced with permission from [74], Copyright Optical Society of America, 2016.

4. Polarization Modulation in Graphene Metasurfaces

Polarization manipulation, usually achieved by birefringence materials, total internal reflection, optical gratings, and the Faraday effect, is instrumental in a wide range of optical applications, such as telecommunications [75], imaging [76], sensing [77], polarimetry [78], and spectroscopy [79]. As compared to those conventional methods, metasurfaces have shown exceptional capabilities for flexible polarization control in a planar, ultrathin, and integrable manner. Introduction of graphene into the metasurfaces opens an exciting route to dynamically and actively taming the polarization in a desired manner. The basic idea is an anisotropic subwavelength pattern made of graphene or hybrid graphene/metal that supports two plasmonic eigenmodes along two orthogonal polarization directions. Therefore, the polarization control generally utilizes the plasmonic property of graphene in Figure 1b. By altering the relative magnitude and phase delay between the two eigenmodes via geometric design and dynamical modulation of graphene conductivity, several outstanding features are proposed, such as wave plates with tunable working frequency, switching between two polarization states, or variation of the polarization along a continuous path in the Poincare sphere, all without readjusting the geometry of the metasurface. The review of this part is classified by the functionality as quarter-wave plates, half-wave plates, polarizers, and general polarization control.

4.1. Graphene-Based Quarter-Wave Plate

The quarter-wave plate (QWP) converts a linear polarization into a circular one, by engineering the two orthogonal eigenmodes with equal amplitude and 90° phase delay. Figure 8a is an

asymmetric graphene nanocross with distinct plasmon resonances along the two arms [80]. The 45°
linearly-polarized (LP) beam is changed into a circularly-polarized (CP) beam at 7.92 μm when the
graphene Fermi level is 0.75 eV. The working wavelength is blue-shifted with the further increase of the
graphene Fermi level. However, the conversion efficiency is low, and the quarter-wave plate is strictly
satisfied at a single frequency. Thus, the bandwidth is very narrow (estimated as 1% from [80]) once a
certain biasing voltage is applied. Figure 8b enlarges the bandwidth to 40% due to the interaction of
the top graphene grating and the bottom gold grating with opposite phase retardation dispersion [81].
Variation of the carrier density changes the graphene from transparent to conductive, leading to a
polarization switch from co-polarized LP to CP with $E_F = 0$ and 0.5 eV.

The form birefringence of the graphene grating is further enhanced in Figure 8c by introducing
a periodic gradient instead of the binary pattern [82]. The increased birefringence shifts the working
frequency to a lower value. The bandwidth of the quarter-wave plate relative to the center frequency
is therefore increased as compared to the binary design. In order to dynamically move the working
frequency in a wide bandwidth, liquid crystal (LC) is integrated into the graphene grating with
additional electrical modulation of the birefringence. The theoretical study indicates by electrically
controlling the liquid crystal director angle that the dynamic bandwidth of such hybrid quarter-wave
plate achieves 78% centered at 1.15 THz. The design in Figure 8d has shown similar dynamic bandwidth
by independently biasing the top graphene pattern and the bottom graphene film [83]. The bottom
layer composed of seven layers of graphene acts as a reflector with a tunable reflection phase, which
is used to compensate the difference of phase delay due to the frequency shift. Furthermore, the
efficiency is very high (~70%) by working in reflection mode.

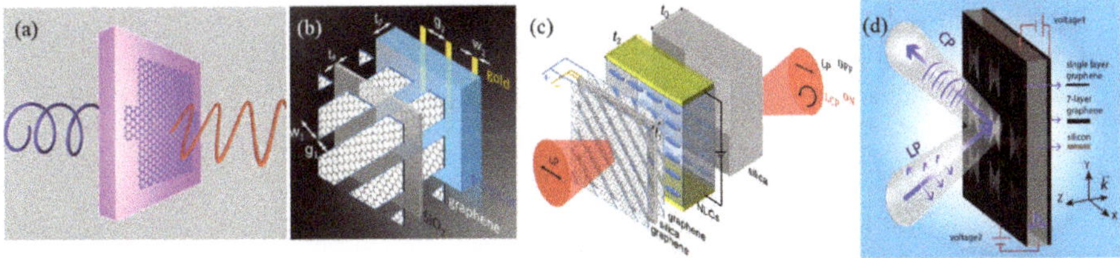

Figure 8. Graphene metasurfaces as active quarter-wave plates. (**a**) Graphene asymmetric nanocross for
LP to CP conversion. The 45° LP beam is converted into a CP beam at 7.92 μm when the graphene Fermi
level is 0.75 eV. The operation frequency blue shifts with the increase of the Fermi level. Reproduced
with permission from [80], Copyright Optical Society of America, 2013. (**b**) Hybrid metasurface
composed of graphene sandwich grating and gold grating separated by a polymer spacer. The form
birefringence in graphene grating and gold grating adds up to offer a constant phase delay of 90°
between two eigenmodes over a wide bandwidth. Reproduced with permission from [81], Copyright
The authors, 2015. (**c**) Graphene grating sandwich with well-designed distance and in-plane gradient
on top of the LC layer. The spatial gradient of the graphene grating increases the form birefringence,
and the electrical control of the LC molecule direction leads the QWP to have a dynamic bandwidth
of 78%. Reproduced with permission from [82], Copyright The authors, 2018. (**d**) A graphene sheet
patterned with butterfly holes and backed by seven layers of graphene with separate biasing for
wideband LP to CP conversion. The bottom seven layers of graphene show a tunable reflection phase in
order to compensate the difference of the phase delay due to the frequency shift over a wide bandwidth.
Reproduced with permission from [83], Copyright The authors, 2017.

4.2. Graphene-Based Half-Wave Plate

Half-wave plates (HWPs) as the cross-polarization converter require 180° phase retardation and a
comparable intensity of the two eigenmodes. Different from the QWP, the HWP is a real challenge
through a single layer of anisotropic graphene pattern. As mentioned in Section 3.1, each eigenmode is
an electric dipole with 180° or less phase shift across the resonance. Spectral shifting of two dipole
resonances from the orthogonal directions of the anisotropic element leads to a phase difference of

less than $180°$. Therefore, the monolayer design working in the transmission mode only rotates the LP wave by a few degrees [84], and almost all the graphene HWPs work in the reflection mode with the anisotropic pattern backed with a metallic mirror at a proper distance. Table 1 summarizes the key features of these HWPs working in THz and mid-IR (all theoretical designs), from which we can generalize the rule of designing graphene HWP to meet the needs of different applications.

The first three designs [85–87] in Table 1 share the same L-shaped graphene pattern (G pattern) or the complementary slot. We found that by properly selecting the arm length and width, the performance of the HWP can be narrowband [85], dual-frequency [86], or broadband [87], depending on the spectral separation of the two eigenmodes and their quality factors. Besides the L-shaped design, the graphene layer is patterned into various anisotropic shapes, from which the sinusoidal [88] and ϕ-shaped designs [89] also gain very wide bandwidth. Here, the bandwidth is specific to the static bandwidth when a fixed Fermi level is given to graphene. It is estimated with the polarization conversion ratio (PCR) above 0.5, where PCR is defined as $|R_{cross}|^2/(|R_{cross}|^2 + |R_{co}|^2)$. R_{cross} and R_{co} are the reflection coefficients of the cross- and co-polarized beams. The maximum PCR in all cases is close to one, indicating a complete polarization conversion. The carrier relaxation time τ used in the modelings varies from 0.5 ps to 1 ps, and it should be at least larger than 0.02 ps to maintain the polarization conversion effect [90]. The Fermi level is relatively high, close to 1 eV in most cases. An exception is [91], where the graphene Fermi level is 0 eV. The birefringence in this study originates from two layers: the I-shaped metallic resonator and the graphene ribbon with orthogonal orientations, leading to a broad operation bandwidth of around 96%. Although all the designs show very high PCR above 90%, this does not account for the absorption. The peak efficiency listed in Table 1 is the maximum value of $|R_{cross}|^2$, which indicates the net power into the cross-polarization. Due to the different relaxation times and different intensities of the field localization, the net efficiency varies from 50 to 90%. Still, these HWPs are efficient due to the blocking of the transmission channel.

Table 1. Comparison of active half-wave plates made of graphene reflective metasurfaces. G, graphene.

Ref.	Structure	Freq.	Static Bandwidth	τ E_F	Peak Eff.	Lattice Size	Thickness	Dynamic Tunability
[85]	L-shaped G pattern + back mirror	38.9 THz	$\sim 3\%$	~0.9 ps 0.9 eV	$\sim 76\%$	0.02λ	0.23λ	$\sim 13\%$ dynamic bandwidth with E_F from 0.7 to 0.9 eV
[86]	L-shaped G slot + back mirror	31.4 THz 41.3 THz	$\sim 5\%$ $\sim 3\%$	1 ps 0.9 eV	$\sim 80\%$	0.016λ	0.28λ	$\sim 30\%$ and $\sim 20\%$ dynamic bandwidth with E_F from 0.7 to 1 eV
[87]	L-shaped G pattern + back mirror	6.65 THz	$\sim 50\%$	1 ps 0.9 eV	$\sim 80\%$	0.17λ	0.33λ	$\sim 69\%$ dynamic bandwidth with E_F from 0.7 to 1 eV
[92]	rectangle G holes + back mirror	46.8 THz	$\sim 5\%$	0.5 ps 1 eV	$\sim 49\%$	0.03λ	0.25λ	$\sim 32\%$ dynamic bandwidth with E_F from 0.6 to 1 eV
[93]	elliptical G pattern + back mirror	22.5 THz	$\sim 9\%$	1 ps 0.9 eV	$\sim 83\%$	0.015λ	0.16λ	$\sim 28\%$ dynamic bandwidth with E_F from 0.6 to 0.9 eV
[91]	I-shaped metallic resonator + G ribbons + back mirror	0.67 THz	$\sim 96\%$	1 ps 0 eV	$\sim 90\%$	0.27λ	0.25λ	E_F from 0 to 0.6 eV, polarization varies from cross-LP to ellipse to CP
[90]	H-shaped G holes + back mirror	35.7 THz	11%	0.5 ps 1 eV	$\sim 70\%$	0.02λ	0.26λ	$\sim 37\%$ dynamic bandwidth with E_F from 0.6 to 1 eV
[94]	Cross G pattern + back mirror	7.7 THz	25.8%	1 ps 1 eV	$\sim 80\%$	0.05λ	0.25λ	$\sim 77\%$ dynamic bandwidth with E_F from 0.5 to 1 eV
[88]	Sinusoidal G holes + back mirror	1.6 THz	47%	1 ps 0.4 eV	$\sim 60\%$	0.09λ	0.26λ	Shift of the bandwidth with E_F is small relative to the static bandwidth
[89]	ϕ-shaped G pattern + back mirror	5.98 THz	41.9%	0.6 ps 0.6 eV	$\sim 80\%$	0.04λ	0.25λ	$\sim 88\%$ dynamic bandwidth with E_F from 0.4 to 1 eV

The lattice size is apparently one or two orders of magnitude smaller than the vacuum wavelength taking advantage of the strong localization of graphene plasmons. As the localization increases with frequency, the lattice size is smaller in higher frequency designs. Due to such a small feature size, all the proposed designs are very robust to the variation of the incidence angle up to 40° or more [88,90,93,94]. This is the merit of the graphene HWP compared to the metallic counterpart besides electrical tunability. Interestingly, the thickness of the dielectric spacer layer separating the graphene pattern and the back mirror is always around 0.25λ if the thickness means the single optical path taking the refractive index of the dielectric into account. The round trip through the dielectric with the 180° phase delay upon back reflection enables the enhancement of interference in the reflection side.

Finally, the dynamic tunability of the graphene-based HWP through the electrical control of the Fermi level is discussed in the last column of Table 1. Generally, an increase of E_F leads to reduced loss and stronger plasmonic resonance with blue-shifted resonance frequency and thus helps increase the PCR. Accordingly, a decrease of E_F causes lower PCR and narrower static bandwidth. The lower limit of the dynamic bandwidth is set by PCR > 0.5. In addition, E_F above 1 eV is avoided considering the breakdown of the capacitor and the extraordinarily high biasing voltage, which sets the upper limit of the dynamic bandwidth. The designs in [87,89,94] show a broad dynamic bandwidth around 70~90%. Table 1 indicates that the design with a wide static bandwidth is more likely to have a wide dynamic bandwidth. Please note that all the data with the ~ sign are estimated from the plots and without the ~ sign are the exact ones mentioned in the literature.

4.3. Graphene-Based Polarizer

In addition to engineering the phase retardation of the two eigenmodes, graphene metasurfaces can be tailored to filter selectively one eigenmode while eliminating the orthogonal one so as to be a polarizer, or in other words a polarization-selective surface. A straightforward example is an anisotropic graphene resonator such as Figure 9a supporting two spectrally-separated resonances along orthogonal directions [95]. By working at a frequency where one polarization is on resonance and the other is off resonance, the on-resonance state is strongly reflected, while reflection of the off-resonance state is fairly weak (Figure 9b). Such polarization selectivity is switched on only when the graphene is doped to support the plasmonic resonance.

Besides the switchable function, graphene endows the polarizer with frequency tunability and the active controllable isolation depth of the polarization states with the elaborate designs in Figure 9c,e, respectively. In Figure 9c, the L-shaped trace is inserted into orthogonal-oriented graphene strips to couple the resonances in the two directions such that the whole structure exhibits the CP selectivity [96]. Left circular polarization (LCP) and right circular polarization (RCP) excitation causes different current distributions in the sandwich element with different resonance frequencies, such that one polarization is blocked by strong resonances and the other goes through with small insertion loss. Figure 9d shows that the operation frequency can be dynamically shifted with the variation of the graphene Fermi level.

The chiral metamaterials have been studied with enhanced circular dichroism (CD) [97] and optical activity (OA) [98]. When the graphene layer is incorporated as a loss-controllable medium into the chiral resonators in Figure 9e, the radiation loss of the resonator is controlled by the voltage [99]. Underdamped to overdamped phase transition for the RCP wave is clearly observed in the experiment, while the off-resonance LCP wave experiences negligible modulation (Figure 9f). Therefore, at the critical coupling condition, a giant CD with a 45-dB isolation is experimentally achieved. Different from the plasmonic feature of patterned graphene in most of the polarization controllers, here the tunable loss of the monolayer graphene is the key to tune the CD originating from the chiral meta-atoms.

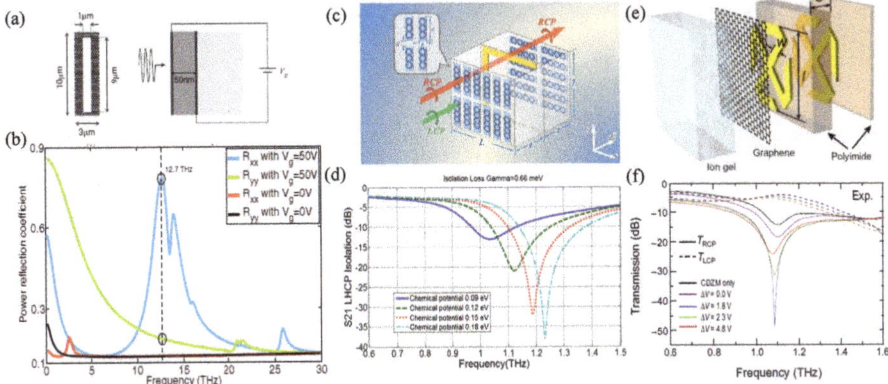

Figure 9. Graphene metasurfaces as active polarizers. (**a**) Rectangular slots in the graphene layer with artificial birefringence and its biasing configuration. (**b**) Different lengths along the x and y directions lead to different resonance frequencies. At the frequency of 12.7 THz, x polarization is on resonance and strongly reflected, while y polarization is off resonance and weakly reflected. The x polarization is filtered upon reflection. Reproduced with permission from [95], Copyright American Physical Society, 2012. (**c**) Two layers of orthogonally-orientated graphene strips sandwiching an L-shaped metallic resonator. The asymmetric L resonator couples to the orthogonal graphene strips, leading the LCP and RCP beams to different resonance frequencies, such that one polarization is blocked by strong resonances and the other goes through with small insertion loss. (**d**) Variation of the LCP transmission with the graphene Fermi level leads to a frequency-tunable polarizer. Reproduced with permisson from [96], Copyright Optical Society of America, 2015. (**e**) Graphene as a loss-tunable material attached to the bilayer of a conjugated double-Z chiral metamaterial for active control of the radiation loss. (**f**) Measured transmission spectra for LCP and RCP waves with different gate voltages. The RCP wave experiences underdamped to overdamped phase transition, while the off-resonance LCP wave experiences negligible modulation. The isolation depth is 45 dB at the critical coupling condition and can be actively controlled via biasing. Reproduced with permission from [99], Copyright The authors, 2017.

4.4. General Polarization Control in Graphene Metasurfaces

In general, the output beam through graphene metasurfaces may exhibit elliptical polarization, where the tilt angle and/or the ellipticity can be actively engineered with the variation of the graphene Fermi level. Multiple polarization states are flexibly switchable with proper biasing voltage at high modulation speed in a single structure, which is highly demanding for polarization encoding and multiplexing.

In 2017, Shvets' group experimentally demonstrated an active graphene metasurface that converts a linear polarization into an elliptical one with controllable tilt angle and ellipticity [100]. The configuration is a C-bar-shaped anisotropic metallic metasurface integrated with a graphene layer (Figure 10a), which induces a strong Fano resonance only when the polarization is along the bar direction (y direction). Therefore, the reflection spectra of the y-polarized beam were well controlled by the voltage due to the intense interaction of the in-plane electric field with graphene, while the x polarization is unaffected, as shown by the measurement results in Figure 10b. At a specific wavelength of 7.72 μm, the tilt angle can be modulated with constant ellipticity when the biasing voltages are −200 V, 0 V, and 250 V in Figure 10c.

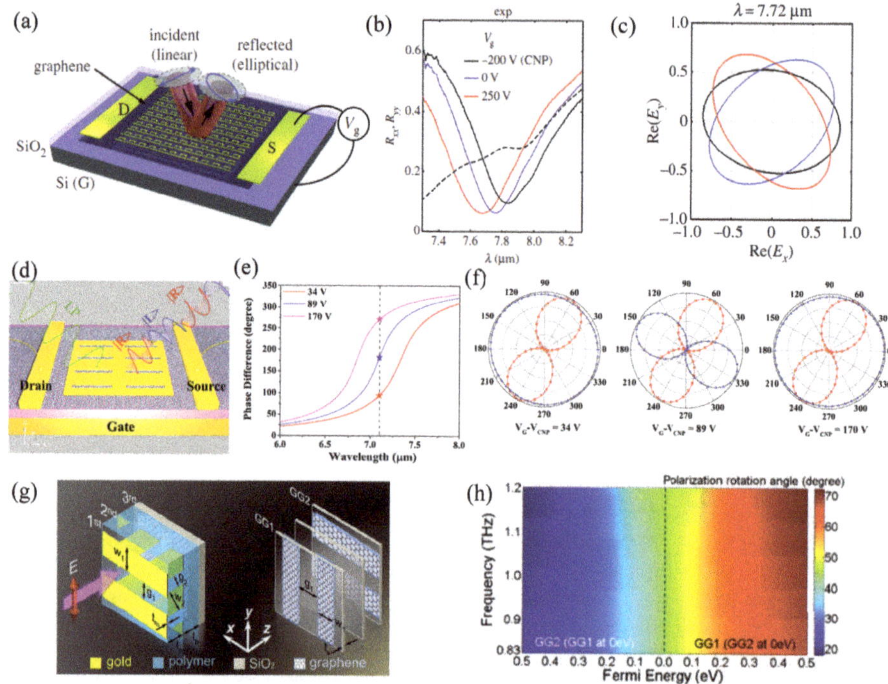

Figure 10. Continuous or multi-state active polarization engineering. (**a**) Graphene integrated with an anisotropic metasurface for linear to elliptical polarization conversion upon reflection. (**b**) Reflection spectra of x- and y-polarized beams with different gate biasing voltages. The design induces a strong Fano resonance only when the polarization is along the bar direction, i.e., the y direction. The Fano resonance dip in y polarization shifts with the voltage, and the x polarization is unaffected. (**c**) The tilt angle is adjustable with constant ellipticity by changing the biasing voltage as −200 V, 0 V, and 250 V at 7.72 μm. Reproduced with permission from [100], Copyright The authors, 2017. (**d**) Graphene-loaded slot-shaped metasurface for polarization conversion due to strong resonance along the short edge and weak interaction along the long edge. (**e**) Reflection phase difference between x and y polarization states under different voltages. The phase differences of 90°, 180°, and 270° are obtained at 7.1 μm with proper voltage applied. (**f**) The 61° linear polarized light is changed to LCP, cross-LP, and RCP when the voltage is 34 V, 89 V, and 170 V relative to the charge neutral point (CNP) for encoding and multiplexing applications. Reproduced with permission from [101], Copyright WILEY-VCH Verlag GmbH & Co. KGaA, Weinheim, 2015. (**g**) Unit cell of a tunable polarization rotator composed of 45°-rotated bi-layer metallic gratings and orthogonal bi-layer graphene gratings. The polarization rotation is enhanced by the Fabry–Perot interference between the top and bottom gratings and dynamically changed via modification of the graphene conductivity (**h**) The polarization rotation angle varies continuously from 20° to 45° and 45° to 70° by sequentially tuning the Fermi level of the orthogonal graphene gratings with a high transmission efficiency of above 75% in a wide band between 0.83 and 1.2 THz. Reproduced with permission from [102], Copyright Elsevier Ltd., 2018.

Similarly, by utilizing another anisotropic metasurface, namely the rectangle slot array in Figure 10d, the polarization conversion comes from strong resonance along the short edge and weak interaction along the long edge [101]. The voltage variation shifts the resonance frequency, as well as the phase difference between the orthogonal LP modes. Figure 10e shows that the phase differences of 90°, 180°, and 270° are obtained at 7.1 μm with the corresponding voltage applied, which are used to convert an LP mode into LCP, cross-LP, and RCP, respectively (Figure 10f). The three polarization states can be used for encoding binary data and polarization multiplexing with any desired time-domain sequence. In addition, graphene gratings and metallic gratings are stacked with proper orientation in Figure 10g to rotate the polarization direction of LP beam dynamically from 20° to 70° with a high transmission efficiency of above 75% in a wide band between 0.83 and 1.2 THz [102], taking advantage of the Fabry-Perot interference and the dynamic graphene conductivity. Such controllable optical activity without chiral material holds good prospects in biology and spectroscopy.

5. Discussion and Conclusions

If the metasurfaces are classified as two categories: graphene-integrated metallic resonator array and patterned graphene resonator array, most of the experimental validations until now fall into the first case. This is not only because of the relatively mature transfer technique [103] of graphene without further patterning steps, but also due to the simple biasing at a single location. A recent study [104] concluded that the graphene/metal hybrid structure, as compared to the graphene-only structure, has a higher quality factor and extinction ratio, with the performance maintained better during the dynamic modulation. In contrast, patterning and independent biasing of the isolated graphene inclusion remains a big challenge, especially at mid-IR, where the feature size is in the nanometer scale.

The relaxation time of carriers in most of the CVD-grown graphene layer is around 10~30 fs when transferred to different substrates [28,52,56,100]. We notice that this value is one or two orders higher in most of the theoretical designs, though high mobility of 10^4 cm^2/Vs has been reported before [40]. It is predicted that the small scattering time of tens of femtoseconds will introduce strong absorption and significantly reduced the range of phase response [88–90]. Therefore, improving the quality of graphene remains a key step to boosting the efficiency of active metasurfaces. The designs robust to the graphene loss or utilizing the lossy feature are more favorable in practice. Other practical considerations include ways to increase the carrier density at low gate voltage by using the ion-gel top gate [105] or the bi-layer graphene capacitor [45].

The development of graphene metasurfaces is generally following the track of metallic metasurfaces. The initially-proposed monolayer metallic metasurfaces usually suffer from poor efficiency and limited phase response range. The introduction of the back mirror enhances the interaction of light with the metallic resonators for efficient wavefront shaping in reflection mode. The geometric phase is then successfully used for wideband phase manipulation, and the layered configuration is helpful to improve the efficiency in transmission mode. All of these progresses have been witnessed in the development of graphene metasurfaces. However, most of the graphene configurations stay at the stage of dynamic phase shifters. To continue following the track of metallic metasurfaces, we envision graphene designs with spatially- and temporally-variant phase profiles for complicated wavefront shaping beyond focusing and bending, such as dynamic hologram imaging and reconfigurable structured beam generation, where the static metallic counterparts have been widely studied.

With the recent trend from plasmonic to dielectric metasurfaces due to programmable electric and magnetic responses, graphene-dielectric hybrid design has become one of the most promising directions for active wave engineering with improved efficiency and increased flexibility. Initial studies include enhanced transmission modulation [106,107], tunable electromagnetically-induced transparency [108], quarter-wave plates [109], and very recently, experimentally-demonstrated tunable absorbers [110].

In summary, the recent progress of graphene metasurfaces is overviewed with emphasis on the active phase and polarization control by relating the basic material property to the tunable device functionality. With the introduction of graphene, the metasurface is vested with tunable operation frequency, controllable efficiency, or switchable multiple functionalities. Additionally, graphene also extends the operation frequency of metallic metasurfaces into the sub-THz regime where metals as perfect electric conductors do not interact much with light, and even the near-IR and visible region where the interband transition dominates the conductivity [111]. Although most of the reviewed studies stay in the theoretical designs and numerical validations, we have noticed that active wavefront shaping, such as anomalous reflection [52] and focusing [53], and electrical tuning of the polarization states [99,100] have been proven by experiments in the past two years. We believe more and more active devices based on graphene will become feasible with the advances of the fabrication technique in the near future, to enrich the optical components and to address technical challenges for the THz and mid-IR technology.

Author Contributions: Writing, original draft preparation, J.C.; writing, review and editing, J.C., F.F., and S.C.; supervision, S.C.

Funding: This research was funded by the National Natural Science Foundation of China No. 61805123 and No. 61831012, the State's Key Project of Research and Development Plan No. 2017YFA0701000 and No. 2016YFC0101002, and the Fundamental Research Funds for the Central Universities, Natural Science Foundation of Tianjin No. 18JCQNJC02200.

Conflicts of Interest: The authors declare no conflict of interest.

References

1. He, Q.; Sun, S.; Xiao, S.; Zhou, L. High-Efficiency Metasurfaces: Principles, Realizations, and Applications. *Adv. Opt. Mater.* **2018**, *6*, 1800415. [CrossRef]
2. Kamali, S.M.; Arbabi, E.; Arbabi, A.; Faraon, A. A review of dielectric optical metasurfaces for wavefront control. *Nanophotonics* **2018**, *7*, 1041–1068. [CrossRef]
3. Chang, S.; Guo, X.; Ni, X. Optical metasurfaces: Progress and applications. *Annu. Rev. Mater. Res.* **2018**, *48*, 279–302. [CrossRef]
4. Kotlyar, V.V.; Nalimov, A.G.; Stafeev, S.S.; Hu, C.; O'Faolain, L.; Kotlyar, M.V.; Gibson, D.; Song, S. Thin high numerical aperture metalens. *Opt. Express* **2017**, *25*, 8158–8167. [CrossRef]
5. Paniagua-Domínguez, R.; Yu, Y.F.; Khaidarov, E.; Choi, S.M.; Leong, V.; Bakker, R.M.; Liang, X.N.; Fu, Y.H.; Valuckas, V.; Krivitsky, L.A.; et al. A Metalens with a Near-Unity Numerical Aperture. *Nano Lett.* **2017**, *18*, 2124–2132. [CrossRef] [PubMed]
6. Liang, H.; Lin, Q.; Xie, X.; Sun, Q.; Wang, Y.; Zhou, L.; Liu, L.; Yu, X.; Zhou, J.; Krauss, T.F.; et al. Ultrahigh Numerical Aperture Metalens at Visible Wavelengths. *Nano Lett.* **2018**, *18*, 4460–4466. [CrossRef] [PubMed]
7. Chen, W.T.; Yang, K.Y.; Wang, C.M.; Huang, Y.W.; Sun, G.; Chiang, I.D.; Liao, C.Y.; Hsu, W.L.; Lin, H.T.; Sun, S.; et al. High-efficiency broadband meta-hologram with polarization-controlled dual images. *Nano Lett.* **2013**, *14*, 225–230. [CrossRef] [PubMed]
8. Wen, D.; Yue, F.; Li, G.; Zheng, G.; Chan, K.; Chen, S.; Chen, M.; Li, K.F.; Wong, P.W.H.; Cheah, K.W.; et al. Helicity multiplexed broadband metasurface holograms. *Nat. Commun.* **2015**, *6*, 8241. [CrossRef]
9. Arbabi, A.; Horie, Y.; Bagheri, M.; Faraon, A. Dielectric metasurfaces for complete control of phase and polarization with subwavelength spatial resolution and high transmission. *Nat. Nanotechnol.* **2015**, *10*, 937–943. [CrossRef]
10. Li, Z.; Qiao, P.; Dong, G.; Shi, Y.; Bi, K.; Yang, X.; Meng, X. Polarization-multiplexed broadband hologram on all-dielectric metasurface. *EPL Eur. Lett.* **2018**, *124*, 14003. [CrossRef]
11. Deng, Z.; Cao, Y.; Li, X.; Wang, G. Multifunctional metasurface: From extraordinary optical transmission to extraordinary optical diffraction in a single structure. *Photonics Res.* **2018**, *6*, 443–450. [CrossRef]
12. Forouzmand, A.; Salary, M.M.; Inampudi, S.; Mosallaei, H. A Tunable Multigate Indium-Tin-Oxide-Assisted All-Dielectric Metasurface. *Adv. Opt. Mater.* **2018**, *6*, 1701275. [CrossRef]
13. Cheng, J.; Inampudi, S.; Mosallaei, H. Optimization-based dielectric metasurfaces for angle-selective multifunctional beam deflection. *Sci. Rep.* **2017**, *7*, 12228. [CrossRef]
14. Zhang, F.; Pu, M.; Li, X.; Gao, P.; Ma, X.; Luo, J.; Yu, H.; Luo, X. All-Dielectric Metasurfaces for Simultaneous Giant Circular Asymmetric Transmission and Wavefront Shaping Based on Asymmetric Photonic Spin–Orbit Interactions. *Adv. Funct. Mater.* **2017**, *27*, 1704295. [CrossRef]
15. Li, G.; Wu, L.; Li, K.F.; Chen, S.; Schlickriede, C.; Xu, Z.; Huang, S.; Li, W.D.; Liu, Y.J.; Pun, E.Y.B.; et al. Nonlinear Metasurface for Simultaneous Control of Spin and Orbital Angular Momentum in Second Harmonic Generation. *Nano Lett.* **2017**, *17*, 7974–7979. [CrossRef] [PubMed]
16. Geim, A.K.; Novoselov, K.S. The rise of graphene. *Nat. Mater.* **2007**, *6*, 183–91. [CrossRef] [PubMed]

17. Novoselov, K.S.; Fal, V.; Colombo, L.; Gellert, P.; Schwab, M.; Kim, K. A roadmap for graphene. *Nature* **2012**, *490*, 192–200. [CrossRef]

18. Savo, S.; Shrekenhamer, D.; Padilla, W.J. Liquid Crystal Metamaterial Absorber Spatial Light Modulator for THz Applications. *Adv. Opt. Mater.* **2014**, *2*, 275–279. [CrossRef]

19. Yu, J.P.; Chen, S.; Fan, F.; Cheng, J.R.; Xu, S.T.; Wang, X.H.; Chang, S.J. Tunable terahertz wave-plate based on dual-frequency liquid crystal controlled by alternating electric field. *Opt. Express* **2018**, *26*, 663–673. [CrossRef] [PubMed]

20. Raeis-Hosseini, N.; Rho, J. Metasurfaces based on phase-change material as a reconfigurable platform for multifunctional devices. *Materials* **2017**, *10*, 1046. [CrossRef]

21. Chen, Z.C.; Rahmani, M.; Gong, Y.D.; Chong, C.T.; Hong, M.H. Realization of Variable Three-Dimensional Terahertz Metamaterial Tubes for Passive Resonance Tunability. *Adv. Mater.* **2012**, *24*, OP143–OP147.

22. Li, J.; Shah, C.M.; Withayachumnankul, W.; Ung, S.Y.; Mitchell, A.; Sriram, S.; Bhaskaran, M.; Chang, S.; Abbott, D. Mechanically tunable terahertz metamaterials. *Appl. Phys. Lett.* **2013**, *102*, 4773–252. [CrossRef]

23. Xu, S.T.; Mou, L.L.; Fan, F.; Chen, S.; Zhao, Z.; Xiang, D.; de Andrade, M.J.; Liu, Z.; Chang, S.J. Mechanical modulation of terahertz wave via buckled carbon nanotube sheets. *Opt. Express* **2018**, *26*, 28738–28750. [CrossRef] [PubMed]

24. Yao, Y.; Kats, M.A.; Genevet, P.; Yu, N.; Song, Y.; Kong, J.; Capasso, F. Broad Electrical Tuning of Graphene-Loaded Plasmonic Antennas. *Nano Lett.* **2013**, *13*, 1257–1264. [CrossRef] [PubMed]

25. Liu, P.Q.; Luxmoore, I.J.; Mikhailov, S.A.; Savostianova, N.A.; Valmorra, F.; Faist, J.; Nash, G.R. Highly tunable hybrid metamaterials employing split-ring resonators strongly coupled to graphene surface plasmons. *Nat. Commun.* **2015**, *6*, 8969. [CrossRef] [PubMed]

26. Yao, Y.; Shankar, R.; Kats, M.A.; Song, Y.; Kong, J.; Loncar, M.; Capasso, F. Electrically tunable metasurface perfect absorbers for ultrathin mid-infrared optical modulators. *Nano Lett.* **2014**, *14*, 6526–6532. [CrossRef] [PubMed]

27. Ju, L.; Geng, B.; Horng, J.; Girit, C.; Martin, M.; Hao, Z.; Bechtel, H.A.; Liang, X.; Zettl, A.; Shen, Y.R. Graphene plasmonics for tunable terahertz metamaterials. *Nat. Nanotechnol.* **2011**, *6*, 630–634. [CrossRef]

28. Lee, S.H.; Choi, M.; Kim, T.T.; Lee, S.; Liu, M.; Yin, X.; Choi, H.K.; Lee, S.S.; Choi, C.G.; Choi, S.Y.; et al. Switching terahertz waves with gate-controlled active graphene metamaterials. *Nat. Mater.* **2012**, *11*, 936. [CrossRef]

29. Vasić, B.; Jakovljević, M.M.; Isić, G.; Gajić, R. Tunable metamaterials based on split ring resonators and doped graphene. *Appl. Phys. Lett.* **2013**, *103*, 165113. [CrossRef]

30. Fang, Z.; Thongrattanasiri, S.; Schlather, A.; Liu, Z.; Ma, L.; Wang, Y.; Ajayan, P.M.; Nordlander, P.; Halas, N.J.; de Abajo, F.J.G. Gated tunability and hybridization of localized plasmons in nanostructured graphene. *ACS Nano* **2013**, *7*, 2388–2395. [CrossRef]

31. Dabidian, N.; Kholmanov, I.; Khanikaev, A.B.; Tatar, K.; Trendafilov, S.; Mousavi, S.H.; Magnuson, C.; Ruoff, R.S.; Shvets, G. Electrical Switching of Infrared Light Using Graphene Integration with Plasmonic Fano Resonant Metasurfaces. *ACS Photonics* **2015**, *2*, 216–227. [CrossRef]

32. Thongrattanasiri, S.; Koppens, F.H.L.; de Abajo, F.J.G. Complete optical absorption in periodically patterned graphene. *Phys. Rev. Lett.* **2012**, *108*, 047401. [CrossRef] [PubMed]

33. Kakenov, N.; Takan, T.; Ozkan, V.A.; Balci, O.; Polat, E.O.; Altan, H.; Kocabas, C. Graphene-enabled electrically controlled terahertz spatial light modulators. *Opt. Lett.* **2015**, *40*, 1984–1987. [CrossRef] [PubMed]

34. Liu, C.; Bai, Y.; Zhou, J.; Zhao, Q.; Qiao, L. A Review of Graphene Plasmons and its Combination with Metasurface. *J. Korean Ceram. Soc.* **2017**, *54*, 349–365. [CrossRef]

35. Nemati, A.; Wang, Q.; Hong, M.; Teng, J. Tunable and reconfigurable metasurfaces and metadevices. *Opto-Electron. Adv.* **2018**, *1*, 180009. [CrossRef]

36. Gusynin, V.; Sharapov, S.; Carbotte, J. Magneto-optical conductivity in graphene. *J. Phys. Condens. Matter* **2006**, *19*, 026222. [CrossRef]

37. Falkovsky, L.; Pershoguba, S. Optical far-infrared properties of a graphene monolayer and multilayer. *Phys. Rev. B* **2007**, *76*, 153410. [CrossRef]

38. Hanson, G.W. Dyadic Green's functions and guided surface waves for a surface conductivity model of graphene. *J. Appl. Phys.* **2008**, *103*, 064302. [CrossRef]

39. Li, X.; Magnuson, C.W.; Venugopal, A.; Tromp, R.M.; Hannon, J.B.; Vogel, E.M.; Colombo, L.; Ruoff, R.S. Large-area graphene single crystals grown by low-pressure chemical vapor deposition of methane on copper. *J. Am. Chem. Soc.* **2011**, *133*, 2816–2819. [CrossRef]

40. Novoselov, K.S.; Geim, A.K.; Morozov, S.V.; Jiang, D.; Zhang, Y.; Dubonos, S.V.; Grigorieva, I.V.; Firsov, A.A. Electric field effect in atomically thin carbon films. *Science* **2004**, *306*, 666–669. [CrossRef]

41. Castro, E.V.; Ochoa, H.; Katsnelson, M.; Gorbachev, R.; Elias, D.; Novoselov, K.; Geim, A.; Guinea, F. Limits on charge carrier mobility in suspended graphene due to flexural phonons. *Phys. Rev. Lett.* **2010**, *105*, 266601. [CrossRef] [PubMed]

42. Jablan, M.; Buljan, H.; Soljačić, M. Plasmonics in graphene at infrared frequencies. *Phys. Rev. B* **2009**, *80*, 245435. [CrossRef]

43. Cheng, J.; Wang, W.L.; Mosallaei, H.; Kaxiras, E. Surface plasmon engineering in graphene functionalized with organic molecules: A multiscale theoretical investigation. *Nano Lett.* **2013**, *14*, 50–56. [CrossRef]

44. Nair, R.R.; Blake, P.; Grigorenko, A.N.; Novoselov, K.S.; Booth, T.J.; Stauber, T.; Peres, N.M.; Geim, A.K. Fine structure constant defines visual transparency of graphene. *Science* **2008**, *320*, 1308–1308. [CrossRef] [PubMed]

45. Balci, O.; Kakenov, N.; Karademir, E.; Balci, S.; Cakmakyapan, S.; Polat, E.O.; Caglayan, H.; Özbay, E.; Kocabas, C. Electrically switchable metadevices via graphene. *Sci. Adv.* **2018**, *4*, eaao1749. [CrossRef] [PubMed]

46. Lu, F.; Liu, B.; Shen, S. Infrared wavefront control based on graphene metasurfaces. *Adv. Opt. Mater.* **2014**, *2*, 794–799. [CrossRef]

47. Li, Z.; Yao, K.; Xia, F.; Shen, S.; Tian, J.; Liu, Y. Graphene plasmonic metasurfaces to steer infrared light. *Sci. Rep.* **2015**, *5*, 12423. [CrossRef]

48. AbdollahRamezani, S.; Arik, K.; Farajollahi, S.; Khavasi, A.; Kavehvash, Z. Beam manipulating by gate-tunable graphene-based metasurfaces. *Opt. Lett.* **2015**, *40*, 5383–5386. [CrossRef]

49. Ding, X.; Monticone, F.; Zhang, K.; Zhang, L.; Gao, D.; Burokur, S.N.; de Lustrac, A.; Wu, Q.; Qiu, C.W.; Alù, A. Ultrathin Pancharatnam–Berry Metasurface with Maximal Cross-Polarization Efficiency. *Adv. Mater.* **2015**, *27*, 1195–1200. [CrossRef]

50. Lin, D.; Fan, P.; Hasman, E.; Brongersma, M.L. Dielectric gradient metasurface optical elements. *Science* **2014**, *345*, 298–302. [CrossRef]

51. Cheng, H.; Chen, S.Q.; Yu, P.; Liu, W.W.; Li, Z.C.; Li, J.X.; Xie, B.Y.; Tian, J.G. Dynamically tunable broadband infrared anomalous refraction based on graphene metasurfaces. *Adv. Opt. Mater.* **2015**, *3*, 1744–1749. [CrossRef]

52. Kim, T.T.; Kim, H.; Kenney, M.; Park, H.S.; Kim, H.D.; Min, B.; Zhang, S. Amplitude Modulation of Anomalously Refracted Terahertz Waves with Gated-Graphene Metasurfaces. *Adv. Opt. Mater.* **2018**, *6*, 1700507. [CrossRef]

53. Liu, W.G.; Hu, B.; Huang, Z.D.; Guan, H.Y.; Li, H.T.; Wang, X.K.; Zhang, Y.; Yin, H.X.; Xiong, X.L.; Liu, J.; et al. Graphene-enabled electrically controlled terahertz meta-lens. *Photonics Res.* **2018**, *6*, 703–708. [CrossRef]

54. Huang, Z.D.; Hu, B.; Liu, W.G.; Liu, J.; Wang, Y.T. Dynamical tuning of terahertz meta-lens assisted by graphene. *JOSA B* **2017**, *34*, 1848–1854. [CrossRef]

55. Miao, Z.; Wu, Q.; Li, X.; He, Q.; Ding, K.; An, Z.; Zhang, Y.; Zhou, L. Widely tunable terahertz phase modulation with gate-controlled graphene metasurfaces. *Phys. Rev. X* **2015**, *5*, 041027. [CrossRef]

56. Sherrott, M.C.; Hon, P.W.C.; Fountaine, K.T.; Garcia, J.C.; Ponti, S.M.; Brar, V.W.; Sweatlock, L.A.; Atwater, H.A. Experimental demonstration of >230 degrees phase modulation in gate-tunable graphene—Gold reconfigurable mid-infrared metasurfaces. *Nano Lett.* **2017**, *17*, 3027–3034. [CrossRef] [PubMed]

57. Carrasco, E.; Tamagnone, M.; Mosig, J.R.; Low, T.; Perruisseau-Carrier, J. Gate-controlled mid-infrared light bending with aperiodic graphene nanoribbons array. *Nanotechnology* **2015**, *26*, 134002. [CrossRef]

58. Luo, L.; Wang, K.; Guo, K.; Shen, F.; Zhang, X.; Yin, Z.; Guo, Z. Tunable manipulation of terahertz wavefront based on graphene metasurfaces. *J. Opt.* **2017**, *19*, 115104. [CrossRef]

59. Yao, W.; Tang, L.; Wang, J.; Ji, C.; Wei, X.; Jiang, Y. Spectrally and Spatially Tunable Terahertz Metasurface Lens Based on Graphene Surface Plasmons. *IEEE Photonics J.* **2018**, *10*, 1–8. [CrossRef]

60. Biswas, S.R.; Gutiérrez, C.E.; Nemilentsau, A.; Lee, I.H.; Oh, S.H.; Avouris, P.; Low, T. Tunable Graphene Metasurface Reflectarray for Cloaking, Illusion, and Focusing. *Phys. Rev. Appl.* **2018**, *9*, 034021. [CrossRef]

61. Ding, P.; Li, Y.; Shao, L.; Tian, X.; Wang, J.; Fan, C. Graphene aperture-based metalens for dynamic focusing of terahertz waves. *Opt. Express* **2018**, *26*, 28038–28050. [CrossRef] [PubMed]

62. Liu, L.; Zarate, Y.; Hattori, H.T.; Neshev, D.N.; Shadrivov, I.V.; Powell, D.A. Terahertz focusing of multiple wavelengths by graphene metasurfaces. *Appl. Phys. Lett.* **2016**, *108*, 031106. [CrossRef]

63. Zhao, H.; Chen, Z.; Su, F.; Ren, G.; Liu, F.; Yao, J. Terahertz wavefront manipulating by double-layer graphene ribbons metasurface. *Opt. Commun.* **2017**, *402*, 523–526. [CrossRef]

64. Ma, W.; Huang, Z.; Bai, X.; Zhan, P.; Liu, Y. Dual-band light focusing using stacked graphene metasurfaces. *ACS Photonics* **2017**, *4*, 1770–1775. [CrossRef]

65. Vakil, A.; Engheta, N. Transformation optics using graphene. *Science* **2011**, *332*, 1291–1294. [CrossRef] [PubMed]

66. Vakil, A.; Engheta, N. Fourier optics on graphene. *Phys. Rev. B* **2012**, *85*, 075434. [CrossRef]

67. Nasari, H.; Abrishamian, M.S. Electrically tunable graded index planar lens based on graphene. *J. Appl. Phys.* **2014**, *116*, 083106. [CrossRef]

68. Du, L.; Tang, D. Manipulating propagating graphene plasmons at near field by shaped graphene nano-vacancies. *JOSA A* **2014**, *31*, 691–695. [CrossRef]

69. Gao, W.; Shi, G.; Jin, Z.; Shu, J.; Zhang, Q.; Vajtai, R.; Ajayan, P.M.; Kono, J.; Xu, Q. Excitation and active control of propagating surface plasmon polaritons in graphene. *Nano Lett.* **2013**, *13*, 3698–3702. [CrossRef]

70. Fei, Z.; Rodin, A.S.; Andreev, G.O.; Bao, W.; McLeod, A.S.; Wagner, M.; Zhang, L.M.; Zhao, Z.; Thiemens, M.; Dominguez, G.; et al. Gate-tuning of graphene plasmons revealed by infrared nano-imaging. *Nature* **2012**, *487*, 82–85. [CrossRef]

71. Chen, J.; Badioli, M.; Alonso-González, P.; Thongrattanasiri, S.; Huth, F.; Osmond, J.; Spasenović, M.; Centeno, A.; Pesquera, A.; Godignon, P.; et al. Optical nano-imaging of gate-tunable graphene plasmons. *Nature* **2012**, *487*, 77–81. [CrossRef] [PubMed]

72. Esquius-Morote, M.; Gómez-Díaz, J.S.; Perruisseau-Carrier, J. Sinusoidally modulated graphene leaky-wave antenna for electronic beamscanning at THz. *IEEE Trans. Terahertz Sci. Technol.* **2014**, *4*, 116–122. [CrossRef]

73. Chen, M.; Fan, F.; Wu, P.; Zhang, H.; Chang, S. Active graphene plasmonic grating for terahertz beam scanning device. *Opt. Commun.* **2015**, *348*, 66–70. [CrossRef]

74. Cheng, J.; Jafar-Zanjani, S.; Mosallaei, H. Real-time two-dimensional beam steering with gate-tunable materials: a theoretical investigation. *Appl. Opt.* **2016**, *55*, 6137–6144. [CrossRef] [PubMed]

75. Miller, S.; Chynoweth, A. *Optical Fiber Telecommunications*; Academic Press: New York, NY, USA, **1979**.

76. Morris, H.R.; Hoyt, C.C.; Miller, P.; Treado, P.J. Liquid crystal tunable filter Raman chemical imaging. *Appl. Spectrosc.* **1996**, *50*, 805–811. [CrossRef]

77. Langley, D.P.; Balaur, E.; Hwang, Y.; Sadatnajafi, C.; Abbey, B. Optical Chemical Barcoding Based on Polarization Controlled Plasmonic Nanopixels. *Adv. Funct. Mater.* **2018**, *28*, 1704842. [CrossRef]

78. Losurdo, M.; Bergmair, M.; Bruno, G.; Cattelan, D.; Cobet, C.; de Martino, A.; Fleischer, K.; Dohcevic-Mitrovic, Z.; Esser, N.; Galliet, M.; et al. Spectroscopic ellipsometry and polarimetry for materials and systems analysis at the nanometer scale: State-of-the-art, potential, and perspectives. *J. Nanopart. Res.* **2009**, *11*, 1521–1554. [CrossRef] [PubMed]

79. Oron, D.; Dudovich, N.; Silberberg, Y. Femtosecond phase-and-polarization control for background-free coherent anti-Stokes Raman spectroscopy. *Phys. Rev. Lett.* **2003**, *90*, 213902. [CrossRef]

80. Cheng, H.; Chen, S.Q.; Yu, P.; Li, J.X.; Deng, L.; Tian, J.G. Mid-infrared tunable optical polarization converter composed of asymmetric graphene nanocrosses. *Opt. Lett.* **2013**, *38*, 1567–1569. [CrossRef]

81. Zhang, Y.; Feng, Y.J.; Zhu, B.; Zhao, J.M.; Jiang, T. Switchable quarter-wave plate with graphene based metamaterial for broadband terahertz wave manipulation. *Opt. Express* **2015**, *23*, 27230–27239. [CrossRef]

82. Ji, Y.Y.; Fan, F.; Wang, X.H.; Chang, S.J. Broadband controllable terahertz quarter-wave plate based on graphene gratings with liquid crystals. *Opt. Express* **2018**, *26*, 12852–12862. [CrossRef] [PubMed]

83. Gao, X.; Yang, W.L.; Cao, W.P.; Chen, M.; Jiang, Y.N.; Yu, X.H.; Li, H.O. Bandwidth broadening of a graphene-based circular polarization converter by phase compensation. *Opt. Express* **2017**, *25*, 23945–23954. [CrossRef] [PubMed]

84. Zhu, L.; Fan, Y.H.; Wu, S.; Yu, L.Z.; Zhang, K.Y.; Zhang, Y. Electrical control of terahertz polarization by graphene microstructure. *Opt. Commun.* **2015**, *346*, 120–123. [CrossRef]

85. Cheng, H.; Chen, S.; Yu, P.; Li, J.; Xie, B.; Li, Z.; Tian, J. Dynamically tunable broadband mid-infrared cross polarization converter based on graphene metamaterial. *Appl. Phys. Lett.* **2013**, *103*, 151107. [CrossRef]

86. Ding, J.; Arigong, B.; Ren, H.; Shao, J.; Zhou, M.; Lin, Y.K.; Zhang, H.L. Mid-Infrared Tunable Dual-Frequency Cross Polarization Converters Using Graphene-Based L-Shaped Nanoslot Array. *Plasmonics* **2015**, *10*, 351–356. [CrossRef]

87. Guo, T.; Argyropoulos, C. Broadband polarizers based on graphene metasurfaces. *Opt. Lett.* **2016**, *41*, 5592–5595. [CrossRef] [PubMed]

88. Zhu, J.F.; Li, S.F.; Deng, L.; Zhang, C.; Yang, Y.; Zhu, H.B. Broadband tunable terahertz polarization converter based on a sinusoidally-slotted graphene metamaterial. *Opt. Mater. Express* **2018**, *8*, 1164–1173. [CrossRef]

89. Yadav, V.S.; Ghosh, S.K.; Bhattacharyya, S.; Das, S. Graphene-based metasurface for a tunable broadband terahertz cross-polarization converter over a wide angle of incidence. *Appl. Opt.* **2018**, *57*, 8720–8726. [CrossRef] [PubMed]

90. Chen, M.; Chang, L.Z.; Gao, X.; Chen, H.; Wang, C.Y.; Xiao, X.F.; Zhao, D.P. Wideband Tunable Cross Polarization Converter Based on a Graphene Metasurface with a Hollow-Carved "H" Array. *IEEE Photonics J.* **2017**, *9*, 1–11. [CrossRef]

91. Yu, X.Y.; Gao, X.; Qiao, W.; Wen, L.L.; Yang, W.L. Broadband Tunable Polarization Converter Realized by Graphene-Based Metamaterial. *IEEE Photonics Technol. Lett.* **2016**, *28*, 2399–2402. [CrossRef]

92. Yang, C.; Luo, Y.; Guo, J.; Pu, Y.; He, D.; Jiang, Y.D.; Xu, J.; Liu, Z.J. Wideband tunable mid-infrared cross polarization converter using rectangle-shape perforated graphene. *Opt. Express* **2016**, *24*, 16913–16922. [CrossRef] [PubMed]

93. Chen, M.; Sun, W.; Cai, J.J.; Chang, L.Z.; Xiao, X.F. Frequency-Tunable Mid-Infrared Cross Polarization Converters Based on Graphene Metasurface. *Plasmonics* **2017**, *12*, 699–705. [CrossRef]

94. Luo, S.W.; Li, B.; Yu, A.L.; Gao, J.; Wang, X.B.; Zuo, D.L. Broadband tunable terahertz polarization converter based on graphene metamaterial. *Opt. Commun.* **2018**, *413*, 184–189. [CrossRef]

95. Fallahi, A.; Perruisseau-Carrier, J. Design of Tunable Biperiodic Graphene Metasurfaces. *Phys. Rev. B* **2012**, *86*, 4608–4619. [CrossRef]

96. Li, Y.Z.; Zhao, J.M.; Lin, H.; Milne, W.; Hao, Y. Tunable circular polarization selective surfaces for low-THz applications using patterned graphene. *Opt. Express* **2015**, *23*, 7227–7236. [CrossRef] [PubMed]

97. Yu, Y.; Yang, Z.Y.; Li, S.X.; Zhao, M. Higher extinction ratio circular polarizers with hetero-structured double-helical metamaterials. *Opt. Express* **2011**, *19*, 10886–10894. [CrossRef]

98. Kim, T.T.; Oh, S.S.; Park, H.S.; Zhao, R.; Kim, S.H.; Choi, W.; Min, B.; Hess, O. Optical activity enhanced by strong inter-molecular coupling in planar chiral metamaterials. *Sci. Rep.* **2014**, *4*, 5864. [CrossRef] [PubMed]

99. Kim, T.T.; Oh, S.S.; Kim, H.D.; Park, H.S.; Hess, O.; Min, B.; Zhang, S. Electrical access to critical coupling of circularly polarized waves in graphene chiral metamaterials. *Sci. Adv.* **2017**, *3*, e1701377. [CrossRef]

100. Dutta-Gupta, S.; Dabidian, N.; Kholmanov, I.; Belkin, M.A.; Shvets, G. Electrical tuning of the polarization state of light using graphene-integrated anisotropic metasurfaces. *Philos. Trans. R. Soc. A* **2017**, *375*, 20160061. [CrossRef]

101. Li, J.X.; Yu, P.; Cheng, H.; Liu, W.W.; Li, Z.; Xie, B.Y.; Chen, S.Q.; Tian, J.G. Optical Polarization Encoding Using Graphene-Loaded Plasmonic Metasurfaces. *Adv. Opt. Mater.* **2016**, *4*, 91–98. [CrossRef]

102. Zhang, Y.; Feng, Y.J.; Jiang, T.; Cao, J.; Zhao, J.M.; Zhu, B. Tunable broadband polarization rotator in terahertz frequency based on graphene metamaterial. *Carbon* **2018**, *133*, 170–175. [CrossRef]

103. Li, X.S.; Cai, W.W.; An, J.; Kim, S.; Nah, J.; Yang, D.X.; Piner, R.; Velamakanni, A.; Jung, I.; Tutuc, E. Large-area synthesis of high-quality and uniform graphene films on copper foils. *Science* **2009**, *324*, 1312–1314. [CrossRef]

104. Arezoomandan, S.; Quispe, H.O.C.; Ramey, N.; Nieves, C.A.; Sensale-Rodriguez, B. Graphene-based reconfigurable terahertz plasmonics and metamaterials. *Carbon* **2017**, *112*, 177–184. [CrossRef]

105. Hu, H.; Zhai, F.; Hu, D.B.; Li, Z.J.; Bai, B.; Yang, X.X.; Dai, Q. Broadly tunable graphene plasmons using an ion-gel top gate with low control voltage. *Nanoscale* **2015**, *7*, 19493–19500. [CrossRef]

106. Argyropoulos, C. Enhanced transmission modulation based on dielectric metasurfaces loaded with graphene. *Opt. Express* **2015**, *23*, 23787–23797. [CrossRef] [PubMed]

107. Wu, C.H.; Arju, N.; Kelp, G.; Fan, J.A.; Dominguez, J.; Gonzales, E.; Tutuc, E.; Brener, I.; Shvets, G. Spectrally Selective Chiral Silicon Metasurfaces Based on Infrared Fano Resonances. *Nat. Commun.* **2014**, *5*, 3892. [CrossRef]

108. Zhu, L.; Dong, L.; Guo, J.; Meng, F.Y.; Wu, Q. Tunable electromagnetically induced transparency in hybrid graphene/all-dielectric metamaterial. *Appl. Phys. A* **2017**, *123*, 192. [CrossRef]

109. Owiti, E.O.; Yang, H.N.; Liu, P.; Ominde, C.F.; Sun, X.D. Polarization Converter with Controllable Birefringence Based on Hybrid All-Dielectric-Graphene Metasurface. *Nanoscale Res. Lett.* **2018**, *13*, 38. [CrossRef]
110. Arezoomandan, S.; Quispe, H.C.; Chanana, A.; Gopalan, P.; Banerji, S.; Nahata, A.; Sensale-Rodriguez, B. Graphene-dielectric Integrated Terahertz Metasurfaces. *Semicond. Sci. Technol.* **2018**, *33*, 104007. [CrossRef]
111. Kong, X.T.; Khan, A.A.; Kidambi, P.R.; Deng, S.; Yetisen, A.K.; Dlubak, B.; Hiralal, P.; Montelongo, Y.; Bowen, J.; Xavier, S.; et al. Graphene-based ultrathin flat lenses. *ACS Photonics* **2015**, *2*, 200–207. [CrossRef]

MDPI
St. Alban-Anlage 66
4052 Basel
Switzerland
Tel. +41 61 683 77 34
Fax +41 61 302 89 18
www.mdpi.com

Nanomaterials Editorial Office
E-mail: nanomaterials@mdpi.com
www.mdpi.com/journal/nanomaterials